VOLUME ONE HUNDRED AND FOURTEEN

PROGRESS IN MOLECULAR BIOLOGY AND TRANSLATIONAL SCIENCE

G Protein-Coupled Receptors in Energy Homeostasis and Obesity Pathogenesis

VOLUME ONE HUNDRED AND FOURTEEN

PROGRESS IN MOLECULAR BIOLOGY AND TRANSLATIONAL SCIENCE

G Protein-Coupled Receptors in Energy Homeostasis and Obesity Pathogenesis

Edited by

YA-XIONG TAO
Department of Anatomy, Physiology and Pharmacology, College of Veterinary Medicine, Auburn University, Auburn, Alabama USA

ELSEVIER

AMSTERDAM • BOSTON • HEIDELBERG • LONDON
NEW YORK • OXFORD • PARIS • SAN DIEGO
SAN FRANCISCO • SINGAPORE • SYDNEY • TOKYO
Academic Press is an imprint of Elsevier

Academic Press is an imprint of Elsevier
The Boulevard, Langford Lane, Kidlington, Oxford, OX51GB, UK
32, Jamestown Road, London NW1 7BY, UK
Radarweg 29, PO Box 211, 1000 AE Amsterdam, The Netherlands
225 Wyman Street, Waltham, MA 02451, USA
525 B Street, Suite 1900, San Diego, CA 92101-4495, USA

First edition 2013

Library of Congress Cataloging-in-Publication Data
A catalog record for this book is available from the Library of Congress

British Library Cataloguing in Publication Data
A catalogue record for this book is available from the British Library

ISBN: 978-0-12-386933-3
ISSN: 1877-1173

For information on all Academic Press publications
visit our website at store.elsevier.com

Printed and bound by CPI Group (UK) Ltd, Croydon, CR0 4YY

Transferred to Digital Printing, 2013

CONTENTS

CONTRIBUTORS

Benson T. Akingbemi
Department of Anatomy, Physiology and Pharmacology, Auburn University, Auburn, Alabama, USA

Karima Begriche
Department of Metabolism and Aging, The Scripps Research Institute, Jupiter, Florida, USA, and Inserm U991, Foie, Métabolisme et Cancer, 2 avenue du Pr Léon Bernard, Rennes, France

Andrew A. Butler
Department of Metabolism and Aging, The Scripps Research Institute, Jupiter, Florida, USA

Clemencé Girardet
Department of Metabolism and Aging, The Scripps Research Institute, Jupiter, Florida, USA

Anke Hinney
Department of Child and Adolescent Psychiatry, Universitätsklinikum, University of Duisburg-Essen, Essen, Germany

Qingsheng Huang
Research Center for Gene and Cell Engineering, Institute of Biomedicine and Biotechnology, Shenzhen Institute of Advance Technology, Chinese Academy of Sciences, Shenzhen, China

Nadja Knoll
Department of Child and Adolescent Psychiatry, Universitätsklinikum, University of Duisburg-Essen, Essen, Germany

Shiva Priya Dharshan Senthil Kumar
Cellular, Molecular and Structural Biology, Miami University, Oxford, Ohio, USA

Lei Li
Research Center for Gene and Cell Engineering, Institute of Biomedicine and Biotechnology, Shenzhen Institute of Advance Technology, Chinese Academy of Sciences, Shenzhen, China

Yin Li
Department of Physiology and Pathophysiology, School of Basic Medical Sciences, Peking University, and Key Laboratory of Molecular Cardiovascular Science, Ministry of Education, Beijing, China

Ziru Li
Department of Physiology and Pathophysiology, School of Basic Medical Sciences, Peking University, and Key Laboratory of Molecular Cardiovascular Science, Ministry of Education, Beijing, China

Xian Liu
Cellular, Molecular and Structural Biology, Miami University, Oxford, Ohio, USA

Patricia McDonald
Department of Molecular Therapeutics, The Scripps Research Institute, Jupiter, Florida, USA

Xiu-Lei Mo
Department of Anatomy, Physiology and Pharmacology, College of Veterinary Medicine, Auburn University, Auburn, Alabama, USA

Jian Peng
Department of Animal Nutrition and Feed Science, College of Animal Science and Technology, Huazhong Agricultural University, Wuhan, People's Republic of China

Pei-Gen Ren
Innovative Pharmacology and Biotherapy Pre-clinical Test Public Service Platform of Shenzhen, China

Ayman I. Sayegh
Gastroenterology Laboratory, Department of Biomedical Sciences, College of Veterinary Medicine, Tuskegee University, Tuskegee, Alabama, USA

Haifei Shi
Department of Biology, Center for Physiology and Neuroscience, and Cellular, Molecular and Structural Biology, Miami University, Oxford, Ohio, USA

Ya-Xiong Tao
Department of Anatomy, Physiology and Pharmacology, College of Veterinary Medicine, Auburn University, Auburn, Alabama, USA

Anna-Lena Volckmar
Department of Child and Adolescent Psychiatry, Universitätsklinikum, University of Duisburg-Essen, Essen, Germany

Hong-Kui Wei
Department of Anatomy, Physiology and Pharmacology, College of Veterinary Medicine, Auburn University, Auburn, Alabama, USA, and Department of Animal Nutrition and Feed Science, College of Animal Science and Technology, Huazhong Agricultural University, Wuhan, People's Republic of China

Jun Xie
Freshwater Fisheries Research Center, Chinese Academy of Fishery Sciences, Wuxi, Jiangsu, People's Republic of China

Zong-Hui Yuan
Department of Basic Veterinary Medicine, College of Animal Science and Veterinary Medicine, Huazhong Agricultural University, Wuhan, Hubei, People's Republic of China

Jian V. Zhang
Research Center for Gene and Cell Engineering, Institute of Biomedicine and Biotechnology, Shenzhen Institute of Advance Technology, Chinese Academy of Sciences, Shenzhen, and Innovative Pharmacology and Biotherapy Pre-clinical Test Public Service Platform of Shenzhen, China

Weizhen Zhang
Department of Physiology and Pathophysiology, School of Basic Medical Sciences, Peking University; Key Laboratory of Molecular Cardiovascular Science, Ministry of Education, Beijing, China, and Department of Surgery, University of Michigan Medical Center, Ann Arbor, Michigan, USA

PREFACE

Obesity is an epidemic in the industrialized countries and is increasing at an alarming rate in developing countries. Obesity is much more than a cosmetic problem. It is associated with a number of important diseases, including diabetes mellitus, hypertension, cardiovascular disease, and certain types of cancer. Therefore, obesity represents an enormous health, economic, and social burden. Current therapeutic options for obesity are far from ideal in terms of efficacy and side effects. Novel pharmacotherapy is urgently needed.

Obesity is caused by the imbalance of energy homeostasis, with energy intake exceeding energy expenditure. Genetic, environmental, and psychosocial factors contribute to the pathogenesis of obesity. This volume summarizes the current status in studies of a number of G protein-coupled receptors that have been shown to be prominent regulators of energy homeostasis and promising targets for obesity treatment. Some of these receptors also cause monogenic obesity. G protein-coupled receptors have proved to be excellent drug targets, accounting for a significant share of current therapeutics. Studies on G protein-coupled receptors are also advancing at an extremely fast pace, with two leaders in the field, Robert J. Lefkowitz and Brian Kobilka, awarded Nobel Prize in Chemistry in 2012. The chapters in this volume illustrate the rapid progress in these areas of research.

I thank all the authors for taking time out of their busy schedules to contribute to this volume. It was a privilege to review their chapters. I learned a lot from their summary of their exciting areas of research. Their outstanding contributions to this volume provide comprehensive overviews of the state of the art in the particular research topics that they are working on.

As this volume goes to print, I want to express my sincere appreciation to Dr. P. Michael Conn, the Series Editor, for his guidance, encouragement, and support. His trust and friendship give me confidence in undertaking this project. I also thank my colleagues at Elsevier, including Lisa Tickner, May Ann Zimmerman, Delsy Retchagar, Sarah Latham, Sarah E. Lay, Poornisha Meganathan, and Helene Kabes, for their tremendous help and patience in the various stages of this project.

Finally, I am grateful to my family for their understanding and support when I was spending long hours in the office working on this volume.

Ya-Xiong Tao
Auburn, Alabama

CHAPTER ONE

G Protein-Coupled Receptors as Regulators of Energy Homeostasis

Ya-Xiong Tao[*], **Zong-Hui Yuan**[†], **Jun Xie**[‡]

[*]Department of Anatomy, Physiology and Pharmacology, College of Veterinary Medicine, Auburn University, Auburn, Alabama, USA

[†]Department of Basic Veterinary Medicine, College of Animal Science and Veterinary Medicine, Huazhong Agricultural University, Wuhan, Hubei, People's Republic of China

[‡]Freshwater Fisheries Research Center, Chinese Academy of Fishery Sciences, Wuxi, Jiangsu, People's Republic of China

Contents

Abstract

G protein-coupled receptors (GPCRs) are versatile regulators of physiological processes. They are also important drug targets. Many of the molecules controlling energy homeostasis act through GPCRs. This article summarizes the regulators of energy homeostasis in the central nervous system: those secreted by the gastrointestinal peptides and those secreted by the endocrine pancreas. Some examples of orphan GPCRs are also given. The regulation of energy homeostasis is conserved in other mammals, including those species relevant in veterinary medicine, and fish. Finally, the genetics of human obesity is briefly summarized. Genetic susceptibility in the current obesogenic environment is likely causing the obesity pandemic. A better understanding of the regulation of energy homeostasis will lead to novel pharmacotherapy for obesity treatment.

Progress in Molecular Biology and Translational Science, Volume 114
ISSN 1877-1173
http://dx.doi.org/10.1016/B978-0-12-386933-3.00001-7

ABBREVIATIONS

AgRP Agouti-related peptide
AP area postrema
BBS Bardet–Biedl syndrome
BDNF brain-derived neurotrophic factor
BMI body mass index
BRS-3 bombesin receptor subtype-3
cAMP cyclic AMP
CART cocaine- and amphetamine-regulated transcript
CCK cholecystokinin
CeA central nucleus of the amygdala
CRH corticotropin-releasing hormone
DAG diacylglycerol
DPP-IV dipeptidyl peptidase IV
FTO the fat mass and obesity-related gene
GHRH growth hormone-releasing hormone
GI gastrointestinal
GLP-1 glucagon-like peptide-1
GLP-1R glucagon-like peptide-1 receptor
GPCR G protein-coupled receptor
GRP gastrin-releasing peptide
GWAS genome-wide association studies
ICV intracerebroventricular
KO knockout
MC3R melanocortin-3 receptor
MC4R melanocortin-4 receptor
MCH melanin-concentrating hormone
MSH melanocyte-stimulating hormone
NPY neuropeptide Y
NTS nucleus of the tractus solitaries
OXM oxyntomodulin
PC prohormone convertases
PKA protein kinase A
PKC protein kinase C
POMC proopiomelanocortin
PP pancreatic polypeptide
PYY peptide tyrosine tyrosine
TrkB tropomyosin-related kinase B

1. INTRODUCTION TO G PROTEIN-COUPLED RECEPTORS

G protein-coupled receptors (GPCRs) comprise the largest family of membrane receptors in the mammalian genome. In humans, there are 799 human GPCR sequences and 342 unique functional nonolfactory GPCRs identified from the genome.[1] GPCRs are versatile regulators of every aspect of the body's functions. It is estimated that 50% of current drugs target GPCRs, amounting to $50 billion in sales annually.[2]

Extensive analysis of GPCR sequences and mutagenesis studies suggested that GPCRs share a common topology consisting of seven transmembrane domains connected with alternating extracellular and intracellular loops, with the N-terminus located extracellularly and the C-terminus located intracellularly.[3] The crystal structures of GPCRs, first that of rhodopsin, and more recently, for those of increasing numbers of GPCRs that bind diffusible ligands, have confirmed this hypothesis.[4–12] These crystal structures further revealed an additional α-helix that is perpendicular to the other seven transmembrane domains after transmembrane domain 7, which is called Helix 8.

Diverse extracellular signals serve as the ligands for GPCRs, including light, odorants, ions, amines, nucleotides, fatty acids, peptides, and large glycoproteins.[13] The binding of these ligands causes conformational changes in GPCRs resulting in the exchange of GTP for GDP on the α-subunit of the heterotrimeric G protein and dissociation of the α-subunit from the βγ-dimer. The GTP-bound α-subunits activate several enzymes resulting in changes in levels of several second messengers. For example, the stimulatory G protein Gs will increase the enzyme activity of adenylyl cyclase causing intracellular levels of the second messenger cyclic AMP (cAMP) to rise whereas the inhibitory G protein Gi will have the opposite effect, decreasing intracellular cAMP levels. The activation of Gq will increase intracellular inositol phosphate and diacylglycerol (DAG) as well as calcium ion levels.

The second messengers will activate downstream protein kinase. cAMP will activate protein kinase A (PKA, also known as cAMP-dependent protein kinase) whereas inositol phosphate, DAG, and calcium ion will activate protein kinase C (PKC). PKA is a holoenzyme consisting of two regulatory and two catalytic subunits. When cellular levels of cAMP are low, PKA is catalytically inactive. When the intracellular cAMP levels increase, cAMP binds to the two binding sites on each of the regulatory subunits (with a total of four molecules of cAMP bound). This binding results in conformational change in the enzyme releasing the catalytic subunits. The free catalytic subunits will phosphorylate other proteins on either serine or threonine residues, changing their activities. This pathway is quick, on the time scale of minutes. There is also a slower pathway affecting protein synthesis that takes hours to days. In this pathway, PKA phosphorylates cAMP response element-binding protein, a transcription factor that binds to the cAMP response element in target genes, changing the transcription and subsequent synthesis of target proteins.

The PKC is in fact a family of enzymes, consisting of 15 members in humans.[14] Depending on their requirements of second messengers, these isozymes are divided into three subfamilies: conventional (or classical), including α, β, and γ, that require Ca^{2+}, DAG, and phosphatidylserine for activation; novel, including δ, δ_1, δ_2, δ_3, ε, η, and θ, that require DAG but not Ca^{2+} for activation; and atypical, including ι, ξ, PK-N1, PK-N2, and PK-N3, that require neither DAG nor Ca^{2+} for activation. Different from PKA, PKC members are single polypeptides consisting of a regulatory domain and a catalytic domain tethered together by a hinge region.

Although drugs targeting GPCR already account for a significant share of current market, they only target 20 or so of the 799 GPCRs.[15] Therefore, there is enormous potential in exploiting the remaining nonolfactory GPCRs as novel therapeutics. Obesity is an area that is very actively pursued for novel therapeutic modalities. Several antiobesity drugs that were used but currently discontinued act on GPCRs or change the ligand concentration of GPCRs, including amphetamine, d-fenfluramine, rimonabant, an antagonist for the endocannabinoid receptor CB1. Some of these drugs (such as amphetamine) are discontinued due to their potential for abuse; some (such as d-fenfluramine and rimonabant) are discontinued due to serious side effects.

2. OBESITY AND CURRENT TREATMENTS

Obesity in adults is defined by a body mass index (BMI, body weight in kilograms divided by height in meters squared) of $\geq$30 kg/m^2, whereas a BMI of 25 and 29.9 is defined as overweight. Obesity is not just a cosmetic problem with its consequent social and psychological effects on the obese.[16] Obesity, especially abdominal obesity frequently found in middle-aged people, is associated with a number of important health disorders including type 2 diabetes mellitus, nonalcoholic fatty liver disease, dyslipidemia, hypertension, arteriosclerosis, cardiovascular disease, certain types of cancer (such as endometrial, postmenopausal breast, kidney, prostate and colon cancer), polycystic ovarian syndrome, and sleep apnea.[17] Therefore obesity should be considered as a disease.[18]

Obesity has become an epidemic in developed countries and is increasing at an alarming rate in developing countries, rapidly becoming a critical health problem. In the United States, nearly 70% of adults are considered as overweight or obese compared with less than 25% 40 years ago.[19] According to World Health Organization, globally, more than 1 billion adults are overweight and at least 300 million adults are obese (http://www.who.int/dietphysicalactivity/media/en/gsfs_obesity.pdf, accessed 20 August 2012). According to this report, "current obesity levels range from below 5% in China, Japan, and certain African nations, to over 75% in urban Samoa. But even in relatively low prevalence countries like China, rates are almost 20% in some cities." The report also pointed out that the obesity epidemic is not restricted to developed countries. Developing countries often have faster increases than in the developed countries. Another survey showed that the prevalence of overweight, general obesity, and abdominal obesity among Chinese adults has increased significantly from 1993 to 2009.[20]

Childhood obesity is also becoming a significant public health crisis. It was estimated that 10% of the world's school-aged children are carrying excess body fat, with a quarter of these children obese.[21] The prevalence is dramatically higher in developed countries, but it is rising significantly in most parts of the world. These overweight children will be more likely to have obesity-associated diseases such as type 2 diabetes and heart disease before or during early adulthood.[21] For white men aged 20–30 years with a severe level of obesity (BMI >45), the maximum years of life lost is 13, representing a 22% reduction in life expectancy.[22] Because of the effect of obesity on longevity, it was suggested that the steady rise in life expectancy during the past two centuries in the United States might soon come to an end, with a potential decline in life expectancy in the United States in the twenty-first century.[23,24]

In addition to the health burden, obesity is associated with significant economical costs. Obesity imposes a substantial drain on health care resources, accounting for ~6% of adult medical expenditures in the United States.[25] A paper published in 2004 reported that annual medical expenditures that can be attributed to obesity are estimated at $75 billion in 2003 in the United States, with state-level estimates ranging from $87 million in Wyoming to $7.7 billion in California.[25] The majority of these costs are financed by taxpayers,[26] with roughly half financed by Medicare and Medicaid.[25] Even after taking into account the shortened lifespan for the obese, the lifetime costs of obesity are positive and generally increase with increasing BMI.[27]

Current treatments for obesity include lifestyle intervention such as increased exercise, dieting, and behavioral counseling, pharmacological methods such as intestinal lipase inhibitors (orlistat), as well as bariatric surgery for the morbidly obese (BMI of 40 or more) patients. For the vast majority of severely obese patients, lifestyle intervention can achieve short-term success, but often fail over the long term due to the body's defense of body fat mass, representing perhaps the single largest obstacle in preventing weight regain.[28,29] Pharmacotherapy, including sibutramine, a serotonin–norepinephrine reuptake inhibitor that reduces appetite, and orlistat, a triacylglycerol lipase inhibitor that reduces dietary fat absorption, is appropriate for some patients as an adjunct to lifestyle interventions to facilitate weight loss and prevent weight regain.[30] However, existing pharmacotherapy has limited effectiveness (mean weight loss of 4.45 kg for sibutramine and 2.89 kg for orlistat at 12 months, according to a meta-analysis [31]) and an imperfect safety record.[32] Indeed, sibutramine was withdrawn in Europe and North America in 2010 when a large cardiovascular outcome trial found long-term sibutramine treatment is linked to increased risk of nonfatal myocardial infarction and nonfatal stroke in patients with preexisting cardiovascular problems.[33]

Bariatric surgery, reserved for patients who have a BMI of 40 or more or for those who have a BMI of 35 or more with coexisting medical conditions,[34] is the most effective treatment for morbid obesity, resulting in 61.2% of weight loss on average.[35] It also reduces obesity-related comorbidities, especially type 2 diabetes, independent of weight loss.[35–37] In one meta-analysis, it was shown that diabetes is completely resolved in 76.8% of patients and resolved or improved in 86.0%.[35] In patients with uncontrolled diabetes, bariatric surgery eliminates the need for medications in some patients and markedly reduces the need for drug treatment in others, representing a potential strategy for management of these patients.[38,39] The long-term efficacy of bariatric surgery has also been established.[40,41] These studies showed that bariatric surgery could be more effective than either standard or intensive medical treatment in the management of type 2 diabetes. Bariatric surgery can even prevent the development of type 2 diabetes in obese subjects.[42] However, only a small fraction of the qualifying obese population undergoes these procedures because of the significant risks around and after operation, like any surgery, and the high cost.[32,43,44] In a large clinical trial with operations performed by the best surgeons, the overall rate of death for bariatric surgery was 0.3%, and major complications occurred in 4.1% of the patients.[45]

3. REGULATION OF ENERGY HOMEOSTASIS IN THE CENTRAL NERVOUS SYSTEM

Energy homeostasis refers to the matching of energy intake with energy expenditure over a prolonged period in adult animals. Therefore there is no significant change in body weight. For example, when animals are forcibly overfed (or underfed) for a short term and then allowed to feed *ad libitum*, they return to their initial weights with dramatic precision.[29] Caloric restriction, especially short term, is associated with a compensatory decrease in energy expenditure, contributing to the difficulty of maintaining the weight loss by dieting.[46] Energy intake is intake through the form of food and drinks. Energy expenditure includes resting energy expenditure (basal metabolism), which represents the energy needed to support minimal daily functions, accounting for two-thirds of total daily energy expenditure, and physical activity.

The maintenance of energy balance is achieved by incorporating both long-term adiposity and short-term satiety signals. The two major adiposity signals are leptin and insulin.[28,47–49] The discovery of leptin by Douglas Coleman and Jeffrey Friedman was a landmark event in modern biomedical research, for which they received the Albert Lasker Basic Medical Research Award in 2010.[50] Leptin is a hormone produced by adipocytes[51] that informs the brain about the status of energy stores in the organism through a negative feedback loop. In mammals, leptin decreases food intake through inhibiting the orexigenic neuropeptide Y (NPY)/Agouti-related peptide (AgRP) and activating anorexigenic proopiomelanocortin (POMC)/cocaine- and amphetamine-regulated transcript (CART) neurons (reviewed in Refs. 47,52,53). These two types of neurons express leptin receptors.[54] Leptin, by binding to the leptin receptors in the arcuate nucleus, stimulates arcuate nucleus POMC gene expression.[55,56] AgRP is an antagonist of the melanocortin-4 receptor (MC4R), and POMC-derived α-melanocyte-stimulating hormone (MSH) is an agonist of the MC4R. MC4R signaling is an important mediator of leptin's effects on energy homeostasis.[57] Deficiencies in leptin signaling are responsible for the profound genetic obesity present in rodent obesity models, including *ob/ob* mice (leptin deficiency),[51] *db/db* mice (leptin receptor deficiency),[58] and Zucker *fa/fa* rat (leptin receptor deficiency)[59].

After the cloning of leptin gene, it was thought that leptin would be the wonder drug that could cure obesity. However, disappointing efficacy in clinical trials[60] dashed the initial enthusiasm. Indeed, circulating leptin levels are proportional to body fat mass, therefore obese humans, rather than having low leptin levels (that could be corrected by leptin administration), have very high circulating leptin levels.[61–63] Therefore human obesity is associated with leptin resistance rather than leptin deficiency.[61,64,65] Several mechanisms for leptin resistance have been proposed, such as impaired transport of leptin across the blood–brain barrier, endoplasmic reticulum stress, inflammation, perturbations in developmental programming, and leptin-signaling defects.[66,67] The exact mechanisms of leptin resistance are not clear yet, but they are very actively studied at the cellular, physiological, and behavioral levels (for recent reviews, see Refs. 67,68).

Feeding behavior is finely tuned by a complex interplay of numerous molecules, including neurotransmitters, neuropeptides, and hormones, and the hypothalamus plays an important role in the integration of these signals.[69] Table 1.1 lists some of the neuropeptides involved in regulating food intake. The hypothalamus receives and integrates neural, metabolic, and hormonal signals from the periphery and central nervous system (CNS) to control energy homeostasis and body weight. In addition to hypothalamus, different brain areas, including the brainstem, are involved in responding to these molecules and are interconnected, activating either the orexigenic or anorexigenic pathways.

There are several nuclei in the hypothalamus, including the arcuate nucleus, the paraventricular nucleus, the dorsomedial nucleus, and the preoptic area, that are critical for regulating energy homeostasis. These nuclei perform different functions. Arcuate nucleus neurons respond to a wide array of hormones and nutrients, including adiposity signals (leptin and insulin), gastrointestinal (GI) hormones (such as ghrelin), gonadal steroids, and glucose.[70] There are transport mechanisms that carry these molecules through the blood–brain barrier. In addition, some arcuate neurons have cell bodies in the median eminence or have projections to the median eminence, where there is no blood–brain barrier. Neurons in other parts of the hypothalamus, such as the paraventricular nucleus and the dorsomedial nucleus are also critical for regulating energy homeostasis, for example, in mediating leptin

Table 1.1 Regulation of appetite by neuropeptides acting on GPCRs

Neuropeptide	Receptor	Effects on food intake
NPY	Y1R, Y5R	↑
α-MSH	MC4R	↓
AgRP	MC4R	↑
CART	Unknown	↓
CRH	CRH-R	↓
Neurotensin	Neurotensin receptor 1	↓
Orexin	Orexin receptors	↑
MCH	MCH-R	↑
Galanin	Galanin receptors	↑

action.[71] The preoptic area controls energy expenditure and core body temperature regulation.[72]

In addition to hypothalamus, central regulation of energy balance also includes the midbrain[73] and the brainstem. For example, the ventral tegmental area and striatum, which comprise the core of the mesolimbic dopamine system, are largely responsible for mediating the hedonic or rewarding value of palatable food such as fast food and soft drinks.[71] These reward circuits are integrated with the circuits controlling energy homeostasis to generate signals for food intake,[29] for example, deciding whether you are going to eat the dessert after you are already full with the main courses. The brainstem is also intimately involved in regulating energy homeostasis. The dorsovagal complex comprises: (1) the area postrema (AP), a circumventricular organ lacking a functional blood–brain barrier therefore capable of sensing chemical changes in the blood; (2) the nucleus of the tractus solitarius (NTS), a center that integrates GI distension, mechanosensory, and other inputs from the viscera; and (3) the dorsal motor nucleus of the vagus, a center that integrates motor and secretory drive to the viscera. NTS receives and integrates vagal afferent satiation, blood-borne energy status signals (such as glucose and fatty acid levels), and GI peptides secreted during a meal (see below), and send signals to the dorsal motor nucleus of the vagus to maintain energy balance.[74] Several peptides, including those secreted in response to meals, such as amylin, are sensed at AP to terminate the meal, generating anorexic responses (see below). Administration of melanin-concentrating hormone (MCH) into the fourth ventricle targeting the NTS decreases body temperature without changing physical activity and food intake, suggesting that MCH inhibits energy expenditure through hindbrain.[75]

Neurons in the NTS also contain elements of leptinergic and melanocortinergic signaling systems, potentially also integrating leptin and melanocortin signals to control food intake.[76] For example, in addition to its production in the arcuate nucleus of the hypothalamus and pituitary gland (where it is cleaved to adrenocorticotropin), POMC is also prominently expressed in the NTS in the hindbrain.[77,78] The MC4R is also expressed abundantly in the dorsal motor nucleus of the vagus[79] and these receptors regulate food intake and energy expenditure similar to the receptors in the hypothalamus, decreasing food intake and increasing energy expenditure when activated.[80–84] Injection of MC4R ligands into the fourth ventricle near the dorsal vagal complex affects meal size, though not meal frequency.[85] Brainstem melanocortin system is also involved in mediating cholecystokinin (CCK) inhibition of food intake.[86]

4. REGULATION OF ENERGY HOMEOSTASIS BY THE GI PEPTIDES

The GI tract sends signals to the brain to control energy homeostasis. These hormones produced by the GI tract convey information on the nutritional status and/or energy storage to the CNS. The CNS processing of GI signals secreted in response to ingested food determines meal size (satiation) and meal frequency (satiety).[74] Important hormones produced by the GI tract include ghrelin that is reviewed in this volume by Weizhen Zhang and colleagues,[87] obestatin summarized in this volume by Jian Zhang and colleagues,[88] and CCK and gastrin-releasing peptide (GRP) that are reviewed in this volume by Sayegh.[89,90] There are several other hormones produced by the GI tract that exert important effects on energy balance, forming the gut–brain axis (Table 1.2).

Peptide tyrosine tyrosine (PYY) is produced by the *enteroendocrine* L cells in the distal gut that decreases food intake (reviewed in Ref. 91). Two forms of PYY, PYY_{1-36} and PYY_{3-36}, are released postprandially, with the latter form consisting of the predominant form, generated by cleavage PYY_{1-36} of the tyrosine proline residues at the N-terminus by dipeptidyl peptidase IV (DPP-IV), an aminopeptidase that cleaves dipeptides from the N-terminus of proteins containing an alanine or proline at the second position. Postprandial plasma PYY levels are proportional to the amounts of calories consumed. These PYY molecules signal through binding to the G protein-coupled NPY

Table 1.2 Regulation of appetite by gastrointestinal peptides acting on GPCRs

GI peptide	Receptor	Effect on food intake
PYY_{1-36}	Y1R, Y5R	↑
PYY_{3-36}	Y2R	↓
PP	Y4R	↓
GLP-1	GLP1R	↓
Oxyntomodulin	GLP1R?	↓
Ghrelin	GHSR	↑
CCK	CCK-R	↓
Gastrin-releasing peptide	GRP-R/BB2R	↓

receptors (Y1, Y2, Y4, Y5, and Y6).[92] Of the five subtypes of NPY receptors, PYY_{1-36} bind to Y1 and Y5 receptors stimulating appetite, generating an orexigenic signal, whereas PYY_{3-36} binds selectively to Y2 receptor enhancing satiety, generating an anorexigenic signal.[93]

Preproglucagon is a preprohormone that undergoes tissue-specific post-translational processing generating a number of important hormones. In the pancreas, the major hormone produced is glucagon, whereas in the intestine, it is processed into glucagon-like peptide-1 (GLP-1), GLP-2, and oxyntomodulin (OXM). The 29-amino acid glucagon produced in α-cells in the pancreas is a well-known hormone increasing plasma glucose level, counteracting the action of insulin produced by β-cells in the pancreas (its action on regulating energy homeostasis will be reviewed in Section 5). Prohormone convertase (PC) 2 is suggested to be essential for the processing of proglucagon to glucagon. In the gut enteroendocrine L cells, proglucagon is processed to glicentin, OXM, GLP-1, and GLP-2. PC1/3 is responsible for the processing of proglucagon to these peptides. DPP-IV is also involved in degradation of these peptides, resulting in a half-life of about 1 min for GLP-1.[94] Glucagon, GLP-1, and GLP-2 all have their own receptors (all are GPCRs) whereas OXM can bind to and activate both glucagon and GLP-1 receptor (GLP-1R). For a comprehensive review on these receptors, the reader is referred to Ref. 95.

GLP-1R was first cloned from a rat pancreatic islet cDNA library[96] and the human *GLP-1R* gene was isolated and mapped to chromosome 6 band p21.1, in 1993.[97] GLP-1R is coupled to multiple G proteins, including Gs, Gq/11, and Gi1/2, changing cellular levels of second messengers such as cAMP, calcium, and DAG.

In addition to various peripheral tissues such as the intestine and pancreatic islets, GLP-1R is also highly expressed in several regions of the brain,[98] including nuclei in the hypothalamus critically involved in regulating food intake (such as the arcuate nucleus, paraventricular nucleus, and supraoptic nucleus) and hindbrain.[99–101] Intracerebroventricular (ICV) administration of GLP-1 potently inhibits feeding in fasted rats, which is blocked by the exendin (9–39),[102] a specific GLP-1R antagonist.[103] Exendin (9–39) also potentiates the orexigenic effect of NPY.[102] Immunohistochemistry experiments showed that ICV injection of GLP-1 induces *c-fos* expression exclusively in the paraventricular nucleus of the hypothalamus and central nucleus of the amygdala, and exendin (9–39) blocks this neuronal activation.[102] Direct injection of GLP-1 into the fourth ventricle decreases food intake without inducing conditioned taste aversion, suggesting that brainstem

GLP-1R is also involved in the anorexic effect of GLP-1.[104] ICV administration of GLP-1 induces conditioned taste aversion.[104,105] Intraperitoneal administration of GLP-1 either does not affect food intake[100,106] or decreases food intake.[107] Chronic administration of a long-acting injectable GLP-1 derivative, NN2211, inhibits food intake and decreases body weight in both lean and obese rats.[108] The vagal afferents, brainstem, and direct CNS action all contribute to the effects of circulating GLP-1 on food intake.[107,109–111] In humans, intravenous GLP-1 infusions decrease food intake and enhance satiety dose dependently.[112,113] This inhibitory effect of GLP-1 on food intake and appetite has been confirmed in obese men[114] and type 2 diabetics.[115] A meta-analysis showed that GLP-1 infusion decreases *ad libitum* energy intake dose dependently in both lean and overweight subjects.[116] A reduced gastric emptying rate may contribute to the increased satiety induced by GLP-1.[116] Indeed, weight loss is frequently observed in diabetic patients treated with GLP-1R agonists.[117]

The OXM contains the glucagon molecule (1–29) elongated at the C-terminus by eight amino acids. In addition to its originally described function of stimulating acid secretion from the oxyntic glands in the stomach,[118] OXM also stimulates insulin secretion[119], enhances intestinal glucose uptake,[120] inhibits gastric emptying,[121] and decreases basal as well as postprandial gastric acid secretion.[121–123] More relevant to the present discussion, it has been shown to be a potent anorexigenic peptide with similar potency as GLP-1. Both peripheral and central administration of OXM has been shown to decrease food intake in rodents.[124–128] Chronic administration of OXM results in decreased body weight and adiposity, even when food was restricted, suggesting that increased energy expenditure was also involved.[125] In a double-blind, randomized, controlled trial, subcutaneous administration of OXM decreases energy intake, increases energy expenditure, resulting in decreased body weight in healthy overweight and obese human volunteers.[129,130] Subjects treated with OXM have decreased leptin and increased adiponectin levels, consistent with loss of white adipose tissue.[129] Both glucagon receptor and GLP-1R are involved in mediating the weight-lowering actions of OXM.[127,128] Pharmacologically, OXM has weak affinity for both glucagon receptor and GLP-1R.[131]

5. REGULATION OF ENERGY HOMEOSTASIS BY PEPTIDES FROM ENDOCRINE PANCREAS

Several peptides released from endocrine pancreas that act on GPCRs, including glucagon, pancreatic polypeptide (PP), and amylin, also induce satiety. Insulin is another critical regulator of energy homeostasis secreted by the endocrine pancreas.[132–134] However, the insulin receptor, widely distributed in the CNS,[135] is a tyrosine kinase receptor, not a GPCR, the focus of this volume.

Studies dating back several decades already demonstrated clearly that in rodents and humans, glucagon decreases food intake and increases energy expenditure[136–140] (reviewed in Refs. 141–143). Glucagon can abolish gastric contractions and diminish hunger sensation.[144] Intraperitoneal injections of glucagon antibodies to fasted rats at the onset of the first meal increase meal size and meal duration, suggesting that glucagon is necessary for the normal termination of meals.[145] ICV administration of glucagon also decreases food intake in chickens[146] and sheep.[147] These results suggest that glucagon can act directly on the CNS to inhibit food intake, consistent with expression of glucagon receptor in the brain,[148] although the hepatic branch of the abdominal vagus is likely critical for relaying the signal to the brain.[143] Partly through activation of the sympathetic nervous system, glucagon also modulates thermogenesis and energy expenditure.[143] Studies with glucagon receptor KO mice showed that these mice have normal food intake, energy expenditure, and body weight, but decreased adiposity and leptin levels, suggesting glucagon action may be involved in the regulation of whole body composition.[149] When challenged with high-fat diet, glucagon receptor KO mice have decreased food intake and body weight.[150] Two important studies highlighted potential clinical utility of dual agonism of both glucagon and GLP-1Rs , suggesting potential benefits of activating glucagon receptor as a novel therapeutics for metabolic syndrome.[151,152]

PP, first isolated from chicken pancreas[153] and then in mammals,[154] is secreted by the PP cells in islets distinct from those storing insulin, glucagon, or somatostatin, during meals. Blood PP levels are directly proportional to the amount of calories ingested. PP is a 36-amino acid peptide, the newest member in a family of peptides that also includes NPY and PYY.[92,155] Of the five NPY receptors, PP preferentially binds to Y4 receptor. Through activation of Y4 receptor in hindbrain,[156] PP decreases food intake in both mice[157,158] and men,[159,160] including obese models such as *ob/ob* mice and patients with Prader–Willie syndrome, the latter being characterized by early-onset morbid obesity and a deficiency in both basal and meal-stimulated PP secretion. Peripherally administered PP decreases gene expression of the orexigenic NPY, orexin, and ghrelin and increases the expression of anorexigenic urocortin. Energy expenditure is also increased, with increased sympathetic nerve activity and oxygen consumption.[158] Due to both decreased food intake and increased energy expenditure, repeated administration of PP results in decreased body weight.[158] Transgenic animal models that overexpress PP confirmed the inhibitory effect of PP on food intake and body weight. The mice with increased plasma PP are hypophagic and lean.[161] In addition, PP secretion is diminished in obese subjects and increased in patients with anorexia nervosa.[162,163] It is interesting to note that when administered centrally, PP is orexigenic, mimicking the action of NPY.[164–166] These data suggest that PP is involved in regulating energy balance in both experimental animals and men.

Amylin or islet amyloid polypeptide, a 37-amino acid peptide found in extracellular amyloid deposits in some pancreatic islets of type 2 diabetic patients,[167] is co-secreted with insulin from the β-cells in the pancreatic islet. Amylin has characteristics of both short-term (satiety) and long-term (adiposity) signals in its role in regulating food intake.[141] Similar to PP, amylin is also secreted during meals proportional to the amount of calories ingested.[168] Nutrient secretagogues and peptide modulators can directly act on β-cells to alter amylin secretion.[169]

Several groups have reported an anorexigenic effect of amylin in rodents administered either centrally or peripherally, primarily by reducing meal size without producing conditioned taste aversion, decreasing body weight with chronic administration.[170–176] The anorexigenic effect is as potent as that of CCK, and when administered together, amylin and CCK have a synergistic effect on inhibiting food intake; ineffective doses of amylin and CCK, when combined, result in near-maximal inhibition of food intake.[177] It was suggested that since amylin and CCK are both secreted in response to mixed meals, the synergy between them could indicate a shared role in physiological appetite control.[177] Administration of amylin antagonists increases food intake by increasing both meal size and meal frequency, suggesting that endogenous amylin plays a physiological role in regulating food intake.[178–180]

AP senses circulating amylin. Peripheral administration of amylin results in strong neuronal activation in the AP and NTS, as well as the lateral parabrachial nucleus and the central nucleus of the amygdala (CeA), but not in the paraventricular nucleus.[181] In rats with lesions of the AP/NTS, the anorectic effect of amylin was completely eliminated or significantly attenuated, suggesting that the anorectic effect of amylin is primarily through its direct action at the AP/NTS.[182–184] In these rats, amylin-induced c-Fos expression in the NTS, the lateral parabrachial nucleus, and the CeA was blunted, suggesting that the activation of these structures is mediated by AP.[185] In rats with electrolytic lesions in the lateral parabrachial nucleus, the anorectic effect of a low-dose amylin was significantly reduced, and the amylin-induced neuronal activation in the CeA was completely abolished, suggesting that the lateral parabrachial nucleus lesion disrupted the signal transduction pathway between the AP/NTS and CeA.[186]

Taken together, these studies suggest that the well-established neuronal axis regulating ingestive behavior comprising the AP, the NTS, the lateral parabrachial nucleus, and the CeA, mediates amylin's satiating effect.[186]

Synthetic analog of amylin developed by Amylin Pharmaceuticals, pramlintide (Symlin®), causes significant reductions in meal size and cravings for food and body weight in obese humans.[187] Further clinical trials suggested that amylin is not efficacious enough to justify as a monotherapy.[188,189] However, in both rats and humans, amylin (acting in the hindbrain) and leptin (acting in the hypothalamus) coadministration shows significant synergistic effects on body weight reduction, primarily due to decreased food intake, suggesting that amylin agonism restores leptin responsiveness in diet-induced obesity.[190–192] Amylin agonist together with leptin agonist was in Phase 2 clinical trials by Amylin (acquired by Bristol-Myers Squibb) and Takeda Pharmaceutical Company Limited for treating human obesity; however, it was suspended due to a safety issue in 2011.[193]

6. REGULATION OF ENERGY HOMEOSTASIS BY ORPHAN GPCRs

Of the ~800 GPCRs in humans, more than 300 have been deorphanized during the past two decades.[194] Some of the important GPCRs in regulating energy homeostasis, such as the MCH receptor,[195–199] orexin receptors,[200] and ghrelin receptor,[201] were deorphanized by reverse pharmacology. The more recently deorphanized fatty acid receptors are also important for regulating energy homeostasis and metabolism. Among them, GPR120 is reviewed in this volume by Mo *et al.*[202] Despite the tremendous success, there are still ~100 chemosensory GPCRs remaining as orphan receptors with no endogenous ligands identified yet.[194] A number of these GPCRs also regulate energy homeostasis and therefore potential therapeutic targets for obesity treatment. The following are just two examples.

GPR21, located on chromosome 9 in human and chromosome 2 in mouse, is an orphan GPCR that is widely expressed in the mouse with the highest expression in the brain (including hypothalamus) and spleen.[203–205] Two groups recently independently generated *Gpr21* knockout mice. *Gpr21* deficient mice are resistant to high-fat diet-induced obesity due to increased activity and energy expenditure.[204,205] No difference in food intake, respiratory exchange ratio, and thermogenesis was observed between the KO mice and the wild-type littermates.[204] The KO mice have decreased fat percentage, decreased inflammation markers, and improved glucose tolerance and increased insulin sensitivity.[204,205] GPR21 is highly expressed in macrophage. Macrophage-specific *Gpr21* knockout mice are protected from high-fat diet-induced adipose tissue inflammation and have improved insulin sensitivity.[205]

Another orphan GPCR, bombesin receptor subtype-3 (BRS-3), in the same family as neuromedin B receptor and GRP receptor,[206] has also shown to be important for maintaining normal energy balance. Despite significant homology, BRS-3 does not respond to neuromedin B or GRP and the endogenous ligand for BRS-3 has not been identified. Mice lacking BRS-3 develop mild obesity with hyperphagia, reduced metabolic rate, and increased feeding efficiency.[207] These mice also have hypertension and impairment of glucose metabolism.[207] In the hypothalami of BRS-3 KO mice, expression of MCH receptor and MCH are upregulated (with increased hyperphagic response to MCH compared to wild-type littermates), which contributes to hyperphagia,[208] and hyperphagia is a major factor contributing to obesity and hyperinsulinemia in the BRS-3 KO mice.[209] After the onset of obesity, there is central leptin resistance.[208] Administration of synthetic BRS-3 antagonist increases food intake and body weight, whereas administration of synthetic agonists increases metabolic rate and reduces food intake and body weight, with no significant tachyphylaxis.[210] Additional data showed that the role of BRS-3 in energy homeostasis complements several well-known pathways, including leptin, NPY, MC4R, and CB1R.[210] An orally active, potent, and selective BRS-3 agonist, MK-5046, also demonstrates antiobesity efficacy in both rodents and dogs.[211] These studies clearly showed that BRS-3 agonists represent a potential approach for obesity treatment, either alone or in combination with drugs targeting other pathways.[210,211] A human genetic study in Japanese subjects did not find any mutation or polymorphism in the coding region of BRS-3, suggesting that mutation of this gene is not a major cause of obesity.[212]

7. REGULATION OF ENERGY HOMEOSTASIS BY GPCRs IN DOMESTIC ANIMALS

The goal in farm animals such as in pigs, cows, and sheep is to increase growth rate and feed efficiency. Therefore the studies of feed intake and feed efficiency are important for maximal financial gains for the farmers. Increased food intake can increase growth rate and meat production. It has also been proposed that some of these larger animals are better models than rodents for modeling human obesity, for example, the similar metabolic features, cardiovascular systems, and proportional organ sizes between pigs and humans.[213–215] Hence, studies on agricultural animals can benefit both animal agriculture and human medicine. The findings from reports published on these species are mostly consistent with results obtained in experimental animals such as rodents and clinical studies in humans. However, there are some notable exceptions. For example, ghrelin is known to be a potent stimulator of feeding in both rodents and humans (reviewed in this volume by Li *et al.*[87]). However, in sheep, although ghrelin can stimulate postprandial rise in growth hormone levels (similar to rodents and humans), it cannot stimulate voluntary food intake, whether administered intracerebroventricularly or intravenously, despite the fact that there is a preprandial increase in plasma ghrelin levels in sheep on programmed feeding[216].

Similarly, in the melanocortin system, there are also some important differences (reviewed in Ref. 53). Although *in vitro*, SHU9119 and AgRP act as antagonists at the porcine MC4R with similar affinities as human MC4R,[217,218] *in vivo*, these antagonists cannot stimulate feed intake, whereas ICV administration of superpotent agonist does decrease feed intake.[217] We have cloned porcine melanocortin-3 receptor (MC3R) and showed that it has the same ranking in both binding affinities and potencies as human MC3R, when multiple ligands were used in binding and signaling experiments.[219] No *in vivo* data have been published to show whether the MC3R has similar functions in regulating feed intake and fat storage in pigs. Different from ghrelin, in sheep, the melanocortin system is a critical component of appetite regulation.[53,220]

In companion animals, similar obesity epidemic is found as in humans, with 25–40% of owned dogs and cats considered to be obese,[221–223] with similar associated health consequences such as type 2 diabetes mellitus.[224] All these conditions greatly impact quality of life and owner's financial situations. Unfortunately, pet obesity is also a very difficult condition to treat and only one pharmacological weight-reduction agent has recently been approved for treating dog obesity.

Very limited studies on the regulation of energy homeostasis have been done in companion animals. Canine leptin has been cloned,[225] and similar to other mammals, plasma leptin concentration is positively correlated with adiposity, and therefore there is a good index of adiposity in dogs regardless of breed, age, and sex,[226,227] whereas canine adiponectin is negatively associated with adiposity.[228] Subcutaneous or intrathecal administration of recombinant human leptin decreases food intake, body weight, and body fat in dogs.[229]

With the MC4R, one naturally occurring variant, V213F, has been identified.[230,231] We showed that this variant does not cause any overt functional defects compared with wild-type canine MC4R.[232] The cloned canine MC4R has similar pharmacological properties as the human MC4R.[232] However, the *in vivo* functions of the MC4R have not been reported.

8. REGULATION OF ENERGY HOMEOSTASIS BY GPCRs IN LOWER VERTEBRATES

Because of the economic importance of several lower vertebrates, regulation of energy balance has been actively studied by scientists in both academia and industry. The goal of these studies is to promote positive energy intake with faster growth and shorter time to market. Energy efficiency is also relevant. A few examples of these studies are summarized below.

There are numerous studies in chicks on the regulation of energy homeostasis due to the economic importance of poultry industry. Similar mechanisms, neural pathways, and neuroanatomic sites are responsible for the regulation of feed intake and energy expenditure in birds as in mammals (reviewed in Refs. 233,234). Most of the time, findings in mammals on food intake and energy expenditure can be repeated in birds with some exceptions. The anorexigenic molecules include leptin, α-MSH, corticotropin-releasing hormone (CRH), ghrelin, growth-hormone-releasing hormone (GHRH), CCK, urotensin I/urocortin, GLP-1, serotonin, dopamine, among others, whereas the orexigenic molecules include NPY, AgRP, MCH, orexins, PP, galanin, β-endorphin, endocannabinoid, and others.[233] As it can be seen from this list, most of the molecules serve the same role in birds as in mammals. There are several notable exceptions. The cloning of chicken leptin is very controversial.[235,236] Some investigators suggested that chicken leptin gene originally cloned[237,238] does not exist in chicken genome.[236,239,240] Whether there is a leptin gene in chicks or whether indeed the leptin originally cloned was correct, or the real leptin gene remains to be identified and cloned is still hotly debated. However, the existence of leptin receptor genes in poultry species is firmly established, including chicken,[241] duck, turkey,[242] and goose.[243] The avian leptin receptor activates the same signaling system as mammalian leptin receptor. Exogenous administration of murine or chicken leptin through ICV, intravenous, intraperitoneal, or intramuscular routes, has been found to reduce feed intake[244,245] in chickens as well as in wild birds.[246] Leptin receptor polymorphisms have been reported to be associated with adiposity in chickens.[247]

Another notable difference between the mammalian and avian system in energy balance regulation is ghrelin and GHRH. Both ghrelin and GHRH are potent orexigenic molecules in mammals, but they are anorexigenic in birds. ICV administration of (rat and chicken) ghrelin and GHRH inhibits food intake in neonatal chicks.[248,249]

The melanocortins inhibit feed intake in birds, similar to mammals (reviewed in Ref. 53). A single intravenous injection of melanotan-II into feed-restricted birds leads to attenuation of food intake upon exposure to feeding *ad libitum* in both broiler and layer chickens and feed restriction significantly reduces POMC mRNA level in both types of chickens,[250] similar to findings in mammals. ICV injection of α-MSH but not chicken β- or γ-MSH decreases feed intake in chickens.[251] The effect of α-MSH is blocked by an MC4R-specific antagonist, suggesting the involvement of MC4R in the decreased feed intake.[251] Analysis of the differentially expressed genes in two extremely divergent lines of chickens in body weight showed that there are no significant differences in gene expression for AgRP, MC4R, POMC, and melanocortin-5 receptor genes.[252] Only the expression of MC3R mRNA is significantly higher in male chickens from the low body weight line compared to those from the high body weight line.[252]

Fishery is an important economy in many countries, including developing countries like China. Therefore there are extensive studies on the regulation of feeding in fish. Fish can also be used as another model system, in addition to rodents, for studying basic biological mechanism. Goldfish has been used extensively in studies on feeding regulation by giants in fish endocrinology like Richard Peter and his former associates. Zebrafish and pufferfish are also being increasingly used because their genome sequences are already published. For example, Cone and colleagues have been using zebrafish as a model for studying the melanocortin regulation of energy homeostasis.[253,254]

Earlier studies using either electrical stimulation or lesioning of specific brain regions confirmed that fish hypothalamus is also critical for regulating food intake (reviewed in Ref. 255). Leptin has been identified from several teleosts.[256–261] Although the homology between fish leptin and mammalian leptin is very low—about 15–30% in amino acid sequences, the predicted tertiary structure of fish leptin is very similar to mammalian leptin, with the typical four α-helix bundle structure of the class-I α-helical cytokines. Two cysteine residues for disulfide bond conserved in the mammalian leptin exist in identical locations in fish leptin. Duplicate leptin genes are a common feature of teleosts and the two leptin isoforms in a particular species also share a low homology, for example, at 24% in zebrafish,[259] although Perciformes may only have one leptin gene.[262] The major site of expression for leptin is liver, different from the adipose tissue in mammals.[256,262] Similarly, fish leptin receptor has low homology with mammalian leptin receptors, but they have similar gene structures.[263,264] An early study showed that green sunfish responds to murine leptin injections by increasing fat metabolism, although there is no difference in loss of total weight and body fat with chronic leptin treatment, likely due to the low homology between the murine and green sunfish leptin; therefore, murine leptin cannot activate sunfish leptin receptor optimally.[265] Similarly, peripheral administration of human leptin to immature coho salmon under both fed and fasted conditions, resulting in high circulating levels of human leptin, does not change energy homeostasis significantly.[266] However, in striped bass, intraperitoneal injection of human leptin suppresses appetite.[262] ICV injection of the rat leptin fragment 22–56 does not change food intake in channel catfish.[267] However, in goldfish, ICV or intraperitoneal administration of murine leptin significantly decreases food intake, partly through modulating the orexigenic effects of NPY and orexin.[69,268] Using homologous protein, leptin has been shown to have an acute effect on the regulation of food intake, energy expenditure, and lipid metabolism in grass carp, although chronically it is not effective.[260]

Results from studies on the goldfish have shown that most of the neuropeptides and their receptors important in regulating feeding in mammals are also present in fish, with similar expression patterns. Although most of the neuropeptides have similar actions on fish energy homeostasis, there are some examples of dramatic differences. For example, MCH is a known orexigenic peptide with data from different methodologies consistent with this conclusion. However, in goldfish, ICV administration of MCH of either fish or human origin dose dependently decreases feeding, whereas ICV administration of anti-MCH antiserum to neutralize the endogenous MCH significantly increases food intake.[269,270] Fasting decreases MCH-like immunoreactivity in the dorsal part of the nucleus recessus lateralis, a region of the goldfish hypothalamus.[270] It should be mentioned that transgenic medaka that overexpresses MCH has a lightened body color but normal feeding behavior and growth.[271]

The pathways for the anorexigenic and orexigenic neuropeptides in feeding regulation have been studied in teleosts, primarily by administering receptor antagonists (reviewed in Ref. 272). For example, in goldfish, preinjection of a Y1 receptor antagonist, BIBP-3226, completely blocks ghrelin-induced feeding, suggesting that ghrelin-induced feeding is mediated by NPY released in ghrelin-responsive neurons that then activates NPY receptor.[273] α-MSH and MC4R likely mediate the anorexigenic action of MCH in the goldfish mentioned above because MC4R antagonist HS024 blocks the MCH action on feeding.[274,275] The MC4R does not mediate the anorexigenic action of another hormone, chicken gonadotropin-releasing hormone II, in goldfish.[276] From similar studies like these, it was suggested that in goldfish, CRH mediates the anorexigenic actions of neuromedin U, pituitary adenylate cyclase-activating polypeptide, vasoactive intestinal polypeptide, and α-MSH, whereas in the orexigenic pathway, NPY, ghrelin, and orexin have a number of reciprocal interactions among them.[272]

Although most of the neuropeptides and their receptors in fish follow the discovery of their counterparts in mammals, MCH was indeed first described in fish as a hormone mediating color change in chum salmon[277] by concentrating melanin granules in the melanophores to lighten the skin. Subsequently, the mammalian MCH was shown to be a potent orexigenic hormone, playing a major role in the control of feeding and energy homeostasis[278] (reviewed in Ref. 279). In fish, the MCH has the opposite role in feeding regulation compared to mammals (see above).

Because the enormous diversity in the feeding habits and responses to nutritional and environmental changes of different species of fishes display, there are likely species-specific mechanisms regulating feeding.[280]

9. GENETICS OF HUMAN OBESITY

Obesity is a complex disorder with contributions from genetic, behavioral, psychological, and environmental factors. Studies have shown that both environmental factors and genetics contribute to human obesity. O'Rahilly and colleagues elegantly summarized the earlier studies on heritability of weight from following up of twins and adopted children, with a heritability of fat mass of between 40% and 70%.[281] A higher concordance of 0.7–0.9 between monozygotic twins, who share 100% of their genes, compared with a lower concordance of 0.35–0.45 between dizygotic twins, who share about 50% of their genes, was observed from these studies.[281,282]

Since the mid-1990s, mutations in a number of genes have been found to cause monogenic obesity. These genes include leptin,[283,284] leptin receptor,[285] *POMC*,[286] *PC1*,[287] *Sim1*,[288] *MC3R*[289–294] (reviewed in Ref. 295), *MC4R*[296–298] (reviewed earlier in Refs. 53,299,300, and comprehensively in this volume by Hinney *et al.*[301]), brain-derived neurotrophic factor (*BDNF*),[302] and tropomyosin-related kinase B (*TrkB*), the cognate receptor of BDNF.[303] Very recently, it was shown that retinoic acid-induced 1 haploinsufficiency, through downstream pathways including that of BDNF, may contribute to the significant early-onset obesity in patients with Smith–Magenis syndrome.[304,305] Most of the molecules are part of the leptin-melanocortin circuit, including leptin, leptin receptor, PC1, the enzyme that cleaves POMC into melanocortins, POMC, the precursor of the melanocortins α- and β-MSH (both important in maintaining human energy balance, see also Refs. 306,307), and the neural melanocortin receptors, MC3R and MC4R. Sim1 is important for development of paraventricular nucleus,[308] an area with abundant expression of the MC4R. BDNF and its receptor TrkB are important downstream mediators of MC4R action, implicated in regulating food intake.[309,310] Retinoic acid-induced 1 regulates BDNF expression. BDNF haploinsufficiency in patients with the Wilms' tumor, aniridia, genitourinary anomalies, and mental retardation (WAGR) syndrome is associated with early-onset obesity.[311] These data highlight the critical importance of the leptin-melanocortin circuit in maintaining energy homeostasis, not just in experimental animals such as rodents, but also in humans.

Studies of the monogenic obesity caused by mutations in these genes have led to a better understanding of the regulation of energy homeostasis. It also led to some immediate benefits, especially with the leptin-deficient patients, where administration of recombinant human leptin results in normalization of hyperphagia, a reduction in body weight and fat mass, and normal pubertal development.[282,312–314] Potential therapeutic approaches for patients with *MC4R* mutations, including pharmacological chaperones and novel synthetic ligands, are also investigated.[53,315–317]

In addition to the monogenic nonsyndromic obesity, syndromic obesity can also be caused by genetic mutations. For example, with Bardet–Biedl syndrome (BBS), a disorder characterized by multiple clinical phenotypes that include, in addition to obesity, pigmentary retinal dystrophy, postaxial polydactyly, developmental delay, and renal malformations.[318] Compared to non-BBS subjects with similar BMI, BBS subjects usually have higher adiposity, especially abdominal visceral fat.[319] Mutations in at least 12 BBS genes have been shown to cause this disease, including obesity.[318,320–322] Patients heterozygous for BBS gene mutations have increased risk of developing obesity than noncarriers, but without displaying other BBS phenotypes.[323] Four single-nucleotide polymorphisms in three genes (*BBS2*, *BBS4*, and *BBS6*) were found to be associated with common obesity in French Caucasians who do not have BBS syndrome.[321] A knockin mouse model for the most common mutation in *BBS1* gene (M390R) was generated.[324] The mutant mice are obese and hyperleptinemic, with increased food intake and decreased locomotor activity.[324] Although the 9+2 arrangement of axonemal microtubules was intact in the ependymal cell cilia that line the enlarged third ventricle of the brains from the knockin mice, some of the cilia were elongated or with abnormally swollen distal ends.[324] Defective cilia in POMC neurons have been shown to cause obesity.[325] Both the mouse models and BBS patients are hyperleptinemic, suggesting leptin resistance. BBS patients have markedly elevated serum leptin compared with matched obese controls, suggesting they may have more severe leptin resistance than non-BBS obese individuals.[319] Recent studies showed that the BBS proteins are important for leptin receptor trafficking, and leptin resistance caused by the loss of BBS genes could be due to abnormal leptin receptor handling in a subset of leptin-responsive neurons (POMC neurons)[326,327] (reviewed in Ref. 322). BBS1 protein interacts directly with the signaling isoform of leptin receptor but not the nonsignaling isoform, and the C-terminal cytoplasmic domain that is unique to signaling isoform is sufficient to mediate the interaction with BBS1 [327].

Alström syndrome is another autosomal recessive genetic disorder with childhood truncal obesity and its associated metabolic dysfunction in adulthood. It is caused solely by mutations in *ALMS1*, a large and complex gene with 23 exons and encoding a protein of 4169 amino acids. ALMS1 is also a ciliary protein and the phenotypes in Alström syndrome subjects overlap with those of other ciliopathies such as BBS. A mouse line with a spontaneous mutation in the ortholog of human ALMS1, *Fat Aussie*, is born with a normal birth weight but has hyperphagia and becomes severely obese.[328] Studies on the pathogenic mechanisms underlying obesity in BBS and Alström syndrome should not only lead to better treatments for these patients, they can also lead to a better understanding of the pathophysiology and potential treatment of common obesity.

There are also tremendous progresses in understanding the genetics of more common nonsyndromic forms of obesity, primarily of onset in adolescence or adulthood.[329–333] Indeed, the widely popular Obesity Gene Map update incorporating published results up to the end of October 2005 showed that there are putative loci on all chromosomes except Y, with single-gene mutations in 11 different genes, 50 loci related to Mendelian syndromes relevant to human obesity mapped, and causal genes or strong candidates identified for most of these syndromes, 253 QTLs for obesity-related phenotypes from 61 genome-wide scans, and associations between DNA sequence variation and obesity phenotypes in 127 candidate genes.[334] There is no comprehensive update for the past 7 years because of the dramatic increase in the field making it almost impossible.

Unlike the monogenic form of obesity, the effects of any individual genetic variant identified in these studies are small to very small, explaining only 1–2% of the BMI variance.[333] Proving the association of these variants with obesity phenotype is very challenging, necessitating the involvement of populations of sufficient size. Recent genome-wide association studies (GWAS) have identified common polymorphisms at numerous loci, each with very modest contributions to adiposity.[335–337] The strongest signal from the GWAS studies is the association with variants within *FTO* (the fat mass and obesity-related gene),[338–340] with individuals homozygous for the at-risk A allele weighing ~3 kg more than individuals with the low-risk T allele.[341] Genes identified from these GWAS studies will likely provide new insights into human body weight regulation,[336,337,342] for example, with the *FTO*.[341,343,344]

Genetics, epigenetics, and environment interact to cause obesity.[345] Even for monogenic form of obesity, environment is important for the presentation of obesity phenotype.[346] The current obesogenic environment, including the availability of energy-dense fast foods and sweetened soft drinks at low prices, and the sedentary lifestyle at work (mostly sitting in front of a computer screen, rather than laboring in a farm field) and home (TV viewing, mostly), together with genetic susceptibility, are the causes of current obesity pandemic. A better understanding of gene–environment interaction is critical for behavioral interventions to be successful in the prevention of obesity.

10. SUMMARY

Tremendous progresses have been made in our understanding of regulation of energy balance during the past several decades. These studies identified adiposity and satiety signals maintaining energy homeostasis over the long term and short term, respectively. The pathways for these signals are also being elucidated. Many of these regulators act through GPCRs, proven therapeutic targets. Although limited success has been obtained so far in turning the monumental knowledge on regulation of energy balance into pharmacotherapy for the obese patients, "the future looks bright for development of new and better antiobesity medications to be used with diet and exercise to facilitate substantial weight loss[347]". Indeed, the recent approval by the U.S. Food and Drug Administration of two new antiobesity drugs targeting GPCR or transporters affecting the release of neurotransmitters acting on GPCRs—Belviq (locaserin, targeting the 5HT2C receptor to induce satiety) and Qysmia (Qnexa, a combination of topiramate and phentermine, targeting the norepinephrine, dopamine, and serotonin systems to reduce hunger),[348] and a number of drug candidates in research or clinical trials targeting a number of GPCRs, including Y1R, Y2R, Y4R, Y5R, 5-HT2C, MC4R, amylin receptor, glucagon receptor, GLP-1R, ghrelin receptor, and MCH receptor [193]—suggests that this optimism is justified.

ACKNOWLEDGMENTS

Our studies on the melanocortin-3 and melanocortin-4 receptors were supported by the American Diabetes Association Grant 1-12-BS-212, National Institutes of Health Grant R15-DK077213, and Animal Health and Diseases Research Program as well as Interdisciplinary Grant from Auburn University College of Veterinary Medicine, and Intramural Grant Program of Auburn University (to Y.-X. T.).

REFERENCES

1. Fredriksson R, Lagerstrom MC, Lundin LG, Schioth HB. The G-protein-coupled receptors in the human genome form five main families. Phylogenetic analysis, paralogon groups, and fingerprints. *Mol Pharmacol* 2003;**63**:1256–72.
2. Heimann A, Ferro E. New approaches for G-protein coupled receptor ligands identification. *Annu Rev Biomed Sci* 2009;**11**:T95–T101.
3. Baldwin JM. The probable arrangement of the helices in G protein-coupled receptors. *EMBO J* 1993;**12**:1693–703.
4. Palczewski K, Kumasaka T, Hori T, Behnke CA, Motoshima H, Fox BA, et al. Crystal structure of rhodopsin: a G protein-coupled receptor. *Science* 2000;**289**:739–45.
5. Rasmussen SG, Choi HJ, Rosenbaum DM, Kobilka TS, Thian FS, Edwards PC, et al. Crystal structure of the human β_2 adrenergic G-protein-coupled receptor. *Nature* 2007;**450**:383–7.
6. Cherezov V, Rosenbaum DM, Hanson MA, Rasmussen SG, Thian FS, Kobilka TS, et al. High-resolution crystal structure of an engineered human β_2-adrenergic G protein-coupled receptor. *Science* 2007;**318**:1258–65.
7. Rosenbaum DM, Cherezov V, Hanson MA, Rasmussen SG, Thian FS, Kobilka TS, et al. GPCR engineering yields high-resolution structural insights into β_2-adrenergic receptor function. *Science* 2007;**318**:1266–73.
8. Jaakola VP, Griffith MT, Hanson MA, Cherezov V, Chien EY, Lane JR, et al. The 2.6 angstrom crystal structure of a human A_{2A} adenosine receptor bound to an antagonist. *Science* 2008;**322**:1211–7.
9. Chien EY, Liu W, Zhao Q, Katritch V, Han GW, Hanson MA, et al. Structure of the human dopamine D3 receptor in complex with a D2/D3 selective antagonist. *Science* 2010;**330**:1091–5.

10. Wu B, Chien EY, Mol CD, Fenalti G, Liu W, Katritch V, et al. Structures of the CXCR4 chemokine GPCR with small-molecule and cyclic peptide antagonists. *Science* 2010;**330**:1066–71.
11. Xu F, Wu H, Katritch V, Han GW, Jacobson KA, Gao ZG, et al. Structure of an agonist-bound human A2A adenosine receptor. *Science* 2011;**332**:322–7.
12. Hanson MA, Roth CB, Jo E, Griffith MT, Scott FL, Reinhart G, et al. Crystal structure of a lipid G protein-coupled receptor. *Science* 2012;**335**:851–5.
13. Bockaert J, Pin JP. Molecular tinkering of G protein-coupled receptors: an evolutionary success. *EMBO J* 1999;**18**:1723–9.
14. Mellor H, Parker PJ. The extended protein kinase C superfamily. *Biochem J* 1998;**332** (Pt 2):281–92.
15. Overington JP, Al-Lazikani B, Hopkins AL. How many drug targets are there? *Nat Rev Drug Discov* 2006;**5**:993–6.
16. Friedman JM. Modern science versus the stigma of obesity. *Nat Med* 2004;**10**:563–9.
17. Desvergne B, Michalik L, Wahli W. Be fit or be sick: peroxisome proliferator-activated receptors are down the road. *Mol Endocrinol* 2004;**18**:1321–32.
18. Allison DB, Downey M, Atkinson RL, Billington CJ, Bray GA, Eckel RH, et al. Obesity as a disease: a white paper on evidence and arguments commissioned by the Council of the Obesity Society. *Obesity (Silver Spring)* 2008;**16**:1161–77.
19. Flegal KM, Carroll MD, Ogden CL, Curtin LR. Prevalence and trends in obesity among US adults, 1999–2008. *JAMA* 2010;**303**:235–41.
20. Xi B, Liang Y, He T, Reilly KH, Hu Y, Wang Q, et al. Secular trends in the prevalence of general and abdominal obesity among Chinese adults, 1993–2009. *Obes Rev* 2012;**13**:287–96.
21. Lobstein T, Baur L, Uauy R. Obesity in children and young people: a crisis in public health. *Obes Rev* 2004;**5**(Suppl. 1):4–85.
22. Fontaine KR, Redden DT, Wang C, Westfall AO, Allison DB. Years of life lost due to obesity. *J Am Med Assoc* 2003;**289**:187–93.
23. Olshansky SJ, Passaro DJ, Hershow RC, Layden J, Carnes BA, Brody J, et al. A potential decline in life expectancy in the United States in the 21st century. *N Engl J Med* 2005;**352**:1138–45.
24. Finkelstein EA, Brown DS, Wrage LA, Allaire BT, Hoerger TJ. Individual and aggregate years-of-life-lost associated with overweight and obesity. *Obesity (Silver Spring)* 2010;**18**:333–9.
25. Finkelstein EA, Fiebelkorn IC, Wang G. State-level estimates of annual medical expenditures attributable to obesity. *Obes Res* 2004;**12**:18–24.
26. Finkelstein EA, Ruhm CJ, Kosa KM. Economic causes and consequences of obesity. *Annu Rev Pub Health* 2005;**26**:239–57.
27. Finkelstein EA, Trogdon JG, Brown DS, Allaire BT, Dellea PS, Kamal-Bahl SJ. The lifetime medical cost burden of overweight and obesity: implications for obesity prevention. *Obesity (Silver Spring)* 2008;**16**:1843–8.
28. Woods SC, Schwartz MW, Baskin DG, Seeley RJ. Food intake and the regulation of body weight. *Annu Rev Psychol* 2000;**51**:255–77.
29. Guyenet SJ, Schwartz MW. Clinical review: regulation of food intake, energy balance, and body fat mass: implications for the pathogenesis and treatment of obesity. *J Clin Endocrinol Metab* 2012;**97**:745–55.
30. Eckel RH. Clinical practice. Nonsurgical management of obesity in adults. *N Engl J Med* 2008;**358**:1941–50.
31. Li Z, Maglione M, Tu W, Mojica W, Arterburn D, Shugarman LR, et al. Meta-analysis: pharmacologic treatment of obesity. *Ann Int Med* 2005;**142**:532–46.
32. Nguyen N, Champion JK, Ponce J, Quebbemann B, Patterson E, Pham B, et al. A review of unmet needs in obesity management. *Obes Surg* 2012;**22**:956–66.

33. James WP, Caterson ID, Coutinho W, Finer N, Van Gaal LF, Maggioni AP, et al. Effect of sibutramine on cardiovascular outcomes in overweight and obese subjects. *N Engl J Med* 2010;**363**:905–17.
34. Gastrointestinal surgery for severe obesity: National Institutes of Health Consensus Development Conference Statement. *Am J Clin Nutr* 1992;**55**:615S–619S.
35. Buchwald H, Avidor Y, Braunwald E, Jensen MD, Pories W, Fahrbach K, et al. Bariatric surgery: a systematic review and meta-analysis. *J Am Med Assoc* 2004;**292**: 1724–37.
36. Chambers AP, Jessen L, Ryan KK, Sisley S, Wilson-Perez HE, Stefater MA, et al. Weight-independent changes in blood glucose homeostasis after gastric bypass or vertical sleeve gastrectomy in rats. *Gastroenterology* 2011;**141**:950–8.
37. Stefater MA, Wilson-Perez HE, Chambers AP, Sandoval DA, Seeley RJ. All bariatric surgeries are not created equal: insights from mechanistic comparisons. *Endocr Rev* 2012;**33**:595–622.
38. Schauer PR, Kashyap SR, Wolski K, Brethauer SA, Kirwan JP, Pothier CE, et al. Bariatric surgery versus intensive medical therapy in obese patients with diabetes. *N Engl J Med* 2012;**366**:1567–76.
39. Mingrone G, Panunzi S, De Gaetano A, Guidone C, Iaconelli A, Leccesi L, et al. Bariatric surgery versus conventional medical therapy for type 2 diabetes. *N Engl J Med* 2012;**366**:1577–85.
40. Sjostrom L, Narbro K, Sjostrom CD, Karason K, Larsson B, Wedel H, et al. Effects of bariatric surgery on mortality in Swedish obese subjects. *N Engl J Med* 2007;**357**:741–52.
41. Adams TD, Gress RE, Smith SC, Halverson RC, Simper SC, Rosamond WD, et al. Long-term mortality after gastric bypass surgery. *N Engl J Med* 2007;**357**:753–61.
42. Carlsson LM, Peltonen M, Ahlin S, Anveden A, Bouchard C, Carlsson B, et al. Bariatric surgery and prevention of type 2 diabetes in Swedish obese subjects. *N Engl J Med* 2012;**367**:695–704.
43. DeMaria EJ. Bariatric surgery for morbid obesity. *N Engl J Med* 2007;**356**:2176–83.
44. Terranova L, Busetto L, Vestri A, Zappa MA. Bariatric surgery: cost-effectiveness and budget impact. *Obes Surg* 2012;**22**:646–53.
45. Flum DR, Belle SH, King WC, Wahed AS, Berk P, Chapman W, et al. Perioperative safety in the longitudinal assessment of bariatric surgery. *N Engl J Med* 2009;**361**:445–54.
46. Schwartz A, Doucet E. Relative changes in resting energy expenditure during weight loss: a systematic review. *Obes Rev* 2010;**11**:531–47.
47. Schwartz MW, Woods SC, Porte Jr D, Seeley RJ, Baskin DG. Central nervous system control of food intake. *Nature* 2000;**404**:661–71.
48. Niswender KD, Schwartz MW. Insulin and leptin revisited: adiposity signals with overlapping physiological and intracellular signaling capabilities. *Front Neuroendocrinol* 2003;**24**:1–10.
49. Morton GJ, Cummings DE, Baskin DG, Barsh GS, Schwartz MW. Central nervous system control of food intake and body weight. *Nature* 2006;**443**:289–95.
50. Flier JS, Maratos-Flier E. Lasker lauds leptin. *Cell* 2010;**143**:9–12.
51. Zhang Y, Proenca R, Maffei M, Barone M, Leopold L, Friedman JM. Positional cloning of the mouse obese gene and its human homologue. *Nature* 1994; **372**:425–32.
52. Cone RD. Anatomy and regulation of the central melanocortin system. *Nat Neurosci* 2005;**8**:571–8.
53. Tao YX. The melanocortin-4 receptor: physiology, pharmacology, and pathophysiology. *Endocr Rev* 2010;**31**:506–43.
54. Cheung CC, Clifton DK, Steiner RA. Proopiomelanocortin neurons are direct targets for leptin in the hypothalamus. *Endocrinology* 1997;**138**:4489–92.

55. Schwartz MW, Seeley RJ, Woods SC, Weigle DS, Campfield LA, Burn P, et al. Leptin increases hypothalamic pro-opiomelanocortin mRNA expression in the rostral arcuate nucleus. *Diabetes* 1997;**46**:2119–23.
56. Thornton JE, Cheung CC, Clifton DK, Steiner RA. Regulation of hypothalamic proopiomelanocortin mRNA by leptin in ob/ob mice. *Endocrinology* 1997;**138**: 5063–6.
57. Seeley RJ, Yagaloff KA, Fisher SL, Burn P, Thiele TE, van Dijk G, et al. Melanocortin receptors in leptin effects. *Nature* 1997;**390**:349.
58. Chen H, Charlat O, Tartaglia LA, Woolf EA, Weng X, Ellis SJ, et al. Evidence that the diabetes gene encodes the leptin receptor: identification of a mutation in the leptin receptor gene in db/db mice. *Cell* 1996;**84**:491–5.
59. Phillips MS, Liu Q, Hammond HA, Dugan V, Hey PJ, Caskey CJ, et al. Leptin receptor missense mutation in the fatty Zucker rat. *Nat Genet* 1996;**13**:18–9.
60. Heymsfield SB, Greenberg AS, Fujioka K, Dixon RM, Kushner R, Hunt T, et al. Recombinant leptin for weight loss in obese and lean adults: a randomized, controlled, dose-escalation trial. *JAMA* 1999;**282**:1568–75.
61. Maffei M, Halaas J, Ravussin E, Pratley RE, Lee GH, Zhang Y, et al. Leptin levels in human and rodent: measurement of plasma leptin and ob RNA in obese and weight-reduced subjects. *Nat Med* 1995;**1**:1155–61.
62. Considine RV, Sinha MK, Heiman ML, Kriauciunas A, Stephens TW, Nyce MR, et al. Serum immunoreactive-leptin concentrations in normal-weight and obese humans. *N Engl J Med* 1996;**334**:292–5.
63. Rosenbaum M, Nicolson M, Hirsch J, Heymsfield SB, Gallagher D, Chu F, et al. Effects of gender, body composition, and menopause on plasma concentrations of leptin. *J Clin Endocrinol Metab* 1996;**81**:3424–7.
64. Rohner-Jeanrenaud F, Jeanrenaud B. Obesity, leptin, and the brain. *N Engl J Med* 1996;**334**:324–5.
65. Caro JF, Kolaczynski JW, Nyce MR, Ohannesian JP, Opentanova I, Goldman WH, et al. Decreased cerebrospinal-fluid/serum leptin ratio in obesity: a possible mechanism for leptin resistance. *Lancet* 1996;**348**:159–61.
66. Myers MG, Cowley MA, Munzberg H. Mechanisms of leptin action and leptin resistance. *Annu Rev Physiol* 2008;**70**:537–56.
67. St-Pierre J, Tremblay ML. Modulation of leptin resistance by protein tyrosine phosphatases. *Cell Metab* 2012;**15**:292–7.
68. Myers Jr MG, Leibel RL, Seeley RJ, Schwartz MW. Obesity and leptin resistance: distinguishing cause from effect. *Trends Endocrinol Metab* 2010;**21**:643–51.
69. de Pedro N, Martinez-Alvarez R, Delgado MJ. Acute and chronic leptin reduces food intake and body weight in goldfish (*Carassius auratus*). *J Endocrinol* 2006;**188**:513–20.
70. Cone RD, Cowley MA, Butler AA, Fan W, Marks DL, Low MJ. The arcuate nucleus as a conduit for diverse signals relevant to energy homeostasis. *Int J Obes Relat Metab Disord* 2001;**25**(Suppl. 5):S63–S67.
71. Myers Jr MG, Munzberg H, Leinninger GM, Leshan RL. The geometry of leptin action in the brain: more complicated than a simple ARC. *Cell Metab* 2009;**9**:117–23.
72. Nakamura K, Morrison SF. A thermosensory pathway that controls body temperature. *Nat Neurosci* 2008;**11**:62–71.
73. Morley JE. Neuropeptide regulation of appetite and weight. *Endocr Rev* 1987;**8**:256–87.
74. Grill HJ, Hayes MR. The nucleus tractus solitarius: a portal for visceral afferent signal processing, energy status assessment and integration of their combined effects on food intake. *Int J Obes (Lond)* 2009;**33**(Suppl. 1):S11–S15.
75. Zheng H, Patterson LM, Morrison C, Banfield BW, Randall JA, Browning KN, et al. Melanin concentrating hormone innervation of caudal brainstem areas involved in gastrointestinal functions and energy balance. *Neuroscience* 2005;**135**:611–25.

76. Schwartz GJ. Integrative capacity of the caudal brainstem in the control of food intake. *Philos Trans R Soc Lond B Biol Sci* 2006;**361**:1275–80.
77. Palkovits M, Mezey E, Eskay RL. Pro-opiomelanocortin-derived peptides (ACTH/ beta-endorphin/alpha-MSH) in brainstem baroreceptor areas of the rat. *Brain Res* 1987;**436**:323–38.
78. Bronstein DM, Schafer MK, Watson SJ, Akil H. Evidence that beta-endorphin is synthesized in cells in the nucleus tractus solitarius: detection of POMC mRNA. *Brain Res* 1992;**587**:269–75.
79. Mountjoy KG, Mortrud MT, Low MJ, Simerly RB, Cone RD. Localization of the melanocortin-4 receptor (MC4-R) in neuroendocrine and autonomic control circuits in the brain. *Mol Endocrinol* 1994;**8**:1298–308.
80. Grill HJ, Ginsberg AB, Seeley RJ, Kaplan JM. Brainstem application of melanocortin receptor ligands produces long-lasting effects on feeding and body weight. *J Neurosci* 1998;**18**:10128–35.
81. Brown KS, Gentry RM, Rowland NE. Central injection in rats of alpha-melanocyte-stimulating hormone analog: effects on food intake and brain Fos. *Regul Pept* 1998;**78**:89–94.
82. Williams DL, Kaplan JM, Grill HJ. The role of the dorsal vagal complex and the vagus nerve in feeding effects of melanocortin-3/4 receptor stimulation. *Endocrinology* 2000;**141**:1332–7.
83. Skibicka KP, Grill HJ. Energetic responses are triggered by caudal brainstem melanocortin receptor stimulation and mediated by local sympathetic effector circuits. *Endocrinology* 2008;**149**:3605–16.
84. Skibicka KP, Grill HJ. Hindbrain leptin stimulation induces anorexia and hyperthermia mediated by hindbrain melanocortin receptors. *Endocrinology* 2009;**150**:1705–11.
85. Zheng H, Patterson LM, Phifer CB, Berthoud HR. Brain stem melanocortinergic modulation of meal size and identification of hypothalamic POMC projections. *Am J Physiol Regul Integr Comp Physiol* 2005;**289**:R247–R258.
86. Fan W, Ellacott KL, Halatchev IG, Takahashi K, Yu P, Cone RD. Cholecystokinin-mediated suppression of feeding involves the brainstem melanocortin system. *Nat Neurosci* 2004;**7**:335–6.
87. Li ZR, Li Y, Zhang WZ. Ghrelin receptor in energy homeostasis and obesity pathogenesis. *Prog Mol Biol Transl Sci* 2013;**114**:46–88.
88. Li L, Huang QS, Ren PG, Zhang JV. Obestatin receptor in energy homeostasis and obesity pathogenesis. *Prog Mol Biol Transl Sci* 2013;**114**:89–108.
89. Sayegh A. The role of cholecystokinin receptors in the short-term control of food intake. *Prog Mol Biol Transl Sci* 2013;**114**:277–316.
90. Sayegh A. The role of bombesin and bombesin-related peptides in the short-term control of food intake. *Prog Mol Biol Transl Sci* 2013;**114**:343–71.
91. Zac-Varghese S, De Silva A, Bloom SR. Translational studies on PYY as a novel target in obesity. *Curr Opin Pharmacol* 2011;**11**:582–5.
92. Michel MC, Beck-Sickinger A, Cox H, Doods HN, Herzog H, Larhammar D, et al. XVI. International Union of Pharmacology recommendations for the nomenclature of neuropeptide Y, peptide YY, and pancreatic polypeptide receptors. *Pharmacol Rev* 1998;**50**:143–50.
93. Batterham RL, Cowley MA, Small CJ, Herzog H, Cohen MA, Dakin CL, et al. Gut hormone PYY(3-36) physiologically inhibits food intake. *Nature* 2002; **418**:650–4.
94. Deacon CF, Nauck MA, Toft-Nielsen M, Pridal L, Willms B, Holst JJ. Both subcutaneously and intravenously administered glucagon-like peptide I are rapidly degraded from the NH2-terminus in type II diabetic patients and in healthy subjects. *Diabetes* 1995;**44**:1126–31.

95. Mayo KE, Miller LJ, Bataille D, Dalle S, Goke B, Thorens B, et al. International Union of Pharmacology. XXXV. The glucagon receptor family. *Pharmacol Rev* 2003;**55**:167–94.
96. Thorens B. Expression cloning of the pancreatic beta cell receptor for the gluco-incretin hormone glucagon-like peptide 1. *Proc Natl Acad Sci USA* 1992;**89**:8641–5.
97. Stoffel M, Espinosa 3rd R, Le Beau MM, Bell GI. Human glucagon-like peptide-1 receptor gene. Localization to chromosome band 6p21 by fluorescence in situ hybridization and linkage of a highly polymorphic simple tandem repeat DNA polymorphism to other markers on chromosome 6. *Diabetes* 1993;**42**:1215–8.
98. Campos RV, Lee YC, Drucker DJ. Divergent tissue-specific and developmental expression of receptors for glucagon and glucagon-like peptide-1 in the mouse. *Endocrinology* 1994;**134**:2156–64.
99. Kanse SM, Kreymann B, Ghatei MA, Bloom SR. Identification and characterization of glucagon-like peptide-1 7-36 amide-binding sites in the rat brain and lung. *FEBS Lett* 1988;**241**:209–12.
100. Navarro M, Rodriquez de Fonseca F, Alvarez E, Chowen JA, Zueco JA, Gomez R, et al. Colocalization of glucagon-like peptide-1 (GLP-1) receptors, glucose transporter GLUT-2, and glucokinase mRNAs in rat hypothalamic cells: evidence for a role of GLP-1 receptor agonists as an inhibitory signal for food and water intake. *J Neurochem* 1996;**67**:1982–91.
101. Shughrue PJ, Lane MV, Merchenthaler I. Glucagon-like peptide-1 receptor (GLP1-R) mRNA in the rat hypothalamus. *Endocrinology* 1996;**137**:5159–62.
102. Turton MD, O'Shea D, Gunn I, Beak SA, Edwards CM, Meeran K, et al. A role for glucagon-like peptide-1 in the central regulation of feeding. *Nature* 1996;**379**:69–72.
103. Goke R, Fehmann HC, Linn T, Schmidt H, Krause M, Eng J, et al. Exendin-4 is a high potency agonist and truncated exendin-(9-39)-amide an antagonist at the glucagon-like peptide 1-(7-36)-amide receptor of insulin-secreting beta-cells. *J Biol Chem* 1993;**268**: 19650–5.
104. Kinzig KP, D'Alessio DA, Seeley RJ. The diverse roles of specific GLP-1 receptors in the control of food intake and the response to visceral illness. *J Neurosci* 2002;**22**: 10470–6.
105. Thiele TE, Van Dijk G, Campfield LA, Smith FJ, Burn P, Woods SC, et al. Central infusion of GLP-1, but not leptin, produces conditioned taste aversions in rats. *Am J Physiol* 1997;**272**:R726–R730.
106. Tang-Christensen M, Larsen PJ, Goke R, Fink-Jensen A, Jessop DS, Moller M, et al. Central administration of GLP-1-(7-36) amide inhibits food and water intake in rats. *Am J Physiol* 1996;**271**:R848–R856.
107. Abbott CR, Monteiro M, Small CJ, Sajedi A, Smith KL, Parkinson JR, et al. The inhibitory effects of peripheral administration of peptide YY(3-36) and glucagon-like peptide-1 on food intake are attenuated by ablation of the vagal-brainstem-hypothalamic pathway. *Brain Res* 2005;**1044**:127–31.
108. Larsen PJ, Fledelius C, Knudsen LB, Tang-Christensen M. Systemic administration of the long-acting GLP-1 derivative NN2211 induces lasting and reversible weight loss in both normal and obese rats. *Diabetes* 2001;**50**:2530–9.
109. Yamamoto H, Kishi T, Lee CE, Choi BJ, Fang H, Hollenberg AN, et al. Glucagon-like peptide-1-responsive catecholamine neurons in the area postrema link peripheral glucagon-like peptide-1 with central autonomic control sites. *J Neurosci* 2003;**23**:2939–46.
110. Kanoski SE, Fortin SM, Arnold M, Grill HJ, Hayes MR. Peripheral and central GLP-1 receptor populations mediate the anorectic effects of peripherally administered GLP-1 receptor agonists, liraglutide and exendin-4. *Endocrinology* 2011;**152**:3103–12.
111. Baraboi ED, St-Pierre DH, Shooner J, Timofeeva E, Richard D. Brain activation following peripheral administration of the GLP-1 receptor agonist exendin-4. *Am J Physiol Regul Integr Comp Physiol* 2011;**301**:R1011–R1024.

112. Flint A, Raben A, Astrup A, Holst JJ. Glucagon-like peptide 1 promotes satiety and suppresses energy intake in humans. *J Clin Invest* 1998;**101**:515–20.
113. Gutzwiller JP, Goke B, Drewe J, Hildebrand P, Ketterer S, Handschin D, et al. Glucagon-like peptide-1: a potent regulator of food intake in humans. *Gut* 1999;**44**:81–6.
114. Naslund E, Barkeling B, King N, Gutniak M, Blundell JE, Holst JJ, et al. Energy intake and appetite are suppressed by glucagon-like peptide-1 (GLP-1) in obese men. *Int J Obes Relat Metab Disord* 1999;**23**:304–11.
115. Gutzwiller JP, Drewe J, Goke B, Schmidt H, Rohrer B, Lareida J, et al. Glucagon-like peptide-1 promotes satiety and reduces food intake in patients with diabetes mellitus type 2. *Am J Physiol* 1999;**276**:R1541–R1544.
116. Verdich C, Flint A, Gutzwiller JP, Naslund E, Beglinger C, Hellstrom PM, et al. A meta-analysis of the effect of glucagon-like peptide-1 (7-36) amide on ad libitum energy intake in humans. *J Clin Endocrinol Metab* 2001;**86**:4382–9.
117. Vella A. Mechanism of action of DPP-4 inhibitors—new insights. *J Clin Endocrinol Metab* 2012;**97**:2626–8.
118. Bataille D, Gespach C, Tatemoto K, Marie JC, Coudray AM, Rosselin G, et al. Bioactive enteroglucagon (oxyntomodulin): present knowledge on its chemical structure and its biological activities. *Peptides* 1981;**2**(Suppl. 2):41–4.
119. Jarrousse C, Bataille D, Jeanrenaud B. A pure enteroglucagon, oxyntomodulin (glucagon 37), stimulates insulin release in perfused rat pancreas. *Endocrinology* 1984;**115**:102–5.
120. Collie NL, Zhu Z, Jordan S, Reeve Jr JR. Oxyntomodulin stimulates intestinal glucose uptake in rats. *Gastroenterology* 1997;**112**:1961–70.
121. Schjoldager B, Mortensen PE, Myhre J, Christiansen J, Holst JJ. Oxyntomodulin from distal gut. Role in regulation of gastric and pancreatic functions. *Digest Dis Sci* 1989;**34**:1411–9.
122. Jarrousse C, Niel H, Audousset-Puech MP, Martinez J, Bataille D. Oxyntomodulin and its C-terminal octapeptide inhibit liquid meal-stimulated acid secretion. *Peptides* 1986;**7**(Suppl. 1):253–6.
123. Jarrousse C, Carles-Bonnet C, Niel H, Sabatier R, Audousset-Puech MP, Blache P, et al. Inhibition of gastric acid secretion by oxyntomodulin and its 19-37 fragment in the conscious rat. *Am J Physiol* 1993;**264**:G816–G823.
124. Dakin CL, Gunn I, Small CJ, Edwards CM, Hay DL, Smith DM, et al. Oxyntomodulin inhibits food intake in the rat. *Endocrinology* 2001;**142**:4244–50.
125. Dakin CL, Small CJ, Park AJ, Seth A, Ghatei MA, Bloom SR. Repeated ICV administration of oxyntomodulin causes a greater reduction in body weight gain than in pair-fed rats. *Am J Physiol Endocrinol Metab* 2002;**283**:E1173–E1177.
126. Dakin CL, Small CJ, Batterham RL, Neary NM, Cohen MA, Patterson M, et al. Peripheral oxyntomodulin reduces food intake and body weight gain in rats. *Endocrinology* 2004;**145**:2687–95.
127. Baggio LL, Huang Q, Brown TJ, Drucker DJ. Oxyntomodulin and glucagon-like peptide-1 differentially regulate murine food intake and energy expenditure. *Gastroenterology* 2004;**127**:546–58.
128. Kosinski JR, Hubert J, Carrington PE, Chicchi GG, Mu J, Miller C, et al. The glucagon receptor is involved in mediating the body weight-lowering effects of oxyntomodulin. *Obesity (Silver Spring)* 2012;**20**:1566–71.
129. Wynne K, Park AJ, Small CJ, Patterson M, Ellis SM, Murphy KG, et al. Subcutaneous oxyntomodulin reduces body weight in overweight and obese subjects: a double-blind, randomized, controlled trial. *Diabetes* 2005;**54**:2390–5.
130. Wynne K, Park AJ, Small CJ, Meeran K, Ghatei MA, Frost GS, et al. Oxyntomodulin increases energy expenditure in addition to decreasing energy intake in overweight and obese humans: a randomised controlled trial. *Int J Obes (Lond)* 2006;**30**:1729–36.

131. Gros L, Thorens B, Bataille D, Kervran A. Glucagon-like peptide-1-(7-36) amide, oxyntomodulin, and glucagon interact with a common receptor in a somatostatin-secreting cell line. *Endocrinology* 1993;**133**:631–8.
132. Woods SC, Lotter EC, McKay LD, Porte Jr D. Chronic intracerebroventricular infusion of insulin reduces food intake and body weight of baboons. *Nature* 1979;**282**:503–5.
133. Bruning JC, Gautam D, Burks DJ, Gillette J, Schubert M, Orban PC, et al. Role of brain insulin receptor in control of body weight and reproduction. *Science* 2000;**289**:2122–5.
134. Obici S, Feng Z, Karkanias G, Baskin DG, Rossetti L. Decreasing hypothalamic insulin receptors causes hyperphagia and insulin resistance in rats. *Nat Neurosci* 2002;**5**:566–72.
135. Havrankova J, Roth J, Brownstein M. Insulin receptors are widely distributed in the central nervous system of the rat. *Nature* 1978;**272**:827–9.
136. Davidson IW, Salter JM, Best CH. Calorigenic action of glucagon. *Nature* 1957;**180**:1124.
137. Schulman JL, Carleton JL, Whitney G, Whitehorn JC. Effect of glucagon on food intake and body weight in man. *J Appl Physiol* 1957;**11**:419–21.
138. Penick SB, Hinkle Jr LE. Depression of food intake induced in healthy subjects by glucagon. *N Engl J Med* 1961;**264**:893–7.
139. Inokuchi A, Oomura Y, Nishimura H. Effect of intracerebroventricularly infused glucagon on feeding behavior. *Physiol Behav* 1984;**33**:397–400.
140. Nair KS. Hyperglucagonemia increases resting metabolic rate in man during insulin deficiency. *J Clin Endocrinol Metab* 1987;**64**:896–901.
141. Woods SC, Lutz TA, Geary N, Langhans W. Pancreatic signals controlling food intake; insulin, glucagon and amylin. *Philos Trans R Soc Lond B Biol Sci* 2006;**361**:1219–35.
142. Heppner KM, Habegger KM, Day J, Pfluger PT, Perez-Tilve D, Ward B, et al. Glucagon regulation of energy metabolism. *Physiol Behav* 2010;**100**:545–8.
143. Habegger KM, Heppner KM, Geary N, Bartness TJ, DiMarchi R, Tschop MH. The metabolic actions of glucagon revisited. *Nat Rev Endocrinol* 2010;**6**:689–97.
144. Stunkard AJ, Van Itallie TB, Reis BB. The mechanism of satiety: effect of glucagon on gastric hunger contractions in man. *Proc Soc Exp Biol Med* 1955;**89**:258–61.
145. Langhans W, Zeiger U, Scharrer E, Geary N. Stimulation of feeding in rats by intraperitoneal injection of antibodies to glucagon. *Science* 1982;**218**:894–6.
146. Honda K, Kamisoyama H, Saito N, Kurose Y, Sugahara K, Hasegawa S. Central administration of glucagon suppresses food intake in chicks. *Neurosci Lett* 2007;**416**:198–201.
147. Kurose Y, Kamisoyama H, Honda K, Azuma Y, Sugahara K, Hasegawa S, et al. Effects of central administration of glucagon on feed intake and endocrine responses in sheep. *Anim Sci J* 2009;**80**:686–90.
148. Svoboda M, Tastenoy M, Vertongen P, Robberecht P. Relative quantitative analysis of glucagon receptor mRNA in rat tissues. *Mol Cell Endocrinol* 1994;**105**:131–7.
149. Gelling RW, Du XQ, Dichmann DS, Romer J, Huang H, Cui L, et al. Lower blood glucose, hyperglucagonemia, and pancreatic alpha cell hyperplasia in glucagon receptor knockout mice. *Proc Natl Acad Sci USA* 2003;**100**:1438–43.
150. Conarello SL, Jiang G, Mu J, Li Z, Woods J, Zycband E, et al. Glucagon receptor knockout mice are resistant to diet-induced obesity and streptozotocin-mediated beta cell loss and hyperglycaemia. *Diabetologia* 2007;**50**:142–50.
151. Day JW, Ottaway N, Patterson JT, Gelfanov V, Smiley D, Gidda J, et al. A new glucagon and GLP-1 co-agonist eliminates obesity in rodents. *Nat Chem Biol* 2009;**5**:749–57.
152. Pocai A, Carrington PE, Adams JR, Wright M, Eiermann G, Zhu L, et al. Glucagon-like peptide 1/glucagon receptor dual agonism reverses obesity in mice. *Diabetes* 2009;**58**:2258–66.

153. Kimmel JR, Pollock HG, Hazelwood RL. Isolation and characterization of chicken insulin. *Endocrinology* 1968;**83**:1323–30.
154. Lin TM, Chance RE. Candidate hormones of the gut. VI. Bovine pancreatic polypeptide (BPP) and avian pancreatic polypeptide (APP). *Gastroenterology* 1974;**67**:737–8.
155. Larhammar D. Evolution of neuropeptide Y, peptide YY and pancreatic polypeptide. *Regul Pept* 1996;**62**:1–11.
156. Whitcomb DC, Taylor IL, Vigna SR. Characterization of saturable binding sites for circulating pancreatic polypeptide in rat brain. *Am J Physiol* 1990;**259**:G687–G691.
157. Malaisse-Lagae F, Carpentier JL, Patel YC, Malaisse WJ, Orci L. Pancreatic polypeptide: a possible role in the regulation of food intake in the mouse. Hypothesis. *Experientia* 1977;**33**:915–7.
158. Asakawa A, Inui A, Yuzuriha H, Ueno N, Katsuura G, Fujimiya M, et al. Characterization of the effects of pancreatic polypeptide in the regulation of energy balance. *Gastroenterology* 2003;**124**:1325–36.
159. Berntson GG, Zipf WB, O'Dorisio TM, Hoffman JA, Chance RE. Pancreatic polypeptide infusions reduce food intake in Prader–Willi syndrome. *Peptides* 1993;**14**: 497–503.
160. Batterham RL, Le Roux CW, Cohen MA, Park AJ, Ellis SM, Patterson M, et al. Pancreatic polypeptide reduces appetite and food intake in humans. *J Clin Endocrinol Metab* 2003;**88**:3989–92.
161. Ueno N, Inui A, Iwamoto M, Kaga T, Asakawa A, Okita M, et al. Decreased food intake and body weight in pancreatic polypeptide-overexpressing mice. *Gastroenterology* 1999;**117**:1427–32.
162. Marco J, Zulueta MA, Correas I, Villanueva ML. Reduced pancreatic polypeptide secretion in obese subjects. *J Clin Endocrinol Metab* 1980;**50**:744–7.
163. Lassmann V, Vague P, Vialettes B, Simon MC. Low plasma levels of pancreatic polypeptide in obesity. *Diabetes* 1980;**29**:428–30.
164. Clark JT, Kalra PS, Crowley WR, Kalra SP. Neuropeptide Y and human pancreatic polypeptide stimulate feeding behavior in rats. *Endocrinology* 1984;**115**:427–9.
165. Flynn MC, Turrin NP, Plata-Salaman CR, Ffrench-Mullen JM. Feeding response to neuropeptide Y-related compounds in rats treated with Y5 receptor antisense or sense phosphothio-oligodeoxynucleotide. *Physiol Behav* 1999;**66**:881–4.
166. Campbell RE, Smith MS, Allen SE, Grayson BE, Ffrench-Mullen JM, Grove KL. Orexin neurons express a functional pancreatic polypeptide Y4 receptor. *J Neurosci* 2003;**23**:1487–97.
167. Cooper GJ, Willis AC, Clark A, Turner RC, Sim RB, Reid KB. Purification and characterization of a peptide from amyloid-rich pancreases of type 2 diabetic patients. *Proc Natl Acad Sci USA* 1987;**84**:8628–32.
168. Butler PC, Chou J, Carter WB, Wang YN, Bu BH, Chang D, et al. Effects of meal ingestion on plasma amylin concentration in NIDDM and nondiabetic humans. *Diabetes* 1990;**39**:752–6.
169. Moore CX, Cooper GJ. Co-secretion of amylin and insulin from cultured islet beta-cells: modulation by nutrient secretagogues, islet hormones and hypoglycemic agents. *Biochem Biophys Res Commun* 1991;**179**:1–9.
170. Chance WT, Balasubramaniam A, Zhang FS, Wimalawansa SJ, Fischer JE. Anorexia following the intrahypothalamic administration of amylin. *Brain Res* 1991;**539**:352–4.
171. Morley JE, Flood JF. Amylin decreases food intake in mice. *Peptides* 1991;**12**:865–9.
172. Morley JE, Flood JF, Horowitz M, Morley PM, Walter MJ. Modulation of food intake by peripherally administered amylin. *Am J Physiol* 1994;**267**:R178–R184.
173. Lutz TA, Del Prete E, Scharrer E. Reduction of food intake in rats by intraperitoneal injection of low doses of amylin. *Physiol Behav* 1994;**55**:891–5.

174. Lutz TA, Geary N, Szabady MM, Del Prete E, Scharrer E. Amylin decreases meal size in rats. *Physiol Behav* 1995;**58**:1197–202.
175. Arnelo U, Blevins JE, Larsson J, Permert J, Westermark P, Reidelberger RD, et al. Effects of acute and chronic infusion of islet amyloid polypeptide on food intake in rats. *Scan J Gastroenterol* 1996;**31**:83–9.
176. Rushing PA, Seeley RJ, Air EL, Lutz TA, Woods SC. Acute 3rd-ventricular amylin infusion potently reduces food intake but does not produce aversive consequences. *Peptides* 2002;**23**:985–8.
177. Bhavsar S, Watkins J, Young A. Synergy between amylin and cholecystokinin for inhibition of food intake in mice. *Physiol Behav* 1998;**64**:557–61.
178. Grabler V, Lutz TA. Chronic infusion of the amylin antagonist AC 187 increases feeding in Zucker fa/fa rats but not in lean controls. *Physiol Behav* 2004;**81**:481–8.
179. Mollet A, Gilg S, Riediger T, Lutz TA. Infusion of the amylin antagonist AC 187 into the area postrema increases food intake in rats. *Physiol Behav* 2004;**81**:149–55.
180. Reidelberger RD, Haver AC, Arnelo U, Smith DD, Schaffert CS, Permert J. Amylin receptor blockade stimulates food intake in rats. *Am J Physiol Regul Integr Comp Physiol* 2004;**287**:R568–R574.
181. Rowland NE, Crews EC, Gentry RM. Comparison of Fos induced in rat brain by GLP-1 and amylin. *Regul Pept* 1997;**71**:171–4.
182. Lutz TA, Senn M, Althaus J, Del Prete E, Ehrensperger F, Scharrer E. Lesion of the area postrema/nucleus of the solitary tract (AP/NTS) attenuates the anorectic effects of amylin and calcitonin gene-related peptide (CGRP) in rats. *Peptides* 1998;**19**:309–17.
183. Rowland NE, Richmond RM. Area postrema and the anorectic actions of dexfenfluramine and amylin. *Brain Res* 1999;**820**:86–91.
184. Lutz TA, Mollet A, Rushing PA, Riediger T, Scharrer E. The anorectic effect of a chronic peripheral infusion of amylin is abolished in area postrema/nucleus of the solitary tract (AP/NTS) lesioned rats. *Int J Obes Relat Metab Disord* 2001;**25**:1005–11.
185. Riediger T, Zuend D, Becskei C, Lutz TA. The anorectic hormone amylin contributes to feeding-related changes of neuronal activity in key structures of the gut-brain axis. *Am J Physiol Regul Integr Comp Physiol* 2004;**286**:R114–R122.
186. Becskei C, Grabler V, Edwards GL, Riediger T, Lutz TA. Lesion of the lateral parabrachial nucleus attenuates the anorectic effect of peripheral amylin and CCK. *Brain Res* 2007;**1162**:76–84.
187. Smith SR, Blundell JE, Burns C, Ellero C, Schroeder BE, Kesty NC, et al. Pramlintide treatment reduces 24-h caloric intake and meal sizes and improves control of eating in obese subjects: a 6-wk translational research study. *Am J Physiol Endocrinol Metab* 2007;**293**:E620–E627.
188. Aronne L, Fujioka K, Aroda V, Chen K, Halseth A, Kesty NC, et al. Progressive reduction in body weight after treatment with the amylin analog pramlintide in obese subjects: a phase 2, randomized, placebo-controlled, dose-escalation study. *J Clin Endocrinol Metab* 2007;**92**:2977–83.
189. Smith SR, Aronne LJ, Burns CM, Kesty NC, Halseth AE, Weyer C. Sustained weight loss following 12-month pramlintide treatment as an adjunct to lifestyle intervention in obesity. *Diabetes Care* 2008;**31**:1816–23.
190. Roth JD, Roland BL, Cole RL, Trevaskis JL, Weyer C, Koda JE, et al. Leptin responsiveness restored by amylin agonism in diet-induced obesity: evidence from nonclinical and clinical studies. *Proc Natl Acad Sci USA* 2008;**105**:7257–62.
191. Trevaskis JL, Coffey T, Cole R, Lei C, Wittmer C, Walsh B, et al. Amylin-mediated restoration of leptin responsiveness in diet-induced obesity: magnitude and mechanisms. *Endocrinology* 2008;**149**:5679–87.

192. Ravussin E, Smith SR, Mitchell JA, Shringarpure R, Shan K, Maier H, et al. Enhanced weight loss with pramlintide/metreleptin: an integrated neurohormonal approach to obesity pharmacotherapy. *Obesity (Silver Spring)* 2009;**17**:1736–43.
193. Heal DJ, Gosden J, Smith SL. What is the prognosis for new centrally-acting anti-obesity drugs? *Neuropharmacology* 2012;**63**:132–46.
194. Chung S, Funakoshi T, Civelli O. Orphan GPCR research. *Br J Pharmacol* 2008;**153** (Suppl. 1):S339–S346.
195. Chambers J, Ames RS, Bergsma D, Muir A, Fitzgerald LR, Hervieu G, et al. Melanin-concentrating hormone is the cognate ligand for the orphan G-protein-coupled receptor SLC-1. *Nature* 1999;**400**:261–5.
196. Saito Y, Nothacker HP, Wang Z, Lin SH, Leslie F, Civelli O. Molecular characterization of the melanin-concentrating-hormone receptor. *Nature* 1999;**400**:265–9.
197. Lembo PM, Grazzini E, Cao J, Hubatsch DA, Pelletier M, Hoffert C, et al. The receptor for the orexigenic peptide melanin-concentrating hormone is a G-protein-coupled receptor. *Nat Cell Biol* 1999;**1**:267–71.
198. Bachner D, Kreienkamp H, Weise C, Buck F, Richter D. Identification of melanin concentrating hormone (MCH) as the natural ligand for the orphan somatostatin-like receptor 1 (SLC-1). *FEBS Lett* 1999;**457**:522–4.
199. Shimomura Y, Mori M, Sugo T, Ishibashi Y, Abe M, Kurokawa T, et al. Isolation and identification of melanin-concentrating hormone as the endogenous ligand of the SLC-1 receptor. *Biochem Biophys Res Commun* 1999;**261**:622–6.
200. Sakurai T, Amemiya A, Ishii M, Matsuzaki I, Chemelli RM, Tanaka H, et al. Orexins and orexin receptors: a family of hypothalamic neuropeptides and G protein-coupled receptors that regulate feeding behavior. *Cell* 1998;**92**:573–85.
201. Kojima M, Hosoda H, Date Y, Nakazato M, Matsuo H, Kangawa K. Ghrelin is a growth-hormone-releasing acylated peptide from stomach. *Nature* 1999;**402**: 656–60.
202. Mo XL, Wei HK, Peng J, Tao YX. Free fatty acid receptor GPR120 and pathogenesis of obesity and type 2 diabetes mellitus. *Prog Mol Biol Transl Sci* 2013;**114**:251–76.
203. O'Dowd BF, Nguyen T, Jung BP, Marchese A, Cheng R, Heng HH, et al. Cloning and chromosomal mapping of four putative novel human G-protein-coupled receptor genes. *Gene* 1997;**187**:75–81.
204. Gardner J, Wu S, Ling L, Danao J, Li Y, Yeh WC, et al. G-protein-coupled receptor GPR21 knockout mice display improved glucose tolerance and increased insulin response. *Biochem Biophys Res Commun* 2012;**418**:1–5.
205. Osborn O, Oh da Y, McNelis J, Sanchez-Alavez M, Talukdar S, Lu M, et al. G protein-coupled receptor 21 deletion improves insulin sensitivity in diet-induced obese mice. *J Clin Invest* 2012;**122**:2444–53.
206. Jensen RT, Battey JF, Spindel ER, Benya RV. International Union of Pharmacology. LXVIII. Mammalian bombesin receptors: nomenclature, distribution, pharmacology, signaling, and functions in normal and disease states. *Pharmacol Rev* 2008;**60**:1–42.
207. Ohki-Hamazaki H, Watase K, Yamamoto K, Ogura H, Yamano M, Yamada K, et al. Mice lacking bombesin receptor subtype-3 develop metabolic defects and obesity. *Nature* 1997;**390**:165–9.
208. Maekawa F, Quah HM, Tanaka K, Ohki-Hamazaki H. Leptin resistance and enhancement of feeding facilitation by melanin-concentrating hormone in mice lacking bombesin receptor subtype-3. *Diabetes* 2004;**53**:570–6.
209. Ladenheim EE, Hamilton NL, Behles RR, Bi S, Hampton LL, Battey JF, et al. Factors contributing to obesity in bombesin receptor subtype-3-deficient mice. *Endocrinology* 2008;**149**:971–8.

210. Guan XM, Chen H, Dobbelaar PH, Dong Y, Fong TM, Gagen K, et al. Regulation of energy homeostasis by bombesin receptor subtype-3: selective receptor agonists for the treatment of obesity. *Cell Metab* 2010;**11**:101–12.
211. Guan XM, Metzger JM, Yang L, Raustad KA, Wang SP, Spann SK, et al. Antiobesity effect of MK-5046, a novel bombesin receptor subtype-3 agonist. *J Pharmacol Exp Ther* 2011;**336**:356–64.
212. Hotta K, Matsukawa Y, Nishida M, Kotani K, Takahashi M, Kuriyama H, et al. Mutation in bombesin receptor subtype-3 gene is not a major cause of obesity in the Japanese. *Horm Metab Res* 2000;**32**:33–4.
213. Allen MS, Bradford BJ, Harvatine KJ. The cow as a model to study food intake regulation. *Annu Rev Nutr* 2005;**25**:523–47.
214. Spurlock ME, Gabler NK. The development of porcine models of obesity and the metabolic syndrome. *J Nutr* 2008;**138**:397–402.
215. Clouard C, Meunier-Salaun MC, Val-Laillet D. Food preferences and aversions in human health and nutrition: how can pigs help the biomedical research? *Animal* 2012;**6**:118–36.
216. Iqbal J, Kurose Y, Canny B, Clarke IJ. Effects of central infusion of ghrelin on food intake and plasma levels of growth hormone, luteinizing hormone, prolactin, and cortisol secretion in sheep. *Endocrinology* 2006;**147**:510–9.
217. Barb CR, Robertson AS, Barrett JB, Kraeling RR, Houseknecht KL. The role of melanocortin-3 and -4 receptor in regulating appetite, energy homeostasis and neuroendocrine function in the pig. *J Endocrinol* 2004;**181**:39–52.
218. Fan ZC, Sartin JL, Tao YX. Pharmacological analyses of two naturally occurring porcine melanocortin-4 receptor mutations in domestic pigs. *Domest Anim Endocrinol* 2008;**34**:383–90.
219. Fan ZC, Sartin JL, Tao YX. Molecular cloning and pharmacological characterization of porcine melanocortin-3 receptor. *J Endocrinol* 2008;**196**:139–48.
220. Sartin JL, Wagner CG, Marks DL, Daniel JA, McMahon CD, Obese FY, et al. Melanocortin-4 receptor in sheep: a potential site for therapeutic intervention in disease models. *Domest Anim Endocrinol* 2005;**29**:446–55.
221. German AJ. The growing problem of obesity in dogs and cats. *J Nutr* 2006;**136**: 1940S–1946S.
222. Kushner RF, Blatner DJ, Jewell DE, Rudloff K. The PPET Study: people and pets exercising together. *Obesity (Silver Spring)* 2006;**14**:1762–70.
223. Gossellin J, Wren JA, Sunderland SJ. Canine obesity: an overview. *J Vet Pharmacol Ther* 2007;**30**(Suppl. 1):1–10.
224. Hoenig M. The cat as a model for human obesity and diabetes. *J Diabetes Sci Technol* 2012;**6**:525–33.
225. Iwase M, Kimura K, Sasaki N, Komagome R, Ishioka K, Morimatsu M, et al. Canine leptin: cDNA cloning, expression and activity of recombinant protein. *Res Vet Sci* 2000;**68**:109–14.
226. Ishioka K, Soliman MM, Sagawa M, Nakadomo F, Shibata H, Honjoh T, et al. Experimental and clinical studies on plasma leptin in obese dogs. *J Vet Med Sci* 2002;**64**:349–53.
227. Ishioka K, Hosoya K, Kitagawa H, Shibata H, Honjoh T, Kimura K, et al. Plasma leptin concentration in dogs: effects of body condition score, age, gender and breeds. *Res Vet Sci* 2007;**82**:11–5.
228. Ishioka K, Omachi A, Sagawa M, Shibata H, Honjoh T, Kimura K, et al. Canine adiponectin: cDNA structure, mRNA expression in adipose tissues and reduced plasma levels in obesity. *Res Vet Sci* 2006;**80**:127–32.
229. LeBel C, Bourdeau A, Lau D, Hunt P. Biologic response to peripheral and central administration of recombinant human leptin in dogs. *Obes Res* 1999;**7**:577–85.

230. Skorczyk A, Stachowiak M, Szczerbal I, Klukowska-Roetzler J, Schelling C, Dolf G, et al. Polymorphism and chromosomal location of the MC4R (melanocortin-4 receptor) gene in the dog and red fox. *Gene* 2007;**392**:247–52.
231. van den Berg L, van den Berg SM, Martens EECP, Hazewinke HAW, Dijkshoorn NA, Delemarre-van de Waal HA, et al. Analysis of variation in the melanocortin-4 receptor gene (mc4r) in golden retriever dogs. *Anim Genet* 2010;**41**:557.
232. Yan J, Tao YX. Pharmacological characterization of canine melancortin-4 receptor and its natural variant V213F. *Domest Anim Endocrinol* 2011;**41**:91–7.
233. Richards MP. Genetic regulation of feed intake and energy balance in poultry. *Poult Sci* 2003;**82**:907–16.
234. Boswell T. Regulation of energy balance in birds by the neuroendocrine hypothalamus. *J Poult Sci* 2005;**42**:161–81.
235. Scanes CG. Absolute and relative standards—the case of leptin in poultry: first do no harm. *Poult Sci* 2008;**87**:1927–8.
236. Sharp PJ, Dunn IC, Waddington D, Boswell T. Chicken leptin. *Gen Comp Endocrinol* 2008;**158**:2–4.
237. Taouis M, Chen JW, Daviaud C, Dupont J, Derouet M, Simon J. Cloning the chicken leptin gene. *Gene* 1998;**208**:239–42.
238. Ashwell CM, Czerwinski SM, Brocht DM, McMurtry JP. Hormonal regulation of leptin expression in broiler chickens. *Am J Physiol* 1999;**276**:R226–R232.
239. Friedman-Einat M, Boswell T, Horev G, Girishvarma G, Dunn IC, Talbot RT, et al. The chicken leptin gene: has it been cloned? *Gen Comp Endocrinol* 1999; **115**:354–63.
240. Yosefi S, Hen G, Rosenblum CI, Cerasale DJ, Beaulieu M, Criscuolo F, et al. Lack of leptin activity in blood samples of Adelie penguin and bar-tailed godwit. *J Endocrinol* 2010;**207**:113–22.
241. Horev G, Einat P, Aharoni T, Eshdat Y, Friedman-Einat M. Molecular cloning and properties of the chicken leptin-receptor (CLEPR) gene. *Mol Cell Endocrinol* 2000;**162**:95–106.
242. Richards MP, Poch SM. Molecular cloning and expression of the turkey leptin receptor gene. *Comp Biochem Physiol B Biochem Mol Biol* 2003;**136**:833–47.
243. Wang F, Lu L, Yuan H, Tian Y, Li J, Shen J, et al. Molecular cloning, expression, and regulation of goose leptin receptor gene in adipocytes. *Mol Cell Biochem* 2011;**353**: 267–74.
244. Denbow DM, Meade S, Robertson A, McMurtry JP, Richards M, Ashwell C. Leptin-induced decrease in food intake in chickens. *Physiol Behav* 2000;**69**:359–62.
245. Dridi S, Raver N, Gussakovsky EE, Derouet M, Picard M, Gertler A, et al. Biological activities of recombinant chicken leptin C4S analog compared with unmodified leptins. *Am J Physiol Endocrinol Metab* 2000;**279**:E116–E123.
246. Lohmus M, Sundstrom LF, El Halawani M, Silverin B. Leptin depresses food intake in great tits (Parus major). *Gen Comp Endocrinol* 2003;**131**:57–61.
247. Lei M, Luo C, Peng X, Fang M, Nie Q, Zhang D, et al. Polymorphism of growth-correlated genes associated with fatness and muscle fiber traits in chickens. *Poult Sci* 2007;**86**:835–42.
248. Furuse M, Tachibana T, Ohgushi A, Ando R, Yoshimatsu T, Denbow DM. Intracerebroventricular injection of ghrelin and growth hormone releasing factor inhibits food intake in neonatal chicks. *Neurosci Lett* 2001;**301**:123–6.
249. Saito ES, Kaiya H, Takagi T, Yamasaki I, Denbow DM, Kangawa K, et al. Chicken ghrelin and growth hormone-releasing peptide-2 inhibit food intake of neonatal chicks. *Eur J Pharmacol* 2002;**453**:75–9.
250. Hen G, Yosefi S, Simchaev V, Shinder D, Hruby VJ, Friedman-Einat M. The melanocortin circuit in obese and lean strains of chicks. *J Endocrinol* 2006;**190**:527–35.

251. Saneyasu T, Honda K, Kamisoyama H, Nakayama Y, Ikegami K, Hasegawa S. Alpha-melanocyte stimulating hormone plays an important role in the regulation of food intake by the central melanocortin system in chicks. *Peptides* 2011;**32**:996–1000.
252. Ka S, Albert FW, Denbow DM, Paabo S, Siegel PB, Andersson L, et al. Differentially expressed genes in hypothalamus in relation to genomic regions under selection in two chicken lines resulting from divergent selection for high or low body weight. *Neurogenetics* 2011;**12**:211–21.
253. Song Y, Golling G, Thacker TL, Cone RD. Agouti-related protein (AGRP) is conserved and regulated by metabolic state in the zebrafish, *Danio rerio*. *Endocrine* 2003;**22**:257–65.
254. Song Y, Cone RD. Creation of a genetic model of obesity in a teleost. *FASEB J* 2007;**21**:2042–9.
255. Volkoff H, Canosa LF, Unniappan S, Cerda-Reverter JM, Bernier NJ, Kelly SP, et al. Neuropeptides and the control of food intake in fish. *Gen Comp Endocrinol* 2005;**142**:3–19.
256. Kurokawa T, Uji S, Suzuki T. Identification of cDNA coding for a homologue to mammalian leptin from pufferfish, *Takifugu rubripes*. *Peptides* 2005;**26**:745–50.
257. Huising MO, Geven EJ, Kruiswijk CP, Nabuurs SB, Stolte EH, Spanings FA, et al. Increased leptin expression in common Carp (*Cyprinus carpio*) after food intake but not after fasting or feeding to satiation. *Endocrinology* 2006;**147**:5786–97.
258. Kurokawa T, Murashita K. Genomic characterization of multiple leptin genes and a leptin receptor gene in the Japanese medaka, Oryzias latipes. *Gen Comp Endocrinol* 2009;**161**:229–37.
259. Gorissen M, Bernier NJ, Nabuurs SB, Flik G, Huising MO. Two divergent leptin paralogues in zebrafish (*Danio rerio*) that originate early in teleostean evolution. *J Endocrinol* 2009;**201**:329–39.
260. Li GG, Liang XF, Xie Q, Li G, Yu Y, Lai K. Gene structure, recombinant expression and functional characterization of grass carp leptin. *Gen Comp Endocrinol* 2010;**166**:117–27.
261. Froiland E, Murashita K, Jorgensen EH, Kurokawa T. Leptin and ghrelin in anadromous Arctic charr: cloning and change in expressions during a seasonal feeding cycle. *Gen Comp Endocrinol* 2010;**165**:136–43.
262. Won ET, Baltzegar DA, Picha ME, Borski RJ. Cloning and characterization of leptin in a Perciform fish, the striped bass (*Morone saxatilis*): control of feeding and regulation by nutritional state. *Gen Comp Endocrinol* 2012;**178**:98–107.
263. Kurokawa T, Murashita K, Suzuki T, Uji S. Genomic characterization and tissue distribution of leptin receptor and leptin receptor overlapping transcript genes in the pufferfish, *Takifugu rubripes*. *Gen Comp Endocrinol* 2008;**158**:108–14.
264. Ronnestad I, Nilsen TO, Murashita K, Angotzi AR, Gamst Moen AG, Stefansson SO, et al. Leptin and leptin receptor genes in Atlantic salmon: cloning, phylogeny, tissue distribution and expression correlated to long-term feeding status. *Gen Comp Endocrinol* 2010;**168**:55–70.
265. Londraville RL, Duvall CS. Murine leptin injections increase intracellular fatty acid-binding protein in green sunfish (*Lepomis cyanellus*). *Gen Comp Endocrinol* 2002;**129**:56–62.
266. Baker DM, Larsen DA, Swanson P, Dickhoff WW. Long-term peripheral treatment of immature coho salmon (*Oncorhynchus kisutch*) with human leptin has no clear physiologic effect. *Gen Comp Endocrinol* 2000;**118**:134–8.
267. Silverstein JT, Plisetskaya EM. The effects of NPY and insulin on food intake regulation in fish. *Am Zool* 2000;**40**:296–308.
268. Volkoff H, Eykelbosh AJ, Peter RE. Role of leptin in the control of feeding of goldfish Carassius auratus: interactions with cholecystokinin, neuropeptide Y and orexin A, and modulation by fasting. *Brain Res* 2003;**972**:90–109.

269. Matsuda K, Shimakura S, Maruyama K, Miura T, Uchiyama M, Kawauchi H, et al. Central administration of melanin-concentrating hormone (MCH) suppresses food intake, but not locomotor activity, in the goldfish, Carassius auratus. *Neurosci Lett* 2006;**399**:259–63.
270. Matsuda K, Shimakura S, Miura T, Maruyama K, Uchiyama M, Kawauchi H, et al. Feeding-induced changes of melanin-concentrating hormone (MCH)-like immunoreactivity in goldfish brain. *Cell Tissue Res* 2007;**328**:375–82.
271. Kinoshita M, Morita T, Toyohara H, Hirata T, Sakaguchi M, Ono M, et al. Transgenic medaka overexpressing a melanin-concentrating hormone exhibit lightened body color but no remarkable abnormality. *Mar Biotechnol (NY)* 2001;**3**:536–43.
272. Matsuda K. Recent advances in the regulation of feeding behavior by neuropeptides in fish. *Ann N Y Acad Sci* 2009;**1163**:241–50.
273. Miura T, Maruyama K, Shimakura S, Kaiya H, Uchiyama M, Kangawa K, et al. Neuropeptide Y mediates ghrelin-induced feeding in the goldfish, *Carassius auratus*. *Neurosci Lett* 2006;**407**:279–83.
274. Shimakura S, Kojima K, Nakamachi T, Kageyama H, Uchiyama M, Shioda S, et al. Neuronal interaction between melanin-concentrating hormone- and α-melanocyte-stimulating hormone-containing neurons in the goldfish hypothalamus. *Peptides* 2008;**29**:1432–40.
275. Shimakura S, Miura T, Maruyama K, Nakamachi T, Uchiyama M, Kageyama H, et al. Alpha-melanocyte-stimulating hormone mediates melanin-concentrating hormone-induced anorexigenic action in goldfish. *Horm Behav* 2008;**53**:323–8.
276. Matsuda K, Nakamura K, Shimakura S, Miura T, Kageyama H, Uchiyama M, et al. Inhibitory effect of chicken gonadotropin-releasing hormone II on food intake in the goldfish, *Carassius auratus*. *Horm Behav* 2008;**54**:83–9.
277. Kawauchi H, Kawazoe I, Tsubokawa M, Kishida M, Baker BI. Characterization of melanin-concentrating hormone in chum salmon pituitaries. *Nature* 1983; **305**:321–3.
278. Qu D, Ludwig DS, Gammeltoft S, Piper M, Pelleymounter MA, Cullen MJ, et al. A role for melanin-concentrating hormone in the central regulation of feeding behaviour. *Nature* 1996;**380**:243–7.
279. Pissios P, Bradley RL, Maratos-Flier E. Expanding the scales: the multiple roles of MCH in regulating energy balance and other biological functions. *Endocr Rev* 2006;**27**:606–20.
280. Volkoff H, Hoskins LJ, Tuziak SM. Influence of intrinsic signals and environmental cues on the endocrine control of feeding in fish: potential application in aquaculture. *Gen Comp Endocrinol* 2010;**167**:352–9.
281. O'Rahilly S, Farooqi IS, Yeo GS, Challis BG. Minireview: human obesity-lessons from monogenic disorders. *Endocrinology* 2003;**144**:3757–64.
282. Farooqi S, O'Rahilly S. Genetics of obesity in humans. *Endocr Rev* 2006;**27**:710–8.
283. Montague CT, Farooqi IS, Whitehead JP, Soos MA, Rau H, Wareham NJ, et al. Congenital leptin deficiency is associated with severe early-onset obesity in humans. *Nature* 1997;**387**:903–8.
284. Strobel A, Issad T, Camoin L, Ozata M, Strosberg AD. A leptin missense mutation associated with hypogonadism and morbid obesity. *Nat Genet* 1998;**18**:213–5.
285. Clement K, Vaisse C, Lahlou N, Cabrol S, Pelloux V, Cassuto D, et al. A mutation in the human leptin receptor gene causes obesity and pituitary dysfunction. *Nature* 1998;**392**:398–401.
286. Krude H, Biebermann H, Luck W, Horn R, Brabant G, Gruters A. Severe early-onset obesity, adrenal insufficiency and red hair pigmentation caused by POMC mutations in humans. *Nat Genet* 1998;**19**:155–7.

287. Jackson RS, Creemers JW, Ohagi S, Raffin-Sanson ML, Sanders L, Montague CT, et al. Obesity and impaired prohormone processing associated with mutations in the human prohormone convertase 1 gene. *Nat Genet* 1997;**16**:303–6.
288. Holder Jr JL, Butte NF, Zinn AR. Profound obesity associated with a balanced translocation that disrupts the SIM1 gene. *Hum Mol Genet* 2000;**9**:101–8.
289. Lee YS, Poh LK, Loke KY. A novel melanocortin 3 receptor gene (MC3R) mutation associated with severe obesity. *J Clin Endocrinol Metab* 2002;**87**:1423–6.
290. Tao YX, Segaloff DL. Functional characterization of melanocortin-3 receptor variants identify a loss-of-function mutation involving an amino acid critical for G protein-coupled receptor activation. *J Clin Endocrinol Metab* 2004;**89**:3936–42.
291. Tao YX. Functional characterization of novel melanocortin-3 receptor mutations identified from obese subjects. *Biochim Biophys Acta* 2007;**1772**:1167–74.
292. Zegers D, Beckers S, de Freitas F, Peeters AV, Mertens IL, Verhulst SL, et al. Identification of three novel genetic variants in the melanocortin-3 receptor of obese children. *Obesity (Silver Spring)* 2010;**19**:152–9.
293. Mencarelli M, Dubern B, Alili R, Maestrini S, Benajiba L, Tagliaferri M, et al. Rare melanocortin-3 receptor mutations with in vitro functional consequences are associated with human obesity. *Hum Mol Genet* 2011;**20**:392–9.
294. Yang F, Tao YX. Functional characterization of nine novel naturally occurring human melanocortin-3 receptor mutations. *Biochim Biophys Acta* 2012;**1822**:1752–61.
295. Tao YX. Mutations in the melanocortin-3 receptor (MC3R) gene: impact on human obesity or adiposity. *Curr Opin Investig Drugs* 2010;**11**:1092–6.
296. Yeo GS, Farooqi IS, Aminian S, Halsall DJ, Stanhope RG, O'Rahilly S. A frameshift mutation in MC4R associated with dominantly inherited human obesity. *Nat Genet* 1998;**20**:111–2.
297. Vaisse C, Clement K, Guy-Grand B, Froguel P. A frameshift mutation in human MC4R is associated with a dominant form of obesity. *Nat Genet* 1998;**20**:113–4.
298. Hinney A, Schmidt A, Nottebom K, Heibult O, Becker I, Ziegler A, et al. Several mutations in the melanocortin-4 receptor gene including a nonsense and a frameshift mutation associated with dominantly inherited obesity in humans. *J Clin Endocrinol Metab* 1999;**84**:1483–6.
299. Tao YX. Molecular mechanisms of the neural melanocortin receptor dysfunction in severe early onset obesity. *Mol Cell Endocrinol* 2005;**239**:1–14.
300. Tao YX. Mutations in melanocortin-4 receptor and human obesity. *Prog Mol Biol Transl Sci* 2009;**88**:173–204.
301. Hinney A, Volckmar A-L, Knoll N. Melanocortin-4 receptor in energy homeostasis and obesity pathogenesis. *Prog Mol Biol Transl Sci* 2013;**114**:147–92.
302. Gray J, Yeo GS, Cox JJ, Morton J, Adlam AL, Keogh JM, et al. Hyperphagia, severe obesity, impaired cognitive function, and hyperactivity associated with functional loss of one copy of the brain-derived neurotrophic factor (BDNF) gene. *Diabetes* 2006;**55**:3366–71.
303. Yeo GS, Connie Hung CC, Rochford J, Keogh J, Gray J, Sivaramakrishnan S, et al. A de novo mutation affecting human TrkB associated with severe obesity and developmental delay. *Nat Neurosci* 2004;**7**:1187–9.
304. Burns B, Schmidt K, Williams SR, Kim S, Girirajan S, Elsea SH. Rai1 haploinsufficiency causes reduced Bdnf expression resulting in hyperphagia, obesity and altered fat distribution in mice and humans with no evidence of metabolic syndrome. *Hum Mol Genet* 2010;**19**:4026–42.
305. Rosas-Vargas H, Martinez-Ezquerro JD, Bienvenu T. Brain-derived neurotrophic factor, food intake regulation, and obesity. *Arch Med Res* 2011;**42**:482–94.
306. Biebermann H, Castaneda TR, van Landeghem F, von Deimling A, Escher F, Brabant G, et al. A role for β-melanocyte-stimulating hormone in human body-weight regulation. *Cell Metab* 2006;**3**:141–6.

307. Lee YS, Challis BG, Thompson DA, Yeo GS, Keogh JM, Madonna ME, et al. A POMC variant implicates β-melanocyte-stimulating hormone in the control of human energy balance. *Cell Metab* 2006;**3**:135–40.
308. Michaud JL, Rosenquist T, May NR, Fan CM. Development of neuroendocrine lineages requires the bHLH-PAS transcription factor SIM1. *Genes Dev* 1998; **12**:3264–75.
309. Lebrun B, Bariohay B, Moyse E, Jean A. Brain-derived neurotrophic factor (BDNF) and food intake regulation: a minireview. *Auto Neurosci* 2006;**126–127**:30–8.
310. Xu B, Goulding EH, Zang K, Cepoi D, Cone RD, Jones KR, et al. Brain-derived neurotrophic factor regulates energy balance downstream of melanocortin-4 receptor. *Nat Neurosci* 2003;**6**:736–42.
311. Han JC, Liu QR, Jones M, Levinn RL, Menzie CM, Jefferson-George KS, et al. Brain-derived neurotrophic factor and obesity in the WAGR syndrome. *N Engl J Med* 2008;**359**:918–27.
312. Farooqi IS, Jebb SA, Langmack G, Lawrence E, Cheetham CH, Prentice AM, et al. Effects of recombinant leptin therapy in a child with congenital leptin deficiency. *N Engl J Med* 1999;**341**:879–84.
313. Farooqi IS, Matarese G, Lord GM, Keogh JM, Lawrence E, Agwu C, et al. Beneficial effects of leptin on obesity, T cell hyporesponsiveness, and neuroendocrine/metabolic dysfunction of human congenital leptin deficiency. *J Clin Invest* 2002;**110**:1093–103.
314. Licinio J, Caglayan S, Ozata M, Yildiz BO, de Miranda PB, O'Kirwan F, et al. Phenotypic effects of leptin replacement on morbid obesity, diabetes mellitus, hypogonadism, and behavior in leptin-deficient adults. *Proc Natl Acad Sci USA* 2004;**101**:4531–6.
315. Xiang Z, Pogozheva ID, Sorenson NB, Wilczynski AM, Holder JR, Litherland SA, et al. Peptide and small molecules rescue the functional activity and agonist potency of dysfunctional human melanocortin-4 receptor polymorphisms. *Biochemistry* 2007;**46**:8273–87.
316. Fan ZC, Tao YX. Functional characterization and pharmacological rescue of melanocortin-4 receptor mutations identified from obese patients. *J Cell Mol Med* 2009;**13**:3268–82.
317. Rene P, Le Gouill C, Pogozheva ID, Lee G, Mosberg HI, Farooqi IS, et al. Pharmacological chaperones restore function to MC4R mutants responsible for severe early-onset obesity. *J Pharmacol Exp Ther* 2010;**335**:520–32.
318. Katsanis N, Ansley SJ, Badano JL, Eichers ER, Lewis RA, Hoskins BE, et al. Triallelic inheritance in Bardet-Biedl syndrome, a Mendelian recessive disorder. *Science* 2001; **293**:2256–9.
319. Feuillan PP, Ng D, Han JC, Sapp JC, Wetsch K, Spaulding E, et al. Patients with Bardet–Biedl syndrome have hyperleptinemia suggestive of leptin resistance. *J Clin Endocrinol Metab* 2011;**96**:E528–E535.
320. Badano JL, Ansley SJ, Leitch CC, Lewis RA, Lupski JR, Katsanis N. Identification of a novel Bardet-Biedl syndrome protein, BBS7, that shares structural features with BBS1 and BBS2. *Am J Hum Genet* 2003;**72**:650–8.
321. Benzinou M, Walley A, Lobbens S, Charles MA, Jouret B, Fumeron F, et al. Bardet-Biedl syndrome gene variants are associated with both childhood and adult common obesity in French Caucasians. *Diabetes* 2006;**55**:2876–82.
322. Guo DF, Rahmouni K. Molecular basis of the obesity associated with Bardet–Biedl syndrome. *Trends Endocrinol Metab* 2011;**22**:286–93.
323. Croft JB, Morrell D, Chase CL, Swift M. Obesity in heterozygous carriers of the gene for the Bardet–Biedl syndrome. *Am J Med Genet* 1995;**55**:12–5.
324. Davis RE, Swiderski RE, Rahmouni K, Nishimura DY, Mullins RF, Agassandian K, et al. A knockin mouse model of the Bardet–Biedl syndrome 1 M390R mutation has

cilia defects, ventriculomegaly, retinopathy, and obesity. *Proc Natl Acad Sci USA* 2007;**104**:19422–7.

325. Davenport JR, Watts AJ, Roper VC, Croyle MJ, van Groen T, Wyss JM, et al. Disruption of intraflagellar transport in adult mice leads to obesity and slow-onset cystic kidney disease. *Curr Biol* 2007;**17**:1586–94.
326. Rahmouni K, Fath MA, Seo S, Thedens DR, Berry CJ, Weiss R, et al. Leptin resistance contributes to obesity and hypertension in mouse models of Bardet–Biedl syndrome. *J Clin Invest* 2008;**118**:1458–67.
327. Seo S, Guo DF, Bugge K, Morgan DA, Rahmouni K, Sheffield VC. Requirement of Bardet–Biedl syndrome proteins for leptin receptor signaling. *Hum Mol Genet* 2009;**18**:1323–31.
328. Arsov T, Silva DG, O'Bryan MK, Sainsbury A, Lee NJ, Kennedy C, et al. Fat aussie—a new Alstrom syndrome mouse showing a critical role for ALMS1 in obesity, diabetes, and spermatogenesis. *Mol Endocrinol* 2006;**20**:1610–22.
329. Barsh GS, Farooqi IS, O'Rahilly S. Genetics of body-weight regulation. *Nature* 2000;**404**:644–51.
330. Bell CG, Walley AJ, Froguel P. The genetics of human obesity. *Nat Rev Genet* 2005;**6**:221–34.
331. Walley AJ, Asher JE, Froguel P. The genetic contribution to non-syndromic human obesity. *Nat Rev Genet* 2009;**10**:431–42.
332. Hinney A, Vogel CI, Hebebrand J. From monogenic to polygenic obesity: recent advances. *Eur Child Adolesc Psychiatry* 2010;**19**:297–310.
333. Hebebrand J, Volckmar AL, Knoll N, Hinney A. Chipping away the 'missing heritability': GIANT steps forward in the molecular elucidation of obesity—but still lots to go. *Obes Facts* 2010;**3**:294–303.
334. Rankinen T, Zuberi A, Chagnon YC, Weisnagel SJ, Argyropoulos G, Walts B, et al. The human obesity gene map: the 2005 update. *Obesity (Silver Spring)* 2006;**14**: 529–644.
335. Thorleifsson G, Walters GB, Gudbjartsson DF, Steinthorsdottir V, Sulem P, Helgadottir A, et al. Genome-wide association yields new sequence variants at seven loci that associate with measures of obesity. *Nat Genet* 2009;**41**:18–24.
336. Willer CJ, Speliotes EK, Loos RJ, Li S, Lindgren CM, Heid IM, et al. Six new loci associated with body mass index highlight a neuronal influence on body weight regulation. *Nat Genet* 2009;**41**:25–34.
337. Speliotes EK, Willer CJ, Berndt SI, Monda KL, Thorleifsson G, Jackson AU, et al. Association analyses of 249,796 individuals reveal 18 new loci associated with body mass index. *Nat Genet* 2010;**42**:937–48.
338. Frayling TM, Timpson NJ, Weedon MN, Zeggini E, Freathy RM, Lindgren CM, et al. A common variant in the FTO gene is associated with body mass index and predisposes to childhood and adult obesity. *Science* 2007;**316**:889–94.
339. Dina C, Meyre D, Gallina S, Durand E, Korner A, Jacobson P, et al. Variation in FTO contributes to childhood obesity and severe adult obesity. *Nat Genet* 2007;**39**:724–6.
340. Scuteri A, Sanna S, Chen WM, Uda M, Albai G, Strait J, et al. Genome-wide association scan shows genetic variants in the FTO gene are associated with obesity-related traits. *PLoS Genet* 2007;**3**:e115.
341. Church C, Moir L, McMurray F, Girard C, Banks GT, Teboul L, et al. Overexpression of Fto leads to increased food intake and results in obesity. *Nat Genet* 2010;**42**:1086–92.
342. McCarthy MI. Genomics, type 2 diabetes, and obesity. *N Engl J Med* 2010;**363**: 2339–50.
343. Fawcett KA, Barroso I. The genetics of obesity: FTO leads the way. *Trends Genet* 2010;**26**:266–74.

344. Larder R, Cheung MK, Tung YC, Yeo GS, Coll AP. Where to go with FTO? *Trends Endocrinol Metab* 2011;**22**:53–9.
345. Heitmann BL, Westerterp KR, Loos RJ, Sorensen TI, O'Dea K, McLean P, et al. Obesity: lessons from evolution and the environment. *Obes Rev* 2012;http://dx.doi.org/10.1111/j.467-789X.2012.01007.x.
346. Stutzmann F, Tan K, Vatin V, Dina C, Jouret B, Tichet J, et al. Prevalence of melanocortin-4 receptor deficiency in Europeans and their age-dependent penetrance in multigenerational pedigrees. *Diabetes* 2008;**57**:2511–8.
347. Foster-Schubert KE, Cummings DE. Emerging therapeutic strategies for obesity. *Endocr Rev* 2006;**27**:779–93.
348. The fat of thematter. *Nat Med* 2012;**18**:1155.

CHAPTER TWO

Ghrelin Receptor in Energy Homeostasis and Obesity Pathogenesis

Ziru Li[*,†], **Yin Li**[*,†], **Weizhen Zhang**[*,†,‡]
[*]Department of Physiology and Pathophysiology, School of Basic Medical Sciences, Peking University, Beijing, China
[†]Key Laboratory of Molecular Cardiovascular Science, Ministry of Education, Beijing, China
[‡]Department of Surgery, University of Michigan Medical Center, Ann Arbor, Michigan, USA

Contents

Progress in Molecular Biology and Translational Science, Volume 114
ISSN 1877-1173
http://dx.doi.org/10.1016/B978-0-12-386933-3.00002-9

Abstract

The ghrelin receptor, also known as growth hormone secretagogue receptor (GHS-R), was identified in porcine and rat anterior pituitary membranes, where the synthetic secretagogue MK-0677 causes amplified pulsatile growth hormone (GH) release. In addition to its function in the stimulation of GH secretion, ghrelin, the natural ligand of ghrelin receptor is now recognized as a peptide hormone with fundamental influence on energy homeostasis. Despite the potential existence of multiple subtypes of ghrelin receptor, the effects of ghrelin on energy metabolism, obesity, and diabetes are mediated by its classical receptor GHS-R1a, whose activation requires the n-octanoylation of ghrelin. Here we review the current understanding of the role of the ghrelin receptor in the regulation of energy homeostasis. An overview of the ghrelin receptor is presented first, followed by the discussion on its effects on food intake, glucose homeostasis, and lipid metabolism. Finally, potential strategies for treating obesity and diabetes via manipulation of the ghrelin/ghrelin receptor axis are explored.

1. OVERVIEW OF THE GHRELIN RECEPTOR

1.1. Discovery of the ghrelin receptor

1.1.1 Development of growth hormone secretagogue

Growth hormone (GH) was first identified for its notable effect on longitudinal growth. Subsequent research revealed that the GH has anabolic effects on protein, lipid, and carbohydrate metabolism.[1] GH replacement therapy, using recombinant GH, is therefore used to treat individuals with short stature due to a variety of conditions.[2,3] However, GH replacement therapy suffers from significant drawbacks such as low bioavailability and side effects. Moreover, most GH-deficient individuals exhibit a secretory defect rather than a primary deficiency in GH production. Research seeking a better drug to replace GH was therefore extensively active in the 1980s and 1990s.

In 1984, a synthetic hexapeptide, His-D-Trp-Ala-Trp-D-Phe-Lys-NH2 (GHRP-6), was identified by Bowers and colleagues. This hexapeptide was shown to potently stimulate GH release *in vitro* and *in vivo* by an unknown mechanism.[4] Because of its poor oral bioavailability (0.3%) and short half-life (20 min) in human serum,[5] GHRP-6 was selected only as a structural model to design a nonpeptide mimetic. Based on the structure–activity relationships (SARs) of GHRP-6, the nonpeptidyl growth hormone secretagogue (GHS) L-692,429 was identified by Smith *et al.* in 1993.[6] This nonpeptidyl GHS synergizes with GHRP-6 to stimulate GH release and cAMP production, accompanied by a significant increase in intracellular calcium concentration ($[Ca^{2+}]_i$), indicating that this nonpeptidyl GHS acts through a distinct signal transduction pathway.[6] In 1995, a potent oral GHS L-163,191 (MK-0677) was reported by Patchett *et al.* This agent was found to have excellent oral bioavailability and specificity in its release of GH, without significant effect on plasma levels of other hormones such as aldosterone, luteinizing hormone, thyroxine, and prolactin.[7]

1.1.2 Identification of GHS-R

Due to its potential use in treating GH deficiency-related diseases, GH secretion became a research focus in the 1990s. Secretion of GH is mainly regulated by GH-releasing hormone (GHRH) and somatostatin, which have stimulatory and inhibitory effects, respectively.[8,9] GHRH acts on the GHRH receptor by increasing intracellular cAMP, which serves as a second messenger; in contrast, GHSs stimulate phospholipase C, resulting in an increase in inositol trisphosphate (IP3) and intracellular Ca^{2+}.[7] The proceeding research on synthetic peptides and nonpeptides that stimulate GH release revealed a distinct third pathway in addition to GHRH and somatostatin, suggesting the presence of a unique novel receptor.[4] This novel receptor was identified in 1996 by Pong *et al.* in porcine and rat anterior pituitary membranes using MK-0677 as a ligand.[10] Binding of the GHS with this receptor is Mg^{2+} dependent and could be inhibited by stable guanosine triphosphate (GTP-γ-S), indicating that it is a G-protein-coupled receptor (GPCR).[10] Activation of this GPCR results in the stimulation of amplified pulsatile GH release. This GPCR was therefore recognized as a novel neuroendocrine pathway for the control of pulsatile GH release. Using the expression-cloning strategy in Xenopus oocytes, the full-length sequence of this receptor was revealed by Howard and colleagues and named growth hormone secretagogue receptor (GHS-R).[11]

1.2. Natural ligands for the ghrelin receptor

Through a process of reverse pharmacology, the endogenous ligand for GHS-R was identified from gastric extracts and named "ghrelin" by Kojima *et al.* in 1999.[12] Using the intracellular calcium concentration in CHO cells expressing GHS-R as a functional readout, Kojima *et al.* screened several tissue extracts from brain, lung, heart, kidney, stomach, and intestine.[12] Unexpectedly, they discovered that the active agonist for the distinct ghrelin receptor exits in the stomach instead of the brain. The active component was purified by successive chromatography including gel filtration, ion exchange high-performance liquid chromatography (HPLC), and reverse-phase HPLC.[12] The amino acid sequence of ghrelin in the rat was reported as

GSSFLSPEHQKAQQRKESKKPPAKLQPR, in which the third serine (Ser^3) is *n*-octanoylated. This 28 amino acids peptide was named ghrelin based on the word root "ghre" meaning "growth" in proto-Indo-European language, to depict its ability to stimulate GH release.[12] The octanoylation of the third serine is necessary for its activity. Ghrelin is the only known peptide modified by a fatty acid, and it shares no structural homology with other GHSs. Cloning of human ghrelin, using primers based on rat ghrelin cDNA under low stringency conditions, reveals that only two amino acids are different between human and rat ghrelin.[12] Ghrelin is thus highly conserved between species.

A second endogenous ligand for GHS-R was subsequently purified from rat gastric extracts, and named des-Gln^{14}-Ghrelin. The amino acid sequence for this ligand is identical to that of ghrelin, except that the 14th glutamine is missing.[13] Again the *n*-octanoyl modification on the third serine of this peptide is necessary for its impact on GH release. Analysis of the genomic structure reveals that des-Gln^{14}-ghrelin is processed from the ghrelin gene by alternative splicing. Thus, there are at least two ghrelin peptides in the stomach that stimulate the GH release through GHS-R.

Despite its high concentration in the blood, the physiologic role of unacylated ghrelin (UAG or des-acyl ghrelin) has not been conclusively determined. Studies by Cassoni *et al.* and Muccioli *et al.* have shown that UAG can displace radiolabeled ghrelin from membrane extracts of cell lines lacking expression of GHS-R mRNA.[14,15] Additionally, UAG has been shown to inhibit ghrelin-induced glucose release in primary hepatocytes.[16] In contrast, studies by Gauna *et al.*, using CHO-K1 cells expressing human GHS-R, demonstrate that UAG is a full agonist for the ghrelin receptor with an EC between 1.6 and 2 μM.[17] In addition, ghrelin receptor antagonists such as BIM28163 and [D-lys^3] GHRP6 block UAG-induced $[Ca^{2+}]_i$ signaling with almost similar potency to their effect on ghrelin activation of the ghrelin receptor.[17] These findings raise the question of whether UAG is one of the ligands for the ghrelin receptor, or whether there exists another unidentified ghrelin receptor subtype.

Other possible natural ligands, including adenosine and cortistatin, have also been considered. Although adenosine is able to induce an increase in $[Ca^{2+}]_i$ in cell lines expressing recombinant human GHS-R1a, it is not considered a natural ligand for the ghrelin receptor as it has no effect on GH release.[18] Cortistatin, a natural neuropeptide with high structural homology to somatostatin, has also been reported to compete with ^{125}I-Tyr-Ala-hexarelin for human pituitary binding sites in a binding affinity comparable to that of hexarelin and ghrelin. In addition, cortistatin and ghrelin exhibit similar binding affinity in displacing ^{125}I-Tyr-Ala-hexarelin.[19] Whether cortistatin is a natural ligand for ghrelin receptor remains to be explored.

1.3. Variants of the ghrelin receptor

The GHS-R gene encodes two transcripts that are translated into GHS-R1a and GHS-R1b. Only GHS-R1a has been convincingly demonstrated to be functionally active upon binding with its natural ligand ghrelin, and is often referred to as the ghrelin receptor. We will therefore use the terms "ghrelin receptor" and "GHS-R1a" interchangeably throughout the rest of this review. GHS-R1b is the product of a read-through into an intron, producing an in-frame stop codon. The predicted translation of GHS-R1b is therefore a truncated product of GHS-R gene, containing only the first five transmembrane (TM) domains of the ghrelin receptor at the N-terminus and lacking the TM6 and TM7 segments. Although GHS-R1b mRNA is expressed in a wide range of tissues,[20] whether it is translated into protein is unclear. In addition, neither ghrelin nor GHS binds to GHS-R1b protein *in vitro*.[21] The functional significance of GHS-1b is therefore not apparent.

Based on a variety of functional and binding studies, the existence of a novel subtype of ghrelin receptor distinct from GHS-R1a is proposed to exist in peripheral tissues such as cardiomyocytes, endothelium, skeletal muscle, and tumor cells. A radiolabeled peptidyl GHS has been shown to bind to a more diverse array of tissues than ghrelin or a nonpeptidyl GHS. Discrepancy exists between the reported actions of neutral ghrelin receptor antagonists in the regulation of GH release and body weight,[22,23] suggesting the presence of an unknown ghrelin receptor that modulates ghrelin actions on weight gain other than its function on the stimulation of GH. In addition, binding sites of des-acyl ghrelin have been detected in certain cells such as myocytes. However, whether this unknown subtype of ghrelin receptor is an alternative receptor for ghrelin or des-acyl ghrelin requires further investigation. Until the novel ghrelin receptor subtype is cloned, the possibility of an alternative ghrelin receptor or putative des-acyl ghrelin receptor remains speculative.

Variations in the ghrelin receptor gene have been linked to obesity, eating disorders, hypertriglyceridemia, and insulin resistance.[24–26] A comprehensive analysis of single nucleotide polymorphisms (SNPs) and haplotype structure across the entire ghrelin receptor region (99.3 kb) has identified a linkage disequilibrium block consisting of five SNPs that shows linkage with body mass index in an intra-familial segregation study among 178 obese White pedigrees, as well as association with obesity at the population level among 1418 whites.[26] Such common variants in ghrelin receptor were also shown to be associated with obesity in a French case–control study of 602 subjects, while such an association was not identified in a German study of 888 individuals.[27]

The effects of ghrelin receptor variants on the pathogenesis of obesity have been reported. A mutation screening by Gjesing *et al.* identified the -151C/T mutation, located in the ghrelin receptor promoter, as the most likely functional variant. This mutation increases the transcriptional activity of the ghrelin receptor gene, probably due to the introduction of a highly specific SP-1-like site (such as Egr1 or Ets1) predicted to selectively bind the -151T mutant promoter. As the ghrelin receptor has high constitutive activity at baseline, any increase in receptor transcription and translation related to the -151C/T mutation is proposed to increase ghrelin-independent receptor signaling. Consequently, this mutation may lead to increased appetite and decreased energy expenditure at the organism level.[28]

Rare deleterious mutations in the ghrelin receptor gene have been associated with short stature in humans.[29,30] Carriers of a change from alanine 204 to glutamate (A204E) are three standard deviations smaller than average for height and weight.[29] The 204 glutamate variant is characterized by a decrease in cell membrane receptor density but no change in receptor affinity for the agonist.[29] Another nonsynonymous ghrelin receptor variation at amino acid 279 has been identified in one heterozygous obese child with short normal stature, but its function is yet unknown.[30]

Other naturally occurring variants of the human ghrelin receptor, including I134T, V160M, A204E, and F279L, might affect basal activity and receptor expression.[31] Each of the missense mutations in the ghrelin receptor is characterized by a decrease in receptor expression compared with the wild-type value. Individuals who express these variants may also show diminished receptor-mediated function. While the ghrelin receptor has a high level of ligand-independent basal activity exceeding 50% of the ghrelin-induced maximum, the V160M, A204E, and F279L variants show reduced or abolished basal function.[31]

1.4. Antagonists for the ghrelin receptor

In addition to the agonists described above, a variety of low molecular weight ghrelin receptor antagonists have been investigated as potential therapeutic agents in the treatment of obesity and diabetes. Evaluation of SARs led to the identification of several ghrelin receptor antagonists, including quinazolinone derivates,[32] piperazine-bisamide,[33] and tetralin carboxamide.[34,35]

Ligand-based structure virtual screening and screening of RNA-based compounds have identified other ghrelin receptor antagonists with possible therapeutic actions.

Identification of phenyl or phenoxy groups as optimal substituents at the 6-position of the quinazolinone core led to the discovery of piperidine-substituted quinazolinone derivatives as a new class of small-molecule ghrelin receptor antagonists. To manipulate specific drug properties including selectivity, metabolism, and pharmacokinetics, phenyl groups at the 2-position were replaced by small alkyl substituents. Extensive screening of these chemicals identified two compounds with optimal oral bioavailability, potency, and selectivity as ghrelin receptor antagonists. *In vivo* evaluation of these two compounds demonstrated a significant effect on suppression of food intake and body weight, as well as glucose-lowering effects mediated by glucose-dependent insulin secretion.[32]

A series of piperazine-bisamide analogs originally discovered as the partial agonists of the ghrelin receptor have been modified and converted into potent antagonists through the SAR study. By modifying the phenyl B-ring of the biphenyl moiety to remove the partial agonist activity, a new compound was developed. This agent features high potency, satisfactory pharmacokinetic profile, and sufficient CNS exposure in mdr1a (multi-drug resistance gene) knockout mice.[33] The *ex vivo* study also confirmed this compound as an antagonist for recombinant ghrelin-induced GH release from isolated primary rat pituitary cells.[33]

A full-length analog of human ghrelin, BIM-28163 binds to the ghrelin receptor, but does not induce receptor activation. BIM-28163 functions as a competitive antagonist to the ghrelin receptor, blocking ghrelin-induced GH secretion *in vivo*. However, this antagonist unexpectedly increases food intake and weight gain.[23] The significance of this finding is still unclear, but may indicate the existence of an unknown subtype of ghrelin receptor that modulates ghrelin actions on weight gain independent of its effect on GH release.

Other strategies to potently block ghrelin receptor activation *in vitro* and *in vivo* include the interference of ghrelin action by bio-stable RNA-based compounds termed RNA Spiegelmers and antibodies against the ghrelin receptor. L-NOX-B11, an RNA Spiegelmer, has been reported to inhibit the action of ghrelin at the receptor level *in vitro* in a low nanomolar range and to suppress ghrelin-induced GH release *in vivo* when injected intravenously.[36] Additionally, vaccination with ghrelin immunoconjugates has been shown to slow weight gain in rats by decreasing feeding efficiency.[37]

1.5. Structure of ghrelin receptor and ligand-binding domains

The ghrelin receptor (GHS-R) gene is located on chromosome 3q26.2 and encodes two transcripts: 1a and 1b, which encode a full-length receptor GHS-R1a (by exon 1 and 2) and a shortened version GHS-R1b (only exon 1), respectively.[38] The GHS-R1a is encoded by a noncontiguous open reading frame of about 1.1 kb; approximately 2 kb of nonencoding intronic sequence divides the open reading frame into an amino-terminal segment ending at leucine-265 encoding TM 1-5 and a carboxyl-terminal segment encoding TM6/TM7. Moreover, the intron contains a stop codon that can explain the production of GHS-R1b mRNA. Both sequences are identical from the Met translation site to Leu at the 265th amino acid position.[11] Comparison of the predicted human, rat, pig, and sheep GHS-R1a amino acid sequences reveals 91.8–95.6% sequence homology.[39]

Human GHS-R1a consists of 366 amino acids with a molecular mass of approximately 41 kDa[11] and belongs to family A of GPCRs.[40,41] It spans the membrane with seven α-helix hydrophobic domains connected by three intra- and extracellular domains, beginning with an extracellular N-terminal domain and ending with an intracellular C-terminal domain.[41] The N-terminal domain forms a β-hairpin structure and the TM domains form a round calyx-like structure that is attributed to Pro residues in the center of the TM helices. TM3 occupies the central position among the TM segments and TM5 is the most peripheral.[42] The ligand activation domains are in the second TM domain (TM2) and the third TM domain (TM3). Moreover, GHS-R1a possesses three conserved residues, Glu-Arg-Tyr, at the intracellular end of TM3 in position 140–142 (ERY/DRY motif), which are critical for the isomerization between the active and inactive conformation. Additionally, two conserved cysteine residues (Cys116 and Cys198) on extracellular loops 1 and 2 form a disulfide bond.[41,43] These key amino acid residues are essential for binding and activation by different ligands, and have been evolutionarily conserved for 400 million years, highlighting their importance in fundamental physiological processes.[39]

GHS-R1b contains 298 amino acids corresponding to the first five TM domains encoded by exon 1, plus a unique 24 amino acid tail encoded by an alternatively spliced intron sequence. GHS-R1b neither binds nor responds to ghrelin or GHS stimulation.[21] Since GHS-R1b is comprehensively expressed in various tissues,[20] it is reasonable to assume that this receptor possesses some unidentified biological function. A report in 2007 indicated that GHS-R1b decreases the cell surface expression of GHS-R1a and acts as a repressor of the constitutive activity of GHS-R1a when overexpressed in

HEK-293 cells.[44] This finding suggests that GHS-R1b may act as an endogenous modulator for GHS-R1a constitutive activity.

GHS-R1a is highly expressed in the hypothalamus and pituitary, consistent with ghrelin's functional role in the control of appetite and GH release. Ligand binding to GHS-R1a is believed to stabilize the active conformation. The main binding pocket for small amines is deep in the cavity created by the TM domains. According to the general model based on the β_2-adrenergic receptor and rhodopsin, ligand binding results in an alteration of receptor molecular structure. This change is characterized by a reciprocal rearrangement of the α-helices with vertical seesaw movements of TM6 and TM7 around their central proline residues, which act as a pivot.[45] Hence, the intracellular ends of TM6 and TM7 move away from the center of the receptor toward TM3, exposing the sites subsequently recognized by G-proteins and β-arrestin.[45] The "toggle switch model," as it is known, is applicable to GHS-R1a, with a binding domain for the natural ligand ghrelin involving six amino acids located in TM3, TM6, or TM7.[45] In addition, binding of the natural ligand ghrelin with GHS-R1a requires the ligand to interact with one pocket formed by polar amino acids in TM2/TM3 and one formed by nonpolar amino acids in TM5/TM6.[42] The synthetic peptidyl and nonpeptidyl GHSs share a common binding pocket in the TM3 region of GHS-R1a, although there are other distinct binding sites in the receptor that seem to be selective for particular classes of agonists.[46] In contrast, the inverse agonist D-Arg^1-D-Phe^5-D-Trp^{7,9}-Leu^{11}-substance P requires a wider binding pocket, dispersed across the main binding crevice.[41]

Small peptides bind to extracellular epitopes of GHS-R1a as well. The basic amine common to peptidyl (GHRP-6) and nonpeptidyl (MK-0677) GHS establishes an electrostatic interaction with Glu124 in the TM3 domain.[46] Substitution of glutamine for glutamic acid [Q124E mutant] in human GHS-R1a eliminates its function. Furthermore, mutation of Arg283 in TM6, which interacts with Glu124, abolishes both constitutive and agonist-induced signaling.[46,47] In addition, disruption of the disulfide bond between Cys116 and Cys198 in the extracellular portion of the receptor [C116A mutant] completely abolishes the activity of all agonists.[39,46]

1.6. Constitutive activity of ghrelin receptor

Evidence for the constitutive activity of the ghrelin receptor and its physiologic implications is steadily growing.[48] The high basal activity of ghrelin receptor has been postulated to play a role in the regulation of both food intake and growth.[28,49] When overexpressed in COS-7 cells, the ghrelin receptor possesses a constitutive activity in terms of the turnover of IP3, which is approximately 50% of the maximal agonist-induced activity.[45,47]

The ligand-binding domains of the ghrelin receptor are characterized by a cluster of aromatic amino acids in TM3, TM6, and TM7.[50,51] The molecular basis of its constitutive activity relates therefore to three aromatic residues located in TM6 and TM7, namely PheVI:16, PheVII:06, and PheVII:09. This region promotes the formation of a hydrophobic core between TM6 and TM7, which ensures proper docking of the extracellular end of TM7 into TM6, mimicking agonist activation and stabilizing the receptor in active conformation.[52]

Currently, the physiological relevance of this high constitutive activity is not fully understood. In HEK-293 cells, transfection of ghrelin receptor leads to constitutively stimulated cAMP response element and sterol response element activity.[47] In contrast, the ghrelin receptor demonstrates no constitutive activity in the pituitary cell line RC-4B/ C40, indicating that its constitutive activity may be turned off or dependent on the cellular context.[53] D-Arg^{1}-D-Phe^{5}-D-$Trp^{7,9}$-Leu^{11}-substance P is recognized as the only known ligand blocking GHS-R1a constitutive activity.[45,47] Based on these observations, efforts are in progress to develop drugs that inhibit ghrelin receptor constitutive activity as a treatment for obesity. Preliminary results have demonstrated that drugs functioning in this manner might reduce the craving for desserts and intermeal snacks by increasing sensitivity to inhibitory signals.[52]

1.7. Internalization of ghrelin receptor

In the normal healthy organism, ghrelin receptor desensitization and endocytosis govern the ability of cells to respond to ghrelin, thereby regulating intracellular signaling to avoid permanent stimulation of target cells. Receptor resensitization determines the frequency of the response to ghrelin. In addition, the composition of the plasma membrane affects the responsiveness of cells to metabolically important hormones such as ghrelin and insulin. Deficiencies in this attenuation system may lead to an uncontrolled or defective stimulation of target cells, alteration of their intracellular signaling, and subsequent pathological changes.

Kinetic studies of ghrelin receptor internalization based on radioligand-binding experiments demonstrate that the ghrelin receptor is internalized in a time-dependent manner with peaks at approximately 20 min after ghrelin stimulation. Once the ligand–receptor complex is internalized into intracellular vesicles, ghrelin receptors are sorted into endosomes to be recycled back to the membrane.[54] About 360 min after agonist removal, the level of ghrelin receptors on the cell surface recovers to almost 100% of the original level. This process is not affected by cycloheximide, suggesting that ghrelin receptors are replenished from endosomes rather than *de novo* receptor synthesis. Study of fluorescence associated with GHS-R1a-EGFP in CHO cells demonstrates that ghrelin receptors reappear on the membrane with similar kinetics after a ghrelin pulse. Furthermore, fluorescence emitted by the EGFP-labeled GHS-R1a in cells colocalizes with the endosome marker early endosomal protein 1 but not with the lysosomal marker cathepsin. Thus, most ghrelin receptors appearing at the cell surface are derived from endocytosed receptors with completely restored binding capacity and functionality.[54]

Functional studies also demonstrate the theory of ghrelin receptor recycling. GH response to two consecutive pulses of ghrelin is blunted when pulses are separated by a short interval of 60 min. However, GH response retains its initial amplitude when the second pulse is administered after 180, 240, or 360 min. The slower recycling (3–6 h) of this receptor compared to other GPCRs is determined by the stability of the ghrelin receptor/β-arrestin complex during clathrin-mediated endocytosis. This complex appears to dictate the profile of ghrelin receptor resensitization.[55]

Lipid and plasma membrane composition also influence ghrelin receptor activation and desensitization, as demonstrated by the effect of oligounsaturated fatty acids (OFAs) in a study by Delhanty *et al.*[56] OFAs disrupt plasma membrane structure, rendering the membrane more fluid. Exposing ghrelin receptor expressing cells to the 18-carbon OFAs such as oleic and linoleic acids for a prolonged period (96 h) causes significantly increased receptor sensitivity to ghrelin.[56] The inhibitory effects of ghrelin pretreatment on subsequent ghrelin responsiveness are blunted, suggesting that OFAs suppress desensitization of ghrelin receptor. Furthermore, it has been found that OFA treatment suppresses the internalization of the ghrelin receptor,[56] likely explaining OFA-induced refractoriness to ligand-induced desensitization. The altered responsiveness of the ghrelin receptor under conditions that alter membrane cholesterol suggests the involvement of lipid rafts. Together, these findings indicate that the membrane composition is very important for ghrelin receptor activation and desensitization, and might at least partially explain the mechanism by which OFAs and cholesterol affect ghrelin's activity *in vivo*.

1.8. Interaction between the ghrelin receptor and other receptors

Cross talk between different GPCRs has been well demonstrated, often resulting in amplification of intracellular signaling, but occasionally reversing the response. The concept that GPCRs exist as functional homo- and heterodimers is gaining acceptance based on the evidence of their physiological role in receptor trafficking, cellular signaling, and pharmacological function.[57–60] Under physiologic conditions, cells are stimulated by individual agonists or several agonists simultaneously according to the temporal pattern. Activation of a signaling pathway is then evaluated to determine its potential for interaction with pathways activated by other receptors. This process ensures exchange of information between individual signaling pathways and provides a molecular basis for their cooperation.[61–63] Receptor–receptor interactions potentially stabilize specific conformations and lead to coupling with discrete effectors, resulting in heterodimer-specific signal transduction.

1.8.1 The ghrelin receptor and the GHRH receptor

In rat primary pituitary cell cultures, MK-0677 has been shown to amplify GH release by augmenting GHRH-induced cAMP accumulation.[21] Because pituitary somatotrophs coexpress the ghrelin receptor and the GHRH receptor endogenously, ghrelin mimetics amplify GHRH-induced cAMP accumulation by a mechanism involving ghrelin receptor-mediated activation of PKC.[6,64,65] A similar mechanism applies when the ghrelin receptor and the GHRH receptor are cotransfected in HEK293 cells. Ghrelin significantly augments GHRH-induced cAMP accumulation in these cells. The well-characterized ghrelin receptor signal transduction pathway consisting of $G\alpha_{q/11}$, phospholipase C, diacylglycerol (DAG), and PKC is apparently involved because the synergistic effect is blocked by a specific PKC inhibitor (Bis), but not by PTX (pertussis toxin). Therefore, at least in the case of the ghrelin receptor and GHRH-R, experiments in HEK293 cells validate what occurs in native cells.

1.8.2 Ghrelin receptor and dopamine receptors

1.8.2.1 Ghrelin receptor and D1-R (dopamine receptor subtype-1)

In most cell types, ghrelin alone activates the ghrelin receptor to transduce its signal through $G\alpha_{q/11}$ and phospholipase C without the involvement of adenylyl cyclase. Even in HEK293 cells coexpressing the ghrelin receptor and D1-R, ghrelin alone does not change cAMP levels. However, when both receptors are stimulated by treating the cells with both ghrelin and dopamine, ghrelin causes augmentation of dopamine-induced cAMP accumulation. Similarly, both GHRH and dopamine activate their respective receptors in HEK293 cells to cause increased cAMP accumulation; in each case coactivation of the ghrelin receptor by ghrelin augments cAMP accumulation through distinct mechanisms.

The interaction between the ghrelin receptor and D1-R is further supported by the colocalization of these two receptor proteins. Immunoprecipitation of cell lysates from cells expressing ghrelin and D1-R using a ghrelin receptor antibody, followed by Western blots with D1-R antibody, demonstrates the formation of [ghrelin receptor/D1-R]$_2$ heterodimers in the presence of dopamine and ghrelin.

As for the detailed intracellular mechanism of interaction between these two receptors, a switch in G-protein coupling of the ghrelin receptor from $G\alpha_{q/11}$ to $G\alpha_{i/s}$ upon agonist-induced formation of ghrelin receptor/D1-R heterodimers has been proposed. The ghrelin receptor predominantly couples to $G\alpha_{q/11}$, while D1-R typically signals through $G\alpha_s$ to activate the AC2 isozyme. Upon coactivation by ghrelin and dopamine, ghrelin receptor and D1-R form a heterodimer that subsequently induces a conformational change in the ghrelin receptor. This conformational change has been demonstrated to result in coupling of the ghrelin receptor with $G\alpha_i$ protein, releasing $\beta\gamma$ subunits that associate with AC2, thereby amplifying AC2 activity. $G\alpha_s$ subunits are often assumed to dominate in the regulation of AC, while the contributions of $G\beta\gamma$ subunits have been largely unappreciated.[66] A recent study by Sunahara *et al.* has clearly demonstrated that the $\beta\gamma$ subunits are powerful modulators of AC activity that can be either stimulatory for AC2, AC4, and AC7 or inhibitory for AC1 and AC8. Different from the interaction between the ghrelin receptor and GHRH-R, the synergism between the ghrelin receptor and D1-R is not inhibited by Bis, a PKC inhibitor. Instead, PTX treatment significantly inhibits ghrelin amplification of dopamine-induced cAMP accumulation while demonstrating no significant effect on basal and dopamine-induced cAMP accumulation in the absence of ghrelin.[67]

1.8.2.2 The ghrelin receptor and D2-R (dopamine receptor subtype-2)

As a member of family A GPCR, D2-R typically transmits dopamine signals through $G\alpha_{i/o}$ coupling, resulting in inhibition of adenylyl cyclase activity and subsequent decrease in cAMP level.[68] However, ghrelin receptor/D2-R heterodimers induce $G\beta\gamma$ signaling and PLC-dependent mobilization of

Ca^{2+} from intracellular stores,[69] while dopamine (or a selective D2-R agonist) has no such effect in cells expressing only D2-R. Although the ghrelin receptor typically couples to $G\alpha_{q/11}$,[11] application of $G\alpha_{q/11}$ siRNA in cells coexpressing GHS-R1a and D2-R has no effect on dopamine-induced mobilization of $[Ca^{2+}]_i$. Furthermore, inhibition of PKC signaling blocks ghrelin receptor signal transduction,[21] but demonstrates no effect on dopamine-induced mobilization of $[Ca^{2+}]_i$ in cells coexpressing these two receptors.[69] These observations preclude basal activity of GHS-R1a as an explanation for modification of D2-R signal transduction. Instead, interactions between the ghrelin receptor and D2-R have been proposed to alter canonical D2-R signal transduction resulting in dopamine-induced $[Ca^{2+}]_i$ mobilization.

As further evidence of this mechanism, formation of ghrelin receptor/D2-Rheteromers has been detected by Tr-FRET methodology. This technique is ideal for detecting homo- and heteromers on cell surfaces at physiological levels of GPCR expression because of its high sensitivity and high signal-to-noise ratio.[70,71] Tr-FRET assays using SNAP- and CLIP-tagged ghrelin receptor and D2-R clearly shows heteromers formed at equimolar concentrations of ghrelin receptor and D2-R. Moreover, by detecting dopamine-induced mobilization of intracellular Ca^{2+}, the hypothesis that heterodimer function correlates with the Tr-FRET signal produced by ghrelin receptor/D2-R heteromers has been validated.[69] Endogenously formed ghrelin receptor/D2-R heteromers have also been tested in hypothalamic and striatal membrane preparations isolated from wild-type and ghrelin receptor gene knockout mice. Tr-FRET assays show high FRET signals in the hypothalamic membrane of wild-type mice, but not in ghrelin receptor knockout mice, illustrating ghrelin receptor/D2-R heteromer formation.[69]

The formation of the ghrelin receptor/D2-R heteromer is important for the regulation of food intake and energy metabolism. Dopamine signaling through D2-R is critical for regulation of food intake.[72–74] Feeding frequency and volume are significantly reduced when D2-R is activated in hypothalamic neurons.[75] Pharmacologic studies further indicate that increasing dopamine in the lateral hypothalamus induces anorexia, whereas injection of a D2-R antagonist increases food intake,[76] supporting a role for D2-R signaling in the regulation of feeding behavior. However, the anorexic effect of D2-R activation can be blocked by either ghrelin receptor gene deletion or ghrelin receptor antagonism. Administration of cabergoline, a selective D2-R inhibitor, significantly reduces food intake compared to vehicle-treated mice. However, food intake in ghrelin receptor gene knockout mice is unaffected by cabergoline. Furthermore, cabergoline-induced anorexia is blocked by targeting the interaction between the ghrelin receptor and D2-R in wild-type mice with a highly selective ghrelin receptor antagonist (JMV2959).[69]

1.8.3 The ghrelin receptor and melanocortin-3 receptor

Studies by Rediger *et al.*[77] demonstrate mutual signaling interference by the interaction between the ghrelin receptor and melanocortin-3 receptor (MC3R) in hypothalamic neurons. The ghrelin receptor significantly increases melanocortin-induced cAMP signaling, while the interaction with MC3R markedly impairs the ghrelin-induced $G\alpha_{q/11}$ signaling and the agonist-independent basal Ca^{2+}/calmodulin-induced cAMP-responsive-element-binding protein signaling activity of the ghrelin receptor. Interaction of the ghrelin receptor and MC3R therefore leads to enhanced MC3R signaling efficacy (signaling surplus) while simultaneously causing a decreased ghrelin receptor signaling capacity. It is the ligand-independent ghrelin receptor signaling activity that plays a key role in the process of interaction with MC3R. These findings offer new implications for the current established model of hypothalamic body weight regulation.

1.8.4 The ghrelin receptor and the adenosine receptor

Adenosine was originally proposed to represent a third class of ghrelin receptor agonists because of its potentiation of ghrelin receptor-induced IP3 signaling. Recent studies have suggested that the IP3 signaling triggered by adenosine in cells expressing ghrelin receptor may result from the interaction between the ghrelin receptor and adenosine receptor. Treatment of HEK293 or baby hamster kidney (BHK) cells overexpressing ghrelin receptor with adenosine induces intracellular calcium mobilization;[18,78] despite that, adenosine is unable to compete for binding of the ghrelin receptor with either ghrelin or GHSs.[78]

Intracellular calcium mobilization evoked by adenosine is sensitive to cholera toxin and adenylyl cyclase inhibitors, suggesting signaling via a $G\alpha_s$ pathway.[79] Both adenosine A1 (A1R) and A3 receptors (A3R) are predominantly $G\alpha_i$ coupled. However, adenosine A2A (A2AR) and A2B (A2BR) receptors are mainly $G\alpha_s$ coupled.[80] Adenosine binds to and activates the $G\alpha_s$-coupled adenosine receptor A2BR, causing increased cAMP production and activation of protein kinase A (PKA). PKA subsequently phosphorylates and sensitizes IP3Rs to activation by IP3, although additional mechanisms for cAMP-dependent sensitization of IP3Rs may also exist. Crosstalk between the ghrelin receptor and A2BR has been demonstrated by the potentiation of adenosine-induced calcium release achieved by cotransfection of A2BR and the ghrelin receptor. Evidence suggests that adenosine-induced activation of the ghrelin receptor may lead to PKA-mediated phosphorylation and activation of IP3Rs via $G\alpha_s$ signaling.[79] First, adenosine activates A2BR, causing increased cAMP production. Second, ghrelin and GHSs bind to the ghrelin receptor, which leads to activation of $G\alpha_{q/11}$ and PLC. PLC hydrolyses phosphatidylinositol 4,5-bisphosphate to DAG and IP3. Inhibition of the ghrelin receptor blocks calcium release from the intracellular calcium stores in a PLC- and IP3-dependent manner, leaving cAMP levels unaltered.[81] The intracellular calcium mobilization induced by adenosine is significantly suppressed by the IP3R blocker 2-Aminoethoxydiphenyl borate. Third, adenosine does not alter IP3 levels in cells cotransfected with the ghrelin receptor and A2BR. The activation of the adenosine receptor by adenosine could therefore contribute to cAMP-dependent potentiation of IP3-induced Ca^{2+} release caused by ghrelin receptor activation.

1.9. Intracellular signaling pathways

Upon binding with ghrelin, GHS-R1a undergoes a profound change in the TM α helices, which alters the conformation of the intracellular loops and facilitates its interaction with G-proteins. The interaction causes the exchange of GDP bound to the G-protein α subunit for GTP, which activates G-protein subunits and initiates various signaling responses via a series of intracellular molecules.

1.9.1 Calcium mobilization-related signaling

Binding of ghrelin to the ghrelin receptor activates phospholipase C to generate IP3 and DAG, resulting in an increase of intracellular Ca^{2+}. This pathway is consistent with receptor coupling to a $G\alpha q$ subunit.[6] IP3 triggers the release of calcium from IP3-sensitive intracellular calcium stores, whereas DAG is responsible for the activation of PKC. The IP3-dependent calcium rise activates voltage-independent K^+ channels, which transiently hyperpolarizes the cells.[82] The initial calcium current is followed by sustained calcium entry through L- and T-type calcium channels via membrane depolarization.[6,82] In addition to this typical $G\alpha q$/PLC/IP3 pathway, ghrelin may also evoke the intracellular calcium signaling by an alternate pathway. In neuropeptide Y (NPY)-containing neurons, the ghrelin-induced increase in intracellular calcium concentration is dependent on calcium influx through

N-type calcium channel. These channels are activated by the cAMP-PKA signaling pathway following the coupling of the Gαs protein to the ghrelin receptor.[83]

1.9.2 MAPK signaling

In addition to the well-characterized activation of ionic currents, ghrelin also exerts a proliferative activity through the MAP kinase (MAPK) signaling pathway. In human and rat adrenal zona glomerulosa cells, the proliferative effect of ghrelin involves the activation of a tyrosine kinase-dependent MAPK p42/p44 mechanism, and appears to be independent of PKA and PKC.[79,84] In 3T3-L1 preadipocytes, exposure to ghrelin causes a rapid activation of MAPKs, especially ERK1/2. PD98059, an ERK inhibitor, significantly attenuates the mitogenic and anti-apoptotic activities of ghrelin.[85] Pretreatment of cells with a $G\alpha_{i/o}$ inhibitor (PTX), PKC inhibitors (staurosporin and GF109203X), or a PI3K inhibitor (wortmannin) significantly attenuates ghrelin-induced ERK1/2 phosphorylation. These findings suggest that multiple signaling pathways are involved in ghrelin signaling associated with MAPK activation.

1.9.3 AMPK signaling

AMPK plays a pivotal role in the regulation of energy metabolism. Administration of ghrelin leads to an increase in food intake via stimulation of AMPK activity in the hypothalamus. The exact mechanism by which ghrelin regulates AMPK is still unknown, although tumor suppressor LKB1 may be an upstream mediator activating AMPK kinase.[86] Furthermore, the biological effect of ghrelin on AMPK activity is related to fat distribution and metabolism in a tissue-specific manner. In the rat liver, ghrelin inhibits AMPK activity to evoke lipogenic and glucogenic-related gene expression and to increase triglyceride content without changing the mitochondrial oxidative enzyme activities.[87]

1.9.4 PI3K/AKT signaling

Insulin receptor substrate (IRS-1)-associated PI3K activity and Akt phosphorylation are also regulated by ghrelin. In hepatoma cells, ghrelin upregulates IRS-1-associated PI3K activity.[88] On the other hand, ghrelin inhibits Akt kinase activity and partially reverses the downregulation of insulin on phosphoenolpyruvate carboxykinase (PEPCK) mRNA expression, a rate-limiting enzyme of gluconeogenesis that catalyzes the conversion of oxaloacetate to phosphoenolpyruvate.

Ghrelin also stimulates the IRS-1-associated PI3K/Akt pathway in 3T3-L1 preadipocytes, whereas inhibition of PI3K blocks the effects of ghrelin on the proliferation and apoptosis of these cells. Furthermore, ghrelin increases basal and insulin-stimulated glucose transport. The effect of ghrelin on glucose transport can be blocked by coadministration of a PI3K inhibitor.[85] These findings indicate that the PI3K/Akt pathway acts to mediate the effects of ghrelin on 3T3-L1 cells.

1.10. Distribution of the ghrelin receptor

Originally identified in the brain, the ghrelin receptor has been reported to be expressed in both central and peripheral tissues. In the rat brain, strong ghrelin receptor signal has been detected in multiple hypothalamic nuclei including the anteroventralpreoptic nucleus, anterior hypothalamic area, suprachiasmatic nucleus, lateroanterior hypothalamic nucleus, supraoptic nucleus, ventromedial hypothalamic nucleus, arcuate nucleus, paraventricular nucleus, and tuberomammillary nucleus, as well as in the pituitary gland.[89] Expression of the ghrelin receptor in these neuroendocrine centers is consistent with its direct role in the regulation of GH function. In addition to the hypothalamus, mRNA encoding the ghrelin receptor is also expressed in several other discrete regions of the rat brain. For example, specific signals have been detected in the dentate gyrus, CA2 and CA3 regions of the hippocampal formation, thalamic regions, and several nuclei within the brain stem including the pars compacta of the substantia nigra, ventral tegmental area, median and dorsal raphe nuclei, Edinger–Westphal nucleus, laterodorsal tegmental nucleus, and the nuclei of the facial nerve.[89] These observations suggest that the ghrelin receptor may play a role in the control of other neuroendocrine functions besides regulation of GH release. In contrast to the ghrelin receptor GHS-R1a, GHS-R1b signal is negligible in the brain and pituitary.[89] Distribution of the ghrelin receptor in human brain is similar to the rat. Among encephalic regions, the hypothalamus, hippocampal formation, and pituitary are the regions with the most abundant expression of ghrelin receptor mRNA.[89]

In peripheral tissues, ghrelin receptor mRNA has been identified in the stomach and intestine,[90] pancreas,[89] kidney,[91] heart and aorta,[92] adipose tissues,[20] as well as in different human pituitary adenomas,[93,94] and various endocrine neoplasms of the lung,[95] stomach,[96] and pancreas.[94] Distribution of the ghrelin receptor in a variety of peripheral tissues is consistent with the observations of its broader functions beyond the control of GH release and food intake. Existence of a novel ghrelin receptor subtype with a binding profile different from the classical ghrelin receptor GHS-R1a in the peripheral tissues and human thyroid and breast tumors, as well as in related cancer cell lines has been reported but requires further confirmation.[14,97]

1.11. Life without the ghrelin receptor

Deletion of the ghrelin receptor gene demonstrates a significant effect on energy metabolism, rendering the mice resistant to high-fat diet-induced obesity and its associated diabetes, and attenuating age-associated obesity and insulin resistance.[98]

1.11.1 Condition of normal chow diet

There is no significant difference in food intake or body length between ghrelin receptor null mice and littermate wild-type mice fed with a normal chow diet,[99] even though the endogenous GHS-R1a ligand ghrelin fails to stimulate GH release in these ghrelin receptor null mice. Moreover, both bone mineral density and bone mass are comparable in wild-type and ghrelin receptor null littermates, suggesting that the lack of ghrelin receptor does not compromise bone growth. The body weights of ghrelin receptor null mice are modestly lower than that of wild-type mice from 16 to 24 weeks of age. Hence, it is unlikely that ghrelin plays a dominant role in determining growth and body composition.[99]

1.11.2 Condition of high-fat diet

When fed with a high-fat diet, the ghrelin receptor null mice are protected against the full development of diet-induced obesity,[99] which has been demonstrated in a mixed genetic background (C57BL6J:129sv) *ghsr−/−* mice line created by Zigman *et al.* Ghrelin receptor null mice also demonstrate lower feeding efficiency defined as the ratio of weight gain to food intake[100] and lower energetic efficiency defined as the ratio of energy gain to energy intake.[101] These findings suggest that ghrelin receptor null mice store less energy relative to wild-type mice when they consume the same quantity of calories, indicating that increased energy expenditure in ghrelin receptor null mice contributes to their lean phenotype. Interestingly, ghrelin receptor knockout mice maintained on high-fat diet also demonstrate a reduced respiratory quotient (RQ). This observation indicates a shift in metabolic fuel preference toward the utilization of fat as an energy substrate, and reduced locomotor activity especially during the dark cycle that seems contradictory with the decreased body weight.[99] These *ghsr−/−* mice exhibit lean phenotype, lower RQ, and decreased total locomotor activity, suggesting ghrelin's role in energy metabolism may be more extensive and complex than merely increasing appetite.

In addition to the effects on body weight, deletion of the ghrelin receptor leads to lower blood glucose levels in mice maintained on both standard chow diet and high-fat diet.[102] Corresponding insulin levels for the ghrelin receptor null mice are not increased, indicating that ghrelin receptor null mice have enhanced insulin sensitivity as compared to wild-type controls.[99]

1.11.3 Aging-associated obesity

Ablation of the ghrelin receptor improves aging-associated obesity and insulin resistance by reducing adiposity and increasing thermogenesis. There exist two types of adipose tissue: one is energy-storing white adipose tissue (WAT), which stores energy in the form of triglycerides and supplies energy to the body in the form of ATP through lipolysis/β-oxidation; the other is the energy-burning brown adipose tissue (BAT) which is a key organ of nonshivering thermogenesis. WAT, and particularly visceral adipose tissue, is a prominent risk factor for insulin resistance and type 2 diabetes. Visceral WAT is significantly elevated in obese individuals and the elderly.[103–105] On the other hand, BAT plays an important role in energy expenditure by regulating thermogenesis. Thermogenesis in BAT is achieved by activation and expression of uncoupling protein 1 (UCP1) in the mitochondria, which is located in the inner mitochondrial membrane and recruits free fatty acids into the mitochondria to generate heat.[106,107] Both the mass and activity of BAT are severely diminished during aging.[108] Compared with wild-type littermates, old *ghsr−/−* mice exhibit diminished fat mass, increased energy expenditure, and a healthier lipid profile. Resting metabolic rate (RMR) is an important parameter in determining energy balance, which accounts for approximately 60% of total energy expended.[109] *ghsr−/−* mice have a higher RMR with no change in locomotor activity.[98] Thus, the increased energy expenditure observed in old *ghsr−/−* mice is likely due to increased thermogenesis. Unlike young ghrelin receptor null mice, old *ghsr−/−* mice fed normal chow diet have a higher RQ,[98] indicating that they favor carbohydrate as a fuel substrate. Moreover, aging *ghsr−/−* mice have greater relative cumulative frequency, suggesting greater metabolic flexibility consistent with the insulin-sensitive phenotype.[98]

Furthermore, ablation of the ghrelin receptor gene reduces the expression of genes related to glucose/lipid uptake and lipogenesis in WATs, while simultaneously increasing thermogenic gene clusters in BATs.[98] Expression of UCP1 decreases gradually with age, suggesting aging-associated impairment of thermogenesis. Remarkably, higher mRNA and protein expression levels of UCP1 and higher mitochondrial density were detected in old *ghsr−/−* mice, suggesting that ablation of the ghrelin receptor improves thermogenic function and protects the mice against aging-associated decline in thermogenesis.[98]

Collectively, these data demonstrate that deletion of the ghrelin receptor reduces energy storage in WAT, while increasing energy expenditure in BAT during aging. The net result is maintenance of a youthful, insulin-sensitive metabolic state. Such a unique property of ghrelin receptor knockout mice suggests that antagonism of the ghrelin receptor to shift the energy balance from obesogenic to thermogenic may serve as potential strategy for the pharmacologic treatment of obesity and diabetes.[98]

In addition, calorie restriction is the only proven intervention that prolongs lifespan. Low blood glucose, low triglycerides, and low insulin are the characteristic metabolic hallmarks of calorie-restricted subjects and centenarians.[110,111] The findings that *ghsr−/−* mice exhibit all of these youthful health traits suggest that antagonism of the ghrelin receptor may offer great promise for novel interventions to mimic calorie restriction without dietary restriction or exercise.

2. THE GHRELIN RECEPTOR AND ENERGY METABOLISM

Energy homeostasis is determined by the balance between energy intake (food ingestion) and energy expenditure (activity and/or heat production); even a slight imbalance over prolonged periods can lead to changes in body composition[112] and result in the development of obesity, insulin resistance and other serious metabolic disorders.

Ghrelin functions as a key modulator of energy homeostasis, acting via the ghrelin receptor to regulate appetite, select fuel substrate, modulate body weight, and maintain glucose/lipid homeostasis. Ghrelin may also signal the peripheral energy supply status to the brain through the ghrelin receptor, leading to changes in food intake and adiposity.[113,114] Ghrelin and its receptor therefore play an important role in enabling adaptation to changes in nutritional status by controlling metabolic parameters such as energy intake and expenditure.

2.1. Regulation of food intake

As the natural ligand for the ghrelin receptor GHS-R1a, ghrelin is the only known gastrointestinal hormone to stimulate food intake. The orexigenic effect of ghrelin is mainly ascribed to the activation of its receptor localized in the hypothalamic nucleus. The hypothalamic arcuate nucleus, a circumventricular organ with a permeable blood–brain barrier, expresses the ghrelin receptor and is activated following systemic administration of GHSs. Although systemic and intracerebroventricular (ICV) injections of ghrelin both increase plasma GH, ICV administration of ghrelin is more efficient, indicating that ghrelin functions mainly at the hypothalamic level.[113]

ICV administration of ghrelin intensely augments feeding in rodents and increases body weight gain. Antighrelin immunoglobulin G apparently suppresses feeding behavior. The orexigenic effect of ghrelin is dependent on its receptor GHS-R1a in the central nervous system. Ghrelin-induced feeding is suppressed by an antagonist for GHS-R1a, [D-Lys^3]-GHRP-6.[115] Administration of ghrelin significantly increases the expression of c-FOS, a marker of neuronal activation, in regions of primary importance in the regulation of feeding. Within the arcuate nucleus, the ghrelin receptor is mainly expressed on almost all NPY (NPY)/agouti-related peptide (AgRP)-expressing neurons.[116] Only 8% of proopiomelanocortin (POMC) neurons and 20–25% of GHRH mRNA-expressing neurons express the ghrelin receptor.[116] Both antibodies against NPY or AgRP, and antagonists of the NPY receptor have been reported to abolish ghrelin-induced feeding. In addition, the stimulatory effects of ghrelin on these orexigenic neurons are complemented by reduced POMC neuronal activity via inhibitory β-aminobutyric acidergic inputs from NPY/AgRP neurons.[117] Thus, ghrelin may stimulate the food intake not only via activating orexigenic NPY/AgRP neurons, but also by simultaneously suppressing anorexigenic POMC.

2.2. Glucose homeostasis

2.2.1 Effect of exogenous ghrelin

In spite of the controversy, ghrelin may contribute to the modulation of the glucose homeostasis via activation of the ghrelin receptor. There exist differences between the acute effect and long-term effect of exogenous ghrelin. In the short term, exogenous ghrelin induces hyperglycemia and hypoinsulinism in healthy humans and rodents in a GH-independent fashion.[118] The long-term effects of exogenous ghrelin on glucose metabolism vary and may be dependent on the pathway of its administration. Intraperitoneal (ip) injection of ghrelin for 4 days increases plasma glucose concentration. These elevated plasma glucose levels may be attributed to the decreased storage of glycogen in muscle and kidney, while liver glycogen content is unaffected.[119] However, ICV injection of ghrelin for five consecutive days to adult male rats significantly increases serum insulin levels without evoking changes in blood glucose levels.[120]

2.2.2 Effect of endogenous ghrelin

Blocking the effect of endogenous ghrelin by [D-Lys^3]-GHRP-6, an inhibitor of the ghrelin receptor, markedly decreases fasting glucose concentrations over several hours in normal mice. Similarly, the intraperitoneal glucose tolerance test is improved by ghrelin receptor antagonism, indicating that endogenous ghrelin is involved in the regulation of insulin release and blood glucose homeostasis.[121] In *ob/ob* mice, ghrelin receptor antagonism also markedly reduces plasma glucose levels and serum insulin levels, suggesting the amelioration of insulin resistance.[122]

Genetic deletion of the ghrelin receptor (*ghsr−/−*) leads to lower blood glucose levels in mice maintained on a standard chow diet. Interestingly, corresponding insulin levels for the ghrelin receptor null mice are not increased, suggesting that *ghsr−/−* mice have enhanced insulin sensitivity relative to wild-type controls.[99] Because compensatory elevation of blood ghrelin level is observed in the ghrelin receptor null mice, it is rational to propose a reciprocal relationship between ghrelin and insulin. This notion is supported by clinical studies showing that low plasma ghrelin is associated with insulin resistance.[123,124]

2.2.3 Response altered by energy status

Plasma levels of ghrelin and glucose are inversely related. Under conditions of negative energy balance, circulating levels of ghrelin are high; ghrelin levels are much lower with positive energy balance. Fasted mice have higher endogenous ghrelin levels and are refractory to the inhibitory effects of exogenous ghrelin on glucose-stimulated insulin secretion.[125] This effect may result from the saturation of ghrelin-binding sites on the ghrelin receptor by endogenous ghrelin.

Under fasting conditions, ghrelin receptor gene knockout mice fed with normal chow exhibit lower glucose and insulin levels than WT littermates. These observations indicate that the ghrelin receptor is critical for glucose homeostasis, and that deletion of the ghrelin receptor increases insulin sensitivity.[126] However, these ameliorations are absent in high-fat diet-fed mice, indicating that the effects of ghrelin receptor on glucose homeostasis are associated with the energy state of the animals.[126]

2.2.4 Effects on the major target organs for insulin

Mechanisms detailing how ghrelin regulates glucose homeostasis remain largely unknown, but may involve both insulin secretion by pancreatic β cells and insulin sensitivity in its major target organs including skeletal muscle, liver, and adipose tissue.[127]

2.2.4.1 Pancreas

Effects of ghrelin on the development and function of pancreatic β-cells have been demonstrated.

2.2.4.1.1 Effect on the development of β-cells Ghrelin becomes detectable in the fetal pancreas even sooner than in the stomach; after birth, its level is robustly downregulated in pancreas, but increased in stomach. In addition, the number of ghrelin-positive cells is greater in transgenic mouse models lacking functional Nkx2-2, Pax6, or Pax4, with reciprocal β-cell deficiency.[128] Ghrelin-immunoreactive cells have been detected primarily in the ε-cells,[128–130] but may also be present in other islets cells such as α-cells,[121,130] β-cells,[97] PP-cells[130] in rat and human pancreas. All these observations suggest the paracrine role of ghrelin in the development of pancreatic β-cells.

2.2.4.1.2 Effect on insulin secretion In cultured islet cells, acyl ghrelin (10^{-8} M) suppresses both basal insulin secretion[121] and glucose-induced insulin release by the mediation of the ghrelin receptor.[131] Ghrelin knockout mice are resistant to high-fat diet-induced glucose intolerance because of enhanced insulin release. No significant difference in the density and average size of the islets between wild-type and ghrelin knockout mice is observed.[131] In the dissected and perfused rat pancreas, ghrelin markedly inhibits insulin release in response to increasing glucose concentrations, arginine, and carbachol, but has no significant effect on basal insulin release.[132] Moreover, the level of insulin released from the perfused pancreas is significantly increased by blocking the ghrelin receptor or immunoneutralizing endogenous ghrelin. Glucose also enhances insulin release in pancreatic islets isolated from ghrelin-null mice.[131] Overall, these data suggest that ghrelin regulates insulin secretion from pancreatic islet cells.

Glucose stimulation of insulin secretion is tightly associated with cytoplasmic membrane excitability, which is mainly determined by the activity of ATP-sensitive K^+ channels (KATP).[133] The rise in blood glucose induces an increase in β-cell glucose metabolism, resulting in increased ATP/ADP ratio, which inhibits KATP channels. The inhibition of KATP channels then depolarizes the plasma membrane, leading to the opening of voltage-dependent calcium channels and subsequent calcium influx. The resulting increase in intracellular calcium concentration $[Ca^{2+}]_i$ triggers exocytosis of insulin-containing granules.[134]

The Kv channel, regulated mainly by NADPH to NADP ratio,[133] repolarizes the β-cell. Kv channels are thought to repolarize glucose-stimulated action potentials and inhibit Ca^{2+} entry through voltage-gated Ca^{2+} channels, thereby serving as negative regulators of insulin secretion.[135] Glucose-induced insulin release is greater in islets isolated from ghrelin-null mice than in those isolated from wild-type mice. This excess insulin release can be prevented by pretreatment with PTX. Ghrelin has been demonstrated to increase Kv currents, and the Kv channel blocker, tetraethylammonium, blocks the ability of ghrelin to inhibit insulin release. Furthermore, ghrelin treatment suppresses glucose-induced $[Ca^{2+}]_i$ increases in β-cells. When islet β-cells are treated with the antisense oligonucleotide specific for $G\alpha_{i2}$-subunit of G-proteins, the inhibitory effects of ghrelin on $[Ca^{2+}]_i$ and insulin release are abolished. These results indicate that ghrelin suppresses glucose-induced insulin release by attenuation of Ca^{2+} signaling in β-cells via $G\alpha_{i2}$ and Kv channels.[136]

2.2.4.2 Skeletal muscle

Studies by our laboratory suggest that ghrelin may exercise a direct effect on the skeletal muscle to modulate its glucose transport. Presence of the ghrelin receptor mRNA and its function in the skeletal muscle has been validated. Inhibition of mTOR signaling by rapamycin significantly increases circulating ghrelin level in mice,[137] which subsequently affects glucose homeostasis. There exists a positive correlation between circulating plasma acyl ghrelin levels and glucose intolerance induced by rapamycin. Interference of ghrelin receptor by its antagonism ([D-Lys^3] GHRP-6) or gene deletion improves glucose intolerance induced by elevated plasma ghrelin concentration, ameliorates rapamycin-induced impairment in insulin receptor signaling of skeletal muscle by stimulating GLUT4 expression and translocation. Ghrelin decreases the membrane GLUT4 in differentiated C2C12 cells, while ghrelin receptor antagonism significantly increases the amount of GLUT4 under either basal condition or insulin stimulation.[102] The effect of ghrelin on glucose transport seems to be mediated by JNK activity because upregulation of gastrocnemius muscle JNK phosphorylation induced by rapamycin is blocked by ghrelin receptor antagonism or gene deletion.

2.2.4.3 Liver

In vitro experiments show that ghrelin blocks the inhibitory effects of insulin on gluconeogenesis. In the rat or human hepatoma cell line, ghrelin decreases the phosphorylation levels of protein kinase B (PKB/AKT) and glycogen synthase kinase (GSK). Ghrelin also attenuates the effect of insulin on the expression of PEPCK, and upregulates gluconeogenesis in cultured rat hepatoma cells.[88] These observations provide evidence that ghrelin has a direct effect on hepatic glucose metabolism.[88] However, expression of the ghrelin receptor in the liver has not yet been demonstrated. Whether the effect of ghrelin on hepatic glucose metabolism is exerted via the ghrelin receptor thus remains unknown.

2.2.4.4 Adipocytes

The ghrelin receptor has been detected in the epididymal and pericardial adipose tissue deposits. Ghrelin may act synergistically to potentiate insulin-stimulated glucose uptake and may as a result increase the deposition of adipose tissue.[85] In addition, ghrelin treatment increases IRS-1 and AKT phosphorylation. Pretreatment of adipocytes with wortmannin, a PI3K inhibitor, completely blocks the effect of ghrelin on glucose transport and AKT phosphorylation.[85] All these data suggest that the direct action of ghrelin to facilitate insulin-stimulated glucose uptake in adipocytes may contribute to the overall maintenance of glucose homeostasis.

2.3. Lipid metabolism

2.3.1 Central effect

Chronic central administration of ghrelin increases adipose deposition independently of food intake[115,138] in pair-fed animals. Although total energy expenditure and spontaneous physical activity remain unchanged, the RQ is significantly increased, indicating greater carbohydrate metabolism instead of lipid substrate consumption. Further studies demonstrate that central ghrelin stimulates triglyceride (TG) uptake and lipogenesis, while inhibiting lipid oxidation in white adipocytes. The mRNA expression of various fat storage-promoting enzymes such as lipoprotein lipase, acetyl-CoA carboxylase, fatty acid synthase, and stearoyl-CoA desaturase-1 is markedly increased,[138] while expression of the rate-limiting enzymes in fat oxidation such as carnitine palmitoyltransferase-1α is significantly decreased. In brown adipocytes, ICV injection of ghrelin decreases the expression of uncoupling proteins (UCPs) in a dose-dependent manner. This effect is independent of ghrelin-induced hyperphagia and appears to be mediated by the sympathetic nervous system.[138]

2.3.2 Peripheral effect

Peripheral daily administration of ghrelin for 2 weeks causes a significant increase in fat mass as measured by dual energy X-ray absorptiometry.[139] This increase appears to occur only in the visceral fat, but not in the superficial subcutaneous fat depot.[140] As ghrelin receptor null mice are unresponsive to acylated ghrelin administration, it is proposed that this effect is mediated by the ghrelin receptor GHS-R1a. This concept is further supported by the observation that blockade of the ghrelin receptor abolishes the effect of acylated ghrelin on adiposity.[140] *In vitro* experiments clearly demonstrate that ghrelin may increase WAT volume by either stimulating adipogenesis or inhibiting lipolysis and lipid efflux from adipocytes. In cultured epididymal adipocytes, application of ghrelin decreases the expression of ABCG1, the primary mediator of cholesterol efflux [141], and inhibits glycerol release.[142] Liver X receptor α and fatty acid synthase demonstrate no significant change, suggesting that ghrelin may not affect lipogenesis.[143]

Other ghrelin gene products such as des-acyl ghrelin and obestatin have also been reported to stimulate fat acid uptake and inhibit NEFA or glycerol release in adipocytes. The effect of des-acyl ghrelin on fatty acid uptake can be blocked by the ghrelin receptor antagonist [D-Lys^3]-GHRP6, lipid raft disruptor beta-cyclodextrin, PLC inhibitor U73122, and PI3K inhibitor wortmannin,[144] suggesting that des-acyl ghrelin exercises its effect on lipid metabolism in a ghrelin receptor-dependent manner through PI3K/AKT pathway. Since des-acyl ghrelin demonstrates no binding activity to the classical ghrelin receptor GHS-R1a, this observation indicates that it may facilitate the activity of the ghrelin receptor by a yet-to-be-defined mechanism.

Ghrelin-induced lipid accumulation is not limited to WATs. In mouse liver, ghrelin significantly elevates the number of lipid droplets, total lipid area, and triacylglycerol content.[140] In contrast, ghrelin demonstrates no effect on hepatic lipid content in loxTB-GHS-R1a mice.[140] These data suggest that ghrelin induces hepatic steatosis in a GHS-R1a-dependent manner.

2.4. Ghrelin receptor polymorphism and energy metabolism

One of the major physiological functions of the ghrelin receptor appears to be the regulation of food intake and body weight. The ghrelin receptor gene is located within the quantitative trait locus on chromosome 3q26–q29, which has been strongly linked to multiple phenotypes related to obesity and the metabolic syndrome[26] in the same family cohort. Because of its genomic location and biological function, the ghrelin receptor represents an excellent candidate gene that may contribute to the pathophysiology and risk of obesity. This gene has been found to be associated with obesity in a study using the transmission disequilibrium test and a case–control approach in obese subjects with a BMI >32 kg/m^2 and nonobese subjects with a BMI <28 kg/m^2.[26]

Gene expression is regulated by multiple transcription factors binding at multiple sites in or near a gene, and therefore can be affected by multiple SNPs. In fact, it has been demonstrated that gene regulatory elements reside in noncoding and intergene regions,[145,146] which are able to modulate gene expression over long distances.[147] However, the impact of sequence variations in these regions is rarely studied. The study by Baessler *et al.* reveals that SNPs located in the intergenic region of the ghrelin receptor gene demonstrate a stronger association than the SNPs located in the coding region of the gene.[26] This observation suggests that the promoter, regulatory elements or transcriptional initiation of the ghrelin receptor gene could be involved.

A linkage disequilibrium block of five SNPs has been found to be associated with BMI in 1095 individuals from 178 pedigrees with multiple obese members.[26] F279L and A204E mutations are associated with short stature and obesity, respectively.[30] Four naturally occurring missense mutations have been documented in either the National Center for Biotechnology Information SNP database or the literature.[28,49] Two SNPs (rs490683 and rs9819506) in the promoter region of the human ghrelin receptor gene are associated with abnormalities in body weight and BMI.[148]

Thus, genetic variations in the ghrelin receptor gene may change either ghrelin receptor expression or receptor properties and thereby affect appetite regulation and energy metabolism. Altering the response to ghrelin, or impairing the capability to counter-balance inhibitory signals, may consequently increase susceptibility to obesity.[26]

A region of strong linkage disequilibrium has also been detected between five SNPs (rs509035, rs572169, rs519384, rs512692, and rs863441).[26] On the synonymous SNP G477A (rs572169), the minor allele (A) is transmitted more frequently in obese cases than controls, and an additive genetic association with higher BMI has been confirmed in a further study of 1418 Caucasians.[26] Overall, the five SNPs consistently show significant association with obesity. Further studies using full haplotype analysis have identified two haplotypes to be significant. Haplotype 1 is more frequently present in nonobese or "nonsusceptible" individuals, whereas haplotype 2 is more frequently found in obese or "susceptible" individuals.[26] There is a significant relationship between the number of copies of haplotypes 1 and 2 with the qualitative trait obesity in both matched cases and control subjects. Individuals homozygous for the "susceptible" haplotype or lacking the "nonsusceptible" haplotype present more often with obesity and higher BMI than individuals with one or no copy of the respective haplotype.[26]

A study by Mager *et al.* also shows that two SNPs (rs490683 and rs9819506) in the promoter region of the human ghrelin receptor gene are associated with body weight and BMI.[148] Individuals with rs9819506-AA genotype demonstrated the lowest values for body weight and a trend toward the lowest BMI during a 3-year follow-up. Interestingly, subjects with rs490683-CC genotype displayed the highest weight loss values in the whole study group.[148] Moreover, beneficial changes in glucose metabolism could be observed specifically in individuals with rs490683-CC and rs509035-TT genotypes.[148] Rs490683 and rs509035 are in strong linkage disequilibrium, which may further explain the similar results concerning glucose metabolism phenotypes. These results suggest that polymorphisms in the ghrelin receptor gene are longitudinally associated with obesity and glucose metabolism in a prospective study setting. In addition, Chinese women who carry genotype variants of rs2922126 A/A or rs509035 A/A within ghrelin receptor gene have an increased risk of metabolic syndrome.[149] The rs509035 A/A genotype variant may also contribute to a low level of high-density lipoprotein in women.

In another study by Gueorguiev *et al.*, the ghrelin polymorphism rs4864677 and ghrelin receptor polymorphism rs572169 have been associated with obesity. The ghrelin variant rs27647 influences insulin secretion/resistance, whereas the ghrelin variant rs696217 and the GHS-R polymorphism rs2232169 modulate eating behaviors.[27] Ghrelin receptor SNP C447G (rs2232169) is involved in the regulation of eating behavior associated with the overeating phenotype or snacking phenotype, which consequently induces obesity.[27] Although C447G (rs2232169) is conservative and does not cause an amino acid change, it may modulate ghrelin receptor mRNA stability or splicing, resulting in potential enhancement of constitutive activity.[150]

As reported by Matzko *et al.*, roux-en-Y gastric bypass surgery patients who are rs490683 C/C genotype lost about 5% more excess body weight than patients who do not carry this genotype.[151] Moreover, mechanism analysis of SNP rs490683 in an adult mouse hypothalamic cell line has showed that the homozygous C/C genotype attenuates activity of the promoter by approximately 20%.[151] This reduction in ghrelin receptor expression may decrease ghrelin binding and downstream signaling, resulting in diminished appetite and more favorable weight-loss outcomes. Previous study has demonstrated that the "C" allele is associated with a sevenfold attenuation in binding affinity of ghrelin receptor;[148] decreased transcriptional activity of the ghrelin receptor may therefore occur as the result of the alteration of consensus transcription factor-binding sites in the rs490683 C/C genotype.

In conclusion, variations in ghrelin receptor genes have been linked to obesity, disordered eating behavior, hypertriglyceridemia, and insulin resistance. Even though common polymorphisms of ghrelin receptor genes might not be strong contributors to polygenic obesity, they definitely modulate obesity-related traits (insulin secretion/resistance) and eating behavior.

3. STRATEGIES FOR TREATMENT OF OBESITY AND DIABETES

Ghrelin has been emerging as an attractive target for treatment of obesity and diabetes because of its profound effects on food intake and energy metabolism and its unique posttranslational modification by ghrelin *O*-acyl transferase (GOAT), which was identified as a conserved orphan membrane-bound enzyme by two different research groups in 2008.[152,153] Several reviews have already offered detail discussions on this issue. Here, we focus our discussion on the current challenges facing therapeutic strategies for obesity and diabetes.

3.1. Ghrelin receptor antagonists?

The ghrelin/ghrelin receptor system is a potential target for treatment of disturbed weight regulation due to the relevant role of ghrelin in the regulation of energy balance, food intake, and adipogenesis, as well as GH release.[154,155] Targeted at the ghrelin receptor, two different strategies to block its signaling have been developed: antagonizing ghrelin-induced signaling or blocking the high constitutive activity by an inverse agonist.[47] These strategies have led to the evaluation of a group of the GHS-R1a neutral competitive antagonists and inverse agonists.[32,156–158] Unfortunately, no ghrelin antagonist is currently available on the market, mainly due to the defect in long-term efficacy, selectivity, low potency, poor bioavailability, and side effects.[159]

3.2. Alteration of ghrelin sensitivity in obesity and diabetes?

The beneficial effect of ghrelin blockade in reducing appetite and body weight in obesity has been well documented. In rodents, control of appetite and weight loss has been achieved by blocking the action of ghrelin either by receptor antagonism, neutralizing antibodies, or Spiegelmer RNA. Reduction in plasma ghrelin levels has been demonstrated to account for the reduced appetite and weight loss following gastric bypass surgery in obese patients with Prader–Willi syndrome. Despite this therapeutic effect, ghrelin does not appear to play a causative role in the increased appetite in obesity.

Although ghrelin is a potent orexigenic signal for feeding, its plasma levels are commonly lower in obesity; weight loss usually is associated with an increase in the level of ghrelin. Furthermore, alteration in ghrelin sensitivity in obesity has been reported. The NPY/AgRP neuronal response to ghrelin demonstrates an unexpected resistance in diet-induced obese mice.[117,160] Ghrelin resistance is the result of defective NPY/AgRP function, caused by either lower constitutive activity of the ghrelin receptor or fewer receptors available for ghrelin binding at the hypothalamic levels in obese mice. It is currently unclear whether hypothalamic ghrelin resistance is an adaptive response to prevent redundant caloric intake or a pathological consequence of chronic positive energy balance.

Interestingly, ghrelin resistance in diet-induced obesity is not confined to NPY/AgRP neurons, as ghrelin also fails to stimulate GH secretion in diet-induced obese mice. Ghrelin exerts a variety of functions in the brain aside from appetite control, including cognitive function, mood regulation, and protection against neurodegenerative diseases.[160,161] Thus, central ghrelin resistance may potentiate obesity-related cognitive decline. Restoration of ghrelin sensitivity may provide therapeutic options for maintaining healthy aging.

3.3. Targeting the brain or the stomach, a lesson from gastric bypass surgery?

Ghrelin is abundantly synthesized by the specialized X/A like endocrine cells in the gastric mucosa and acts through its receptor mainly expressed on the hypothalamic neurons to stimulate food intake. The search for anti-obesity drugs targeting ghrelin has been therefore focused upon the central nervous system. However, the anatomical structure and location of the hypothalamus pose significant hurdles for therapy targeting this organ. Searching for the peripheral targets is therefore appealing. Gastric bypass surgery is the most effective therapy for morbid obesity. Interestingly, weight reduction after gastric bypass surgery occurs not solely through restriction of food intake after the surgery, but likely also involves the significant decrease of plasma ghrelin. This finding suggests that it may be practical to target the stomach to reduce the production of ghrelin for treatment of obesity. To pursue this goal, an understanding of the cellular and molecular mechanisms by which ghrelin is synthesized, acylated, and secreted is invaluable. Our studies[137,162] demonstrate that the mammalian target of rapamycin (mTOR) selectively regulates the production of gastric ghrelin. An increase in gastric mTOR activity inhibits ghrelin production, while a decrease in gastric mTOR activity stimulates the synthesis of ghrelin. Antagonism of the ghrelin receptor abolishes the rapamycin-induced increase in food intake. Gastric mTOR activity is linked to the regulation of energy intake through ghrelin. Targeting gastric mTOR may therefore provide a novel strategy for treatment of obesity and diabetes.

3.4. Targeting total ghrelin production or acylation of ghrelin?

Deletion of the ghrelin gene renders mice resistant to high-fat diet-induced obesity. These data suggest that targeting the stomach to reduce the production of total ghrelin may be beneficial for weight control. This concept is also supported by our observations that mTOR activity results in the alteration in total ghrelin production and the consequent modulation of the food intake in mice. However, it is worth noting that alteration in the level of total ghrelin may result in the proportional change of acylated ghrelin.

The alternative approach for development of anti-obesity drugs is to target the acylation of ghrelin. Twenty to thirty percent of circulating ghrelin molecules is acylated by an n-octanoic acid,[80] whereas the remainder circulates as UAG. It is widely accepted that the effect of ghrelin on food intake is only exerted by the acylated molecule through its interaction with the ghrelin receptor GHS-R1a. Posttranslational modification of ghrelin by octanoylation of its third amino acid serine residue is therefore key to ghrelin function. The enzyme that mediates octanoylation is GOAT.[152,153] Transcripts for both GOAT and ghrelin are present predominantly in the stomach and pancreas. Genetic disruption of the GOAT gene in mice leads to complete absence of acylated ghrelin in circulation. GOAT is thus suggested as a potential therapeutic target for modulation of food intake and body weight gain. This idea is directly supported by the finding that GO-CoA-Tat, a peptide-based bisubstrate analog that antagonizes GOAT, improves glucose tolerance and reduces weight gain in wild-type mice but not in ghrelin-deficient mice.[163]

Until solid evidence becomes available, it is currently unknown which strategy, that is, targeting total ghrelin or acylation of ghrelin, will provide a better option for treatment of obesity. In addition to the effect on food intake and energy metabolism, ghrelin has been found to regulate gastrointestinal motility and associated sensory functions, to modulate the reproductive and stress axes, mood and emotion, and to affect cardiovascular function. Some of these functions would be important to preserve if aiming to antagonize ghrelin-mediated positive energy balance.

3.5. Long-term efficiency of ghrelin receptor antagonism?

Several strategies have been developed to target the ghrelin for the treatment of obesity. These include blocking its action by receptor antagonism, by neutralizing antibodies, or by Spiegelmer RNA. Both systemic and central administrations of ghrelin receptor antagonists result in decreased body weight and improved glycemic control in obese mice. Neutralization of ghrelin by an RNA Spiegelmer, a synthetic oligonucleotide capable of specific high-affinity binding to bioactive ghrelin, evokes a significant reduction of body weight in a 7-day experiment using mice with high-fat diet-induced obesity. This effect is not observed in ghrelin-deficient mice, suggesting a specific effect of ghrelin neutralization. Antibodies targeting acyl ghrelin have also been demonstrated to significantly reduce food intake.[164,165] Similarly, a vaccination blocking the effects of ghrelin has been reported to reduce relative adiposity, decrease body weight gain, and decrease feeding efficiency without affecting food intake.[37] While these approaches are promising, it is worth noting that the long-term efficacy of these strategies remains questionable. One single report on the long-term use of acyl ghrelin-specific neutralizing antibody for 4 weeks demonstrates no significant effect on body weight and food intake in high-fat diet-induced obesity mice. Moreover, ghrelin vaccination, an efficient approach in animal experiments,[166] had no significant effect on weight control in obese subjects in Phase IIa clinical trial. The exact mechanism for the negligible effect of these approaches with long-term use remains unknown, but may be offset by potential increases in circulating ghrelin due to feedback regulation. Defining novel pathways to inhibit the production of acyl ghrelin, the active form of ghrelin, would shift the rapeutic focus to gastric targets.

4. CONCLUSION

In summary, the ghrelin receptor GHS-R1a is a seven TM GPCR that is widely distributed in the central nervous system and multiple peripheral tissues such as pancreas and adipose tissue. A wide variety of physiological functions have been ascribed to this receptor, whose activation requires a unique octanoylation of the third amino acid serine of its ligand ghrelin. The regulation of food intake and glucose homeostasis by the ghrelin receptor has attracted significant attention from the scientific community. Studies from pharmacological receptor antagonism and ghrelin receptor gene deletion models have established that the ghrelin receptor plays an important role in enabling an animal to adapt to changes in nutritional status by controlling metabolic parameters such as energy intake and expenditure. Therefore, targeting the ghrelin receptor and its ligand ghrelin may provide a therapeutic benefit for people with obesity and diabetes.

ACKNOWLEDGMENTS

This work was supported by grants from the National Natural Science Foundation of China (81030012, 30971434), the Major National Basic Research Program of P. R. China (No. 2010CB912504).

REFERENCES

1. Strobl JS, Thomas MJ. Human growth hormone. *Pharmacol Rev* 1994;**46**:1–34.
2. Chipman JJ. Recent advances in hGH clinical research. *J Pediatr Endocrinol* 1993;**6**: 325–8.
3. Jorgensen JO, Christiansen JS. Brave new senescence: GH in adults. *Lancet* 1993;**341**: 1247–8.
4. Bowers CY, Momany FA, Reynolds GA, Hong A. On the in vitro and in vivo activity of a new synthetic hexapeptide that acts on the pituitary to specifically release growth hormone. *Endocrinology* 1984;**114**:1537–45.
5. Bowers CY, Alster DK, Frentz JM. The growth hormone-releasing activity of a synthetic hexapeptide in normal men and short statured children after oral administration. *J Clin Endocrinol Metab* 1992;**74**:292–8.
6. Smith RG, Cheng K, Schoen WR, Pong SS, Hickey G, Jacks T, et al. A nonpeptidyl growth hormone secretagogue. *Science* 1993;**260**:1640–3.
7. Patchett AA, Nargund RP, Tata JR, Chen MH, Barakat KJ, Johnston DB, et al. Design and biological activities of L-163,191 (MK-0677): a potent, orally active growth hormone secretagogue. *Proc Natl Acad Sci USA* 1995;**92**:7001–5.
8. Katakami H, Arimura A, Frohman LA. Growth hormone (GH)-releasing factor stimulates hypothalamic somatostatin release: an inhibitory feedback effect on GH secretion. *Endocrinology* 1986;**118**:1872–7.
9. Frohman LA, Jansson JO. Growth hormone-releasing hormone. *Endocr Rev* 1986;**7**:223–53.
10. Pong SS, Chaung LY, Dean DC, Nargund RP, Patchett AA, Smith RG. Identification of a new G-protein-linked receptor for growth hormone secretagogues. *Mol Endocrinol* 1996;**10**:57–61.
11. Howard AD, Feighner SD, Cully DF, Arena JP, Liberator PA, Rosenblum CI, et al. A receptor in pituitary and hypothalamus that functions in growth hormone release. *Science* 1996;**273**:974–7.
12. Kojima M, Hosoda H, Date Y, Nakazato M, Matsuo H, Kangawa K. Ghrelin is a growth-hormone-releasing acylated peptide from stomach. *Nature* 1999;**402**:656–60.
13. Hosoda H, Kojima M, Matsuo H, Kangawa K. Purification and characterization of rat des-Gln14-Ghrelin, a second endogenous ligand for the growth hormone secretagogue receptor. *J Biol Chem* 2000;**275**:21995–2000.
14. Cassoni P, Papotti M, Ghe C, Catapano F, Sapino A, Graziani A, et al. Identification, characterization, and biological activity of specific receptors for natural (ghrelin) and synthetic growth hormone secretagogues and analogs in human breast carcinomas and cell lines. *J Clin Endocrinol Metab* 2001;**86**:1738–45.

15. Muccioli G, Pons N, Ghe C, Catapano F, Granata R, Ghigo E. Ghrelin and des-acyl ghrelin both inhibit isoproterenol-induced lipolysis in rat adipocytes via a non-type 1a growth hormone secretagogue receptor. *Eur J Pharmacol* 2004;**498**:27–35.
16. Gauna C, Delhanty PJ, Hofland LJ, Janssen JA, Broglio F, Ross RJ, et al. Ghrelin stimulates, whereas des-octanoyl ghrelin inhibits, glucose output by primary hepatocytes. *J Clin Endocrinol Metab* 2005;**90**:1055–60.
17. Gauna C, van de Zande B, van Kerkwijk A, Themmen AP, van der Lely AJ, Delhanty PJ. Unacylated ghrelin is not a functional antagonist but a full agonist of the type 1a growth hormone secretagogue receptor (GHS-R). *Mol Cell Endocrinol* 2007;**274**:30–4.
18. Tullin S, Hansen BS, Ankersen M, Moller J, Von Cappelen KA, Thim L. Adenosine is an agonist of the growth hormone secretagogue receptor. *Endocrinology* 2000;**141**: 3397–402.
19. Deghenghi R, Papotti M, Ghigo E, Muccioli G. Cortistatin, but not somatostatin, binds to growth hormone secretagogue (GHS) receptors of human pituitary gland. *J Endocrinol Invest* 2001;**24**:RC1–RC3.
20. Gnanapavan S, Kola B, Bustin SA, Morris DG, McGee P, Fairclough P, et al. The tissue distribution of the mRNA of ghrelin and subtypes of its receptor, GHS-R, in humans. *J Clin Endocrinol Metab* 2002;**87**:2988.
21. Smith RG, Van der Ploeg LH, Howard AD, Feighner SD, Cheng K, Hickey GJ, et al. Peptidomimetic regulation of growth hormone secretion. *Endocr Rev* 1997;**18**: 621–45.
22. Moulin A, Demange L, Ryan J, Mousseaux D, Sanchez P, Berge G, et al. New trisubstituted 1,2,4-triazole derivatives as potent ghrelin receptor antagonists. 3. Synthesis and pharmacological in vitro and in vivo evaluations. *J Med Chem* 2008;**51**:689–93.
23. Halem HA, Taylor JE, Dong JZ, Shen Y, Datta R, Abizaid A, et al. A novel growth hormone secretagogue-1a receptor antagonist that blocks ghrelin-induced growth hormone secretion but induces increased body weight gain. *Neuroendocrinology* 2005;**81**:339–49.
24. Baessler A, Fischer M, Mayer B, Koehler M, Wiedmann S, Stark K, et al. Epistatic interaction between haplotypes of the ghrelin ligand and receptor genes influence susceptibility to myocardial infarction and coronary artery disease. *Hum Mol Genet* 2007;**16**: 887–99.
25. den Hoed M, Smeets AJ, Veldhorst MA, Nieuwenhuizen AG, Bouwman FG, Heidema AG, et al. SNP analyses of postprandial responses in (an)orexigenic hormones and feelings of hunger reveal long-term physiological adaptations to facilitate homeostasis. *Int J Obes (Lond)* 2008;**32**:1790–8.
26. Baessler A, Hasinoff MJ, Fischer M, Reinhard W, Sonnenberg GE, Olivier M, et al. Genetic linkage and association of the growth hormone secretagogue receptor (ghrelin receptor) gene in human obesity. *Diabetes* 2005;**54**:259–67.
27. Gueorguiev M, Lecoeur C, Meyre D, Benzinou M, Mein CA, Hinney A, et al. Association studies on ghrelin and ghrelin receptor gene polymorphisms with obesity. *Obesity (Silver Spring)* 2009;**17**:745–54.
28. Gjesing AP, Larsen LH, Torekov SS, Hainerova IA, Kapur R, Johansen A, et al. Family and population-based studies of variation within the ghrelin receptor locus in relation to measures of obesity. *PLoS One* 2010;**5**:e10084.
29. Pantel J, Legendre M, Cabrol S, Hilal L, Hajaji Y, Morisset S, et al. Loss of constitutive activity of the growth hormone secretagogue receptor in familial short stature. *J Clin Invest* 2006;**116**:760–8.
30. Wang HJ, Geller F, Dempfle A, Schauble N, Friedel S, Lichtner P, et al. Ghrelin receptor gene: identification of several sequence variants in extremely obese children and adolescents, healthy normal-weight and underweight students, and children with short normal stature. *J Clin Endocrinol Metab* 2004;**89**:157–62.

31. Liu G, Fortin JP, Beinborn M, Kopin AS. Four missense mutations in the ghrelin receptor result in distinct pharmacological abnormalities. *J Pharmacol Exp Ther* 2007; **322**:1036–43.
32. Rudolph J, Esler WP, O'Connor S, Coish PD, Wickens PL, Brands M, et al. Quinazolinone derivatives as orally available ghrelin receptor antagonists for the treatment of diabetes and obesity. *J Med Chem* 2007;**50**:5202–16.
33. Yu M, Lizarzaburu M, Beckmann H, Connors R, Dai K, Haller K, et al. Identification of piperazine-bisamide GHSR antagonists for the treatment of obesity. *Bioorg Med Chem Lett* 2010;**20**:1758–62.
34. Zhao H, Xin Z, Liu G, Schaefer VG, Falls HD, Kaszubska W, et al. Discovery of tetralin carboxamide growth hormone secretagogue receptor antagonists via scaffold manipulation. *J Med Chem* 2004;**47**:6655–7.
35. Zhao H, Xin Z, Patel JR, Nelson LT, Liu B, Szczepankiewicz BG, et al. Structure-activity relationship studies on tetralin carboxamide growth hormone secretagogue receptor antagonists. *Bioorg Med Chem Lett* 2005;**15**:1825–8.
36. Helmling S, Maasch C, Eulberg D, Buchner K, Schroder W, Lange C, et al. Inhibition of ghrelin action in vitro and in vivo by an RNA-Spiegelmer. *Proc Natl Acad Sci USA* 2004;**101**:13174–9.
37. Zorrilla EP, Iwasaki S, Moss JA, Chang J, Otsuji J, Inoue K, et al. Vaccination against weight gain. *Proc Natl Acad Sci USA* 2006;**103**:13226–31.
38. McKee KK, Palyha OC, Feighner SD, Hreniuk DL, Tan CP, Phillips MS, et al. Molecular analysis of rat pituitary and hypothalamic growth hormone secretagogue receptors. *Mol Endocrinol* 1997;**11**:415–23.
39. Palyha OC, Feighner SD, Tan CP, McKee KK, Hreniuk DL, Gao YD, et al. Ligand activation domain of human orphan growth hormone (GH) secretagogue receptor (GHS-R) conserved from Pufferfish to humans. *Mol Endocrinol* 2000;**14**:160–9.
40. Bockaert J, Pin JP. Molecular tinkering of G protein-coupled receptors: an evolutionary success. *EMBO J* 1999;**18**:1723–9.
41. Schwartz TW, Frimurer TM, Holst B, Rosenkilde MM, Elling CE. Molecular mechanism of 7TM receptor activation—a global toggle switch model. *Annu Rev Pharmacol Toxicol* 2006;**46**:481–519.
42. Pedretti A, Villa M, Pallavicini M, Valoti E, Vistoli G. Construction of human ghrelin receptor (hGHS-R1a) model using a fragmental prediction approach and validation through docking analysis. *J Med Chem* 2006;**49**:3077–85.
43. Petersenn S. Structure and regulation of the growth hormone secretagogue receptor. *Minerva Endocrinol* 2002;**27**:243–56.
44. Chu KM, Chow KB, Leung PK, Lau PN, Chan CB, Cheng CH, et al. Over-expression of the truncated ghrelin receptor polypeptide attenuates the constitutive activation of phosphatidylinositol-specific phospholipase C by ghrelin receptors but has no effect on ghrelin-stimulated extracellular signal-regulated kinase 1/2 activity. *Int J Biochem Cell Biol* 2007;**39**:752–64.
45. Holst B, Lang M, Brandt E, Bach A, Howard A, Frimurer TM, et al. Ghrelin receptor inverse agonists: identification of an active peptide core and its interaction epitopes on the receptor. *Mol Pharmacol* 2006;**70**:936–46.
46. Feighner SD, Howard AD, Prendergast K, Palyha OC, Hreniuk DL, Nargund R, et al. Structural requirements for the activation of the human growth hormone secretagogue receptor by peptide and nonpeptide secretagogues. *Mol Endocrinol* 1998;**12**:137–45.
47. Holst B, Cygankiewicz A, Jensen TH, Ankersen M, Schwartz TW. High constitutive signaling of the ghrelin receptor—identification of a potent inverse agonist. *Mol Endocrinol* 2003;**17**:2201–10.
48. Bond RA, Ijzerman AP. Recent developments in constitutive receptor activity and inverse agonism, and their potential for GPCR drug discovery. *Trends Pharmacol Sci* 2006;**27**:92–6.

49. Holst B, Schwartz TW. Ghrelin receptor mutations—too little height and too much hunger. *J Clin Invest* 2006;**116**:637–41.
50. Holst B, Mokrosinski J, Lang M, Brandt E, Nygaard R, Frimurer TM, et al. Identification of an efficacy switch region in the ghrelin receptor responsible for interchange between agonism and inverse agonism. *J Biol Chem* 2007;**282**:15799–811.
51. Holst B, Frimurer TM, Mokrosinski J, Halkjaer T, Cullberg KB, Underwood CR, et al. Overlapping binding site for the endogenous agonist, small-molecule agonists, and ago-allosteric modulators on the ghrelin receptor. *Mol Pharmacol* 2009;**75**: 44–59.
52. Holst B, Holliday ND, Bach A, Elling CE, Cox HM, Schwartz TW. Common structural basis for constitutive activity of the ghrelin receptor family. *J Biol Chem* 2004;**279**:53806–17.
53. Falls HD, Dayton BD, Fry DG, Ogiela CA, Schaefer VG, Brodjian S, et al. Characterization of ghrelin receptor activity in a rat pituitary cell line RC-4B/C. *J Mol Endocrinol* 2006;**37**:51–62.
54. Camina JP, Carreira MC, El Messari S, Llorens-Cortes C, Smith RG, Casanueva FF. Desensitization and endocytosis mechanisms of ghrelin-activated growth hormone secretagogue receptor 1a. *Endocrinology* 2004;**145**:930–40.
55. Luttrell LM, Lefkowitz RJ. The role of beta-arrestins in the termination and transduction of G-protein-coupled receptor signals. *J Cell Sci* 2002;**115**:455–65.
56. Delhanty PJ, van Kerkwijk A, Huisman M, van de Zande B, Verhoef-Post M, Gauna C, et al. Unsaturated fatty acids prevent desensitization of the human growth hormone secretagogue receptor by blocking its internalization. *Am J Physiol Endocrinol Metab* 2010;**299**:E497–E505.
57. Rios CD, Jordan BA, Gomes I, Devi LA. G-protein-coupled receptor dimerization: modulation of receptor function. *Pharmacol Ther* 2001;**92**:71–87.
58. Milligan G. G protein-coupled receptor dimerization: function and ligand pharmacology. *Mol Pharmacol* 2004;**66**:1–7.
59. Jordan BA, Trapaidze N, Gomes I, Nivarthi R, Devi LA. Oligomerization of opioid receptors with beta 2-adrenergic receptors: a role in trafficking and mitogen-activated protein kinase activation. *Proc Natl Acad Sci USA* 2001;**98**:343–8.
60. AbdAlla S, Lother H, Quitterer U. AT1-receptor heterodimers show enhanced G-protein activation and altered receptor sequestration. *Nature* 2000;**407**:94–8.
61. Gutkind JS. Cell growth control by G protein-coupled receptors: from signal transduction to signal integration. *Oncogene* 1998;**17**:1331–42.
62. Bhalla US, Iyengar R. Emergent properties of networks of biological signaling pathways. *Science* 1999;**283**:381–7.
63. Werry TD, Wilkinson GF, Willars GB. Mechanisms of cross-talk between G-protein-coupled receptors resulting in enhanced release of intracellular Ca2+. *Biochem J* 2003;**374**:281–96.
64. Cheng K, Chan WW, Barreto Jr A, Convey EM, Smith RG. The synergistic effects of His-D-Trp-Ala-Trp-D-Phe-Lys-NH2 on growth hormone (GH)-releasing factor-stimulated GH release and intracellular adenosine 3',5'-monophosphate accumulation in rat primary pituitary cell culture. *Endocrinology* 1989;**124**:2791–8.
65. Cheng K, Chan WW, Butler B, Barreto Jr A, Smith RG. Evidence for a role of protein kinase-C in His-D-Trp-Ala-Trp-D-Phe-Lys-NH2-induced growth hormone release from rat primary pituitary cells. *Endocrinology* 1991;**129**:3337–42.
66. Sunahara RK, Taussig R. Isoforms of mammalian adenylyl cyclase: multiplicities of signaling. *Mol Interv* 2002;**2**:168–84.
67. Jiang H, Betancourt L, Smith RG. Ghrelin amplifies dopamine signaling by cross talk involving formation of growth hormone secretagogue receptor/dopamine receptor subtype 1 heterodimers. *Mol Endocrinol* 2006;**20**:1772–85.

68. Missale C, Nash SR, Robinson SW, Jaber M, Caron MG. Dopamine receptors: from structure to function. *Physiol Rev* 1998;**78**:189–225.
69. Kern A, Albarran-Zeckler R, Walsh HE, Smith RG. Apo-ghrelin receptor forms heteromers with DRD2 in hypothalamic neurons and is essential for anorexigenic effects of DRD2 agonism. *Neuron* 2012;**73**:317–32.
70. Maurel D, Comps-Agrar L, Brock C, Rives ML, Bourrier E, Ayoub MA, et al. Cell-surface protein-protein interaction analysis with time-resolved FRET and snap-tag technologies: application to GPCR oligomerization. *Nat Methods* 2008;**5**:561–7.
71. Albizu L, Cottet M, Kralikova M, Stoev S, Seyer R, Brabet I, et al. Time-resolved FRET between GPCR ligands reveals oligomers in native tissues. *Nat Chem Biol* 2010;**6**:587–94.
72. Fetissov SO, Meguid MM, Sato T, Zhang LH. Expression of dopaminergic receptors in the hypothalamus of lean and obese Zucker rats and food intake. *Am J Physiol Regul Integr Comp Physiol* 2002;**283**:R905–R910.
73. Palmiter RD. Is dopamine a physiologically relevant mediator of feeding behavior? *Trends Neurosci* 2007;**30**:375–81.
74. Volkow ND, Wang GJ, Baler RD. Reward, dopamine and the control of food intake: implications for obesity. *Trends Cogn Sci* 2011;**15**:37–46.
75. Meguid MM, Fetissov SO, Varma M, Sato T, Zhang L, Laviano A, et al. Hypothalamic dopamine and serotonin in the regulation of food intake. *Nutrition* 2000;**16**:843–57.
76. Vucetic Z, Reyes TM. Central dopaminergic circuitry controlling food intake and reward: implications for the regulation of obesity. *Wiley Interdiscip Rev Syst Biol Med* 2010;**2**:577–93.
77. Rediger A, Piechowski CL, Yi CX, Tarnow P, Strotmann R, Gruters A, et al. Mutually opposite signal modulation by hypothalamic heterodimerization of ghrelin and melanocortin-3 receptors. *J Biol Chem* 2011;**286**:39623–31.
78. Smith RG, Griffin PR, Xu Y, Smith AG, Liu K, Calacay J, et al. Adenosine: a partial agonist of the growth hormone secretagogue receptor. *Biochem Biophys Res Commun* 2000;**276**:1306–13.
79. Carreira MC, Camina JP, Smith RG, Casanueva FF. Agonist-specific coupling of growth hormone secretagogue receptor type 1a to different intracellular signaling systems. Role of adenosine. *Neuroendocrinology* 2004;**79**:13–25.
80. Fredholm BB, IJzerman AP, Jacobson KA, Klotz KN, Linden J. International Union of Pharmacology. XXV. Nomenclature and classification of adenosine receptors. *Pharmacol Rev* 2001;**53**:527–52.
81. Hermansson NO, Morgan DG, Drmota T, Larsson N. Adenosine is not a direct GHSR agonist—artificial cross-talk between GHSR and adenosine receptor pathways. *Acta Physiol (Oxf)* 2007;**190**:77–86.
82. Chen C, Zhang J, Vincent JD, Israel JM. Two types of voltage-dependent calcium current in rat somatotrophs are reduced by somatostatin. *J Physiol* 1990;**425**:29–42.
83. Kohno D, Gao HZ, Muroya S, Kikuyama S, Yada T. Ghrelin directly interacts with neuropeptide-Y-containing neurons in the rat arcuate nucleus: Ca2+ signaling via protein kinase A and N-type channel-dependent mechanisms and cross-talk with leptin and orexin. *Diabetes* 2003;**52**:948–56.
84. Mazzocchi G, Neri G, Rucinski M, Rebuffat P, Spinazzi R, Malendowicz LK, et al. Ghrelin enhances the growth of cultured human adrenal zona glomerulosa cells by exerting MAPK-mediated proliferogenic and antiapoptotic effects. *Peptides* 2004;**25**: 1269–77.
85. Kim MS, Yoon CY, Jang PG, Park YJ, Shin CS, Park HS, et al. The mitogenic and anti-apoptotic actions of ghrelin in 3T3-L1 adipocytes. *Mol Endocrinol* 2004;**18**:2291–301.
86. Hardie DG. The AMP-activated protein kinase pathway—new players upstream and downstream. *J Cell Sci* 2004;**117**:5479–87.

87. Barazzoni R, Bosutti A, Stebel M, Cattin MR, Roder E, Visintin L, et al. Ghrelin regulates mitochondrial-lipid metabolism gene expression and tissue fat distribution in liver and skeletal muscle. *Am J Physiol Endocrinol Metab* 2005;**288**:E228–E235.
88. Murata M, Okimura Y, Iida K, Matsumoto M, Sowa H, Kaji H, et al. Ghrelin modulates the downstream molecules of insulin signaling in hepatoma cells. *J Biol Chem* 2002;**277**:5667–74.
89. Guan XM, Yu H, Palyha OC, McKee KK, Feighner SD, Sirinathsinghji DJ, et al. Distribution of mRNA encoding the growth hormone secretagogue receptor in brain and peripheral tissues. *Brain Res Mol Brain Res* 1997;**48**:23–9.
90. Date Y, Kojima M, Hosoda H, Sawaguchi A, Mondal MS, Suganuma T, et al. Ghrelin, a novel growth hormone-releasing acylated peptide, is synthesized in a distinct endocrine cell type in the gastrointestinal tracts of rats and humans. *Endocrinology* 2000;**141**:4255–61.
91. Mori K, Yoshimoto A, Takaya K, Hosoda K, Ariyasu H, Yahata K, et al. Kidney produces a novel acylated peptide, ghrelin. *FEBS Lett* 2000;**486**:213–6.
92. Nagaya N, Uematsu M, Kojima M, Ikeda Y, Yoshihara F, Shimizu W, et al. Chronic administration of ghrelin improves left ventricular dysfunction and attenuates development of cardiac cachexia in rats with heart failure. *Circulation* 2001;**104**:1430–5.
93. Korbonits M, Kojima M, Kangawa K, Grossman AB. Presence of ghrelin in normal and adenomatous human pituitary. *Endocrine* 2001;**14**:101–4.
94. Korbonits M, Bustin SA, Kojima M, Jordan S, Adams EF, Lowe DG, et al. The expression of the growth hormone secretagogue receptor ligand ghrelin in normal and abnormal human pituitary and other neuroendocrine tumors. *J Clin Endocrinol Metab* 2001;**86**:881–7.
95. de Keyzer Y, Lenne F, Bertagna X. Widespread transcription of the growth hormone-releasing peptide receptor gene in neuroendocrine human tumors. *Eur J Endocrinol* 1997;**137**:715–8.
96. Papotti M, Cassoni P, Volante M, Deghenghi R, Muccioli G, Ghigo E. Ghrelin-producing endocrine tumors of the stomach and intestine. *J Clin Endocrinol Metab* 2001;**86**:5052–9.
97. Volante M, Allia E, Gugliotta P, Funaro A, Broglio F, Deghenghi R, et al. Expression of ghrelin and of the GH secretagogue receptor by pancreatic islet cells and related endocrine tumors. *J Clin Endocrinol Metab* 2002;**87**:1300–8.
98. Lin L, Saha PK, Ma X, Henshaw IO, Shao L, Chang BH, et al. Ablation of ghrelin receptor reduces adiposity and improves insulin sensitivity during aging by regulating fat metabolism in white and brown adipose tissues. *Aging Cell* 2011;**10**: 996–1010.
99. Zigman JM, Nakano Y, Coppari R, Balthasar N, Marcus JN, Lee CE, et al. Mice lacking ghrelin receptors resist the development of diet-induced obesity. *J Clin Invest* 2005;**115**:3564–72.
100. Chen AS, Marsh DJ, Trumbauer ME, Frazier EG, Guan XM, Yu H, et al. Inactivation of the mouse melanocortin-3 receptor results in increased fat mass and reduced lean body mass. *Nat Genet* 2000;**26**:97–102.
101. Picard F, Deshaies Y, Lalonde J, Samson P, Richard D. Topiramate reduces energy and fat gains in lean (Fa/?) and obese (fa/fa) Zucker rats. *Obes Res* 2000;**8**:656–63.
102. Xu G, Wang Z, Li Y, Li Z, Tang H, Zhao J, et al. Ghrelin contributes to derangements of glucose metabolism induced by rapamycin in mice. *Diabetologia* 2012;**55**: 1813–23.
103. Gabriely I, Ma XH, Yang XM, Atzmon G, Rajala MW, Berg AH, et al. Removal of visceral fat prevents insulin resistance and glucose intolerance of aging: an adipokine-mediated process? *Diabetes* 2002;**51**:2951–8.
104. Ahima RS. Connecting obesity, aging and diabetes. *Nat Med* 2009;**15**:996–7.

105. Amati F, Dube JJ, Coen PM, Stefanovic-Racic M, Toledo FG, Goodpaster BH. Physical inactivity and obesity underlie the insulin resistance of aging. *Diabetes Care* 2009;**32**:1547–9.
106. Cannon B, Nedergaard J. Brown adipose tissue: function and physiological significance. *Physiol Rev* 2004;**84**:277–359.
107. Nicholls DG, Locke RM. Thermogenic mechanisms in brown fat. *Physiol Rev* 1984;**64**:1–64.
108. Pfannenberg C, Werner MK, Ripkens S, Stef I, Deckert A, Schmadl M, et al. Impact of age on the relationships of brown adipose tissue with sex and adiposity in humans. *Diabetes* 2010;**59**:1789–93.
109. Tentolouris N, Liatis S, Katsilambros N. Sympathetic system activity in obesity and metabolic syndrome. *Ann N Y Acad Sci* 2006;**1083**:129–52.
110. Barzilai N, Gabriely I. The role of fat depletion in the biological benefits of caloric restriction. *J Nutr* 2001;**131**:903S–906S.
111. Barzilai N, Banerjee S, Hawkins M, Chen W, Rossetti L. Caloric restriction reverses hepatic insulin resistance in aging rats by decreasing visceral fat. *J Clin Invest* 1998;**101**:1353–61.
112. Wiskin AE, Davies JH, Wootton SA, Beattie RM. Energy expenditure, nutrition and growth. *Arch Dis Child* 2011;**96**:567–72.
113. Wren AM, Small CJ, Ward HL, Murphy KG, Dakin CL, Taheri S, et al. The novel hypothalamic peptide ghrelin stimulates food intake and growth hormone secretion. *Endocrinology* 2000;**141**:4325–8.
114. Arvat E, Di Vito L, Broglio F, Papotti M, Muccioli G, Dieguez C, et al. Preliminary evidence that Ghrelin, the natural GH secretagogue (GHS)-receptor ligand, strongly stimulates GH secretion in humans. *J Endocrinol Invest* 2000;**23**:493–5.
115. Nakazato M, Murakami N, Date Y, Kojima M, Matsuo H, Kangawa K, et al. A role for ghrelin in the central regulation of feeding. *Nature* 2001;**409**:194–8.
116. Willesen MG, Kristensen P, Romer J. Co-localization of growth hormone secretagogue receptor and NPY mRNA in the arcuate nucleus of the rat. *Neuroendocrinology* 1999;**70**:306–16.
117. Cowley MA, Smith RG, Diano S, Tschop M, Pronchuk N, Grove KL, et al. The distribution and mechanism of action of ghrelin in the CNS demonstrates a novel hypothalamic circuit regulating energy homeostasis. *Neuron* 2003;**37**:649–61.
118. Broglio F, Arvat E, Benso A, Gottero C, Muccioli G, Papotti M, et al. Ghrelin, a natural GH secretagogue produced by the stomach, induces hyperglycemia and reduces insulin secretion in humans. *J Clin Endocrinol Metab* 2001;**86**:5083–6.
119. Nieminen P, Mustonen AM. Effects of peripheral ghrelin on the carbohydrate and lipid metabolism of the tundra vole (Microtus oeconomus). *Gen Comp Endocrinol* 2004; **138**:182–7.
120. Stevanovic D, Nesic D, Milosevic V, Starcevic V, Severs WB. Consummatory behavior and metabolic indicators after central ghrelin injections in rats. *Regul Pept* 2008;**147**: 52–9.
121. Dezaki K, Hosoda H, Kakei M, Hashiguchi S, Watanabe M, Kangawa K, et al. Endogenous ghrelin in pancreatic islets restricts insulin release by attenuating Ca2+ signaling in beta-cells: implication in the glycemic control in rodents. *Diabetes* 2004;**53**:3142–51.
122. Asakawa A, Inui A, Kaga T, Katsuura G, Fujimiya M, Fujino MA, et al. Antagonism of ghrelin receptor reduces food intake and body weight gain in mice. *Gut* 2003;**52**: 947–52.
123. Vestergaard ET, Gormsen LC, Jessen N, Lund S, Hansen TK, Moller N, et al. Ghrelin infusion in humans induces acute insulin resistance and lipolysis independent of growth hormone signaling. *Diabetes* 2008;**57**:3205–10.

124. Pacifico L, Poggiogalle E, Costantino F, Anania C, Ferraro F, Chiarelli F, et al. Acylated and nonacylated ghrelin levels and their associations with insulin resistance in obese and normal weight children with metabolic syndrome. *Eur J Endocrinol* 2009;**161**:861–70.
125. Sun Y, Asnicar M, Saha PK, Chan L, Smith RG. Ablation of ghrelin improves the diabetic but not obese phenotype of ob/ob mice. *Cell Metab* 2006;**3**:379–86.
126. Sun Y, Butte NF, Garcia JM, Smith RG. Characterization of adult ghrelin and ghrelin receptor knockout mice under positive and negative energy balance. *Endocrinology* 2008;**149**:843–50.
127. Petersen KF, Shulman GI. Etiology of insulin resistance. *Am J Med* 2006;**119**:S10–S16.
128. Prado CL, Pugh-Bernard AE, Elghazi L, Sosa-Pineda B, Sussel L. Ghrelin cells replace insulin-producing beta cells in two mouse models of pancreas development. *Proc Natl Acad Sci USA* 2004;**101**:2924–9.
129. Wierup N, Svensson H, Mulder H, Sundler F. The ghrelin cell: a novel developmentally regulated islet cell in the human pancreas. *Regul Pept* 2002;**107**:63–9.
130. Wierup N, Yang S, McEvilly RJ, Mulder H, Sundler F. Ghrelin is expressed in a novel endocrine cell type in developing rat islets and inhibits insulin secretion from INS-1 (832/13) cells. *J Histochem Cytochem* 2004;**52**:301–10.
131. Dezaki K, Sone H, Koizumi M, Nakata M, Kakei M, Nagai H, et al. Blockade of pancreatic islet-derived ghrelin enhances insulin secretion to prevent high-fat diet-induced glucose intolerance. *Diabetes* 2006;**55**:3486–93.
132. Egido EM, Rodriguez-Gallardo J, Silvestre RA, Marco J. Inhibitory effect of ghrelin on insulin and pancreatic somatostatin secretion. *Eur J Endocrinol* 2002;**146**:241–4.
133. Tipparaju SM, Saxena N, Liu SQ, Kumar R, Bhatnagar A. Differential regulation of voltage-gated K+ channels by oxidized and reduced pyridine nucleotide coenzymes. *Am J Physiol Cell Physiol* 2005;**288**:C366–C376.
134. Bryan J, Crane A, Vila-Carriles WH, Babenko AP, Aguilar-Bryan L. Insulin secretagogues, sulfonylurea receptors and K(ATP) channels. *Curr Pharm Des* 2005;**11**:2699–716.
135. Sangiao-Alvarellos S, Cordido F. Effect of ghrelin on glucose-insulin homeostasis: therapeutic implications. *Int J Pept* 2010;**2010**:1–25.
136. Dezaki K, Kakei M, Yada T. Ghrelin uses Galphai2 and activates voltage-dependent K+ channels to attenuate glucose-induced Ca2+ signaling and insulin release in islet beta-cells: novel signal transduction of ghrelin. *Diabetes* 2007;**56**:2319–27.
137. Xu G, Li Y, An W, Li S, Guan Y, Wang N, et al. Gastric mammalian target of rapamycin signaling regulates ghrelin production and food intake. *Endocrinology* 2009;**150**:3637–44.
138. Theander-Carrillo C, Wiedmer P, Cettour-Rose P, Nogueiras R, Perez-Tilve D, Pfluger P, et al. Ghrelin action in the brain controls adipocyte metabolism. *J Clin Invest* 2006;**116**:1983–93.
139. Tschop M, Smiley DL, Heiman ML. Ghrelin induces adiposity in rodents. *Nature* 2000;**407**:908–13.
140. Davies JS, Kotokorpi P, Eccles SR, Barnes SK, Tokarczuk PF, Allen SK, et al. Ghrelin induces abdominal obesity via GHS-R-dependent lipid retention. *Mol Endocrinol* 2009;**23**:914–24.
141. Vaughan AM, Oram JF. ABCG1 redistributes cell cholesterol to domains removable by high density lipoprotein but not by lipid-depleted apolipoproteins. *J Biol Chem* 2005;**280**:30150–7.
142. Tsubone T, Masaki T, Katsuragi I, Tanaka K, Kakuma T, Yoshimatsu H. Ghrelin regulates adiposity in white adipose tissue and UCP1 mRNA expression in brown adipose tissue in mice. *Regul Pept* 2005;**130**:97–103.
143. Schultz JR, Tu H, Luk A, Repa JJ, Medina JC, Li L, et al. Role of LXRs in control of lipogenesis. *Genes Dev* 2000;**14**:2831–8.

144. Miegueu P, St Pierre D, Broglio F, Cianflone K. Effect of desacyl ghrelin, obestatin and related peptides on triglyceride storage, metabolism and GHSR signaling in 3T3-L1 adipocytes. *J Cell Biochem* 2011;**112**:704–14.
145. Pennacchio LA, Rubin EM. Genomic strategies to identify mammalian regulatory sequences. *Nat Rev Genet* 2001;**2**:100–9.
146. Mattick JS. RNA regulation: a new genetics? *Nat Rev Genet* 2004;**5**:316–23.
147. Nobrega MA, Ovcharenko I, Afzal V, Rubin EM. Scanning human gene deserts for long-range enhancers. *Science* 2003;**302**:413.
148. Mager U, Degenhardt T, Pulkkinen L, Kolehmainen M, Tolppanen AM, Lindstrom J, et al. Variations in the ghrelin receptor gene associate with obesity and glucose metabolism in individuals with impaired glucose tolerance. *PLoS One* 2008;**3**:e2941.
149. Li WJ, Zhen YS, Sun K, Xue H, Song XD, Wang YB, et al. Ghrelin receptor gene polymorphisms are associated with female metabolic syndrome in Chinese population. *Chin Med J (Engl)* 2008;**121**:1666–9.
150. Holst B, Schwartz TW. Constitutive ghrelin receptor activity as a signaling set-point in appetite regulation. *Trends Pharmacol Sci* 2004;**25**:113–7.
151. Matzko ME, Argyropoulos G, Wood GC, Chu X, McCarter RJ, Still CD, et al. Association of ghrelin receptor promoter polymorphisms with weight loss following Roux-en-Y gastric bypass surgery. *Obes Surg* 2012;**22**:783–90.
152. Gutierrez JA, Solenberg PJ, Perkins DR, Willency JA, Knierman MD, Jin Z, et al. Ghrelin octanoylation mediated by an orphan lipid transferase. *Proc Natl Acad Sci USA* 2008;**105**:6320–5.
153. Yang J, Brown MS, Liang G, Grishin NV, Goldstein JL. Identification of the acyltransferase that octanoylates ghrelin, an appetite-stimulating peptide hormone. *Cell* 2008;**132**:387–96.
154. Nogueiras R, Perez-Tilve D, Wortley KE, Tschop M. Growth hormone secretagogue (ghrelin-) receptors—a complex drug target for the regulation of body weight. *CNS Neurol Disord Drug Targets* 2006;**5**:335–43.
155. Muller TD, Perez-Tilve D, Tong J, Pfluger PT, Tschop MH. Ghrelin and its potential in the treatment of eating/wasting disorders and cachexia. *J Cachexia Sarcopenia Muscle* 2010;**1**:159–67.
156. Sabbatini FM, Di Fabio R, Corsi M, Cavanni P, Bromidge SM, St-Denis Y, et al. Discovery process and characterization of novel carbohydrazide derivatives as potent and selective GHSR1a antagonists. *ChemMedChem* 2010;**5**:1450–5.
157. Xin Z, Serby MD, Zhao H, Kosogof C, Szczepankiewicz BG, Liu M, et al. Discovery and pharmacological evaluation of growth hormone secretagogue receptor antagonists. *J Med Chem* 2006;**49**:4459–69.
158. Moulin A, Demange L, Berge G, Gagne D, Ryan J, Mousseaux D, et al. Toward potent ghrelin receptor ligands based on trisubstituted 1,2,4-triazole structure. 2. Synthesis and pharmacological in vitro and in vivo evaluations. *J Med Chem* 2007;**50**:5790–806.
159. Schellekens H, Dinan TG, Cryan JF. Lean mean fat reducing "ghrelin" machine: hypothalamic ghrelin and ghrelin receptors as therapeutic targets in obesity. *Neuropharmacology* 2010;**58**:2–16.
160. Briggs DI, Enriori PJ, Lemus MB, Cowley MA, Andrews ZB. Diet-induced obesity causes ghrelin resistance in arcuate NPY/AgRP neurons. *Endocrinology* 2010;**151**:4745–55.
161. Andrews ZB. The extra-hypothalamic actions of ghrelin on neuronal function. *Trends Neurosci* 2011;**34**:31–40.
162. Xu G, Li Y, An W, Zhao J, Xiang X, Ding L, et al. Regulation of gastric hormones by systemic rapamycin. *Peptides* 2010;**31**:2185–92.
163. Barnett BP, Hwang Y, Taylor MS, Kirchner H, Pfluger PT, Bernard V, et al. Glucose and weight control in mice with a designed ghrelin O-acyltransferase inhibitor. *Science* 2010;**330**:1689–92.

164. Mayorov AV, Amara N, Chang JY, Moss JA, Hixon MS, Ruiz DI, et al. Catalytic antibody degradation of ghrelin increases whole-body metabolic rate and reduces refeeding in fasting mice. *Proc Natl Acad Sci USA* 2008;**105**:17487–92.
165. Lu SC, Xu J, Chinookoswong N, Liu S, Steavenson S, Gegg C, et al. An acyl-ghrelin-specific neutralizing antibody inhibits the acute ghrelin-mediated orexigenic effects in mice. *Mol Pharmacol* 2009;**75**:901–7.
166. Kellokoski E, Kummu O, Serpi R, Lehenkari P, Ukkola O, Kesaniemi YA, et al. Ghrelin vaccination decreases plasma MCP-1 level in LDLR(−/−)-mice. *Peptides* 2009;**30**:2292–300.

CHAPTER THREE

Obestatin Receptor in Energy Homeostasis and Obesity Pathogenesis

Jian V. Zhang*,#, Lei Li*,1, Qingsheng Huang*,1, Pei-Gen Ren#

*Research Center for Gene and Cell Engineering, Institute of Biomedicine and Biotechnology, Shenzhen Institute of Advance Technology, Chinese Academy of Sciences, Shenzhen, China
#Innovative Pharmacology and Biotherapy Pre-clinical Test Public Service Platform of Shenzhen, China
[1]These authors contributed equally to this work

Contents

Abstract

Based on the bioinformatic prediction, Zhang and colleagues discovered obestatin, a new 23-amino acid hormone from rat stomach extract encoded by the *ghrelin* gene. Obestatin is present not only in the gastrointestinal tract, but also in the spleen, mammary gland, breast milk, and plasma. Obestatin appears to function as part of a complex gut-brain network whereby hormones and substances from the stomach, intestine and the brain about satiety or hunger. Given the current research regarding the effects of obestatin and its possible cognate receptor(s), this chapter provides the latest review of the physiological and pathological characteristics of this hormone and its possible receptor(s) in energy homeostasis and obesity.

Progress in Molecular Biology and Translational Science, Volume 114
ISSN 1877-1173
http://dx.doi.org/10.1016/B978-0-12-386933-3.00003-0

Ghrelin stimulates food intake and adiposity and thereby increases body weight in rodents after central as well as peripheral administration.[1] Based on the bioinformatic prediction, Zhang and colleagues discovered that, except ghrelin, the *ghrelin* gene encoded another secreted and bioactive peptide. The precursor is encoded by the gene *GHRL*, which is located on chromosome 3, band p25–26,[2] and is cleaved by protease into peptide segments. Zhang and colleagues isolated this new 23-amino acid hormone from rat stomach, and it showed actions opposite to ghrelin on food intake, body weight, and gastric emptying and named it "obestatin".[3] In this chapter, we summarize the possible physiological roles of obestatin and its receptor in energy homeostasis and its possible pathological roles in obesity.

Accumulating publications have confirmed the role of obestatin in regulating energy homeostasis and its association with obesity, among many other physiological functions. Its receptor(s) involved in these functions remains, however, controversial. Two receptors for obestatin have been proposed: GPR39 and GLP-1R.[3,4] Both G protein-coupled receptors (GPCRs) are found to be correlated with the function of obestatin, especially metabolism. Since less evidence demonstrated the exact mechanism, which one is the innate receptor is still unknown.[5]

1. TWO PROPOSED OBESTATIN RECEPTORS

GPR39 was first cloned from a fetal brain cDNA library, by hybridization at low stringency.[6] Due to the sequence similarity, GPR39, along with another GPCR cloned at that time, GPR38, was considered related to the growth hormone secretagogue receptor (GHS-R), whose agonists included synthetic peptides and nonpeptide growth hormone (GH) secretagogues.[6] Later the natural endogenous ligand for GHS-R was discovered, named ghrelin,[7] and GPR39 was speculated to have a ligand of a peptide regulating gastrointestinal functions, since all the other receptors homologous to GHS-R have such ligands.[2] GPR39 had been considered, however, an orphan GPCR until 2005, when Zhang and colleagues discovered obestatin encoded by the *ghrelin* gene. Based on the data of radiolabeled ligand binding assays, it was proposed that the new hormone was a cognate ligand of GPR39.[3] The relationship of obestatin and GPR39 was later challenged by Chartrel, who stated that obestatin did not bind GPR39 and induce any effects on GPR39-transfected cells.[8] In response, Zhang and colleagues found out that the interaction of obestatin and GPR39 could not be reproduced due to the instable bioactivity of iodinated obestatin.[9] Just followed up their response about the bioactivity of obestatin, Zhang and colleagues found out that mono-Iodinated obestatin can bind to GPR39 and GPR39 SNP, but not over-iodinated obestatin [13]. During that time, two groups, Holst *et al.* and Lauwers *et al.*, discovered that zinc ion (Zn^{2+}) is a physiologically relevant agonist of GPR39.[10,11] Both groups emphasized that obestatin did neither bind to GPR39-expressing cells nor display any effect such as cAMP production. Holst *et al.* stated that GPR39 might signal through Gs pathway if Zn^{2+} presented in addition to Gq and G12/13 pathways that were constitutively activated.[10] Lauwers *et al.* showed that the Zn^{2+} response was not affected by obestatin. But there was still some evidence for the role of obestatin as a GPR39 agonist. Moechars *et al.* constructed GPR39-knockout mice and partially confirmed the *in vivo* effects of obestatin depending on GPR39.[12] The knockout mice's differences in energy homeostasis included: accelerated gastric emptying, increased gastric secretion, higher mature body weights and body fat compositions, reduced hyperphagia after fasting, and increased cholesterol levels. Zhang *et al.* demonstrated that *c-fos* expression was induced by obestatin in wild-type mice but not in GPR39 null mice, suggesting that activities of obestatin depended on GPR39.[13] In addition, various studies on bone metabolism[14] and term pregnancy[15] have shown the correlation of expression levels of obestatin and GPR39. Thus, the role of GPR39 remains unresolved.

The other candidate receptor GLP-1R has already gained attention before the discovery of obestatin because it is the receptor for the glucagon-like peptide-1 (GLP-1), which highly stimulates insulin secretion and results in antidiabetogenic effects. The cDNA of GLP-1R was first isolated from a pancreatic islet cDNA library from rats, and its function as a receptor for GLP-1 was confirmed by binding experiments with radiolabeled GLP-1.[16] The cDNA of GLP-1R of human was isolated by several groups in 1993.[17–19] At the same time, two exogenous ligands were identified, which were exendin-4 (Ex-4) and exendin-(9–39).[19] The interation of GLP-1R to obestatin was revealed by Granata *et al.* in 2008.[4] They observed that obestatin bound to GLP-1R and promoted β-cell and islet cell survival. The binding of radiolabeled GLP-1 to the cell membrane was slightly inhibited by motilin, GLP-1, acylated ghrelin, [D-Lys^3]-GHRP-6, and Ex-9, and obestatin dose-dependently inhibited the binding of GLP-1 to the cell membrane.[4] Thus, the binding assay indicated that obestatin and GLP-1 recognized the same binding sites.[4] Furthermore, obestatin did not influence GLP-1 secretion, thus the promotion of cell survival was independent of GLP-1.[4] Favaro *et al.* later confirmed that the anti-apoptotic effects of obestatin and exendin-4 involved PI3K/Akt, ERK1/2, and cAMP/PKA pathways.[20] A study on the metabolic effects of obestatin in rats yielded several negative results that obestatin neither bound to cells overexpressing GLP-1R, nor altered insulin release from islets, where both GLP-1R and GPR39 should be expressed.[21] These controversies on both the obestatin receptor and its related biological functions highlight the urgency of the introduction of study aspects other than cell biology and physiology.

1.1. Phylogenetic relationship of GPR39 and GLP-1R within their GPCR families

GPR39 and GLP-1R are both GPCRs. GPCR is a seven-transmembrane protein, with a N-terminal extracellular region and three extracellular loops recognizing a ligand molecule and a C-terminal intracellular region and three intracellular loops interacting with a G protein. Although all GPCRs possess this structural feature, they are results of convergent evolution. GPR39 and GLP-1R belong to two different families sharing little sequence similarity.

GPR39 is characterized as a member of family A, or rhodopsin-like GPCR. The sequence of GPR39 is most related to the ghrelin receptor (or GHS-R), the neurotensin receptors, the motilin receptor (GPR38), and neuromedin-U receptors.[6] The ghrelin receptor is the cognitive receptor for the peptide hormone ghrelin, which is encoded by the same gene of obestatin but with opposite roles to obestatin in the control of food intake.[22] The two neurotensin receptors, NTS-R1 and NTS-R2, are receptors of a tridecapeptide neurotensin.[23,24] Neurotensin is a cleavage segment of a 17 kDa precursor protein, which is expressed in the brain and gut.[24] Neurotensin and its two receptors are correlated to the regulation of dopaminergic neurotransmissions.[24] The motilin receptor was once known as GPR38, cloned at the same time of GPR39.[6,25] Its ligand, the motilin peptide, regulates interdigestive gastrointestinal motility by stimulating rhythmic contraction of the smooth muscle.[25] The two neuromedin-U receptors are NMU-R1 and NMU-R2. Neuromedin-U, similar to motilin, induces smooth muscle contraction. NUM-R1 is expressed in greatest abundance in the gastrointestinal and urogenital systems but at a much lower level within the central nervous system, and NUM-R2 is expressed in a complementary manner, lower in peripheral tissues but higher in the central nervous system.[26]

GLP-1R belongs to GPCR family B, or the secretin-like family, characterized by the three disulfide bonds in the N-terminal domain.[27] Ligands for receptors in this family are large peptides with paracrine activities.[28] By sequence similarity, GLP-1R forms a subgroup with GLP-2R, GCGR (the glucagon receptor), and GIPR (the gastric inhibitory polypeptide receptor).[28] These receptors, together with their ligands, are all essential in maintaining glucose homeostasis. GLP-2R is stimulated by GLP-2, a 33-amino acid peptide cleaved from proglucagon,[29] which also encodes glucagon and GLP-1. GLP-2R does not respond to GLP-1.[29] GCGR recognizes glucagon and plays a role in controlling the production of hepatic glucose and insulin secretion, so as to regulate the level of blood glucose.[30] GIPR is a receptor for GIP,[31] a hormone that stimulates insulin secretion and thus is a potential drug treating type 2 diabetes.[32]

1.2. Gene, expression and structure of GPR39 and GLP-1R

The human *GPR39* gene is located in chromosome 2, band q21–q22,[6] consisting of two exons.[33] The first exon encodes residues 1 to 285, corresponding to the N-terminal region and the first five transmembrane helices, and the second exon encodes residues 286 to 435, corresponding to the last two transmembrane helices and the C-terminal region.[34] Between the two exons there is an intron as large as 200 kb, whose analogue appears in the genes of the ghrelin receptor and the motilin receptor.[34] The second exon of *GPR39* shares a segment of DNA with *LYPD1* gene[34]: GPR39 is encoded by the forward strand of DNA, while LYPD1 is transcribed from the reverse strand. The two-exon structure and *LYPD1* gene-overlapping pattern are conserved among several species.[34] Beside the canonical form of GPR39 (named GPR39-1a), which is 435-amino acid long and has seven-transmembrane helices,[6] a splice variant GPR39-1b has been identified, which is encoded by the first exon and thus truncated after the fifth transmembrane helix.[34]

The maximal promoter activity resides in the 5′-UTR region. Such a region in U-138 cell lines was reported to be from −673 to −14 from the translation initiation codon, while in HepG2 cell from −573 to −14.[34] Although the negative regulatory elements have not yet been determined, there was such an element from −673 to −573 in HepG2 cells. The translation initiation codon is at 469 nucleotides downstream of the transcriptional start site.[34]

As a product of transcription, alternative splicing, and translation, full-length GPR39-1a is a protein with a molecular weight of 52 kDa[15]; another product, the GPR39-1b protein, with a molecular weight of 32 kDa, is identical in sequence to GPR39-1a for residues 1 to 285, and after residue 285, its sequence is Ser-Glu-Ser and end, which is different from the corresponding sequence of GPR39-1a.[35]

GPR39 is expressed in many organs and various tissues, especially in stomach, white adipose tissue, and hypothalamus.[6,34,36] GPR39-1b is widely expressed but may be nonfunctional. Egerod *et al.* pointed out that some studies before might have been misled by the unawareness of GPR39-1b and LYPD1. Both have a region of sequence overlapping with GPR39-1a, the canonical and functional receptor.[34] The level of constitutive activity of GPR39-1a could be affected by the splicing producing GPR39-1b, which may be an implication of the conservation of the two-exon structure and *LYPD1* gene-overlapping pattern.[34] Furthermore, there are GPCRs that have splice variants like GPR39. For the luteinizing hormone receptor, the calcitonin receptor, and α1A-adrenoceptor, the variants with shorter sequences interact, perhaps by forming heterodimers, with the full-length receptors and alter their functions.[35] Because GPR39, similar to related GPCRs such as the ghrelin receptor, keeps a rather high constitutive, ligand-independent activity, the expression of GPR39-1b may be a way to change the signaling of GPR39-1a.

The human *GPL-1R* gene has been mapped to chromosome 6, band p21.1,[37] including 13 exons.[38] Along with the gene, there are two regions with highly polymorphic simple sequence repeat, named CA1 and CA3.[39] The simple tandem repeat DNA polymorphism adopts the form of $(TG)_n$[37] and discriminates many alleles. When these two repeats were identified, 14 alleles for CA1 were observed in African-Americans and six in Caucasians, and 16 alleles for CA3 were observed in African-Americans and eight in Caucasians.[39] Various analyses of allelic frequencies in noninsulin-dependent diabetes mellitus patients negate the hypothesis that inherited defects in *GLP-1R* in African-Americans and Caucasians determine the genetic susceptibility to noninsulin-dependent diabetes mellitus.[39]

In the 5′-UTR of the *GLP-1R* gene, three binding sites for transcript factor Sp1 have been localized.[40] By experiments with constructed mutants, mutation of the two proximal sites, 123 and 58 nucleotides upstream from the transcription start site, were shown to be relevant to the reduction of activity, while the distal one, 339 nucleotides upstream from the transcription start site, to be a *cis*-regulatory-element as a repressor. Both Sp1 and Sp3 could bind to these sites, involved in the regulation of the expression.[40] Besides, there is yet another distal cell-specific silencer element. This element lies between −2285 to −2235 in the 5′-flanking region of the *GLP-1R* gene. The silencer element may function for tissue- and cell-specific expression of GLP-1R, because Sp1 is a ubiquitous transcription factor maintaining basal promoter activity. It was reported that the transcription of the human *GLP-1R* gene was repressed in fibroblasts and pancreatic A-cells, and not pancreatic B- and D-cells.[41]

Structural study on GPCR family B is less bustling than that on GPCR family A. There are no experimentally determined structures of full-length receptors of family B, but structures of the N-terminal extracellular domains of several receptors, including GLP-1R and GIPR. These structures reveal detailed insights of the receptors, for example, the structural mechanism of the recognition between ligands and receptors. In the crystal structure of GPL-1R extracellular domain bound with an antagonist exendin-4 (9–39),[27] the N-terminal domain of GLP-1R embraces the antagonist peptide, forming an amphipathic α-helix, by both hydrophobic and hydrophilic interactions. Similar to other solved structures of GPCR family B, the core of the N-terminal domain of GPL-1R possesses three conserved disulfide bonds and two regions of antiparallel β-sheets.[27] The hydrophobic binding site is defined by several α-helices and a loop linking two antiparallel β-strands. In contrast to a hypothesis[42] suggested by experiments with truncated analogues of exendin-4 and explaining the superior affinity of exendin-4, the Trp cage of exendin-4, which is a segment of nine amino acid residues in the C-terminus of exendin-4, does not participate the in interaction with GLP-1R.[27] The binding pattern of exendin-4 with GLP-1R is similar to that of other GPCRs of family B with their ligands, for example, GIPR and GIP.[43] The crystal structure displaying the binding pattern confirms a two-step model of the binding mechanism[44]: The first step is the C-terminal region of the peptide ligand interacts with the N-terminal extracellular domain of the GPCR, adjusting the orientation and may be secondary structure; the second step is the N-terminal region of the ligand touches the other part of the GPCR, especially the transmembrane helices, leading to signal transduction. Structure of the other portion of GLP-1R is unknown and believed similar to GPCR of family A.[43] There is still the forth disulfide bond formed by residue Cys226 and Cys296, in which Cys296 belongs to the third extracellular helix and is close to the activation pocket because mutation of Cys226 to alanine decreases GLP-1 potency while mutation of Cys296 to alanine in addition restores the lost potency.[45] In the 463-amino acid-long sequence, residues Cys341 and Arg348 are conserved and could be sites of mono-ADP-ribosylation.[46]

GLP-1R was first known as a regulator of insulin secretion and mainly expressed in pancreatic β-cells. In other tissue, GLP-1R mRNA can be identified in lung, stomach, kidney, hypothalamus, brain, and heart, but not in skeletal muscle, liver, and adipose.[47,48]

1.3. Signal transduction across cell membrane for GPR39 and GLP-1R

GPCR mediates signaling by association with G proteins and involves second messengers such as cAMP and calcium ions (Ca^{2+}). Some signaling process of GPR39 might be triggered by obestatin or zinc ions, beside pathways that are thought to be constitutively activated.[10] The obestatin-binding sites have not yet been identified, although a Gs-mediated signaling pathway has been proposed.[3] In cultured preadipocytes, obestatin bound to GPR39 is reported to stimulate the phosphorylation of ERK1/2 and potentially Akt, the downstream signaling partners.[4,13] Obestatin induces the expression of *c-fos*, a transcription factor as a marker of activity in gastrointestinal and other tissues. If tested with *GPR39*-null mice, the obestatin-induced *c-fos* expression was abolished, confirming the role of GPR39 in inducing the action of obestatin.[13] The zinc ion-binding sites are residues His17 and His19 according to mutagenesis experiments.[49] If Zn^{2+} is present, GPR39 might also signal through Gs pathway, in addition to Gq and G12/13 pathways that are constitutively activated.[10] It remains unknown, however, if Zn^{2+} is only an allosteric modulator enhancing the function of another hormone.[10]

GLP-1R is a GPCR for GLP-1,[17] a hormone playing a key role in incretin effect. The term "incretin" is used to denote the fact that a plasma insulin response can be greater to oral glucose load than to an isoglycemic intravenous glucose infusion.[50] Incretin effect is impeded in patients with type 2 diabetes, thus incretin-based therapies, for example, with GLP-1R agonists such as Liraglutide (approved by Europe in 2009), seem promising for the treatment of the disease.[51] By analysis of mice with null mutations, it was demonstrated that loss of GLP-1R action results in impaired insulin response to oral glucose.[52] GLP-1R couples to Gs, Gq/11, and Gi1/2.[53] Via the coupling, it can trigger the activation of phospholipase C[54] and MAPK,[53] as well as adenylyl cyclase,[55] leading to increases of cytosolic calcium. The Lys334-Leu335-Lys336 sequence within the third intracellular loop is a putative site for G protein coupling, for deletion of the sequence abolishes the activation of adenylyl cyclase and insulin secretion.[56–58] In this loop, residues V327, I328, V331, and K334 may be also essential in interaction with G protein and stimulation of cAMP production, as shown in an alanine scanning experiment.[59] These residues were located in one face of an α-helix in a helical wheel model, suggesting that the secondary structure of the third intracellular region consists of an α-helix.[59] In studies with synthetic peptides analogous to this region, the peptide of the entire loop strongly activated pertussis toxin- and cholera toxin-sensitive G proteins.[58] By measurements of GTPase activity stimulated by these analogues, the effect of the third intracellular loop was further divided into two parts: the N-terminal half for activating cholera toxin-sensitive G proteins (Gs), and the C-terminal half for activating pertussis toxin-sensitive G proteins (Gi).[58] This division provides an explanation of the existence of several G protein partners of GLP-1R.

The interaction of obestatin and GLP-1R was first identified by the transduction of survival signals.[4] In the study, obestatin showed specific binding on HIT-T15 cells and INS-1E β-cells, both of which were from islet β-cells and induced the proliferation and survival of serum-starved and cytokine-treated cells. With the treatment of obestatin, intracellular cAMP concentrations in INS-1E cell increased in 30 min and then decreased but still were higher than control at 60 and 90 min.[4] Any one of the antagonist of the Gs protein, inhibitor of adenylyl cyclase, and inhibitor of protein kinase A blocked the anti-apoptotic effect induced by obestatin. In addition, obestatin promoted ERK1/2 and Akt phosphorylation but not Src kinase, suggesting that the anti-apoptotic effect of obestatin on β-cells depended on adenylyl cyclase/cAMP/PKA, PI3K/Akt, and ERK1/2.[4]

2. OBESTATIN AND ITS RECEPTOR(S) IN ENERGY HOMEOSTASIS

Two independent studies revealed obestatin immunoreactivity in cells of the gastric mucosa, myenteric plexus, and in Leydig cells of the testis[60] and in the stomach, duodenum, jejunum, colon, and pancreas[61] of Sprague–Dawley rats, respectively. The obestatin immunoreactivity in human tissue was localized to the mucosa of the gastrointestinal tract, the pancreatic islets, and the epithelial cells in the ducts of mammary glands.[62] The GPR39 immunoreactivity was localized to gastric mucosa, intestinal villi, white adipose tissues, hepatic cords, and kidney tubules of adult mice.[13] The colocalization of immunoreactive obestatin and its receptor GPR39 in the gastric mucosa and intestinal tissues implies its involvement in energy homeostasis.

Obestatin may play a role in energy homeostasis through different physiological processes, like modulating food intake and gastrointestinal motility, GH secretion, and lipid metabolism.

2.1. Food intake

In the first report of obestatin, Zhang et al. showed that the intraperitoneal injection (125 and 1000 nmol/kg) and intracerebroventricular treatment (8 nmol/kg) of obestatin in adult male mice decreased food intake.[3] Results also showed that treatment with obestatin for 5, 6, 7, and 8 days (1000 nmol/kg, three times a day) in adult male rats resulted in suppressed body weight gain.[3] Then many groups repeated its anorectic effects in a number of experimental paradigms after both peripheral and central administration of obestatin. Bresciani *et al.* showed that obestatin inhibited feeding after short-term starvation in young adult male rats.[63] Green *et al.* reported a 43% reduction in food intake after intraperitoneal obestatin injection 4 h prior to an allowed 15 min period of feeding in mice.[64] Brunetti *et al.* also repeated its anorectic effect by intraperitoneal daily injection of obestatin for 12 days. The decrease in food intake and weight gain became evident on day 6 and day 8, respectively, and maintained through to day 12.[65] Sibilia *et al.* reported that acute intracerebroventricular administration of obestatin in 16 h-starved rats induced a trend toward a reduction of food intake but not statistically significant. Chronic intracerebroventricular administration of obestatin for 1 day significantly decreased food intake but this effect vanished thereafter for the next 2–28 days.[66] Based on their study, Sibilia *et al.* concluded that the effect of obestatin on food intake tends to be short-term instead of long-term influence.[66]

Since administration of ghrelin induces potent stimulating effects on food intake, the relation between obestatin and ghrelin in food intake was also studied. Though Zizzari *et al.* found no difference in the food intake at 1, 3, 5, and 18 h postintraperitoneal injection of 1000 nmol/kg obestatin in either 24 h fasted mice or normal diet mice, they showed that obestatin inhibited ghrelin-induced orexigenic effect in normal diet mice.[67]

On the other hand, several researchers were not able to confirm the effect of obestatin on food intake,[21,68–71] thus the role of obestatin as an anorectic hormone was challenged.[72] The reason for the discrepancies is unclear. The different experimental paradigms such as different animal species, methods of administration, and the duration of drug treatment, might have influenced the outcome of the effect of obestatin on food intake in some way. Also, a recent study questioning the purity of synthetic peptides including obestatin might shed a clue on these discrepancies.[73]

2.2. Gastrointestinal motility

Not only food intake Zhang et al. also showed that jejunal contraction was inhibited and obestatin treatment resulted in a sustained suppression of gastric emptying activity.[3] Ataka *et al.* showed that intravenous injection of obestatin inhibited motor activity in the stomach antrum and duodenum of freely moving conscious rats and that CRF type 1 and type 2 receptors in the brain might be involved in these effects of obestatin on gastroduodenal motility.[74] These results were reproduced in another study using freely moving conscious rat models.[75]

Some other studies also addressed the effect of obestatin on ghrelin-induced gastrointestinal motility. Depoortere *et al.* reported that peripheral administration of obestatin showed no effect on fundic strip contractions but cotreatment of obestatin tended to reduce the excitatory effect of ghrelin in mice.[76] However, another study reported that obestatin failed to antagonize ghrelin-induced stimulation of gastroduodenal motility.[74]

In summary, the effect of obestatin on gastrointestinal motility remains controversial. The effects of obestatin in gastric emptying and the contractility of the strips from fundus and small intestine of mice and rats were not reproduced by De Smet *et al.*.[77]

2.3. Growth hormone secretion

Zhang et al. showed that unlike ghrelin, obestatin did not increase *in vitro* GH secretion by cultured rat pituitary cells,[3] which was confirmed by other groups. Bresciani *et al.* showed that obestatin had no effect on GH secretion or GH-stimulated hexarelin production or in spontaneous corticosterone secretion in 10-day-old rats.[63] Nogueiras *et al.* reported no effect of obestatin treatment not only on GH release but also some hypothalamic neuropeptides involved in energy balance regulation.[68] However, Zizzari *et al.* reported obestatin (10 μg/rat)-affected ghrelin-induced GH secretion *in vivo* though obestatin did not affect spontaneous or ghrelin-induced GH release from superfused pituitaries ex vivo or basal GH levels *in vivo*.[67]

2.4. Lipid metabolism

Another physiological role of obestatin might be as a regulator of adipocyte metabolism and adipogenesis. Zhang and colleagues showed that treatment of preadipocytes 3T3-L1 cells with obestatin induced an early-response gene *c-fos* and stimulated ERK1/2 phosphorylation.[13] Then the study by

Gurriarán-Rodríguez *et al.* supported a role of obestatin in promoting adipogenesis and adipocyte metabolism regulation.[78] Obestatin activated Akt and its downstream targets while it inactivated AMPK and decreased ACC phosphorylation in 3T3-L1 cells. Obestatin treatment together with IBMX and DEX was shown to regulate the expression of C/EBPα, C/EBPβ, C/EBPδ, and PPARγ promoting adipogenesis. The involvement of obestatin in cell metabolism and adipogenesis was confirmed *in vivo* in the white adipose tissue obtained after 24 and 72 h continuous subcutaneous infusion of obestatin.[78] Miegueu *et al.* reported that obestatin stimulated cell differentiation and increased fatty acid uptake in a dose-dependent manner while obestatin decreased lipolysis in 3T3-L1 cells.[79] *in vivo* study showed that plasma triglyceride levels were significantly reduced by a chronic 14-day treatment with a stable obestatin analog, N-terminally PEGylated obestatin.[80] Cholesterol levels were increased significantly in the putative obestatin receptor GPR39 null mice compared with wild-type mice.[12] These results support a role of obestatin in lipid metabolism.

2.5. *GPR39* null mice

GPR39 knockout mice may reveal the physiological roles of obestatin/GPR39. Moechars *et al.* showed that *GPR39* null mice showed accelerated gastric emptying, increased volume of gastric secretion, higher mature body weight and body fat composition, and increased cholesterol levels compared with wild type.[12] The report also showed that food intake during day 1 was not different between the two genotypes while food intake after fasting was decreased in *GPR39* null mice.[12] Then Zhang and colleagues further showed that treatment with obestatin induced an early-response gene *c-fos* expression in gastric mucosa of wild-type, but not *GPR39* null, mice, underscoring a mediating role of this receptor in obestatin actions.[13] The altered gastric activity, body weight, body fat composition, and cholesterol levels in *GPR39* null mice might be due to the lost of obestatin biological function due to its missing receptor. All these supported a role of obestatin-GPR39 system in energy homeostasis.

2.6. Obesity pathogenesis

Obesity with its increasing prevalence becomes a global epidemic problem. In the United States, more than 20% of adults are clinically obese as defined by a body mass index of 30 kg/m^2 or higher, and an additional 30% are overweight.[81] As a new proposed anorectic peptide hormone, the association of obestatin with obesity and obesity-related diseases draws great interest ever since its discovery. Considering the involvement of both ghrelin and obestatin in energy homeostasis, both the plasma obestatin level and plasma ghrelin/obestatin ratio were used to assess the possible involvement of obestatin in obesity pathogenesis. Several groups reported decreased plasma obestatin level in obese adults than normal weight control[15,82–86] among which Guo *et al.* detected an increased ghrelin to obestatin ratio and Huda *et al.* detected no difference in ghrelin to obestatin ratio between obese and lean subjects[1] accompanied by decreased obestatin level. On the contrary, Vicennati *et al.* reported a higher obestatin level in obese women compared to control together with a decreased ghrelin to obestatin ratio.[87] Zou *et al.* reported no significant difference in plasma obestatin level between obese and normal weight children with a decreased ghrelin to obestatin ratio.[88]

Due to the heterogeneity in the relationship between plasma obestatin levels and obesity, Zhang *et al.* performed a meta-analysis on the related studies with obestatin hormone level with the published work before December 2009.[89] Concerning obestatin levels in the nine studies with 566 participants, obestatin of the obese groups is lower than that of the normal weight groups at 64.19 pg/mL [95% confidence interval (CI) = 36.11–92.26, $P < 0.01$]. Concerning the ghrelin/obestatin ratio in the five studies with 259 participants, the ratios of obese group were lower than those of the normal weight group although not statistically significant.[89]

Not only the plasma obestatin level was showed to be associated with obesity, but gastric body mucosa obestatin expression was also reported to decrease in overweight and obese patients by immunohistochemistry.[90] The association between obestatin levels (both circulating and gastric level) with obesity all suggested a possible role for stable obestatin level or balanced obestatin/ghrelin ratio in long-term body weight regulation.

The effects of weight loss surgery or exercise on obestatin level might give us another perspective to decipher the relationship between obestatin and obesity. Changes in plasma obestatin levels after weight loss through either Roux-en-Y gastric bypass (RYGB) surgery or weight reduction in summer camp were studied. Roth *et al.* reported that ghrelin concentrations and ghrelin/obestatin ratios were lower in patients 2 years after RYGB surgery than before, while obestatin levels did not change.[91] Martins *et al.* reported that both ghrelin and obestatin plasma levels were significantly elevated in the patients 3 years after the RYGB surgery as compared to the control group and no statistically significant difference in ghrelin/obestatin ratio was detected among the two groups.[92] In the study by Huda *et al.*, obestatin tended to be lower in gastrectomy subjects and their obestatin/ghrelin ratio differed from healthy controls.[86] A 24 h to 7-day time-course study on the effect of short-term anaerobic exercise training was performed in college women though no effect of the exercise on the circulating obestatin levels was observed.[93]

Accumulating evidence piled up for the involvement of obestatin in energy homeostasis and its association with obesity. But more investigation is needed to decipher the detailed mechanism of its physiological and pathological roles including the innate receptor(s) for obestatin, from which we can understand more about the regulation of energy homeostasis and hopefully provide candidate(s) for obesity-related treatment.

ACKNOWLEDGMENTS

This work was supported by the following grants: One-hundred Talent Program of Chinese Academy of Sciences (to J. V. Z.), The National Basic Research Program of China (973 program 2013CB945503), Distinguished Young Scholars Program of The National Natural Science Foundation of China (NSFC 31100850 to J. V. Z.), The Distinguished Young Scholars of Shenzhen City (SW201110059 to J. V. Z.), Shenzhen Engineering Laboratory of Single-molecule Detection and Instrument Development and Innovative Pharmacology and Biotherapy Pre-clinical Test Public Service Platform, Shenzhen.

REFERENCES

1. Inui A. Ghrelin: an orexigenic and somatotrophic signal from the stomach. *Nat Rev* 2001;**2**:551–60.
2. Kojima M, Kangawa K. Ghrelin: structure and function. *Physiol Rev* 2005;**85**:495–522.
3. Zhang JV, Ren PG, Avsian-Kretchmer O, Luo CW, Rauch R, Klein C, et al. Obestatin, a peptide encoded by the ghrelin gene, opposes ghrelin's effects on food intake. *Science* 2005;**310**:996–9.
4. Granata R, Settanni F, Gallo D, Trovato L, Biancone L, Cantaluppi V, et al. Obestatin promotes survival of pancreatic beta-cells and human islets and induces expression of genes involved in the regulation of beta-cell mass and function. *Diabetes* 2008;**57**:967–79.
5. Lacquaniti A, Donato V, Chirico V, Buemi A, Buemi M. Obestatin: an interesting but controversial gut hormone. *Ann Nutr Metab* 2011;**59**:193–9.
6. McKee KK, Tan CP, Palyha OC, Liu J, Feighner SD, Hreniuk DL, et al. Cloning and characterization of two human G protein-coupled receptor genes (GPR38 and GPR39) related to the growth hormone secretagogue and neurotensin receptors. *Genomics* 1997;**46**:426–34.
7. Kojima M, Hosoda H, Matsuo H, Kangawa K. Ghrelin: discovery of the natural endogenous ligand for the growth hormone secretagogue receptor. *Trends Endocrinol Metab* 2001;**12**:118–22.
8. Chartrel N, Alvear-Perez R, Leprince J, Iturrioz X, Reaux-Le Goazigo A, Audinot V, et al. Comment on "Obestatin, a peptide encoded by the ghrelin gene, opposes ghrelin's effects on food intake" *Science* 2007;**315**:766 [author reply 766].

9. Zhang JV, Klein C, Ren PG, Kass S, Donck LV, Moechars D, et al. Response to Comment on "Obestatin, a peptide encoded by the ghrelin gene, opposes ghrelin's effects on food intake" *Science* 2007;**315**:766.
10. Holst B, Egerod KL, Schild E, Vickers SP, Cheetham S, Gerlach LO, et al. GPR39 signaling is stimulated by zinc ions but not by obestatin. *Endocrinology* 2007;**148**:13–20.
11. Lauwers E, Landuyt B, Arckens L, Schoofs L, Luyten W. Obestatin does not activate orphan G protein-coupled receptor GPR39. *Biochem Biophys Res Commun* 2006;**351**:21–5.
12. Moechars D, Depoortere I, Moreaux B, de Smet B, Goris I, Hoskens L, et al. Altered gastrointestinal and metabolic function in the GPR39-obestatin receptor-knockout mouse. *Gastroenterology* 2006;**131**:1131–41.
13. Zhang JV, Jahr H, Luo CW, Klein C, Van Kolen K, Ver Donck L, et al. Obestatin induction of early-response gene expression in gastrointestinal and adipose tissues and the mediatory role of G protein-coupled receptor, GPR39. *Mol Endocrinol* 2008;**22**:1464–75.
14. Pacheco Pantoja EL, Ranganath LR, Gallagher JA, Wilson PJ, Fraser WD. Receptors and effects of gut hormones in three osteoblastic cell lines. *BMC Physiol* 2011;**11**:12.
15. Fontenot E, DeVente JE, Seidel ER. Obestatin and ghrelin in obese and in pregnant women. *Peptides* 2007;**28**:1937–44.
16. Thorens B. Expression cloning of the pancreatic beta cell receptor for the gluco-incretin hormone glucagon-like peptide 1. *Proc Natl Acad Sci USA* 1992;**89**:8641–5.
17. Dillon JS, Tanizawa Y, Wheeler MB, Leng XH, Ligon BB, Rabin DU, et al. Cloning and functional expression of the human glucagon-like peptide-1 (GLP-1) receptor. *Endocrinology* 1993;**133**:1907–10.
18. Graziano MP, Hey PJ, Borkowski D, Chicchi GG, Strader CD. Cloning and functional expression of a human glucagon-like peptide-1 receptor. *Biochem Biophys Res Commun* 1993;**196**:141–6.
19. Thorens B, Porret A, Buhler L, Deng SP, Morel P, Widmann C. Cloning and functional expression of the human islet GLP-1 receptor. Demonstration that exendin-4 is an agonist and exendin-(9–39) an antagonist of the receptor. *Diabetes* 1993;**42**:1678–82.
20. Favaro E, Granata R, Miceli I, Baragli A, Settanni F, Cavallo Perin P, et al. The ghrelin gene products and exendin-4 promote survival of human pancreatic islet endothelial cells in hyperglycaemic conditions, through phosphoinositide 3-kinase/Akt, extracellular signal-related kinase (ERK)1/2 and cAMP/protein kinase A (PKA) signalling pathways. *Diabetologia* 2012;**55**:1058–70.
21. Unniappan S, Speck M, Kieffer TJ. Metabolic effects of chronic obestatin infusion in rats. *Peptides* 2008;**29**:1354–61.
22. Kojima M, Hosoda H, Date Y, Nakazato M, Matsuo H, Kangawa K. Ghrelin is a growth-hormone-releasing acylated peptide from stomach. *Nature* 1999;**402**:656–60.
23. Vita N, Laurent P, Lefort S, Chalon P, Dumont X, Kaghad M, et al. Cloning and expression of a complementary DNA encoding a high affinity human neurotensin receptor. *FEBS Lett* 1993;**317**:139–42.
24. Vita N, Oury-Donat F, Chalon P, Guillemot M, Kaghad M, Bachy A, et al. Neurotensin is an antagonist of the human neurotensin NT2 receptor expressed in Chinese hamster ovary cells. *Eur J Pharmacol* 1998;**360**:265–72.
25. Feighner SD, Tan CP, McKee KK, Palyha OC, Hreniuk DL, Pong SS, et al. Receptor for motilin identified in the human gastrointestinal system. *Science* 1999;**284**:2184–8.
26. Raddatz R, Wilson AE, Artymyshyn R, Bonini JA, Borowsky B, Boteju LW, et al. Identification and characterization of two neuromedin U receptors differentially expressed in peripheral tissues and the central nervous system. *J Biol Chem* 2000;**275**:32452–9.

27. Runge S, Thogersen H, Madsen K, Lau J, Rudolph R. Crystal structure of the ligand-bound glucagon-like peptide-1 receptor extracellular domain. *J Biol Chem* 2008;**283**: 11340–7.
28. Fredriksson R, Lagerstrom MC, Lundin LG, Schioth HB. The G-protein-coupled receptors in the human genome form five main families. Phylogenetic analysis, paralogon groups, and fingerprints. *Mol Pharmacol* 2003;**63**:1256–72.
29. Munroe DG, Gupta AK, Kooshesh F, Vyas TB, Rizkalla G, Wang H, et al. Prototypic G protein-coupled receptor for the intestinotrophic factor glucagon-like peptide 2. *Proc Natl Acad Sci USA* 1999;**96**:1569–73.
30. Fujisawa T, Ikegami H, Yamato E, Takekawa K, Nakagawa Y, Hamada Y, et al. A mutation in the glucagon receptor gene (Gly40Ser): heterogeneity in the association with diabetes mellitus. *Diabetologia* 1995;**38**:983–5.
31. Yamada Y, Hayami T, Nakamura K, Kaisaki PJ, Someya Y, Wang CZ, et al. Human gastric inhibitory polypeptide receptor: cloning of the gene (GIPR) and cDNA. *Genomics* 1995;**29**:773–6.
32. Alana I, Hewage CM, Malthouse JP, Parker JC, Gault VA, O'Harte FP. NMR structure of the glucose-dependent insulinotropic polypeptide fragment, GIP(1–30)amide. *Biochem Biophys Res Commun* 2004;**325**:281–6.
33. Nogueiras R, Tschop M. Biomedicine. Separation of conjoined hormones yields appetite rivals. *Science* 2005;**310**:985–6.
34. Egerod KL, Holst B, Petersen PS, Hansen JB, Mulder J, Hokfelt T, et al. GPR39 splice variants versus antisense gene LYPD1: expression and regulation in gastrointestinal tract, endocrine pancreas, liver, and white adipose tissue. *Mol Endocrinol* 2007;**21**: 1685–98.
35. Popovics P, Stewart AJ. GPR39: a Zn(2+)-activated G protein-coupled receptor that regulates pancreatic, gastrointestinal and neuronal functions. *Cell Mol Life Sci* 2011;**68**: 85–95.
36. Jackson VR, Nothacker HP, Civelli O. GPR39 receptor expression in the mouse brain. *Neuroreport* 2006;**17**:813–6.
37. Stoffel M, Espinosa 3rd R, Le Beau MM, Bell GI. Human glucagon-like peptide-1 receptor gene. Localization to chromosome band 6p21 by fluorescence in situ hybridization and linkage of a highly polymorphic simple tandem repeat DNA polymorphism to other markers on chromosome 6. *Diabetes* 1993;**42**:1215–8.
38. Tokuyama Y, Matsui K, Egashira T, Nozaki O, Ishizuka T, Kanatsuka A. Five missense mutations in glucagon-like peptide 1 receptor gene in Japanese population. *Diabetes Res Clin Pract* 2004;**66**:63–9.
39. Tanizawa Y, Riggs AC, Elbein SC, Whelan A, Donis-Keller H, Permutt MA. Human glucagon-like peptide-1 receptor gene in NIDDM. Identification and use of simple sequence repeat polymorphisms in genetic analysis. *Diabetes* 1994;**43**:752–7.
40. Wildhage I, Trusheim H, Goke B, Lankat-Buttgereit B. Gene expression of the human glucagon-like peptide-1 receptor is regulated by Sp1 and Sp3. *Endocrinology* 1999;**140**: 624–31.
41. Galehshahi FS, Goke B, Lankat-Buttgereit B. A novel silencer element repressing expression of the GLP-1 receptor gene in fibroblasts and pancreatic A-cells, but not in pancreatic B- and D-cells. *Peptides* 2000;**21**:1169–76.
42. Al-Sabah S, Donnelly D. A model for receptor-peptide binding at the glucagon-like peptide-1 (GLP-1) receptor through the analysis of truncated ligands and receptors. *Br J Pharmacol* 2003;**140**:339–46.
43. Parthier C, Kleinschmidt M, Neumann P, Rudolph R, Manhart S, Schlenzig D, et al. Crystal structure of the incretin-bound extracellular domain of a G protein-coupled receptor. *Proc Natl Acad Sci USA* 2007;**104**:13942–7.

44. Lopez de Maturana R, Willshaw A, Kuntzsch A, Rudolph R, Donnelly D. The isolated N-terminal domain of the glucagon-like peptide-1 (GLP-1) receptor binds exendin peptides with much higher affinity than GLP-1. *J Biol Chem* 2003;**278**:10195–200.
45. Mann RJ, Al-Sabah S, de Maturana RL, Sinfield JK, Donnelly D. Functional coupling of Cys-226 and Cys-296 in the glucagon-like peptide-1 (GLP-1) receptor indicates a disulfide bond that is close to the activation pocket. *Peptides* 2010;**31**:2289–93.
46. Dezelak M, Bavec A. Glucagon like-peptide-1 receptor is covalently modified by endogenous mono-ADP-ribosyltransferase. *Mol Biol Rep* 2012;**39**:4375–81.
47. Bullock BP, Heller RS, Habener JF. Tissue distribution of messenger ribonucleic acid encoding the rat glucagon-like peptide-1 receptor. *Endocrinology* 1996;**137**:2968–78.
48. Campos RV, Lee YC, Drucker DJ. Divergent tissue-specific and developmental expression of receptors for glucagon and glucagon-like peptide-1 in the mouse. *Endocrinology* 1994;**134**:2156–64.
49. Storjohann L, Holst B, Schwartz TW. Molecular mechanism of Zn2+ agonism in the extracellular domain of GPR39. *FEBS Lett* 2008;**582**:2583–8.
50. Elrick H, Stimmler L, Hlad Jr CJ, Arai Y. Plasma insulin response to oral and intravenous glucose administration. *J Clin Endocrinol Metab* 1964;**24**:1076–82.
51. Kazakos K. Incretin effect: GLP-1, GIP, DPP4. *Diabetes Res Clin Pract* 2011;**93**(Suppl. 1): S32–S36.
52. Scrocchi LA, Brown TJ, MaClusky N, Brubaker PL, Auerbach AB, Joyner AL, et al. Glucose intolerance but normal satiety in mice with a null mutation in the glucagon-like peptide 1 receptor gene. *Nat Med* 1996;**2**:1254–8.
53. Montrose-Rafizadeh C, Avdonin P, Garant MJ, Rodgers BD, Kole S, Yang H, et al. Pancreatic glucagon-like peptide-1 receptor couples to multiple G proteins and activates mitogen-activated protein kinase pathways in Chinese hamster ovary cells. *Endocrinology* 1999;**140**:1132–40.
54. Wheeler MB, Lu M, Dillon JS, Leng XH, Chen C, Boyd 3rd AE. Functional expression of the rat glucagon-like peptide-I receptor, evidence for coupling to both adenylyl cyclase and phospholipase-C. *Endocrinology* 1993;**133**:57–62.
55. Drucker DJ, Philippe J, Mojsov S, Chick WL, Habener JF. Glucagon-like peptide I stimulates insulin gene expression and increases cyclic AMP levels in a rat islet cell line. *Proc Natl Acad Sci USA* 1987;**84**:3434–8.
56. Salapatek AM, MacDonald PE, Gaisano HY, Wheeler MB. Mutations to the third cytoplasmic domain of the glucagon-like peptide 1 (GLP-1) receptor can functionally uncouple GLP-1-stimulated insulin secretion in HIT-T15 cells. *Mol Endocrinol* 1999;**13**:1305–17.
57. Takhar S, Gyomorey S, Su RC, Mathi SK, Li X, Wheeler MB. The third cytoplasmic domain of the GLP-1[7–36 amide] receptor is required for coupling to the adenylyl cyclase system. *Endocrinology* 1996;**137**:2175–8.
58. Hallbrink M, Holmqvist T, Olsson M, Ostenson CG, Efendic S, Langel U. Different domains in the third intracellular loop of the GLP-1 receptor are responsible for Galpha(s) and Galpha(i)/Galpha(o) activation. *Biochim Biophys Acta* 2001;**1546**:79–86.
59. Mathi SK, Chan Y, Li X, Wheeler MB. Scanning of the glucagon-like peptide-1 receptor localizes G protein-activating determinants primarily to the N terminus of the third intracellular loop. *Mol Endocrinol* 1997;**11**:424–32.
60. Dun SL, Brailoiu GC, Brailoiu E, Yang J, Chang JK, Dun NJ. Distribution and biological activity of obestatin in the rat. *J Endocrinol* 2006;**191**:481–9.
61. Zhao CM, Furnes MW, Stenstrom B, Kulseng B, Chen D. Characterization of obestatin- and ghrelin-producing cells in the gastrointestinal tract and pancreas of rats: an immunohistochemical and electron-microscopic study. *Cell Tissue Res* 2008;**331**: 575–87.

62. Gronberg M, Tsolakis AV, Magnusson L, Janson ET, Saras J. Distribution of obestatin and ghrelin in human tissues: immunoreactive cells in the gastrointestinal tract, pancreas, and mammary glands. *J Histochem Cytochem* 2008;**56**:793–801.
63. Bresciani E, Rapetti D, Dona F, Bulgarelli I, Tamiazzo L, Locatelli V, et al. Obestatin inhibits feeding but does not modulate GH and corticosterone secretion in the rat. *J Endocrinol Invest* 2006;**29**:RC16–RC18.
64. Green BD, Irwin N, Flatt PR. Direct and indirect effects of obestatin peptides on food intake and the regulation of glucose homeostasis and insulin secretion in mice. *Peptides* 2007;**28**:981–7.
65. Brunetti L, Leone S, Orlando G, Recinella L, Ferrante C, Chiavaroli A, et al. Effects of obestatin on feeding and body weight after standard or cafeteria diet in the rat. *Peptides* 2009;**30**:1323–7.
66. Sibilia V, Bresciani E, Lattuada N, Rapetti D, Locatelli V, De Luca V, et al. Intracerebroventricular acute and chronic administration of obestatin minimally affect food intake but not weight gain in the rat. *J Endocrinol Invest* 2006;**29**:RC31–RC34.
67. Zizzari P, Longchamps R, Epelbaum J, Bluet-Pajot MT. Obestatin partially affects ghrelin stimulation of food intake and growth hormone secretion in rodents. *Endocrinology* 2007;**148**:1648–53.
68. Nogueiras R, Pfluger P, Tovar S, Arnold M, Mitchell S, Morris A, et al. Effects of obestatin on energy balance and growth hormone secretion in rodents. *Endocrinology* 2007;**148**:21–6.
69. Seoane LM, Al-Massadi O, Pazos Y, Pagotto U, Casanueva FF. Central obestatin administration does not modify either spontaneous or ghrelin-induced food intake in rats. *J Endocrinol Invest* 2006;**29**:RC13–RC15.
70. Gourcerol G, Coskun T, Craft LS, Mayer JP, Heiman ML, Wang L, et al. Preproghrelin-derived peptide, obestatin, fails to influence food intake in lean or obese rodents. *Obesity* 2007;**15**:2643–52.
71. Kobelt P, Wisser AS, Stengel A, Goebel M, Bannert N, Gourcerol G, et al. Peripheral obestatin has no effect on feeding behavior and brain Fos expression in rodents. *Peptides* 2008;**29**:1018–27.
72. Gourcerol G, St-Pierre DH, Tache Y. Lack of obestatin effects on food intake: should obestatin be renamed ghrelin-associated peptide (GAP)? *Regul Pept* 2007;**141**:1–7.
73. De Spiegeleer B, Vergote V, Pezeshki A, Peremans K, Burvenich C. Impurity profiling quality control testing of synthetic peptides using liquid chromatography-photodiode array-fluorescence and liquid chromatography-electrospray ionization-mass spectrometry: the obestatin case. *Anal Biochem* 2008;**376**:229–34.
74. Ataka K, Inui A, Asakawa A, Kato I, Fujimiya M. Obestatin inhibits motor activity in the antrum and duodenum in the fed state of conscious rats. *Am J Physiol Gastrointest Liver Physiol* 2008;**294**:G1210–G1218.
75. Fujimiya M, Asakawa A, Ataka K, Kato I, Inui A. Different effects of ghrelin, des-acyl ghrelin and obestatin on gastroduodenal motility in conscious rats. *World J Gastroenterol* 2008;**14**:6318–26.
76. Depoortere I, Thijs T, Moechars D, De Smet B, Ver Donck L, Peeters TL. Effect of peripheral obestatin on food intake and gastric emptying in ghrelin-knockout mice. *Br J Pharmacol* 2008;**153**:1550–7.
77. De Smet B, Thijs T, Peeters TL, Depoortere I. Effect of peripheral obestatin on gastric emptying and intestinal contractility in rodents. *Neurogastroenterol Motil* 2007;**19**:211–7.
78. Gurriaran-Rodriguez U, Al-Massadi O, Roca-Rivada A, Crujeiras AB, Gallego R, Pardo M, et al. Obestatin as a regulator of adipocyte metabolism and adipogenesis. *J Cell Mol Med* 2011;**15**:1927–40.

79. Miegueu P, St Pierre D, Broglio F, Cianflone K. Effect of desacyl ghrelin, obestatin and related peptides on triglyceride storage, metabolism and GHSR signaling in 3T3-L1 adipocytes. *J Cell Biochem* 2011;**112**:704–14.
80. Agnew A, Calderwood D, Chevallier OP, Greer B, Grieve DJ, Green BD. Chronic treatment with a stable obestatin analog significantly alters plasma triglyceride levels but fails to influence food intake; fluid intake; body weight; or body composition in rats. *Peptides* 2011;**32**:755–62.
81. Pi-Sunyer FX. The obesity epidemic: pathophysiology and consequences of obesity. *Obes Res* 2002;**10**(Suppl. 2):97S–104S.
82. Guo ZF, Zheng X, Qin YW, Hu JQ, Chen SP, Zhang Z. Circulating preprandial ghrelin to obestatin ratio is increased in human obesity. *J Clin Endocrinol Metab* 2007;**92**:1875–80.
83. Nakahara T, Harada T, Yasuhara D, Shimada N, Amitani H, Sakoguchi T, et al. Plasma obestatin concentrations are negatively correlated with body mass index, insulin resistance index, and plasma leptin concentrations in obesity and anorexia nervosa. *Biol Psychiatry* 2008;**64**:252–5.
84. Zamrazilova H, Hainer V, Sedlackova D, Papezova H, Kunesova M, Bellisle F, et al. Plasma obestatin levels in normal weight, obese and anorectic women. *Physiol Res* 2008;**57**(Suppl. 1):S49–S55.
85. Lippl F, Erdmann J, Lichter N, Tholl S, Wagenpfeil S, Adam O, et al. Relation of plasma obestatin levels to bmi, gender, age and insulin. *Horm Metab Res* 2008;**40**:806–12.
86. Huda MS, Durham BH, Wong SP, Deepak D, Kerrigan D, McCulloch P, et al. Plasma obestatin levels are lower in obese and post-gastrectomy subjects, but do not change in response to a meal. *Int J Obes* 2008;**32**:129–35.
87. Vicennati V, Genghini S, De Iasio R, Pasqui F, Pagotto U, Pasquali R. Circulating obestatin levels and the ghrelin/obestatin ratio in obese women. *Eur J Endocrinol* 2007;**157**:295–301.
88. Zou CC, Liang L, Wang CL, Fu JF, Zhao ZY. The change in ghrelin and obestatin levels in obese children after weight reduction. *Acta Paediatr* 2009;**98**:159–65.
89. Zhang N, Yuan C, Li Z, Li J, Li X, Li C, et al. Meta-analysis of the relationship between obestatin and ghrelin levels and the ghrelin/obestatin ratio with respect to obesity. *Am J Med Sci* 2011;**341**:48–55.
90. Gao XY, Kuang HY, Liu XM, Ma ZB. Decreased gastric body mucosa obestatin expression in overweight and obese patients. *Peptides* 2010;**31**:291–6.
91. Roth CL, Reinehr T, Schernthaner GH, Kopp HP, Kriwanek S, Schernthaner G. Ghrelin and obestatin levels in severely obese women before and after weight loss after Roux-en-Y gastric bypass surgery. *Obes Surg* 2009;**19**:29–35.
92. Martins C, Kjelstrup L, Mostad IL, Kulseng B. Impact of sustained weight loss achieved through Roux-en-Y gastric bypass or a lifestyle intervention on ghrelin, obestatin, and ghrelin/obestatin ratio in morbidly obese patients. *Obes Surg* 2011;**21**:751–8.
93. Manshouri M, Ghanbari-Niaki A, Kraemer RR, Shemshaki A. Time course alterations of plasma obestatin and growth hormone levels in response to short-term anaerobic exercise training in college women. *Appl Physiol Nutr Metab* 2008;**33**:1246–9.

CHAPTER FOUR

Melanocortin-3 Receptors and Metabolic Homeostasis

Karima Begriche[*,†], **Clemencé Girardet**[*], **Patricia McDonald**[‡], **Andrew A. Butler**[*]
[*]Department of Metabolism and Aging, The Scripps Research Institute, Jupiter, Florida, USA
[†]Inserm U991, Foie, Métabolisme et Cancer, 2 avenue du Pr Léon Bernard, Rennes, France
[‡]Department of Molecular Therapeutics, The Scripps Research Institute, Jupiter, Florida, USA

Contents

Abstract

Attenuated activity of the central nervous melanocortin system causes obesity and insulin resistance. Obese rodents treated with melanocortins exhibit improvements in obesity and metabolic homeostasis that are not mutually dependent, suggesting metabolic actions that are independent of weight changes. These responses are generally thought to involve G-protein-coupled receptors expressed in the brain. Melanocortin-4 receptors (MC4Rs) regulate satiety and autonomic nervous system

Progress in Molecular Biology and Translational Science, Volume 114
ISSN 1877-1173
http://dx.doi.org/10.1016/B978-0-12-386933-3.00004-2

and thyroid function. MC3Rs are expressed in hypothalamic and limbic regions involved in controlling ingestive behaviors and autonomic function. *Mc3r−/−* mice exhibit increased adiposity and an accelerated diet-induced obesity. While this phenotype is not dependent on hyperphagia, data on the regulation of food intake by MC3Rs are inconsistent. Recent investigations by our laboratory suggest a unique combination of behavioral and metabolic disorders in *Mc3r−/−* mice. MC3Rs are critical for the expression of the anticipatory response and metabolic homeostasis when food intake occurs outside the normal voluntary rhythms driven by photoperiod. Using a Cre-Lox strategy, we can now investigate MC3Rs expressed in different brain regions and organ systems in the periphery. While focusing on the functions of neural MC3Rs, early results suggest an additional layer of complexity with central and peripheral MC3Rs involved in the defense of body weight.

1. INTRODUCTION

The rise in the prevalence of obesity is a significant public health issue as it is associated with insulin resistance and increased risk for type 2 diabetes and cardiovascular disease. Indeed, less than half of the population in the United States has a healthy body weight (a body mass index between 18.5 and 24.9 kg/m^2).[1] Lifestyle changes causing weight loss and to prevent development of obesity are thus increasingly considered as imperatives. Unfortunately, strong behavioral responses to caloric restriction make adhering to lifestyle changes extremely difficult in obese individuals. Clinical studies also indicate other counter responses to limit weight loss, with reduced resting metabolic rate and physical activity.[2–4] In individuals where the metabolic stress associated with obesity is severe, surgical and/or pharmacological interventions are therefore often required to obtain weight loss. Gastric bypass procedures are highly effective and also rapidly improve metabolic homeostasis.[5] However, the options for pharmacological intervention are limited.

Body weight exhibits remarkable stability over time, even in the severely obese. This suggests that mechanisms exist to sense energy balance and prevent excess weight gain through the control of appetite and energy expenditure and/or nutrient partitioning. A popular concept for "energy homeostasis" has evolved whereby body weight is maintained by neuronal circuits that emanate from the hypothalamus that connect appetite and the autonomic and neuroendocrine control of metabolism with signals of metabolic status.[6–9] As mentioned previously, caloric restriction produces powerful behavioral and metabolic counter responses to limit weight loss. However, the nature of the "anti-obesity" mechanisms beyond the control of appetite in situations of nutrient abundance is still being debated.[10–16]

In situations where calorie-dense diets are freely available, the mechanisms for the defense of body weight are less robust when compared to the powerful counter-regulatory responses to calorie restriction.[17]

It has been known for nearly a century that damage to the hypothalamus or mutations in certain genes can result in obesity.[18–20] Much of the research into the mechanisms that control the balance of energy intake and expenditure over the last two decades has also focused on the hypothalamus. Strong evidence links altered activity of the central nervous melanocortin system that is centered within the hypothalamus with human obesity, and a model for how the system functions to maintain body weight has been proposed.[6,21] Melanocortins regulate both sides of the energy balance equation, regulating satiety and systems that govern energy expenditure. Much has been learnt about the "first order" melanocortin neurons releasing the endogenous melanocortin ligands that are regulated by signals of metabolic state. Considerable attention has also been given to one of melanocortin receptors that are widely expressed in the brain (melanocortin-4 receptor, MC4R). The functions of the other melanocortin receptors expressed in the brain (melanocortin-3 receptor, MC3R) that is also known to affect body weight and composition have received far less attention.

This review begins by providing an overview of our understanding of the neural systems and the involvement of melanocortins in metabolic homeostasis and prevention of obesity. It will then focus on the most poorly understood component of the melanocortin system, the MC3R, in regulating body weight and metabolic homeostasis.

2. THE "LEPTIN-MELANOCORTIN" PATHWAY: A CANONICAL PATHWAY LINKING ENERGY STORES WITH NEURAL OUTPUTS REGULATING ADAPTATION TO ENERGY REQUIREMENTS

The "modern era" of obesity research that began to elucidate the mechanisms orchestrating metabolic homeostasis and weight control originated with mouse genetics that was used to identify and characterize loci associated with hyperphagia and increased adiposity. The characterization of the *obese* (*ob/ob*) and *diabetic* (*db/db*) mice, which are carriers of recessive mutations causing severe obesity, was a landmark in the development of models explaining how body weight might be controlled. This research also provided a fundamental insight into how organisms adapt behavior and metabolism to fasting. Parabiosis experiments performed by Douglas Coleman and colleagues over

40 years ago suggested that *ob/ob* mutant mice lacked a secreted factor produced by a gene on chromosome 6, while *db/db* mice lacked a receptor for a satiety factor likely expressed in the hypothalamus that is produced by a gene on chromosome 4.[22] Positional cloning of the mouse *obese* gene identified leptin as a 16 kDa peptide hormone belonging to the cytokine superfamily secreted from adipocytes.[23] Shortly thereafter an expression cloning strategy identified the leptin receptor as a member of the single transmembrane-spanning cytokine receptor family.[24] Genetic screens of morbidly obese humans subsequently identified homozygous carriers of "loss of function" mutations in genes encoding leptin (*LEP*) or the leptin receptor (*LEPR*).[25] These individuals exhibited a ravenous appetite and severe obesity from early childhood.

Leptin levels in the circulation are proportional to fat mass, while the actions of leptin inhibit food intake and reduce fat.[26] Circulating leptin may therefore function as an input into a negative feedback loop that maintains energy reserves in adipose tissue at stable levels. Heterozygous carriers of null *LEP* alleles exhibit increased adiposity, consistent with a compensatory increase in fat mass to increase leptin levels and thereby raising the "settling point" of adiposity.[27] The discovery of *LEP* and *LEPR* genes has, however, not led to effective obesity therapies owing to leptin resistance in obesity.[28]

Leptin regulates behavioral and metabolic programs that are important for survival during times of caloric insufficiency (Fig. 4.1).[36] The effect of leptin on energy balance is generally "inhibitory" (i.e., suppression of appetite, stimulation of energy expenditure), loss or attenuation of action therefore has the potential to cause weight gain and obesity. Conditional targeting of the *Lepr* gene in mice indicates that the actions of leptin in the central nervous system are sufficient and obligatory to maintain energy balance.[37–42] Leptin secretion in well-fed conditions provides the brain with a signal of caloric sufficiency, attenuating appetite while maintaining energy expenditure. A decline in leptin secretion from adipocytes with fasting signals caloric deprivation,[43] and instigates behavioral programs including increased food seeking and an attenuated response to satiety signals. Leptin-deficiency also promotes neuroendocrine and autonomic responses to instigate metabolic adaptation to caloric restriction (e.g., reducing energy expenditure).[2,44–46] The decline in leptin with fasting and the triggering of a counter response to a negative energy balance are likely the *raison d'etre* for this hormone.[43]

While leptin levels do increase with weight gain, this is considered an indicator of leptin resistance as opposed to an effective counter response for preventing further weight gain. Indeed, desensitization of hypothalamic neurons to leptin, including the primary melanocortin neurons releasing the

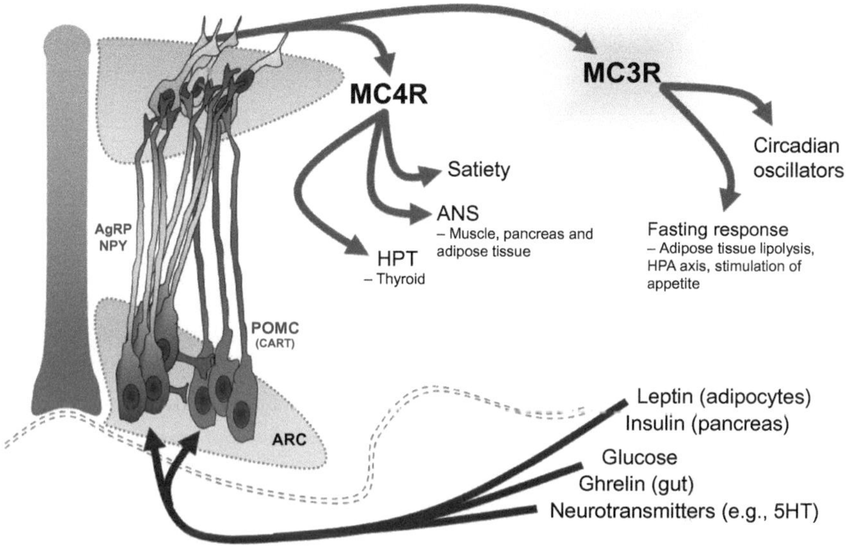

Figure 4.1 Schematic showing the central role played by arcuate (ARC) melanocortin neurons to integrate signals of metabolic status with outputs involved in metabolic homeostasis. The figure shows the two groups of "primary" melanocortin neurons that reside in the ARC and send projections throughout the central nervous system. POMC neurons (red) express the preprohormone processed to produce α- and γ-MSH (not shown). Activation of POMC neurons has been suggested to have "catabolic" actions that limit weight gain by reducing food intake and promoting energy expenditure. AgRP neurons (green) also express neuropeptide Y (NPY); both neuropeptides are orexigenic and have inhibitory effects on outputs that maintain energy expenditure (i.e., ANS and HPT). AgRP and POMC neurons integrate several signals of metabolic state and vigilance. Note, however, that the populations of POMC and AgRP neurons are probably heterogeneous, exhibiting differential responses to some or all of the inputs shown here.[29] Central administration of α-MSH inhibits food intake, a response requiring functional MC4Rs.[30] MC4Rs also regulate the autonomic nervous system (ANS) and the hypothalamo–pituitary–thyroid axis (HPT). Activation of MC4Rs stimulates insulin action to affect glucose disposal and lipid metabolism through effects on the ANS. Activation of MC4Rs also contributes to maintaining normal levels of thyroid activity in fed conditions.[31] The role of MC3Rs in the central nervous system is less clear. However, *Mc3r−/−* mice display reduced activity of circadian oscillators suggesting a role in maintaining rhythms in the central nervous system.[32–34] It was recently reported that *Mc3r−/−* mice also exhibit reduced behavioral and metabolic adaptation to fasting.[35] Stimulation of AgRP expression and appetite is Mc3r dependent, as is the stimulation of adipose tissue lipolysis. (See Color Insert.)

endogenous melanocortin agonists,[47] likely contributes to the deterioration of metabolic homeostasis that is frequently observed in obesity. Several mechanisms have been suggested to explain leptin resistance with weight gain, including attenuation of a saturable transporter system across the blood–brain barrier, increased activity of inhibitory signaling proteins (e.g., protein tyrosine phosphatase 1B, suppressor of cytokine signaling 3), inflammation, and other cellular stress responses linked to insulin resistance.[28] However, results from a recent study suggest that receptor desensitization, which describes the phenomenon whereby the response of the receptor declines over time despite the presence of stimulus, may be critical for the development of leptin resistance. When leptin levels are "clamped" in C57BL/6J (B6) mice subjected to diet-induced obesity, weight gain was normal.[48] Hyperleptinemia therefore does not prevent further weight gain. These mice also did not develop leptin resistance with obesity. Receptor desensitization through upregulation of inhibitory signaling proteins (and perhaps saturation of leptin transport across the blood brain barrier) may therefore be obligatory for the development of leptin resistance. On the other hand, we have also observed that in conditions of extreme obesity, hyperleptinemia may actually exacerbate the obese phenotype. Alleviation of severe hyperleptinemia in B6 mice with obesity due to a combination of reduced MC4R signaling and high fat diet was not associated with further gain, but actually attenuated the obese phenotype.[49]

Mouse genetics also facilitated the neurochemical identification of the first group of hypothalamic neurons involved in mediating the actions of leptin. Mice carrying the dominant agouti alleles *lethal yellow* (A^Y) or *viable yellow* (A^{VY}) had been known for nearly a century to develop an "adult onset" obesity.[50] Positional cloning of the *agouti* locus identified the agouti-signaling protein (ASIP)[51,52] subsequently shown to function as a competitive antagonist for two melanocortin receptors (MC1R, MC4R).[53] The yellow color observed in A^Y and A^{VY} mice results from antagonism of MC1Rs; obesity was hypothesized to result from antagonism of MC4Rs expressed in the central nervous system.[54–56] The similarities in the phenotype of *Mc4r−/−* and A^Y mice (increased longitudinal growth, obesity, insulin resistance, and hyperphagia) supported this hypothesis.[56] Genetic screens using morbidly obese individuals identified homo- and heterozygous carriers of *MC4R* mutations. These individuals have a severe hypherphagic obesity syndrome from a very early age.[57,58] Unlike the case with leptin and its receptor, however, mutations in the *MC4R* gene turned out to be the most common form of monogenic obesity, occurring in 1–6% of obese individuals.

Regulation of satiety is a major factor in the defense of body weight by MC4Rs.[56,59,60] However, as is case the for leptin, studies using rodent models indicate a metabolic component. Prevention ofhyperphagia in *Mc4r*− /− mice only attenuates obesity,[61] while analysis of the response of *Mc4r*− /− mice during the transition to diets with increased fat content suggests deficits in energy expenditure and nutrient partitioning.[60,62–64] MC4Rs in the brain regulate glucose and lipid metabolism via neuroendocrine and autonomic outputs, and their activation promotes energy expenditure during periods of caloric sufficiency.[60,61,64–69] Studies using a Cre-Lox strategy to selectively target *Mc4r* expression in mice suggest that the expression of this receptor in the nervous system is sufficient to restore body weight.[70] These studies also suggest a functional divergence. Expression of Mc4rs in the paraventricular nucleus (PVN) of the hypothalamus is *sufficient* to restore normal regulation of satiety. However, it does not restore the regulation of energy expenditure.[70] On the other hand, deficits in energy expenditure and insulin sensitivity (but not food intake) are rescued by expression of *Mc4r* in preganglionic sympathetic neurons in the hindbrain.[65] These results need to be interpreted with caution, however. For example, several laboratories have reported a role for Mc4rs in the brainstem in the regulation of satiety.[71–76] The actions of these receptors, while important, may thus not be sufficient to overcome loss of Mc4r activity in PVN neurons. Conversely, intracerebroventricular (icv) injections of melanocortin agonists into the hypothalamus can stimulate thermogenesis.[77] One interpretation of these results is a distributed network involving hypothalamic and brainstem sites governing appetite and metabolism. Within this network are critical "nodes" where MC4R activity is sufficient and necessary to regulate one or the other function, but not both. The results from studies using the same genetic strategy to explore leptin signaling in the central nervous system have shown similar outcomes.[78]

Strong evidence from basic research and clinical genetics supporting a role for MC4Rs in preventing obesity in humans has led to several campaigns by pharmaceutical companies and academia to identify small molecule and peptide analogs with activity at this receptor.[79–84] However, the response of obese volunteers to compounds targeting MC4Rs in the clinic has thus far been discouraging due to limited effectiveness and side effects on cardiovascular function.[85] MC4Rs are not, however, the only member of the melanocortin receptor family involved in metabolic homeostasis. *Mc3r*− /− mice display an obese phenotype,[32,62,63,86–90] attenuated behavioral, and metabolic adaptations when food intake is limited both calorically and temporally (to the mid-light cycle),[33,34,91,92] and to fasting.[35] And, although obese *Mc4r−/−* mice treated with melanocortin analogs do not exhibit reduced food intake and weight loss, reductions in fasting insulin and glucose, suggesting improvements in glucose homeostasis, were still observed (Fig. 4.2, Table 4.1).[93] This experiment did not determine the subtype of melanocortin receptor that is involved. There is evidence suggesting that other melanocortin receptors play a role in metabolic homeostasis. MC5Rs expressed in skeletal muscle regulate activity of AMPK, suggesting a role in regulating glucose and fatty acid oxidation.[94] However, with a caveat that human receptors were used to characterize the pharmacology, the melanocortin analog administered to *Mc4r−/−* mice had low affinity for MC5Rs ($K_i > 7\ \mu M$, $EC_{50} = 4\ \mu M$). This suggests that actions involving stimulation of MC3Rs and/or MC1Rs result in improvements in glucose homeostasis in obese mice. As shall be discussed later in the review, data from our laboratory and others suggest that the functions of MC3Rs can impact on glucose homeostasis.[92,95]

While the MC4R has been studied extensively using a combination of mouse genetics and pharmacological tools, far less is known about the functions of MC3Rs expressed in the brain. MC3Rs are expressed in hypothalamic and limbic structures of the central nervous system controlling ingestive behaviors and metabolic homeostasis.[96] The endogenous agonist α-melanocyte-stimulating hormone (α-MSH) exhibits affinity for both receptors.[96,97] Another form of MSH (γ-) exhibits selectivity for the MC3R, suggesting that there may be differential regulation of the two receptors in situations where activity of proopiomelanocortin (POMC) neurons that release the endogenous ligands is increased.[96] A second member of the agouti family, agouti-related peptide (AgRP), is expressed in the brain and binds with high affinity to MC3Rs and MC4Rs[98,99] suggesting parallel regulation in situations where AgRP expression is upregulated. The additive obesity phenotype observed in double *Mc3r−/−.Mc4r−/−* mice suggests that the actions of each receptor affecting body weight involve independent pathways.[87]

While the analysis of *Mc3r* knockout mice suggests functions that impact on adiposity,[32,62,63,86–90] how and where this receptor affects these processes have remained a mystery. Studies analyzing the response of melanocortin receptor knockout mice indicate that MC3Rs are not required for regulating satiety or neuroendocrine and autonomic systems governing metabolism in the periphery.[30,87,100,101] The link between mutations in the human *MC3R* gene and obesity has also not been well established when compared to *MC4R* mutations. These findings may explain in part the lack of

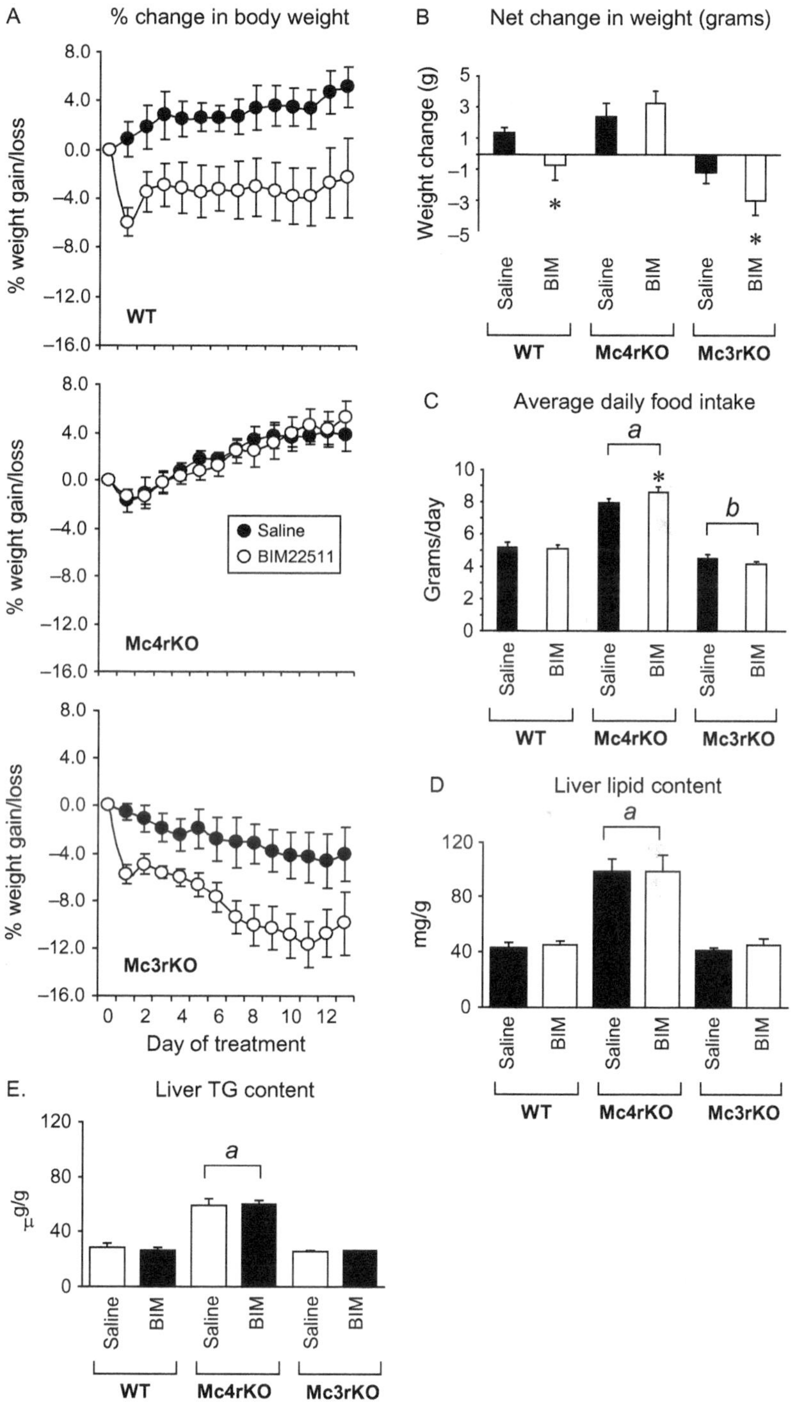
A
% change in body weight
% weight gain/loss
WT
Mc4rKO
Mc3rKO
Saline
BIM22511
Day of treatment
B
Net change in weight (grams)
Weight change (g)
C
Average daily food intake
Grams/day
D
Liver lipid content
mg/g
E.
Liver TG content
μg/g
Saline
BIM
WT
Mc4rKO
Mc3rKO
a
b

enthusiasm shown by the obesity field for investigating this receptor. However, recent observations by our laboratory and others suggest that this receptor may have a critical role in metabolic homeostasis in certain conditions. Specifically, while loss of MC3Rs has minor effects in "nonobesogenic" conditions, the actions of the receptor appear to be critical for metabolic flexibility when mice are challenged with high fat diets. These receptors are also required for both behavioral and metabolic flexibility during times when food availability occurs outside of the normal circadian rhythm. These actions may be related to the regulation of circadian oscillators and the synchronization of circadian rhythms with nutrient availability.[32–34,91,92] Moreover, a recent study by another laboratory reported a critical role for this receptor in the adaptation to fasting, while also suggesting a possible mechanism (mild hypercorticosteronemia) in explaining the obese phenotype.[35]

3. THE CENTRAL NERVOUS MELANOCORTIN SYSTEM

3.1. Multiple ligands with biased agonist and competitive antagonist properties

The melanocortin system is comprised of six known ligands that bind with high affinity to five receptors.[6,102] The cDNA encoding POMC was one of the first to be identified and sequenced in the late 1970s.[103–105] POMC is a precursor polypeptide of 241 amino acids that is processed post-translationally into several peptides including the melanocortins, β-endorphin, and β- and γ-lipotropin. The peptides originally considered to function as melanocortin receptor agonists are adrenocorticotrophic hormone (ACTH) and the three isoforms of MSH (α, β

Figure 4.2 Response of WT, Mc3rKO, and Mc4rKO treated with BIM-22511 for 14 days. (A) Change in body weight as a percent of initial weight in WT (top panel), Mc4rKO (middle panel), or Mc3rKO (bottom panel) treated with BIM-22511 (solid circles) or saline (open circles). Note the absence of a response in Mc4rKO. (B) Net weight gain or loss in mice treated with saline or BIM-22511. The effects of BIM-22511 on body weight are dependent on MC4R activation. (C) Average daily food intake of mice treated with saline or BIM-22511. Mc4rKO were hyperphagic, while Mc3rKO exhibited reduced food intake relative to WT (*a*, $P<0.001$ vs. WT, Mc3rKO; *b*, $P<0.05$ vs. WT). Note that the bars with letters indicate within genotype comparisons (i.e., LSM of saline and BIM treatment groups combined for Mc4rKO, compared to LSM of saline and BIM treatment groups for WT). Repeated measures ANOVA indicated a significant increase in Mc4rKO treated with BIM-22511. $^*P<0.01$ vs. saline within treatment. (D, E) Mc4rKO exhibit hepatic steatosis (elevated liver lipid and triglyceride) relative to WT and Mc3rKO. There was no significant effect of treatment with BIM-22511 on hepatic lipid content, irrespective of genotype. *Reproduced with permission from Kumar* et al., *2009.*[93]

Table 4.1 Blood chemistries of WT, Mc3rKO, and Mc4rKO treated with BIM-22511 compared to controls

Genotype:	**WT**		**Mc3rKO**		**Mc4rKO**	
Treatment:	**Saline**	**BIM-22511**	**Saline**	**BIM-22511**	**Saline**	**BIM-22511**
Glucose (mg/dL)	100±7	129±20	164±22	193±28	153±10	119±15
Insulin (pg/ml)	138±30	145±40	124±21	310±55	3588±812	1899±160*
HOMA-IR	0.9±0.2	1.3±0.5	1.3±0.3	3.5±0.5	33.5±7.0	14.2±3.7*

*$P<0.01$ compared to saline treatment within genotype. Serum was collected after a 4 h fast.

and γ). ACTH is produced by corticotroph cells of the anterior pituitary, which express POMC. In the central nervous system, POMC is processed to produce α-MSH, γ-MSH and (in humans, but not rodents) β-MSH.

The melanocortin system is unique in that soon after the receptors were cloned, high affinity ligands were described with both agonist and antagonist properties. ASIP and AgRP display high affinity for specific melanocortin receptors and were originally reported to function as competitive antagonists[98,53,106–108] against the MSH peptides classically considered to have agonist functions. ASIP was the first to be identified by positional cloning and is expressed in hair follicles and regulates pigmentation by modulating activity of the MC1R.[51–53] Ectopic expression of ASIP due to a spontaneous insertion mutation in the promoter region (e.g., the *lethal yellow* (A^Y/a) mouse) or transgenic expression using a human beta actin promoter produces mice with a yellow coat color. In addition, antagonism of neural MC4Rs results in a hyperphagic obesity syndrome.[54,109,110]

The second member of the agouti family of ligands to be identified (AgRP) functions as an antagonist in cell-based assays using MC3R and MC4R.[98,99,108,111,112] Central administration of AgRP in rodents has the opposite effect to that observed when α-MSH is administered, increasing food intake and altering metabolism to favor deposition of adipose mass.[113] Structurally, AgRP is a 132 residue protein composed of two domains. Residues 1–82 (the N-terminus) are removed by prohormone convertases, while a second domain comprising residues 83–132 is sufficient to induce increased food intake[113] when administered intracerebroventricularly and is the dominant form found in the hypothalamus.[114]

Further investigation has indicated that ASIP and AgRP have functions beyond acting as competitive antagonists. MC4Rs display high basal levels of adenylyl cyclase (AC) activity resulting in high cAMP levels that can be reduced in the presence of AgRP, suggesting "inverse agonist" properties.[115,116] ASIP functions as an MC1R inverse agonist.[117] Mouse genetics also supports the hypothesis that the high level of constitutive activity of melanocortin receptors is physiologically significant. Mice lacking all neural MSH respond to centrally administered AgRP, displaying reduced oxygen consumption, increased respiratory exchange ratio (RER, an indicator of whole body substrate preference), and increased food intake.[118] This observation suggests actions of AgRP at MC3R and MC4R in the central nervous system independent of competitive displacement of MSH. The black coat color phenotype of *Pomc*−/− mice backcrossed onto the C57BL/6J (*a*/*a*) background also suggests that the stimulation of eumelanin (black/brown) pigmentation by MC1R is not dependent on α-MSH.[119] Finally, an examination of the A^Y allele crossed onto mouse lines with varying coat colors suggested that antagonism of MC1R by ASIP is not the equivalent to loss of receptor function[106]; Yang *et al.* interpreted these results as indicating that ASIP could function as a potential MC1R agonist, stimulating signaling through intracellular pathways other than $G_\alpha s$ coupling.

One of the first mechanisms proposed for explaining the high level of constitutive activity of the MC4R involved the long N-terminal extracellular loop functioning as a "tethered" intramolecular ligand.[120] More recently, MSH and AgRP have been suggested to function as biased agonists that regulate the association of MC4R with distinct heterotrimeric G-protein signaling pathways.[121] Activation of MC4R by α-MSH stimulates accumulation of cAMP through increasing coupling with $G_\alpha s$. In contrast, AgRP stimulates the coupling of MC4R expressed in a hypothalamic neuronal cell line (GT1-7) to the $G_\alpha i/o$ subunit, thereby inhibiting adenylate cyclase activity and reducing cellular cAMP levels.

In addition to the melanocortins, "nonclassical" melanocortin ligands were recently identified during a genetic analysis of the *K* locus controlling pigmentation in dogs.[122] The product of the *K* locus was identified as β-defensin103, which regulates coat color by acting as a high affinity ligand for the MC1R. Interestingly, β-defensin103 also binds with low affinity to MC4Rs. The physiological significance of the interaction between defensins and the melanocortin system, beyond influencing coat color variation, remains to be explored and will not be further addressed in this review. Another recent topic in the melanocortin field that will not be discussed in depth here is heterodimerization of melanocortin receptors with other G protein coupled receptors (GPCR) involved in metabolic homeostasis.

For example, MC3Rs form heterodimers with ghrelin receptors (GHSR).[123,124] Heterodimers broaden the selection of ligands that can regulate receptor activity while also modulating post-receptor signaling pathways.

3.2. Neurons expressing the endogenous ligands in the central nervous system: "First order" neurons acting as sensors of metabolic status and vigilance

Neurons expressing the endogenous melanocortin ligands have been observed in two brain regions.[6] Neurons expressing POMC or AgRP are located in the arcuate nucleus of the hypothalamus, sending projections to similar terminal fields in the forebrain, although AgRP neurons may not send projections to the brain stem.[125,126] A smaller population of POMC-expressing neurons is also located in the nucleus of the solitary tract in the brain stem. Deletion of *Pomc* gene in mice results in adrenal insufficiency due to the absence of ACTH and severe obesity.[127] Rescuing *Pomc* expression in the pituitary exacerbates the obese phenotype by restoring glucocorticoid levels.[128] Homozygous carriers of null mutations in the human *POMC* gene have also been identified who display the same characteristics of morbid obesity and adrenal insufficiency.[129] Interestingly, heterozygous carriers of null mutations in the human *POMC* gene also exhibit increased propensity for obesity.[130]

As mentioned before, some of the neurons that release the endogenous ligands for the melanocortin receptors are important central targets for leptin; actions involving the central nervous melanocortin system at least partially mediate the effects of leptin on appetite and metabolism.[131–135] However, these neurons are probably heterogeneous with respect to the kinds of receptors expressed, and integrate multiple inputs from signals of metabolic state including ghrelin,[136–138] insulin,[29,139–142] orexin,[143] serotonin,[66,144–147] and glucose[148,149] (Fig. 4.1). Studies examining hypothalamic neural activity in rats subjected to fasting-refeeding paradigms suggest that these inputs rapidly regulate the activity of arcuate POMC and AgRP neurons.[150–152] In fasted conditions, AgRP neurons are "active" while the activity of POMC neurons is suppressed. In rats allowed access to food for 2 hours, an increase in Fos activity is observed 60 and 90 min after the initiation of feeding in ARC neurons positive for α-MSH staining with a return to normal levels at 180 min. These neurons may therefore integrate signals of long-term energy balance (i.e., leptin and perhaps insulin) with acute meal-related signals.

4. CLONING OF THE MELANOCORTIN RECEPTORS

As key regulators of many biochemical cascades, GPCRs are the largest and most diverse class of cell surface receptors in the human genome and the richest source of drug targets for the pharmaceutical industry. It is estimated that GPCRs are the targets of ~27% of all Food and Drug Administration approved drugs.[153] GPCRs have evolved to respond to a large palette of extracellular signals such as neurotransmitters, hormones, peptides, lipids, biogenic amines, and metabolites. They control and regulate many physiological responses such as cell proliferation, endocrine hormone secretion, learning and memory processes, heart rate, and energy homeostasis.[154] These receptors share a common molecular structure comprising seven hydrophobic transmembrane spanning domains, with an extracellular N-terminus and an intracellular C-terminus linked by three extracellular and three intracellular domains.[151] Based on sequence similarity in the seven transmembrane domains, these receptors have been divided into at least six classes/families: rhodopsin-like receptors (Family A), secretin-like receptors (Family B), metabotropic glutamate receptors (Family C), fungal pheromone receptors (Family D), cyclic AMP receptors (Family E), and frizzled/smoothened receptors.

The first melanocortin receptor (MC1R) was identified using a cDNA library prepared from a human melanoma cell line that exhibited very high levels of MSH binding; the sequence data obtained was used to isolate a second melanocortin receptor (MC2R), which turned out to be the receptor for ACTH.[155] Another three receptors were subsequently identified by several laboratories, and were classified as MC3R, MC4R, and MC5R in the order of their discovery.[96,156–164] Melanocortin receptors belong to the rhodopsin-like family, which is the largest and most intensively studied family of GPCR.[153] The melanocortin receptors were initially reported to be coupled to Gs, with stimulation by agonists leading to increased accumulation of cAMP levels through activation of specific adenylate cyclase isoforms.[165,166]

Of the five melanocortin receptors identified, two (MC3R and MC4R) are expressed in areas of the central nervous system linked to expression of complex behaviors related to feeding and the regulation of metabolic activity. MC4Rs are widely expressed in the brain (~150 nuclei) including the cortex, thalamus, hypothalamus, brainstem, and spinal cord.[55,167] MC4Rs are distributed in many sites involved in autonomic and endocrine function such as the PVN of the hypothalamus, the dorsal motor nucleus of the vagus, and the raphe, consistent with the role of this receptor in feeding and autonomic regulation.[65,168,169] In contrast, *Mc3r* expression measured by *in situ* hybridization was reported in approximately 35 nuclei that are predominantly in hypothalamic and limbic structures; high levels of expression are observed in the ventromedial hypothalamus (VMH), medial habenula, ventral tegmental area, and central linear nucleus of the raphe.[96] *Mc3r* expression has also been reported in peripheral tissues including the immune system, kidney, and areas of the stomach, duodenum, and pancreas.[158]

5. MELANOCORTIN-3 RECEPTORS IN METABOLIC HOMEOSTASIS

5.1. Mutations in the human population

A clear role of the endogenous ligands for the melanocortin receptors in energy balance was reported by Krude *et al.* in 1998, who provided the first description of humans congenitally lacking *POMC* gene products.[129] Patients with *POMC* deficiency developed severe early onset-obesity, hyperphagia, and hypocortisolemia secondary to ACTH deficiency. Homozygous carriers of null mutations in the *POMC* gene therefore require glucocorticoid supplementation.[170] Obesity in *POMC* deficiency is due to loss of agonists regulating MC3R and MC4R, two central receptors known to play an important role in the hypothalamic leptin–melanocortin pathway of body weight regulation. In 2006, Farooqi *et al.* reported that loss of one copy of *POMC* gene (haploinsufficiency) is sufficient to increase risk of obesity in humans.[130] Humans heterozygous for *POMC* null mutation have a body weight toward the upper end of the normal range or are overweight.[130,171] Analysis of mice where one copy of the *Pomc* gene has been deleted also indicates a gene-dosage effect, with increased propensity for diet-induced obesity.[172] Analysis of *POMC* mutations affecting specific melanocortin peptides also provided considerable insights in understanding the relative importance of particular POMC-derived melanocortin ligands in the control of energy balance. Human genetic studies were particularly important to determine the physiological role of β-MSH, a peptide that is not endogenously present in rodents. In 2006, two studies reporting heterozygous missense mutations in the region of POMC encoding β-MSH have suggested a role of this peptide in the hypothalamic control of body weight in humans.[173,174]

Contrasting with these rare monogenic forms of obesity, mutations located in the *MC4R* gene have been shown to be a frequent cause of an early-onset dominant form of obesity. Defects in MC4R signaling account for approximately 1–6 % morbidly obese individuals across diverse ethnic groups.[175–178] Human MC4R deficiency was reported to affect 4% and 5.8% of severely obese French and British populations, respectively.[59,175,179] However, other studies reported a lower incidence of *MC4R* mutations in their respective obese populations. The prevalence of *MC4R* mutations was particularly low in Asian and Scandinavian populations.[178,180] Observation of phenotypes of humans with deficits in MC4R signaling reveal physiological roles of this receptor in the regulation of appetite, body weight, bone mineral content, and cardiovascular function.[59,175,176] Subjects with MC4R deficiency exhibit an obese phenotype characterized by an increase in both fat mass and lean mass from early age.[59] Besides obesity, clinical features of MC4R deficiency include also hyperphagia, which appears in the first year of life, increased bone mineral density and accelerated linear growth probably due to the disproportionate early hyperinsulinemia.[59]

In contrast to *MC4R* mutations, the relevance of *MC3R* mutations to human obesity is less clear and a pathophysiological role of MC3R in human obesity is still debated. To determine whether variations in the *MC3R* gene are associated with obesity, large-scale screening studies were conducted on normal and obese subject cohorts. In 2000, Li *et al.* reported several sequence variants in the coding and 5′ flanking regions of the *MC3R* gene but no significant associations of *MC3R* variants with obesity have been detected.[181] This absence of association between *MC3R* variants and obesity was also observed by other groups. Investigating the prevalence and function of *MC3R* and *MC4R* mutations in two cohorts of North American severely obese and nonobese subjects, Calton *et al.* confirmed that *MC4R* mutations are a significant cause of severe obesity in North American adults whereas *MC3R* mutations are not associated with severe obesity.[182] Recently, an European study conducted by French and Italian investigators reported that rare *MC3R* mutations with *in vitro* functional consequences are associated with human obesity.[183] Collectively these studies indicate that the role of MC3R in regulating body weight is not straightforward, and associations of *MC3R* variants with human obesity are dependent on the ethnicity and the impact of the mutation on receptor functionality. *MC3R* variants have also been associated with subtle changes in onset of weight gain, hyperleptinemia, and hyperinsulinemia.[184,185]

5.2. Targeted deletion of the mouse *Mc3r* gene

The first reports of the phenotype of mice in which the *Mc3r* gene was targeted came in 2000 from Roger Cone's laboratory at The Vollum Institute and a group led by Lex Van der Ploeg at Merck.[86,87] Amgen also produced a *Mc3r* knockout mouse strain that was used in a study reported by a third group at the Pennington Biomedical Research Center.[90] The phenotype of all three strains is very similar, with *Mc3r*−/− mice displaying a moderate obesity syndrome characterized by increased adipose mass and reduced lean mass.[86,87] Chow-fed *Mc3r*−/− mice are hyperinsulinemic suggesting an insulin-resistant phenotype commensurate with a mild increase in adiposity. The hyperinsulinemia in chow-fed conditions that were observed in *Mc3r*−/− mice is modest when compared to mice lacking *Mc4r*.[62,63,89] While the phenotype in the chow-fed condition is modest, *Mc3r*−/− mice display accelerated obesity when fed high-fat diets.[86,87] In these conditions, adiposity, hyperinsulinemia, and impaired glucose tolerance are comparable to that associated with loss of *Mc4r* when using mice backcrossed onto the B6 background.[66]

The results from most of these studies suggest that obesity in *Mc3r*−/− mice does not involve hyperphagia, with altered nutrient partitioning owing to reduced fat oxidation as one possible mechanism.[63] In one study performed by our laboratory where food intake was measured using an automated system (CLAMS unit), there was moderate hyperphagia during the daytime when *Mc3r*−/− mice were first fed a high-fat diet.[63] However, the increased adiposity in *Mc3r*−/− mice in chow-fed conditions is subtle and appears to be stable at least to 6–11 months of age.[87,93] It is therefore unlikely that marked imbalances between energy intake and expenditure occur in this model. The differences in weight gain observed when *Mc3r*−/− and wild type mice are fed a high-fat diet can be more pronounced. However, the differences in energy balance are still technically hard to quantify using indirect calorimetry.[11] Our observations suggest that a subtle imbalance between fat intake and oxidation may be a significant factor in the accelerated diet-induced obesity.[63] Fatty acid oxidation and citrate synthase activity are reduced in skeletal muscle of female *Mc3r*−/− mice, suggesting reduced mitochondrial activity.[62] The physiological significance and mechanisms explaining the differences in mitochondrial activity have, however, not been further explored.

Another potential factor underlying the obese phenotype of *Mc3r*−/− mice is reduced physical activity. *Mc3r*−/− mice can exhibit reduced locomotor behavior in the home-cage and reduced wheel running in the dark phase.[32,63,86,87,186] Reduced physical activity-based or "nonresting" energy expenditure may therefore play a role in the obese phenotype observed in *Mc3r*−/− mice.[32,62] What has yet to be determined is whether reduced locomotor activity is a behavioral disorder (i.e., reduced frequency in the initiation of movement that might represent an adaptation to negative energy balance) or a result of deficits in skeletal muscle oxidative metabolism suggested by measurements of fatty acid oxidation and citrate synthase activity.[62]

A very recent article from Roger Cone's laboratory suggests the interesting hypothesis that MC3Rs are required for adaptation to fasting.[35] The increase in adipose tissue lipolysis, accumulation of triglycerides in the liver, and increased activity of the hypothalamo–pituatary adrenal axis is compromised in *Mc3r−/−* mice. A modest increase in corticosterone in the morning when corticosterone levels are normally low was also proposed as a mechanism for explaining the increased fat mass and reduced lean mass observed in this model. These are important findings, as they are the first to define a real "stimulus-response" function for this receptor. While the site of action and mode of regulation remain to be determined, it is nevertheless interesting to note that the actions of this receptor appear to be dominant in times when AgRP expression is stimulated and POMC activity reduced. Whether this implies that AgRP is acting as a form of "agonist" to regulate the fasting response via modulation of hypothalamic MC3Rs or altered activity of arcuate MC3R/GHSR heterodimers in fasted conditions are involved remains to be determined.

5.3. Response of melanocortin receptor knockout mice to melanocortin analogs

Knockout mice provide an important biological tool to investigate the specificity of action when compounds with high selectivity are not available. They are also a useful tool for investigating "off target" responses in highly selective compounds. Several groups have used *Mc3r−/−* and *Mc4r−/−* mice to dissect the role of each receptor in mediating the response to melanocortin analogs, examining for loss of specific responses. In general, the outcomes from these studies have been inconsistent and, in most cases, have not been that informative with respect to defining the actions of MC3Rs.

Marsh *et al.* reported that *Mc4r−/−* mice do not respond to centrally administered MTII, a potent nonselective analog, indicating that the inhibitory effect of melanocortin agonists is dependent on functional MC4Rs and that MC3Rs cannot compensate.[30] The group at Merck also reported that *Mc4r−/−* mice do not exhibit reduced food intake and increased oxygen consumption in response to peripherally administered MTII. MC4Rs may therefore be required for regulating both aspects of the energy balance equation in situations where energy balance is positive. MC3Rs are also not able to compensate in the absence of MC4Rs. Consistent with this hypothesis, in an experiment designed to examine the response of the melanocortin receptor mutant mice in a more physiological setting, the increase in oxygen consumption associated with hyperphagia is not observed in *Mc4r*−/− mice, but is observed in *Mc3r*−/− mice.[60,64] Again, however, this result needs to be interpreted with caution. It could indicate different responses of secondary neurons expressing either receptor following activation of POMC neurons in response to overfeeding, which has been observed in rats.[187] However, another plausible interpretation would be a "pre-existing" metabolic condition (e.g., reduced sympathetic tone of *Mc4r*−/− mice) that alters the response of peripheral tissues to dietary modifications.

We have also investigated the response of *Mc3r*−/− and *Mc4r*−/− mice to melanocortin receptor agonists,[93] and their response to compounds targeting serotonin receptors expressed by POMC and AgRP neurons.[66,146] The melanocortin analogs provided by Biomeasure (BIM-22493, BIM-22511) are potent (EC_{50} ranging from <1 to 10 nM) agonists for the human MC1R, MC3R, and MC4R, but with low activity at the human MC5R.[93] BIM-22493 is effective when administered by intraperitoneal injection at inhibiting feeding in wild type and *Mc3r*−/− mice, but not in *Mc4r*−/− mice following an overnight fast. A single injection of BIM-22493 also acutely improved glucose tolerance in obese leptin-deficient (Lep^{ob}/Lep^{ob}) mice, indicating that activation of melanocortin receptors rapidly improves glucose disposal. Studies analyzing melanocortin receptor knockout mice treated with serotonin receptor agonists that improve insulin action through regulating POMC neurons suggest that stimulation of MC4Rs on preganglionic autonomic neurons is the most probable "central" mechanism mediating this response.[65,66,188]

We also compared the response of wild type, *Mc3r*−/− and *Mc4r*−/− mice to BIM-22511 administered by osmotic mini-pump over 2 weeks (Fig. 4.2, Table 4.1). Weight loss was observed in wild type and *Mc3r*−/− mice, but was not observed in *Mc4r*−/− mice. Weight loss associated with administration of melanocortin receptor agonists is therefore dependent on functional MC4Rs. Surprisingly there was no significant effect on cumulative food intake in wild type or *Mc3r*−/− mice. However, *Mc4r*−/− mice exhibited a modest increase in food intake. Why no significant effect on food intake was observed in wild type mice is not clear, but could be due to the relatively lean phenotype of the wild type and *Mc3r*−/− mice and modest changes in body weight. This study also used female mice of the B6 background. It would be interesting to repeat these studies with male mice where body weight and adiposity are increased and matched between groups using a high fat diet, as was done for the studies examining the role of melanocortin receptors in response to serotonin receptor agonists.[66]

Another interesting observation made during this experiment was the reduction in fasting insulin and glucose observed in *Mc4r−/−* mice treated with BIM-22511 for 2 weeks. The reduction (50%) was similar in magnitude to that observed in male diet-induced obese mice, but is clearly not dependent on weight loss. This outcome suggests that some of the effects of melanocortin agonists to improve insulin sensitivity are independent of MC4Rs. The improved insulin sensitivity is probably not hepatic, as the severe fatty liver phenotype and increased expression of insulin-responsive genes associated with obesity was not improved by the treatment. This may suggest a specific role for the functions of MC4Rs in maintaining a healthy liver, or that weight loss is important for improving hepatic steatosis.

The agonist used for this study has very low activity at the MC5R, suggesting that these effects involved MC3Rs (and/or perhaps MC1Rs). This effect could be a peripheral response, as stimulation of central MC3Rs in the studies using serotonin agonists was not effective.[66] Future studies exploring whether MC3R agonists can improve insulin sensitivity and the tissues involved are clearly needed. However, a role for MC3Rs in regulating systems that impact on insulin action is not without precedent. *Mc3r−/−* mice fed a high-salt diet develop hypertension and insulin resistance.[95] As discussed in the next section, we have observed what appears to be extrahepatic insulin resistance in *Mc3r−/−* mice subjected to restricted feeding (RF) protocols where food is available in the daytime.[92]

It is important to note, however, that there are discrepancies in the literature. For example, there have been reports indicating differences in the response of melanocortin receptor knockout mice to MTII. Marsh *et al.* observed that AgRP retains an ability to stimulate food intake in *Mc4r−/−* mice when administered icv.,[189] although another group only observed a tendency for response when AgRP was administered over 3 days.[190] One interpretation of these results is that the "agonist" properties of AgRP at the MC3R can have a modest effect to increase food intake. The basis for this interpretation is that no decrease in food intake has been observed in most studies where MC3Rs are stimulated either directly or indirectly[30,66,93,100,146]; simple competitive antagonism by AgRP to inhibit the actions of the α- and γ-MSH at this receptor therefore seems an unlikely explanation for the increase in appetite.

We have observed that PG932, a derivative of SHU9119 (the first functional MC3/4R antagonist reported to increase food intake when administered icv.[54]) and also an MC3/4R antagonist, stimulates food intake when administered peripherally, and that this action is dependent on MC4Rs.[191] Navarro *et al.* observed what appeared to be a stimulatory effect of MTII administered icv. to *Mc4r*−/− mice, while the inhibitory effects of peripherally administered MTII were retained.[192] On the other hand, [194] observed an attenuation of the response of *Mc4r*−/− (and *Mc3r*−/−) mice to centrally administered MTII, with a complete loss of responsiveness in double *Mc3r*−/−.*Mc4r*−/− mice.

It is really not clear how we can reconcile these differences in the response of knockout mice to what is considered the "gold standard" melanocortin agonist used to investigate the behavioral and metabolic responses in rodents (MTII). Of course, technical and environmental variation between laboratory settings could be a factor. However, it is also clear that our understanding of the knowledge of melanocortin receptor signaling and pharmacology remains incomplete. There is a clear need for the development of highly selective compounds with clearly defined agonist/antagonist properties at the MC3R and MC4R. These compounds should also be characterized using mouse receptors, as there can be species specificity in activity.[193] The pharmacology is further complicated by the issue of biased agonism (i.e., ligands of the same receptor that can stabilize distinct receptor conformations and preferentially modulate one signaling pathway to the exclusion of others) and by the heterodimers formed between the MC3R and MC4Rs with other GPCRs that are involved in the regulation of ingestive behaviors and metabolism.

5.4. MC3Rs and the entrainment of rhythms anticipating food presentation: *Mc3r*−/− mice as a form of circadian mutant with reduced sensitivity to caloric cues?

Over the last 5 years, one focus of our laboratory has been to investigate whether MC3Rs are modulators of feeding-related inputs into neurons governing the development of food anticipatory activity (FAA).[32–34,91,92,195] In the circadian research field, scheduled feeding protocols are frequently used to entrain food-seeking behavior in rodents. Indeed, restricting nutrient availability to defined periods at 24 h intervals induces rhythms of locomotor activity and increased wakefulness that anticipate food presentation.[196] In most life forms, the synchronization of rhythms in activity, metabolism, and ingestive behaviors is essential for adaptation to cyclical environmental changes. The ability to predict food intake is an important adaptive feature for surviving competitive environments where nutrients may be limited while also having an important role in the daily maintenance of metabolic homeostasis.[197]

Biological clocks play an important but sill poorly understood role in synchronizing rhythms with caloric intake. The "molecular clock" machinery is broadly expressed in the CNS and in numerous peripheral tissues.[198,199] The core clock machinery responsible for these rhythms is composed of nuclear transcription factors (Bmal1, Clock, Npas2, RORα) and transcription repressors (Per1/2, Cry1/2, Rev-erbα) that establish rhythms of approximately 24 h.[195,200] These transcription factors affect the activity of a large proportion of the transcriptome (approximately 3000 genes) by direct or indirect mechanisms. In mammals, many aspects of behavior and physiology including sleep/wake cycles, locomotor activity, blood pressure, body temperature, hormone secretion, and metabolic pathways exhibit daily rhythmicity under the control of circadian clock.[195,200] Presented first as simple timekeepers, clocks oscillators are now recognized to be essential in metabolic processes and the maintenance of energy homeostasis. Emerging evidence indicates that disturbances of circadian rhythms because of either mutations in clock genes[201,202] or lifestyle modifications[203,204] increase the risk of developing a metabolic syndrome.

The master clock that synchronizes daily rhythms with photoperiod resides in the suprachiasmatic nucleus (SCN) of the hypothalamus. The master clock is able to maintain rhythms even in absence of photic cues. Outside the SCN, other hypothalamic structures have been implicated for maintaining expression of rhythms especially during periods of nutrient insufficiency.[205,206] Clocks residing outside the SCN are rapidly entrained by feeding-related cues that represent the dominant "zeitgeber" or time giver. These clocks in turn send outputs that regulate many enzymatic processes related to energy homeostasis. Because of their properties to sense and rapidly respond to nutritional cues, these oscillator centers located outside of the SCN have been suggested to constitute the Food Entrainable Oscillator (FEO).[207,208] The entrainment to feeding has been proposed to involve clocks distributed in the brain and in peripheral tissues suggesting the existence of several FEO loci. In peripheral tissues, clock oscillators can be rapidly entrained by hormonal signals and restricting feeding.[198] In the liver, clock gene expression is entrained within 2 h even before apparition of FAA suggesting that peripheral oscillators are directly responsive to cellular energy status.[209]

To determine whether MC3Rs regulate inputs into systems governing the expression of rhythms anticipating nutrient intake, wild type and *Mc3r* knockout mice were subjected to a RF protocol. After acclimation to housing conditions where activity can be monitored using running wheels[33,34,92] or infrared beam breaks,[91] access to food was limited to defined intervals with a period length of 24 h. Studies using two models of *Mc3r* deficiency (the original *Mc3r*−/− mouse[86] backcrossed onto the B6 background and the $Mc3r^{TB/TB}$ mouse[32] developed by our laboratory discussed later in this review) where running wheels were used to measure FAA yielded similar results. They suggest that MC3Rs are necessary for the coordinated development of anticipatory activity and increased wakefulness in response to RF.[32,34] Wild type mice subjected to the RF rapidly respond by increasing their activity and vigilance 2 h prior food presentation, while *Mc3r*−/− mice exhibit a significant attenuation of FAA. A similar outcome was observed when *Mc3r*−/− mice housed in constant dark were subjected to the RF protocol.[33]

We have also observed abnormal rhythmic expression of clock genes in the cortex of *Mc3r*−/− mice in *ad libitum* and RF conditions.[32–34] MC3Rs may thus have a critical role in maintaining the activity of clocks distributed throughout the central nervous system. This is an important observation as clocks have been implicated in a range of behavioral disorders including bipolar disorder, depression, mania, sensitivity to drugs, and the formation and consolidation of memories.[210]

A recent paper published in *PLoS One* examined the phenotype of double Lep^{ob}/Lep^{ob}.*Mc3r*−/− mice, and suggested that MC3Rs are not essential for FAA.[211] The authors reported an inhibitory effect of leptin on FAA, as anticipated based on previous studies using rats with mutations in the leptin receptor.[212] Lep^{ob}/Lep^{ob}.*Mc3r*−/− mice actually displayed increased FAA while the inhibitory effects of leptin were preserved. While this result appears at face value to discount a role for MC3Rs in modulating signals governing FAA, these results should be viewed with caution. Bouret and Simerly observed that leptin plays a critical role as a trophic factor in the development of the hypothalamus, stimulating the growth of projections from the arcuate nucleus to other areas (e.g., dorsomedial hypothalamus and PVN).[213] Lep^{ob}/Lep^{ob} mice display marked and permanent disruptions in neural projections from the arcuate. The central nervous melanocortin system is therefore likely also immature in this model, and the results from studies using it are open to debate (including results of our own experiments examining the response of Lep^{ob}/Lep^{ob}.*Mc4r*−/− mice to leptin[135])

MC3Rs are also highly expressed in hypothalamic regions that are critically involved in metabolic homeostasis (e.g., the arcuate and ventromedial nuclei of the hypothalamus[96]). While some studies have reported the importance of clock genes in the expression of food anticipatory behavior, others have questioned this role arguing that the absence of FAA observed in clock mutants is more a consequence of a lethargic and moribund state due to the severe RF paradigm.[214,215] We therefore investigated whether *Mc3r*−/− mice would exhibit evidence of metabolic distress during RF.[92] While *Mc3r*−/− mice fed *ad libitum* exhibited normal glycemia and insulinemia throughout the circadian day, remarkably, they developed a state of insulin resistance and dyslipidemia during RF despite the weight loss promoted by the RF. They also exhibited hyperglycemia, hyperinsulinemia, and glucose intolerance. The expression of genes involved in lipogenesis in the liver was increased as were serum triglycerides. The expression of insulin-responsive lipogenic enzymes correlated with the degree of hyperinsulinism, while the expression of genes involved in gluconeogenesis was reduced. This combination suggests a "mixed insulin resistant" phenotype, where the liver is responding normally to insulin while other tissues responsible for glucose clearance such as skeletal muscle are probably insulin resistant. In contrast, obese *Mc4r*−/− and A^Y/a mice responded to RF by exhibiting the anticipated weight loss and improvements in glucose clearance.[92] *Mc3r*−/− mice thus display a unique combination of physical characteristics when subjected to daytime feeding, exhibiting deteriorating metabolic homeostasis while losing weight.

Mc3r−/− mice also exhibit abnormal rhythmicity in the expression of clock genes in the liver, and abnormal rhythmicity of metabolic oscillators implicated in glucose and lipid metabolism during the RF.[33,92] While this could indicate a role for MC3R to influence rhythms of the liver clock, it is possible that this result is due to feedback/inhibitory effects of abnormalities in metabolic homeostasis that develop in *Mc3r*−/− mice during RF.

5.5. Cell-specific restoration of *Mc3r* expression in LoxTB Mc3r mice: Potential for new insights into the biology of MC3Rs

Two factors have hindered progress in the research of MC3R function. One important problem that is beyond the focus of this article is a scarcity of compounds with specific agonist and antagonist properties. Another issue has been the lack of transgenic approaches using more sophisticated techniques beyond the global knockout approach originally employed to examine MC3R function. To address this issue, we have developed the LoxTB Mc3r mouse using a similar strategy used in the genetic dissection of Mc4r activity.[70] A "transcriptional block" flanked by LoxP site (LoxTB) was inserted into the 5′ UTR of the *Mc3r* gene, and various Cre-expressing transgenic strains were used to remove the LoxTB sequence and restore expression.[32]

Homozygous carriers of the LoxTB *Mc3r* allele exhibit the phenotype observed in all of the previously developed *Mc3r* knockout strains, with reduced lean mass and increased fat mass and normal food intake in chow-fed conditions. They also exhibit the accelerated diet-induced obese phenotype. Crossing LoxTB *Mc3r* mouse with Eiia-Cre mice to remove the LoxTB sequence in early embryogenesis rescued the phenotype. We then used two Cre-expressing lines to answer two questions: the first was to use the Nestin-Cre transgene to investigate whether restoring central MC3R expression would rescue the obese phenotype as observed in the LoxTB *Mc4r* mouse.[70] The second was to examine the function of MC3R expressed in the VMH. The rationale for focusing on the VMH was based on previous studies of expression of the *Mc3r* gene in the rodent brain, and on the abundance of data suggesting functions of VMH neurons in metabolic homeostasis. The dorsomedial VMH (dmVMH) exhibits high levels of *Mc3r* expression relative to other hypothalamic and limbic structures.[96] The VMH is also known to have a role in the regulation of outputs governing glucose homeostasis and that impact on obesity.[216] When the project began, lesioning experiments had suggested a role for the VMH in the expression of FAA.[205,217] More defined studies had suggested that the dmVMH may have an inhibitory role in the expression of FAA,[206] while others have questioned whether any of the hypothalamic nuclei targeted using the lesion approach are required for FAA.[207,218]

The first surprising outcome from these studies was the failure of the Nestin-Cre transgenic to restore a normal body weight phenotype. Expression of MC3R in the brain attenuated the obese phenotype in chow-fed conditions, while having no affect on the obese phenotype when mice were fed a high-fat diet.[32] While this experiment is being repeated, the results nevertheless are startling in suggesting that the functions of peripheral MC3R have some role in the defense of body weight when dietary fat intake is elevated. Ongoing studies are further exploring the phenotype of NesCre. LoxTB Mc3r mice in an attempt to identify the role of peripheral MC3R. We are also investigating whether the expression of FAA is restored in NesCre.LoxTB *Mc3r* mice. While speculative, one possible outcome is that the regulation of circadian-related behaviors will involve "central" MC3R, while the accelerated diet-induced obese phenotype is due to loss of peripheral receptors that impact on metabolic homeostasis.

Sf1Cre.LoxTB *Mc3r* mice exhibited a very modest attenuation of the obese phenotype, while the reduction in lean mass was similar to that observed in LoxTB *Mc3r* mice. There was evidence for improvements in metabolic homeostasis, with fasting hyperinsulinemia normalized suggesting improved insulin sensitivity. Changes in liver gene expression in LoxTB

Mc3r mice suggested altered lipid metabolism, and perhaps fatty liver, were also improved in Sf1Cre.LoxTB *Mc3r* mice. These observations suggest that the functions of MC3Rs expressed in Sf1(+ve) cells can improve metabolic homeostasis in situations of mild obesity. This effect was not due to altered food intake, while analysis using indirect calorimetry suggested altered metabolism. In this group of mice, a significant reduction in the RER was observed in LoxTB Mc3r mice. This suggests a change in substrate preference with less carbohydrate oxidation and increased fat oxidation. LoxTB *Mc3r* mice also exhibited a normal resting energy expenditure (REE) and reductions in "nonresting energy expenditure" (NREE), which was calculated by subtracting REE from total daily energy expenditure.

6. CONCLUSIONS AND FUTURE DIRECTIONS

Over the last decade, technologic advances in genetics, pharmacology, and molecular biology have largely improved our understanding of how the melanocortin system controls energy balance. This system is important in integrating many afferent inputs with behavioral and autonomic responses that adjust food intake and energy expenditure in order to maintain energy homeostasis.

Indeed, melanocortin neurons not only regulate food intake by processing acute and long-term signals related to nutrient intake and adipostatic status but also play a role in regulating energy expenditure, via effects on autonomic outflow to a variety of tissues, as well as effects on endocrine axes. Generation of more refined transgenic animal models harboring disruption of MC3Rs in specific brain areas enabled dissection of neuronal subpopulations that control food intake and/or energy expenditure. Studies conducted in our laboratory have participated in determining the role of MC3R in energy homeostasis. Our results indicate a new homeostatic function of this receptor that does not implicate the classical pathway of reduced food intake and increased energy expenditure to maintain energy balance. Indeed, our studies indicate that MC3Rs could be important mediators in transmitting feeding-related inputs into neurons governing the development of anticipatory activity especially during periods of nutrient insufficiency. MC3R expression is important for maintaining glucose homeostasis and rhythmicity of clock genes known to regulate gluconeogenic and lipogenic enzymes as well as rhythms in glucose and lipid metabolism. Although these findings have contributed to a better understanding of the function of MC3R in energy homeostasis, further investigations targeting reactivation of MC3R in specific brain areas will be useful to identify the neuronal subpopulations of MC3Rs involved in the maintenance of energy homeostasis and circadian rhythms.

ACKNOWLEDGMENTS

Research on the physiological roles of the melanocortin receptors by Andrew Butler's laboratory has been supported by the National Institutes of Health (DK073189), The Pennington Biomedical Research Foundation, and The Scripps Research Institute.

REFERENCES

1. Flegal KM, Carroll MD, Ogden CL, Johnson CL. Prevalence and trends in obesity among US adults, 1999–2000. *JAMA* 2002;**288**(14):1723–7.
2. Rosenbaum M, Goldsmith R, Bloomfield D, Magnano A, Weimer L, Heymsfield S, et al. Low-dose leptin reverses skeletal muscle, autonomic, and neuroendocrine adaptations to maintenance of reduced weight. *J Clin Invest* 2005;**115**(12):3579–86.
3. Leibel RL, Rosenbaum M, Hirsch J. Changes in energy expenditure resulting from altered body weight. *N Engl J Med* 1995;**332**(10):621–8.
4. Martin CK, Heilbronn LK, de Jonge L, DeLany JP, Volaufova J, Anton SD, et al. Effect of calorie restriction on resting metabolic rate and spontaneous physical activity. *Obesity (Silver Spring)* 2007;**15**(12):2964–73.
5. Stefater MA, Wilson-Perez HE, Chambers AP, Sandoval DA, Seeley RJ. All bariatric surgeries are not created equal: insights from mechanistic comparisons. *Endocr Rev* 2012;**33**:595–622.
6. Cone RD. Anatomy and regulation of the central melanocortin system. *Nat Neurosci* 2005;**8**(5):571–8.
7. Myers Jr MG, Munzberg H, Leinninger GM, Leshan RL. The geometry of leptin action in the brain: more complicated than a simple ARC. *Cell Metab* 2009;**9**(2):117–23.
8. Garfield AS, Lam DD, Marston OJ, Przydzial MJ, Heisler LK. Role of central melanocortin pathways in energy homeostasis. *Trends Endocrinol Metab* 2009;**20**(5):203–15.
9. Lowell BB, Spiegelman BM. Towards a molecular understanding of adaptive thermogenesis. *Nature* 2000;**404**(6778):652–60.
10. Tschop MH, Speakman JR, Arch JR, Auwerx J, Bruning JC, Chan L, et al. A guide to analysis of mouse energy metabolism. *Nat Methods* 2011;**9**(1):57–63.
11. Butler AA, Kozak LP. A recurring problem with the analysis of energy expenditure in genetic models expressing lean and obese phenotypes. *Diabetes* 2010;**59**(2):323–9.
12. Ravussin E, Galgani JE. The implication of brown adipose tissue for humans. *Annu Rev Nutr* 2011;**31**:33–47.
13. Kozak LP. Brown fat and the myth of diet-induced thermogenesis. *Cell Metab* 2010;**11**(4):263–7.
14. Cannon B, Nedergaard J. Nonshivering thermogenesis and its adequate measurement in metabolic studies. *J Exp Biol* 2011;**214**(Pt 2):242–53.
15. Flatt JP. Issues and misconceptions about obesity. *Obesity (Silver Spring)* 2011;**19**(4):676–86.
16. Speakman JR, Levitsky DA, Allison DB, Bray MS, de Castro JM, Clegg DJ, et al. Set points, settling points and some alternative models: theoretical options to understand how genes and environments combine to regulate body adiposity. *Dis Model Mech* 2011;**4**(6):733–45.

17. Schwartz MW, Woods SC, Seeley RJ, Barsh GS, Baskin DG, Leibel RL. Is the energy homeostasis system inherently biased toward weight gain? *Diabetes* 2003;**52**(2):232–8.
18. King BM. The rise, fall, and resurrection of the ventromedial hypothalamus in the regulation of feeding behavior and body weight. *Physiol Behav* 2006;**87**(2):221–44.
19. Bray GA, York DA. The MONA LISA hypothesis in the time of leptin. *Recent Prog Horm Res* 1998;**53**:95–117 [discussion 117-8].
20. Friedman JM. A war on obesity, not the obese. *Science* 2003;**299**(5608):856–8.
21. O'Rahilly S. Human genetics illuminates the paths to metabolic disease. *Nature* 2009;**462**(7271):307–14.
22. Coleman DL. A historical perspective on leptin. *Nat Med* 2010;**16**(10):1097–9.
23. Zhang Y, Proenca R, Maffei M, Barone M, Leopold L, Friedman JM. Positional cloning of the mouse obese gene and its human homologue. *Nature* 1994;**372**(6505):425–32.
24. Tartaglia LA, Dembski M, Weng X, Deng N, Culpepper J, Devos R, et al. Identification and expression cloning of a leptin receptor, OB-R. *Cell* 1995;**83**(7):1263–71.
25. Farooqi IS, O'Rahilly S. Mutations in ligands and receptors of the leptin-melanocortin pathway that lead to obesity. *Nat Clin Pract Endocrinol Metab* 2008;**4**(10):569–77.
26. Gautron L, Elmquist JK. Sixteen years and counting: an update on leptin in energy balance. *J Clin Invest* 2011;**121**(6):2087–93.
27. Farooqi IS, Keogh JM, Kamath S, Jones S, Gibson WT, Trussell R, et al. Partial leptin deficiency and human adiposity. *Nature* 2001;**414**(6859):34–5.
28. Myers Jr MG, Leibel RL, Seeley RJ, Schwartz MW. Obesity and leptin resistance: distinguishing cause from effect. *Trends Endocrinol Metab* 2010;**21**(11):643–51.
29. Williams KW, Margatho LO, Lee CE, Choi M, Lee S, Scott MM, et al. Segregation of acute leptin and insulin effects in distinct populations of arcuate proopiomelanocortin neurons. *J Neurosci* 2010;**30**(7):2472–9.
30. Marsh DJ, Hollopeter G, Huszar D, Laufer R, Yagaloff KA, Fisher SL, et al. Response of melanocortin-4 receptor-deficient mice to anorectic and orexigenic peptides. *Nat Genet* 1999;**21**(1):119–22.
31. Lechan RM, Fekete C. The TRH neuron: a hypothalamic integrator of energy metabolism. *Prog Brain Res* 2006;**153**:209–35.
32. Begriche K, Levasseur PR, Zhang J, Rossi J, Skorupa D, Solt LA, et al. Genetic dissection of the functions of the melanocortin-3 receptor, a seven-transmembrane g-protein coupled receptor, suggests roles for central and peripheral receptors in energy homeostasis. *J Biol Chem* 2011;**286**(47):40771–81.
33. Begriche K, Sutton GM, Butler AA. Homeostastic and non-homeostatic functions of melanocortin-3 receptors in the control of energy balance and metabolism. *Physiol Behav* 2011;**104**(4):546–54.
34. Sutton GM, Perez-Tilve D, Nogueiras R, Fang J, Kim JK, Cone RD, et al. The melanocortin-3 receptor is required for entrainment to meal intake. *J Neurosci* 2008;**28**(48):12946–55.
35. Renquist BJ, Murphy JG, Larson EA, Olsen D, Klein RF, Ellacott KL, et al. Melanocortin-3 receptor regulates the normal fasting response. *Proc Natl Acad Sci USA* 2012;**109**(23):E1489–E1498.
36. Friedman JM. Leptin at 14 y of age: an ongoing story. *Am J Clin Nutr* 2009;**89**(3):973S–979S.
37. Asilmaz E, Cohen P, Miyazaki M, Dobrzyn P, Ueki K, Fayzikhodjaeva G, et al. Site and mechanism of leptin action in a rodent form of congenital lipodystrophy. *J Clin Invest* 2004;**113**(3):414–24.
38. Cohen P, Zhao C, Cai X, Montez JM, Rohani SC, Feinstein P, et al. Selective deletion of leptin receptor in neurons leads to obesity. *J Clin Invest* 2001;**108**(8):1113–21.

39. Vong L, Ye C, Yang Z, Choi B, Chua Jr S, Lowell BB. Leptin action on GABAergic neurons prevents obesity and reduces inhibitory tone to POMC neurons. *Neuron* 2011;**71**(1):142–54.
40. Guo K, McMinn JE, Ludwig T, Yu YH, Yang G, Chen L, et al. Disruption of peripheral leptin signaling in mice results in hyperleptinemia without associated metabolic abnormalities. *Endocrinology* 2007;**148**(8):3987–97.
41. de Luca C, Kowalski TJ, Zhang Y, Elmquist JK, Lee C, Kilimann MW, et al. Complete rescue of obesity, diabetes, and infertility in db/db mice by neuron-specific LEPR-B transgenes. *J Clin Invest* 2005;**115**(12):3484–93.
42. Halaas JL, Boozer C, Blair-West J, Fidahusein N, Denton DA, Friedman JM. Physiological response to long-term peripheral and central leptin infusion in lean and obese mice. *Proc Natl Acad Sci USA* 1997;**94**(16):8878–83.
43. Ahima RS, Prabakaran D, Mantzoros C, Qu D, Lowell B, Maratos-Flier E, et al. Role of leptin in the neuroendocrine response to fasting. *Nature* 1996;**382**(6588):250–2.
44. Rosenbaum M, Sy M, Pavlovich K, Leibel RL, Hirsch J. Leptin reverses weight loss-induced changes in regional neural activity responses to visual food stimuli. *J Clin Invest* 2008;**118**(7):2583–91.
45. Rosenbaum M, Murphy EM, Heymsfield SB, Matthews DE, Leibel RL. Low dose leptin administration reverses effects of sustained weight-reduction on energy expenditure and circulating concentrations of thyroid hormones. *J Clin Endocrinol Metab* 2002;**87**(5):2391–4.
46. Galgani JE, Greenway FL, Caglayan S, Wong ML, Licinio J, Ravussin E. Leptin replacement prevents weight loss-induced metabolic adaptation in congenital leptin-deficient patients. *J Clin Endocrinol Metab* 2010;**95**(2):851–5.
47. Enriori PJ, Evans AE, Sinnayah P, Jobst EE, Tonelli-Lemos L, Billes SK, et al. Diet-induced obesity causes severe but reversible leptin resistance in arcuate melanocortin neurons. *Cell Metab* 2007;**5**(3):181–94.
48. Knight ZA, Hannan KS, Greenberg ML, Friedman JM. Hyperleptinemia is required for the development of leptin resistance. *PLoS One* 2010;**5**(6):e11376.
49. Trevaskis JL, Meyer EA, Galgani JE, Butler AA. Counterintuitive effects of double-heterozygous null melanocortin-4 receptor and leptin genes on diet-induced obesity and insulin resistance in C57BL/6J mice. *Endocrinology* 2008;**149**(1):174–84.
50. Yen PM, Chin WW. Molecular mechanisms of dominant negative activity by nuclear hormone receptors. *Mol Endocrinol* 1994;**8**(11):1450–4.
51. Miller MW, Duhl DM, Vrieling H, Cordes SP, Ollmann MM, Winkes BM, et al. Cloning of the mouse agouti gene predicts a secreted protein ubiquitously expressed in mice carrying the lethal yellow mutation. *Genes Dev* 1993;**7**(3):454–67.
52. Bultman SJ, Michaud EJ, Woychik RP. Molecular characterization of the mouse agouti locus. *Cell* 1992;**71**(7):1195–204.
53. Lu D, Willard D, Patel IR, Kadwell S, Overton L, Kost T, et al. Agouti protein is an antagonist of the melanocyte-stimulating-hormone receptor. *Nature* 1994;**371**(6500):799–802.
54. Fan W, Boston BA, Kesterson RA, Hruby VJ, Cone RD. Role of melanocortinergic neurons in feeding and the agouti obesity syndrome. *Nature* 1997;**385**(6612):165–8.
55. Mountjoy KG, Mortrud MT, Low MJ, Simerly RB, Cone RD. Localization of the melanocortin-4 receptor (MC4-R) in neuroendocrine and autonomic control circuits in the brain. *Mol Endocrinol* 1994;**8**(10):1298–308.
56. Huszar D, Lynch CA, Fairchild-Huntress V, Dunmore JH, Fang Q, Berkemeier LR, et al. Targeted disruption of the melanocortin-4 receptor results in obesity in mice. *Cell* 1997;**88**(1):131–41.
57. Vaisse C, Clement K, Guy-Grand B, Froguel P. A frameshift mutation in human MC4R is associated with a dominant form of obesity. *Nat Genet* 1998;**20**(2):113–4.

58. Yeo GS, Farooqi IS, Aminian S, Halsall DJ, Stanhope RG, O'Rahilly S. A frameshift mutation in MC4R associated with dominantly inherited human obesity. *Nat Genet* 1998;**20**(2):111–2.
59. Farooqi IS, Keogh JM, Yeo GS, Lank EJ, Cheetham T, O'Rahilly S. Clinical spectrum of obesity and mutations in the melanocortin 4 receptor gene. *N Engl J Med* 2003;**348** (12):1085–95.
60. Butler AA, Marks DL, Fan W, Kuhn CM, Bartolome M, Cone RD. Melanocortin-4 receptor is required for acute homeostatic responses to increased dietary fat. *Nat Neurosci* 2001;**4**(6):605–11.
61. Ste Marie L, Miura GI, Marsh DJ, Yagaloff K, Palmiter RD. A metabolic defect promotes obesity in mice lacking melanocortin-4 receptors. *Proc Natl Acad Sci USA* 2000;**97**(22):12339–44.
62. Sutton GM, Trevaskis JL, Hulver MW, McMillan RP, Markward NJ, Babin MJ, et al. Diet-genotype interactions in the development of the obese, insulin-resistant phenotype of C57BL/6J mice lacking melanocortin-3 or -4 receptors. *Endocrinology* 2006;**147** (5):2183–96.
63. Butler AA. The melanocortin system and energy balance. *Peptides* 2006;**27**(2):281–90.
64. Albarado DC, McClaine J, Stephens JM, Mynatt RL, Ye J, Bannon AW, et al. Impaired coordination of nutrient intake and substrate oxidation in melanocortin-4 receptor knockout mice. *Endocrinology* 2004;**145**(1):243–52.
65. Rossi J, Balthasar N, Olson D, Scott M, Berglund E, Lee CE, et al. Melanocortin-4 receptors expressed by cholinergic neurons regulate energy balance and glucose homeostasis. *Cell Metab* 2011;**13**(2):195–204.
66. Zhou L, Sutton GM, Rochford JJ, Semple RK, Lam DD, Oksanen LJ, et al. Serotonin 2C receptor agonists improve type 2 diabetes via melanocortin-4 receptor signaling pathways. *Cell Metab* 2007;**6**(5):398–405.
67. Nogueiras R, Wiedmer P, Perez-Tilve D, Veyrat-Durebex C, Keogh JM, Sutton GM, et al. The central melanocortin system directly controls peripheral lipid metabolism. *J Clin Invest* 2007;**117**(11):3475–88.
68. Obici S, Feng Z, Tan J, Liu L, Karkanias G, Rossetti L. Central melanocortin receptors regulate insulin action. *J Clin Invest* 2001;**108**(7):1079–85.
69. Huo L, Gamber K, Greeley S, Silva J, Huntoon N, Leng XH, et al. Leptin-dependent control of glucose balance and locomotor activity by POMC neurons. *Cell Metab* 2009;**9**(6):537–47.
70. Balthasar N, Dalgaard LT, Lee CE, Yu J, Funahashi H, Williams T, et al. Divergence of melanocortin pathways in the control of food intake and energy expenditure. *Cell* 2005;**123**(3):493–505.
71. Wan S, Browning KN, Coleman FH, Sutton G, Zheng H, Butler A, et al. Presynaptic melanocortin-4 receptors on vagal afferent fibers modulate the excitability of rat nucleus tractus solitarius neurons. *J Neurosci* 2008;**28**(19):4957–66.
72. Fan W, Ellacott KL, Halatchev IG, Takahashi K, Yu P, Cone RD. Cholecystokinin-mediated suppression of feeding involves the brainstem melanocortin system. *Nat Neurosci* 2004;**7**(4):335–6.
73. Zheng H, Patterson LM, Phifer CB, Berthoud HR. Brain stem melanocortinergic modulation of meal size and identification of hypothalamic POMC projections. *Am J Physiol Regul Integr Comp Physiol* 2005;**289**(1):R247–R258.
74. Sutton GM, Duos B, Patterson LM, Berthoud HR. Melanocortinergic modulation of cholecystokinin-induced suppression of feeding through extracellular signal-regulated kinase signaling in rat solitary nucleus. *Endocrinology* 2005;**146**(9):3739–47.
75. Williams DL, Kaplan JM, Grill HJ. The role of the dorsal vagal complex and the vagus nerve in feeding effects of melanocortin-3/4 receptor stimulation. *Endocrinology* 2000;**141**(4):1332–7.

76. Grill HJ, Ginsberg AB, Seeley RJ, Kaplan JM. Brainstem application of melanocortin receptor ligands produces long-lasting effects on feeding and body weight. *J Neurosci* 1998;**18**(23):10128–35.
77. Cowley MA, Pronchuk N, Fan W, Dinulescu DM, Colmers WF, Cone RD. Integration of NPY, AGRP, and melanocortin signals in the hypothalamic paraventricular nucleus: evidence of a cellular basis for the adipostat. *Neuron* 1999; **24**(1):155–63.
78. Williams DL, Schwartz MW. Neuroanatomy of body weight control: lessons learned from leptin. *J Clin Invest* 2011;**121**(6):2152–5.
79. Dong JZ, DeOliveira DB, Halem HA, Taylor JE, Roubert P, Plas P, et al. Novel melanocortin-4 receptor agonists that decrease food intake and body weight. *Adv Exp Med Biol* 2009;**611**:485–6.
80. Pantel J, Williams SY, Mi D, Sebag J, Corbin JD, Weaver CD, et al. Development of a high throughput screen for allosteric modulators of melanocortin-4 receptor signaling using a real time cAMP assay. *Eur J Pharmacol* 2011;**660**(1):139–47.
81. Albert JS, Blomberg N, Breeze AL, Brown AJ, Burrows JN, Edwards PD, et al. An integrated approach to fragment-based lead generation: philosophy, strategy and case studies from AstraZeneca's drug discovery programmes. *Curr Top Med Chem* 2007; **7**(16):1600–29.
82. Emmerson PJ, Fisher MJ, Yan LZ, Mayer JP. Melanocortin-4 receptor agonists for the treatment of obesity. *Curr Top Med Chem* 2007;**7**(11):1121–30.
83. Ujjainwalla F, Sebhat IK. Small molecule ligands of the human melanocortin-4 receptor. *Curr Top Med Chem* 2007;**7**(11):1068–84.
84. Hruby VJ, Cai M, Cain J, Nyberg J, Trivedi D. Design of novel melanocortin receptor ligands: multiple receptors, complex pharmacology, the challenge. *Eur J Pharmacol* 2011;**660**(1):88–93.
85. Greenfield JR, Miller JW, Keogh JM, Henning E, Satterwhite JH, Cameron GS, et al. Modulation of blood pressure by central melanocortinergic pathways. *N Engl J Med* 2009;**360**(1):44–52.
86. Butler AA, Kesterson RA, Khong K, Cullen MJ, Pelleymounter MA, Dekoning J, et al. A unique metabolic syndrome causes obesity in the melanocortin-3 receptor-deficient mouse. *Endocrinology* 2000;**141**(9):3518–21.
87. Chen AS, Marsh DJ, Trumbauer ME, Frazier EG, Guan XM, Yu H, et al. Inactivation of the mouse melanocortin-3 receptor results in increased fat mass and reduced lean body mass. *Nat Genet* 2000;**26**(1):97–102.
88. Trevaskis JL, Gawronska-Kozak B, Sutton GM, McNeil M, Stephens JM, Smith SR, et al. Role of adiponectin and inflammation in insulin resistance of Mc3r and Mc4r knockout mice. *Obesity (Silver Spring)* 2007;**15**(11):2664–72.
89. Ellacott KL, Murphy JG, Marks DL, Cone RD. Obesity-induced inflammation in white adipose tissue is attenuated by loss of melanocortin-3 receptor signaling. *Endocrinology* 2007;**148**(12):6186–94.
90. Zhang Y, Kilroy GE, Henagan TM, Prpic-Uhing V, Richards WG, Bannon AW, et al. Targeted deletion of melanocortin receptor subtypes 3 and 4, but not CART, alters nutrient partitioning and compromises behavioral and metabolic responses to leptin. *FASEB J* 2005;**19**(11):1482–91.
91. Begriche K, Maston O, Rossi J, Burke L, McDonald P, Heisler LK, et al. Melanocortin-3 receptors are required for metabolic adaptation to food restriction. *Genes Brain Behav* 2012;**11**(3):291–302.
92. Sutton GM, Begriche K, Kumar KG, Gimble JM, Perez-Tilve D, Nogueiras R, et al. Central nervous system melanocortin-3 receptors are required for synchronizing metabolism during entrainment to restricted feeding during the light cycle. *FASEB J* 2010;**24**(3):862–72.

93. Kumar KG, Sutton GM, Dong JZ, Roubert P, Plas P, Halem HA, et al. Analysis of the therapeutic functions of novel melanocortin receptor agonists in MC3R- and MC4R-deficient C57BL/6J mice. *Peptides* 2009;**30**(10):1892–900.
94. An JJ, Rhee Y, Kim SH, Kim DM, Han DH, Hwang JH, et al. Peripheral effect of alpha-melanocyte-stimulating hormone on fatty acid oxidation in skeletal muscle. *J Biol Chem* 2007;**282**(5):2862–70.
95. Ni XP, Humphreys MH. Abnormal glucose metabolism in hypertensive mice with genetically interrupted gamma-melanocyte stimulating hormone signaling fed a high-sodium diet. *Am J Hypertens* 2008;**21**(12):1284–7.
96. Roselli-Rehfuss L, Mountjoy KG, Robbins LS, Mortrud MT, Low MJ, Tatro JB, et al. Identification of a receptor for gamma melanotropin and other proopiomelanocortin peptides in the hypothalamus and limbic system. *Proc Natl Acad Sci USA* 1993; **90**(19):8856–60.
97. Adan RA, Cone RD, Burbach JP, Gispen WH. Differential effects of melanocortin peptides on neural melanocortin receptors. *Mol Pharmacol* 1994;**46**(6):1182–90.
98. Ollmann MM, Wilson BD, Yang YK, Kerns JA, Chen Y, Gantz I, et al. Antagonism of central melanocortin receptors in vitro and in vivo by agouti-related protein. *Science* 1997;**278**(5335):135–8.
99. Shutter JR, Graham M, Kinsey AC, Scully S, Luthy R, Stark KL. Hypothalamic expression of ART, a novel gene related to agouti, is up-regulated in obese and diabetic mutant mice. *Genes Dev* 1997;**11**(5):593–602.
100. Chen AS, Metzger JM, Trumbauer ME, Guan XM, Yu H, Frazier EG, et al. Role of the melanocortin-4 receptor in metabolic rate and food intake in mice. *Transgenic Res* 2000;**9**(2):145–54.
101. Rahmouni K, Haynes WG, Morgan DA, Mark AL. Role of melanocortin-4 receptors in mediating renal sympathoactivation to leptin and insulin. *J Neurosci* 2003;**23** (14):5998–6004.
102. Gantz I, Fong TM. The melanocortin system. *Am J Physiol Endocrinol Metab* 2003;**284** (3):E468–E474.
103. Nakanishi S, Inoue A, Kita T, Nakamura M, Chang AC, Cohen SN, et al. Nucleotide sequence of cloned cDNA for bovine corticotropin-beta-lipotropin precursor. *Nature* 1979;**278**(5703):423–7.
104. Kita T, Inoue A, Nakanishi S, Numa S. Purification and characterization of the messenger RNA coding for bovine corticotropin/beta-lipotropin precursor. *Eur J Biochem* 1979;**93**(2):213–20.
105. Roberts JL, Seeburg PH, Shine J, Herbert E, Baxter JD, Goodman HM. Corticotropin and beta-endorphin: construction and analysis of recombinant DNA complementary to mRNA for the common precursor. *Proc Natl Acad Sci USA* 1979;**76**(5):2153–7.
106. Ollmann MM, Lamoreux ML, Wilson BD, Barsh GS. Interaction of Agouti protein with the melanocortin 1 receptor in vitro and in vivo. *Genes Dev* 1998;**12**(3):316–30.
107. Yang YK, Ollmann MM, Wilson BD, Dickinson C, Yamada T, Barsh GS, et al. Effects of recombinant agouti-signaling protein on melanocortin action. *Mol Endocrinol* 1997;**11**(3):274–80.
108. Rosenfeld RD, Zeni L, Welcher AA, Narhi LO, Hale C, Marasco J, et al. Biochemical, biophysical, and pharmacological characterization of bacterially expressed human agouti-related protein. *Biochemistry* 1998;**37**(46):16041–52.
109. Klebig ML, Wilkinson JE, Geisler JG, Woychik RP. Ectopic expression of the agouti gene in transgenic mice causes obesity, features of type II diabetes, and yellow fur. *Proc Natl Acad Sci USA* 1995;**92**(11):4728–32.
110. Perry WL, Hustad CM, Swing DA, Jenkins NA, Copeland NG. A transgenic mouse assay for agouti protein activity. *Genetics* 1995;**140**(1):267–74.

111. Tota MR, Smith TS, Mao C, MacNeil T, Mosley RT, Van der Ploeg LH, et al. Molecular interaction of Agouti protein and Agouti-related protein with human melanocortin receptors. *Biochemistry* 1999;**38**(3):897–904.
112. Yang YK, Thompson DA, Dickinson CJ, Wilken J, Barsh GS, Kent SB, et al. Characterization of Agouti-related protein binding to melanocortin receptors. *Mol Endocrinol* 1999;**13**(1):148–55.
113. Rossi M, Kim MS, Morgan DG, Small CJ, Edwards CM, Sunter D, et al. A C-terminal fragment of Agouti-related protein increases feeding and antagonizes the effect of alpha-melanocyte stimulating hormone in vivo. *Endocrinology* 1998;**139**(10):4428–31.
114. Creemers JW, Pritchard LE, Gyte A, Le Rouzic P, Meulemans S, Wardlaw SL, et al. Agouti-related protein is posttranslationally cleaved by proprotein convertase 1 to generate agouti-related protein (AGRP)83-132: interaction between AGRP83-132 and melanocortin receptors cannot be influenced by syndecan-3. *Endocrinology* 2006;**147**(4):1621–31.
115. Nijenhuis WA, Oosterom J, Adan RA. AgRP(83–132) acts as an inverse agonist on the human-melanocortin-4 receptor. *Mol Endocrinol* 2001;**15**(1):164–71.
116. Haskell-Luevano C, Monck EK. Agouti-related protein functions as an inverse agonist at a constitutively active brain melanocortin-4 receptor. *Regul Pept* 2001;**99**(1):1–7.
117. Siegrist W, Drozdz R, Cotti R, Willard DH, Wilkison WO, Eberle AN. Interactions of alpha-melanotropin and agouti on B16 melanoma cells: evidence for inverse agonism of agouti. *J Recept Signal Transduct Res* 1997;**17**(1–3):75–98.
118. Tolle V, Low MJ. In vivo evidence for inverse agonism of Agouti-related peptide in the central nervous system of proopiomelanocortin-deficient mice. *Diabetes* 2008;**57**(1):86–94.
119. Smart JL, Low MJ. Lack of proopiomelanocortin peptides results in obesity and defective adrenal function but normal melanocyte pigmentation in the murine C57BL/6 genetic background. *Ann N Y Acad Sci* 2003;**994**:202–10.
120. Srinivasan S, Lubrano-Berthelier C, Govaerts C, Picard F, Santiago P, Conklin BR, et al. Constitutive activity of the melanocortin-4 receptor is maintained by its N-terminal domain and plays a role in energy homeostasis in humans. *J Clin Invest* 2004;**114**(8):1158–64.
121. Buch TR, Heling D, Damm E, Gudermann T, Breit A. Pertussis toxin-sensitive signaling of melanocortin-4 receptors in hypothalamic GT1-7 cells defines agouti-related protein as a biased agonist. *J Biol Chem* 2009;**284**(39):26411–20.
122. Candille SI, Kaelin CB, Cattanach BM, Yu B, Thompson DA, Nix MA, et al. A b-defensin mutation causes black coat color in domestic dogs. *Science* 2007;**318**(5855):1418–23.
123. Rediger A, Piechowski CL, Habegger K, Gruters A, Krude H, Tschop MH, et al. MC4R dimerization in the paraventricular nucleus and GHSR/MC3R heterodimerization in the arcuate nucleus: is there relevance for body weight regulation. *Neuroendocrinology* 2012;**95**(4):277–88.
124. Rediger A, Piechowski CL, Yi CX, Tarnow P, Strotmann R, Gruters A, et al. Mutually opposite signal modulation by hypothalamic heterodimerization of ghrelin and melanocortin-3 receptors. *J Biol Chem* 2011;**286**(45):39623–31.
125. Bagnol D, Lu XY, Kaelin CB, Day HE, Ollmann M, Gantz I, et al. Anatomy of an endogenous antagonist: relationship between Agouti-related protein and proopiomelanocortin in brain. *J Neurosci* 1999;**19**(18):RC26.
126. Haskell-Luevano C, Chen P, Li C, Chang K, Smith MS, Cameron JL, et al. Characterization of the neuroanatomical distribution of agouti-related protein immunoreactivity in the rhesus monkey and the rat. *Endocrinology* 1999;**140**(3):1408–15.

127. Yaswen L, Diehl N, Brennan MB, Hochgeschwender U. Obesity in the mouse model of pro-opiomelanocortin deficiency responds to peripheral melanocortin. *Nat Med* 1999;**5**(9):1066–70.
128. Smart JL, Tolle V, Low MJ. Glucocorticoids exacerbate obesity and insulin resistance in neuron-specific proopiomelanocortin-deficient mice. *J Clin Invest* 2006;**116**(2):495–505.
129. Krude H, Biebermann H, Luck W, Horn R, Brabant G, Gruters A. Severe early-onset obesity, adrenal insufficiency and red hair pigmentation caused by POMC mutations in humans. *Nat Genet* 1998;**19**(2):155–7.
130. Farooqi IS, Drop S, Clements A, Keogh JM, Biernacka J, Lowenbein S, et al. Heterozygosity for a POMC-null mutation and increased obesity risk in humans. *Diabetes* 2006;**55**(9):2549–53.
131. Cheung CC, Clifton DK, Steiner RA. Proopiomelanocortin neurons are direct targets for leptin in the hypothalamus. *Endocrinology* 1997;**138**(10):4489–92.
132. Wilson BD, Bagnol D, Kaelin CB, Ollmann MM, Gantz I, Watson SJ, et al. Physiological and anatomical circuitry between Agouti-related protein and leptin signaling. *Endocrinology* 1999;**140**(5):2387–97.
133. Seeley RJ, Yagaloff KA, Fisher SL, Burn P, Thiele TE, van Dijk G, et al. Melanocortin receptors in leptin effects. *Nature* 1997;**390**(6658):349.
134. van de Wall E, Leshan R, Xu AW, Balthasar N, Coppari R, Liu SM, et al. Collective and individual functions of leptin receptor modulated neurons controlling metabolism and ingestion. *Endocrinology* 2008;**149**(4):1773–85.
135. Trevaskis JL, Butler AA. Double leptin and melanocortin-4 receptor gene mutations have an additive effect on fat mass and are associated with reduced effects of leptin on weight loss and food intake. *Endocrinology* 2005;**146**(10):4257–65.
136. Andrews ZB, Liu ZW, Walllingford N, Erion DM, Borok E, Friedman JM, et al. UCP2 mediates ghrelin's action on NPY/AgRP neurons by lowering free radicals. *Nature* 2008;**454**(7206):846–51.
137. Chen HY, Trumbauer ME, Chen AS, Weingarth DT, Adams JR, Frazier EG, et al. Orexigenic action of peripheral ghrelin is mediated by neuropeptide Y and agouti-related protein. *Endocrinology* 2004;**145**(6):2607–12.
138. Cowley MA, Smith RG, Diano S, Tschop M, Pronchuk N, Grove KL, et al. The distribution and mechanism of action of ghrelin in the CNS demonstrates a novel hypothalamic circuit regulating energy homeostasis. *Neuron* 2003;**37**(4):649–61.
139. Hill JW, Elias CF, Fukuda M, Williams KW, Berglund ED, Holland WL, et al. Direct insulin and leptin action on pro-opiomelanocortin neurons is required for normal glucose homeostasis and fertility. *Cell Metab* 2010;**11**(4):286–97.
140. Lin HV, Plum L, Ono H, Gutierrez-Juarez R, Shanabrough M, Borok E, et al. Divergent regulation of energy expenditure and hepatic glucose production by insulin receptor in AgRP and POMC neurons. *Diabetes* 2010;**59**(2):337–46.
141. Konner AC, Janoschek R, Plum L, Jordan SD, Rother E, Ma X, et al. Insulin action in AgRP-expressing neurons is required for suppression of hepatic glucose production. *Cell Metab* 2007;**5**(6):438–49.
142. Benoit SC, Air EL, Coolen LM, Strauss R, Jackman A, Clegg DJ, et al. The catabolic action of insulin in the brain is mediated by melanocortins. *J Neurosci* 2002;**22**(20):9048–52.
143. van den Top M, Lee K, Whyment AD, Blanks AM, Spanswick D. Orexigen-sensitive NPY/AgRP pacemaker neurons in the hypothalamic arcuate nucleus. *Nat Neurosci* 2004;**7**(5):493–4.
144. Xu Y, Jones JE, Kohno D, Williams KW, Lee CE, Choi MJ, et al. 5-HT2CRs expressed by pro-opiomelanocortin neurons regulate energy homeostasis. *Neuron* 2008;**60**(4):582–9.

145. Lam DD, Przydzial MJ, Ridley SH, Yeo GS, Rochford JJ, O'Rahilly S, et al. Serotonin 5-HT2C receptor agonist promotes hypophagia via downstream activation of melanocortin 4 receptors. *Endocrinology* 2008;**149**(3):1323–8.
146. Heisler LK, Jobst EE, Sutton GM, Zhou L, Borok E, Thornton-Jones Z, et al. Serotonin reciprocally regulates melanocortin neurons to modulate food intake. *Neuron* 2006;**51**(2):239–49.
147. Heisler LK, Cowley MA, Tecott LH, Fan W, Low MJ, Smart JL, et al. Activation of central melanocortin pathways by fenfluramine. *Science* 2002;**297**(5581):609–11.
148. Ibrahim N, Bosch MA, Smart JL, Qiu J, Rubinstein M, Ronnekleiv OK, et al. Hypothalamic proopiomelanocortin neurons are glucose responsive and express K(ATP) channels. *Endocrinology* 2003;**144**(4):1331–40.
149. Parton LE, Ye CP, Coppari R, Enriori PJ, Choi B, Zhang CY, et al. Glucose sensing by POMC neurons regulates glucose homeostasis and is impaired in obesity. *Nature* 2007;**449**(7159):228–32.
150. Singru PS, Sanchez E, Fekete C, Lechan RM. Importance of melanocortin signaling in refeeding-induced neuronal activation and satiety. *Endocrinology* 2007;**148**(2):638–46.
151. Sanchez E, Singru PS, Acharya R, Bodria M, Fekete C, Zavacki AM, et al. Differential effects of refeeding on melanocortin-responsive neurons in the hypothalamic paraventricular nucleus. *Endocrinology* 2008;**149**(9):4329–35.
152. Johnstone LE, Fong TM, Leng G. Neuronal activation in the hypothalamus and brainstem during feeding in rats. *Cell Metab* 2006;**4**(4):313–21.
153. Williams C, Hill SJ. GPCR signaling: understanding the pathway to successful drug discovery. *Methods Mol Biol* 2009;**552**:39–50.
154. Neves SR, Ram PT, Iyengar R. G protein pathways. *Science* 2002;**296**(5573):1636–9.
155. Mountjoy KG, Robbins LS, Mortrud MT, Cone RD. The cloning of a family of genes that encode the melanocortin receptors. *Science* 1992;**257**(5074):1248–51.
156. Gantz I, Shimoto Y, Konda Y, Miwa H, Dickinson CJ, Yamada T. Molecular cloning, expression, and characterization of a fifth melanocortin receptor. *Biochem Biophys Res Commun* 1994;**200**(3):1214–20.
157. Gantz I, Miwa H, Konda Y, Shimoto Y, Tashiro T, Watson SJ, et al. Molecular cloning, expression, and gene localization of a fourth melanocortin receptor. *J Biol Chem* 1993;**268**(20):15174–9.
158. Gantz I, Konda Y, Tashiro T, Shimoto Y, Miwa H, Munzert G, et al. Molecular cloning of a novel melanocortin receptor. *J Biol Chem* 1993;**268**(11):8246–50.
159. Chhajlani V, Muceniece R, Wikberg JE. Molecular cloning of a novel human melanocortin receptor. *Biochem Biophys Res Commun* 1996;**218**(2):638.
160. Chhajlani V, Muceniece R, Wikberg JE. Molecular cloning of a novel human melanocortin receptor. *Biochem Biophys Res Commun* 1993;**195**(2):866–73.
161. Chhajlani V, Wikberg JE. Molecular cloning and expression of the human melanocyte stimulating hormone receptor cDNA. *FEBS Lett* 1992;**309**(3):417–20.
162. Griffon N, Mignon V, Facchinetti P, Diaz J, Schwartz JC, Sokoloff P. Molecular cloning and characterization of the rat fifth melanocortin receptor. *Biochem Biophys Res Commun* 1994;**200**(2):1007–14.
163. Barrett P, MacDonald A, Helliwell R, Davidson G, Morgan P. Cloning and expression of a new member of the melanocyte-stimulating hormone receptor family. *J Mol Endocrinol* 1994;**12**(2):203–13.
164. Labbe O, Desarnaud F, Eggerickx D, Vassart G, Parmentier M. Molecular cloning of a mouse melanocortin 5 receptor gene widely expressed in peripheral tissues. *Biochemistry* 1994;**33**(15):4543–9.

165. Srinivasan S, Vaisse C, Conklin BR. Engineering the melanocortin-4 receptor to control G(s) signaling in vivo. *Ann N Y Acad Sci* 2003;**994**:225–32.
166. Kim CS, Lee SH, Kim RY, Kim BJ, Li SZ, Lee IH, et al. Identification of domains directing specificity of coupling to G-proteins for the melanocortin MC3 and MC4 receptors. *J Biol Chem* 2002;**277**(35):31310–7.
167. Mountjoy KG. Distribution and function of melanocortin receptors within the brain. *Adv Exp Med Biol* 2010;**681**:29–48.
168. Kishi T, Aschkenasi CJ, Lee CE, Mountjoy KG, Saper CB, Elmquist JK. Expression of melanocortin 4 receptor mRNA in the central nervous system of the rat. *J Comp Neurol* 2003;**457**(3):213–35.
169. Yi CX, Kalsbeek A, Tschop MH. Autonomic MC sets the metabolic tone. *Cell Metab* 2011;**13**(2):121–3.
170. Krude H, Gruters A. Implications of proopiomelanocortin (POMC) mutations in humans: the POMC deficiency syndrome. *Trends Endocrinol Metab* 2000;**11**(1):15–22.
171. Krude H, Biebermann H, Schnabel D, Tansek MZ, Theunissen P, Mullis PE, et al. Obesity due to proopiomelanocortin deficiency: three new cases and treatment trials with thyroid hormone and ACTH4-10. *J Clin Endocrinol Metab* 2003;**88**(10):4633–40.
172. Challis BG, Coll AP, Yeo GS, Pinnock SB, Dickson SL, Thresher RR, et al. Mice lacking pro-opiomelanocortin are sensitive to high-fat feeding but respond normally to the acute anorectic effects of peptide-YY(3–36). *Proc Natl Acad Sci USA* 2004;**101**(13):4695–700.
173. Lee YS, Challis BG, Thompson DA, Yeo GS, Keogh JM, Madonna ME, et al. A POMC variant implicates beta-melanocyte-stimulating hormone in the control of human energy balance. *Cell Metab* 2006;**3**(2):135–40.
174. Biebermann H, Castaneda TR, van Landeghem F, von Deimling A, Escher F, Brabant G, et al. A role for beta-melanocyte-stimulating hormone in human body-weight regulation. *Cell Metab* 2006;**3**(2):141–6.
175. Vaisse C, Clement K, Durand E, Hercberg S, Guy-Grand B, Froguel P. Melanocortin-4 receptor mutations are a frequent and heterogeneous cause of morbid obesity. *J Clin Invest* 2000;**106**(2):253–62.
176. Yeo GS, Lank EJ, Farooqi IS, Keogh J, Challis BG, O'Rahilly S. Mutations in the human melanocortin-4 receptor gene associated with severe familial obesity disrupts receptor function through multiple molecular mechanisms. *Hum Mol Genet* 2003; **12**(5):561–74.
177. Wang CL, Liang L, Wang HJ, Fu JF, Hebebrand J, Hinney A. Several mutations in the melanocortin 4 receptor gene are associated with obesity in Chinese children and adolescents. *J Endocrinol Invest* 2006;**29**(10):894–8.
178. Larsen LH, Echwald SM, Sorensen TI, Andersen T, Wulff BS, Pedersen O. Prevalence of mutations and functional analyses of melanocortin 4 receptor variants identified among 750 men with juvenile-onset obesity. *J Clin Endocrinol Metab* 2005; **90**(1):219–24.
179. Farooqi IS, Yeo GS, Keogh JM, Aminian S, Jebb SA, Butler G, et al. Dominant and recessive inheritance of morbid obesity associated with melanocortin 4 receptor deficiency. *J Clin Invest* 2000;**106**(2):271–9.
180. Ohshiro Y, Sanke T, Ueda K, Shimajiri Y, Nakagawa T, Tsunoda K, et al. Molecular scanning for mutations in the melanocortin-4 receptor gene in obese/diabetic Japanese. *Ann Hum Genet* 1999;**63**(Pt 6):483–7.
181. Li WD, Joo EJ, Furlong EB, Galvin M, Abel K, Bell CJ, et al. Melanocortin 3 receptor (MC3R) gene variants in extremely obese women. *Int J Obes Relat Metab Disord* 2000;**24**(2):206–10.
182. Calton MA, Ersoy BA, Zhang S, Kane JP, Malloy MJ, Pullinger CR, et al. Association of functionally significant Melanocortin-4 but not Melanocortin-3 receptor mutations

with severe adult obesity in a large North American case-control study. *Hum Mol Genet* 2009;**18**(6):1140–7.
183. Mencarelli M, Dubern B, Alili R, Maestrini S, Benajiba L, Tagliaferri M, et al. Rare melanocortin-3 receptor mutations with in vitro functional consequences are associated with human obesity. *Hum Mol Genet* 2011;**20**(2):392–9.
184. Yiannakouris N, Melistas L, Kontogianni M, Heist K, Mantzoros CS. The Val81 missense mutation of the melanocortin 3 receptor gene, but not the 1908c/T nucleotide polymorphism in lamin A/C gene, is associated with hyperleptinemia and hyperinsulinemia in obese Greek caucasians. *J Endocrinol Invest* 2004;**27**(8): 714–20.
185. Schalin-Jantti C, Valli-Jaakola K, Oksanen L, Martelin E, Laitinen K, Krusius T, et al. Melanocortin-3-receptor gene variants in morbid obesity. *Int J Obes Relat Metab Disord* 2003;**27**(1):70–4.
186. Marks DL, Butler AA, Turner R, Brookhart G, Cone RD. Differential role of melanocortin receptor subtypes in cachexia. *Endocrinology* 2003;**144**(4):1513–23.
187. Hagan MM, Rushing PA, Schwartz MW, Yagaloff KA, Burn P, Woods SC, et al. Role of the CNS melanocortin system in the response to overfeeding. *J Neurosci* 1999; **19**(6):2362–7.
188. Xu Y, Berglund ED, Sohn JW, Holland WL, Chuang JC, Fukuda M, et al. 5-HT2CRs expressed by pro-opiomelanocortin neurons regulate insulin sensitivity in liver. *Nat Neurosci* 2010;**13**(12):1457–9.
189. Marsh DJ, Miura GI, Yagaloff KA, Schwartz MW, Barsh GS, Palmiter RD. Effects of neuropeptide Y deficiency on hypothalamic agouti-related protein expression and responsiveness to melanocortin analogues. *Brain Res* 1999;**848**(1–2):66–77.
190. Fekete C, Sarkar S, Rand WM, Harney JW, Emerson CH, Bianco AC, et al. Agouti-related protein (AGRP) has a central inhibitory action on the hypothalamic-pituitary-thyroid (HPT) axis; comparisons between the effect of AGRP and neuropeptide Y on energy homeostasis and the HPT axis. *Endocrinology* 2002;**143**(10): 3846–53.
191. Sutton GM, Josephine Babin M, Gu X, Hruby VJ, Butler AA. A derivative of the melanocortin receptor antagonist SHU9119 (PG932) increases food intake when administered peripherally. *Peptides* 2008;**29**(1):104–11.
192. Navarro M, Lerma-Cabrera JM, Carvajal F, Lowery EG, Cubero I, Thiele TE. Assessment of Voluntary Ethanol Consumption and the Effects of a Melanocortin (MC) Receptor Agonist on Ethanol Intake in Mutant C57BL/6J Mice Lacking the MC-4 Receptor. *Alcohol Clin Exp Res* 2011;**35**(6):1058–66.
193. Joseph CG, Yao H, Scott JW, Sorensen NB, Marnane RN, Mountjoy KG, et al. Gamma-Melanocyte stimulation hormone (gamma-MSH) truncation studies results in the cautionary note that gamma-MSH is not selective for the mouse MC3R over the mouse MC5R. *Peptides* 2010;**31**(12):2304–13.
194. Rowland NE, Schaub JW, Robertson KL, Andreasen A, Haskell-Luevano C. Effect of MTII on food intake and brain c-Fos in melanocortin-3, melanocortin-4, and double MC3 and MC4 receptor knockout mice. *Peptides.* 2010 Dec;**31**(12):2314–7.
195. Begriche K, Sutton GM, Fang J, Butler AA. The role of melanocortin neuronal pathways in circadian biology: a new homeostatic output involving melanocortin-3 receptors? *Obes Rev* 2009;**10**(Suppl. 2):14–24.
196. Mistlberger RE. Circadian food-anticipatory activity: formal models and physiological mechanisms. *Neurosci Biobehav Rev* 1994;**18**(2):171–95.
197. Huang W, Ramsey KM, Marcheva B, Bass J. Circadian rhythms, sleep, and metabolism. *J Clin Invest* 2011;**121**(6):2133–41.
198. Lamia KA, Storch KF, Weitz CJ. Physiological significance of a peripheral tissue circadian clock. *Proc Natl Acad Sci USA* 2008;**105**(39):15172–7.

199. Zvonic S, Floyd ZE, Mynatt RL, Gimble JM. Circadian rhythms and the regulation of metabolic tissue function and energy homeostasis. *Obesity (Silver Spring)* 2007;**15**(3):539–43.
200. Takahashi JS, Hong HK, Ko CH, McDearmon EL. The genetics of mammalian circadian order and disorder: implications for physiology and disease. *Nat Rev Genet* 2008;**9**(10):764–75.
201. Turek FW, Joshu C, Kohsaka A, Lin E, Ivanova G, McDearmon E, et al. Obesity and metabolic syndrome in circadian Clock mutant mice. *Science* 2005;**308**(5724):1043–5.
202. Rudic RD, McNamara P, Curtis AM, Boston RC, Panda S, Hogenesch JB, et al. BMAL1 and CLOCK, two essential components of the circadian clock, are involved in glucose homeostasis. *PLoS Biol* 2004;**2**(11):e377.
203. Scheer FA, Hilton MF, Mantzoros CS, Shea SA. Adverse metabolic and cardiovascular consequences of circadian misalignment. *Proc Natl Acad Sci USA* 2009;**106**(11):4453–8.
204. Arble DM, Bass J, Laposky AD, Vitaterna MH, Turek FW. Circadian timing of food intake contributes to weight gain. *Obesity (Silver Spring)* 2009;**17**(11):2100–2.
205. Choi S, Wong LS, Yamat C, Dallman MF. Hypothalamic ventromedial nuclei amplify circadian rhythms: do they contain a food-entrained endogenous oscillator? *J Neurosci* 1998;**18**(10):3843–52.
206. Gooley JJ, Schomer A, Saper CB. The dorsomedial hypothalamic nucleus is critical for the expression of food-entrainable circadian rhythms. *Nat Neurosci* 2006;**9**(3):398–407.
207. Davidson AJ. Search for the feeding-entrainable circadian oscillator: a complex proposition. *Am J Physiol Regul Integr Comp Physiol* 2006;**290**(6):R1524–R1526.
208. Mistlberger RE, Buijs RM, Challet E, Escobar C, Landry GJ, Kalsbeek A, et al. Standards of evidence in chronobiology: critical review of a report that restoration of Bmal1 expression in the dorsomedial hypothalamus is sufficient to restore circadian food anticipatory rhythms in Bmal1 −/− mice. *J Circadian Rhythms* 2009;**7**:3.
209. Bass J, Takahashi JS. Circadian rhythms: redox redux. *Nature* 2011;**469**(7331):476–8.
210. Menet JS, Rosbash M. When brain clocks lose track of time: cause or consequence of neuropsychiatric disorders. *Curr Opin Neurobiol* 2011;**21**(6):849–57.
211. Ribeiro AC, Ceccarini G, Dupre C, Friedman JM, Pfaff DW, Mark AL. Contrasting effects of leptin on food anticipatory and total locomotor activity. *PLoS One* 2011;**6**(8):e23364.
212. Mistlberger RE, Marchant EG. Enhanced food-anticipatory circadian rhythms in the genetically obese Zucker rat. *Physiol Behav* 1999;**66**(2):329–35.
213. Bouret SG, Draper SJ, Simerly RB. Trophic action of leptin on hypothalamic neurons that regulate feeding. *Science* 2004;**304**(5667):108–10.
214. Landry GJ, Yamakawa GR, Webb IC, Mear RJ, Mistlberger RE. The dorsomedial hypothalamic nucleus is not necessary for the expression of circadian food-anticipatory activity in rats. *J Biol Rhythms* 2007;**22**(6):467–78.
215. Moriya T, Aida R, Kudo T, Akiyama M, Doi M, Hayasaka N, et al. The dorsomedial hypothalamic nucleus is not necessary for food-anticipatory circadian rhythms of behavior, temperature or clock gene expression in mice. *Eur J Neurosci* 2009;**29**(7):1447–60.
216. Kim KW, Sohn JW, Kohno D, Xu Y, Williams K, Elmquist JK. SF-1 in the ventral medial hypothalamic nucleus: a key regulator of homeostasis. *Mol Cell Endocrinol* 2011;**336**(1–2):219–23.
217. Mistlberger RE, Rechtschaffen A. Recovery of anticipatory activity to restricted feeding in rats with ventromedial hypothalamic lesions. *Physiol Behav* 1984;**33**(2):227–35.
218. Mistlberger RE. Neurobiology of food anticipatory circadian rhythms. *Physiol Behav* 2011;**104**:535–45.

CHAPTER FIVE

Melanocortin-4 Receptor in Energy Homeostasis and Obesity Pathogenesis

Anke Hinney, Anna-Lena Volckmar, Nadja Knoll
Department of Child and Adolescent Psychiatry, Universitätsklinikum, University of Duisburg-Essen, Essen, Germany

This chapter is dedicated to the memory of Prof. Dr. Ingrid Schmidt (deceased on July 1, 2012); her encouraging spirit inspired our work at the MC4R in humans. At the Max Planck Institute in Bad Nauheim, she studied Mc4r knock-out pubs just after weaning. Her findings pertaining to early stages of obesity development added vital information also pertaining to the leptinergic–melanocortinergic pathway. She is utterly missed.

Contents

Progress in Molecular Biology and Translational Science, Volume 114
ISSN 1877-1173
http://dx.doi.org/10.1016/B978-0-12-386933-3.00005-4

Abstract

The melanocortin-4 receptor gene (*MC4R*) has intensively been analyzed in molecular genetic obesity research. Decreased MC4R activity leads to obesity. Currently 166 non-synonymous, nonsense, deletion, and frameshift *MC4R* mutations have been described. Vast numbers of these mutations were identified in (extremely) obese individuals. Total or partial loss of MC4R function results from most detected mutations, as shown by *in vitro* analyses. The heterozygote frequency for these mutations in (extremely) obese individuals cumulates to approximately 2–5%. Surprisingly, two polymorphisms in the *MC4R* are associated with a slight protection from obesity. Large study groups were screened to be able to pinpoint these rather small effects on body weight. Recently, the advent of genome-wide association studies led to the discovery of association of polymorphisms in the 3′ region of the *MC4R* with obesity. The functional implication of the finding is still unresolved; an effect on gene expression is the most likely mechanism. Synthetic association could not be detected for the *MC4R* region.

1. THE MELANOCORTIN-4 RECEPTOR GENE (*MC4R*)

A single-exon gene located on chromosome 18q21.3 encodes for the 332 amino acids of the human MC4R.[1] Expression of the *MC4R* is predominantly found in the hypothalamic nucleus paraventricularis (PVN), spinal cord, sympathetic neurons, brainstem, and penis.[2] The G-protein-coupled MC4R, comprising seven transmembrane domains, has a central role in the leptinergic melanocortineric pathway and consequently in energy homeostasis. Leptin, which is a satiety factor that is secreted from adipocytes, activates the expression of pro-opiomelanocortin (POMC) in neurons of the arcuate nucleus. α-MSH, a POMC cleavage product, is an agonist at the MC4R. Activation of MC4R leads to an anorexigenic state in which energy intake and catabolic processes are decreased. Low leptin levels trigger increased agouti-related protein (AgPR) levels. As an inverse agonist, it suppresses activation of MC4R that in turn leads to orexigenic stimuli.[3] The central role of the MC4R in the regulation of the energy homeostasis resulted in the in-depth elucidation of its role in human obesity.

2. FROM RODENT *MC4R* MODELS TO HUMAN OBESITY

MC4R became a focus of molecular genetic research in obesity when the first knockout mouse model was published in 1997.[4] Mice lacking *Mc4r* are hyperphagic, hyperinsulinaemic, extremely obese, and have an increased linear growth, while mice heterozygous for a null *Mc4r* allele exhibit weight gain intermediate to that seen in wild-type and homozygous mutant littermates. Hence, a codominant form of obesity is transferred by a reduced MC4R activity. Lower energy expenditure leads to a faster weight gain in knockout mice than in wild-type littermates that consume the same amount of food.[5] In contrast, in *Mc4r*-deficient mice at weaning hyperphagia, not hypometabolism is the primary disturbance initiating excessive fat deposition.[6]

Rescue of *Mc4r* expression in the PVN of knockout mice showed that Mc4r action in the PVN is responsible for food intake. *Mc4r* expression in other brain areas is relevant for energy expenditure.[7] PVN-specific knockdown of *Mc4r* leads to hyperphagia and obesity in rats, thus underscoring the relevance of the Mc4r in the PVN.[8] Voluntary running wheel activity in Mc4r knockout mice slows down the course of obesity[9] suggesting that even for a severe major gene, obesity effect in rodents can be influenced by a change of lifestyle habits.[10] As anecdotic evidence, an overweight heterozygous carrier of a functionally relevant *MC4R* mutation within a large family pedigree mentioned that he "goes running" whenever he feels hungry. Thus, he used a coping strategy that resulted in a body weight in the overweight range, but he was not obese. Therefore, we assumed that data gained from the rat model also apply in humans.[11,12] Recently, the physiological roles of extrahypothalamic Mc4rs were analyzed in mice. It turned out that Mc4rs expressed by cholinergic neurons regulate hepatic glucose production and energy expenditure. These data provided further evidence of the dissociation in pathways mediating the effects of melanocortins on energy balance and glucose homeostasis.[13] A gender effect was also described for the *Mc4r*-deficient mice. In the initially published knockout mice,[4] the obese phenotype was more severe in females; in the second *Mc4r* knockout model, the male mice showed the more extreme phenotype.[14] These opposing observations are remarkable as the disruption of the *Mc4r* was technically very similar in both models.

3. MUTATIONS IN THE *MC4R* CONFER A MAJOR GENE EFFECT FOR THE DEVELOPMENT OF OBESITY

Based on the mouse model, the *MC4R* was analyzed as a candidate gene for human obesity. In 1998, two groups reported the first functionally relevant *MC4R* mutations for obesity.[15,16] Both analyses comprised relatively small study groups. Yeo *et al.*[15] identified an individual who was heterozygous for a deletion of 4-bps at codon 211 (see Table 5.1). The resulting frameshift eventually leads to a premature stop codon leading to a truncated MC4R protein. For the mutation carrier, hyperphagia with constant food seeking was reported.[15] Another frameshift mutation (4-bp insertion at nt732 = Ile245fs, see Table 5.1) was detected in a 35-year-old woman with early-onset obesity who also showed an increased food intake. Within her large family, all carriers of the *MC4R* frameshift mutation were obese, whereas the noncarriers were not obese.[16] These two papers paved the way for a still increasing number of mutation screens in the *MC4R*-coding region in human obesity. In 1999, the first mutation analysis in a large number of obese individuals ($n = 306$) also substantiated the major gene effect for obesity.[40] This study identified the most frequent loss-of-function mutation (Tyr35Stop) known to date. Approximately 0.6% of extremely obese German and Danish[41] individuals carry this mutation heterozygously. By now, 166 mutations that encompass nonsynonymous, nonsense, deletions, and frameshift mutations have been identified mainly in obese individuals (Tables 5.1 and 5.2; e.g., Refs. 22,34,40,50,96). All in all, *MC4R* mutations are detected in 2–6% of (extremely) obese individuals.[50] Mutation carriers are usually heterozygous (codominant inheritance). However, mutations were infrequently identified in homozygous or compound heterozygous carriers; phenotypes of these people were more severe.[11,26,29,35,38,74,94]

4. PHENOTYPES DETECTED IN *MC4R* MUTATION CARRIERS

To depict the phenotype common to all *MC4R* mutation carriers, we genotyped and phenotyped 43 family members of four index patients with *MC4R* mutations, two of these were second-degree cousins. Among the 19 mutation carriers, extreme obesity was the predominant phenotype, but moderate obesity also occurred in some heterozygotes. Surprisingly, other specific phenotypic abnormalities were not detected. Interestingly, female

Table 5.1 Functional analyses for the frameshift, deletion, and nonsense variants in the MC4R

Kind of mutation	Position within the MC4R#	Effect on amino acid sequence	Mutation identification	Function	Classification according to Tao 2005	Functional analyses
Deletion	TM2	Val88_Leu92del (Del 88–92)	17b	LOF	Class III	17b
	TM4	Ile170del	18	LOF	Class III	19
	TM5	Met200Del	20			
Frameshift	N term	Asn17fs (47–48InsG)	21	LOF	Class I	21,22
	N term	Ser19fs	23 (24,25)			
	N term	Gly34fs (codon 34, Ins A)	26			
	TM1	Ser58fs (171delC)	27			
	Extr loop 1	Thr112fs (InsA, FS)	28	LOF	Class IV	29
	TM4	Ile170fs (DelCAT)	30			
	TM5	Leu211fs (211DelCTCT)	15 (30,31)	RF	Class IV	32
	TM5			LOF	Class IV	29
	TM5	Tyr212fs (c.634_635delT)	33	RF	Class I	33
	TM2	Ile245fs (244InsGATT)	16	LOF	Class I	22
	TM6	Ile251fs (c.750-751delGA)	34 (35,36)	LOF	Class III, IV	34
	TM6	Ile251fs (c.750-751delGA)		RF, low cell surface expression, lcw αMSH binding	Class II, IV	37
	Extr loop 3	Cys279fs (279InsGT)	38	LOF	No allocation possible	29,38
	C term	Leu322fs (Codon 320 DelT)	34	RF	Class III, IV	34
	C term	Cys326fs (c.976DelT)	32	RF	Class IV	32

Continued

Table 5.1 Functional analyses for the frameshift, deletion, and nonsense variants in the MC4R—cont'd

Kind of mutation	Position within the MC4R[#]	Effect on amino acid sequence	Mutation identification	Function	Classification according to Tao 2005	Functional analyses
Nonsense	N term	Trp16Stop	39			
	N term	Tyr35Stop	40 (36)	LOF, no cell surface expression	Class I	20,22,30, 33,41,42
	N term	Asp37Stop	43	RF	Class IV	43
	TM1	Gln43Stop	23	LOF	Class I	22
	TM2	Tyr80Stop	22			
	Extr loop 3	Cys277Stop	44	RF, reduced potency for all endogenous agonists	Class IV	42,45
	TM7	Tyr287Stop	29	LOF		29
	TM7	Tyr287Stop		RF, low cell surface expression, low αMSH stimulation	Class II, IV	33,46
	C term	Gln307Stop	30 (47,48)	LOF, no αMSH binding, no cell surface expression	Class II, III	19

All detected MC4R variants in humans sorted by amino acid position and domain, including functional data if available (state of 02/2012). The effect on amino acid sequence is given by the guidelines of Antonarakis,[17] with the original name of the variant given by the authors of first publication in brackets. If a mutation was detected by several groups, additional references are given in brackets in the column "Mutation identification." When functional data were conflicting between different studies, all different classifications are given. Molecular classification of variants according to Tao 2005 (Class I: Defective protein production, Class II: Intracellular retained protein, Class III: Defective ligand binding, Class IV: Defective signaling, Class V: No known defect). Abbreviations are as follows: TM, transmembrane domain; N term, N-terminal extracellular domain; Extr. loop, extracellular loop; C term, C-terminal intracellular domain; WT, wild type; RF, reduced function; LOF, loss of function.

Table 5.2 Functional analyses for the nonsynonymous variants in the *MC4R*

Position within the MC4R[#]	Effect on amino acid sequence	Mutation identification	Function	Classification according to Tao 2005	Functional analyses
Extracellular N-terminal domain	Ser4Phe	33	Like WT	Class V	33
	His6Ala	49	RF, reduced activation, normal affinity	Class IV	49
	Arg7Cys	50	Like WT	Class V	24,50
			RF, low cell surface expression	Class I	51
	Arg7His	52	Like WT	Class V	37
			RF	Class IV	52
	Thr11Ala	53 (39,46)	Like WT	Class V	54
			RF	Class IV	29
	Thr11Ser	21 (42)	Like WT	Class V	21
			RF	Class II	22,52
	Arg18Cys	21 (42)	Like WT	Class V	21
			RF	Class II	22,55
	Arg18His	52	Like WT	Class V	37
			RF	Class IIa	22

Continued

Table 5.2 Functional analyses for the nonsynonymous variants in the *MC4R*—cont'd

Position within the MC4R[#]	Effect on amino acid sequence	Mutation identification	Function	Classification according to Tao 2005	Functional analyses
	Arg18Leu	30 (52)	RF	Class III	37
	Leu23Arg	44	Like WT	Class V	44
	Ser30Phe	40 (31,56)	Like WT	Class V	33,42,57
			RF	Class II, III, IV	30
	Gly32Glu	31	Like WT	Class V	33
			RF	Class IV	31
	Tyr35Cys	58			
	Ser36Tyr	41	Like WT	Class V	59
	Asp37Val		Like WT	Class V	36,42
	Cys40Arg	60			
Transmembrane domain 1	Glu42Lys	61			
	Pro48Ser	62	RF, low αMSH binding	Class III	37
	Val50Met	63	RF, low αMSH binding	Class III	37
	Phe51Leu	53 (64)	Like WT	Class V	54
			RF, low αMSH binding	Class III	24,37
	Leu54Pro	65	RF, less cell surface trafficking, less binding, less cAMP production	Class II, III, IV	65

	Gly55Asp	65	RF, impaired cAMP signaling	Class IV	65
	Gly55Val	65	RF, impaired cAMP signaling	Class IV	65
			LOF		66
	Ser58Cys	63	RF	Class II	42,45,55
	Glu61Lys	31	RF, less cell surface trafficking, less binding, less cAMP production, lower E_{max}, no αMSH binding	Class II, III, IV	33,37,65
			LOF, low basal activity, low activation by αMSH	Class IV	31
	Asn62Ser	38 (29)	RF	Class IV	38
			LOF, no cell surface expression	Class II	42
	Ile69Arg	20	LOF, no αMSH binding, no cell surface expression	Class II, III	19
	Ile69Met	22	LOF	Class I	22
	Ile69Thr	65	RF, less cell surface trafficking, less binding, less cAMP production, low αMSH binding	Class II, III, IV	37,65
	Ala70Thr	50	Like WT	Class V	50
Intracellular loop 1	Lys73Arg	67	Like WT	Class V	67
	His76Arg	18	Like WT	Class V	18
			Gain of function, increased basal activity, increased cAMP plateau		68

Continued

Table 5.2 Functional analyses for the nonsynonymous variants in the *MC4R*—cont'd

Position within the MC4R#	Effect on amino acid sequence	Mutation identification	Function	Classification according to Tao 2005	Functional analyses
			RF, low cell surface expression, increased αMSH response	Class II	19
Transmembrane domain 2	Pro78Leu	40	Reduced E_{max}, increased E_{50}	Class III, IV	34
	Pro78Leu		LOF, no cell surface expression	Class II	42,69
	Met79Ile	70	LOF	Class I	19
	Cys84Arg	24	RF, low cell surface expression	Class II	51
			RF, low cAMP activation	Class III	24
	Ala87Asp	71	LOF	Class IV	71
	Asp90Ala	72	RF	Class IV	72
	Asp90Asn	73	LOF, dominant negativ	Class IV	73
			RF, low cell surface expression, reduced agonist potency, partial agonist stimulation, no αMSH stimulation	Class II, III, IV	37,47
	Ser94Arg	34	LOF, low cell surface expression, no αMSH binding	Class II, III	34,37
	Ser94Asn	18	RF	Class IV	18
			LOF	Class I	19
Transmembrane domain 2	Val95Ile	34	LOF, no cell surface expression	Class I	20,22,34,42
	Asn97Asp	28	RF, reduced potency for all endogenous agonists	Class IV	42,45
			LOF	Class IV	28

	Gly98Arg	74	LOF, reduced cell surface expression, no αMSH binding	Class II, III	37,74
	Thr101Ala	50	RF	Class IV	50
	Thr101Asn	75			
	Ile102Ser	63	RF, normal surface expression, reduced binding and signaling	Class III, IV	42,45,54,63
	Ile102Thr	27	Gain of function, low αMSH binding, increased αMSH response		37
			RF	Class II	22,54,55
			LOF	Class IV	41
	Val103Ile	76 (36,42, 47,61,66, 71,77–80)	Like WT	Class V	43
			Gain of function, low αMSH binding, increased αMSH response		37,67,81
	Leu106Pro	28	RF, reduced potency for all endogenous agonists	Class IV	42,45
			LOF	Class IV	28
Extracellular loop 1	Thr112Met	82 (23,27,30, 38,41,47, 53,64,83)	Like WT	Class V	33,54,67, 71,80,84
			RF, reduced potency for one or more encogenous	Class IV	42,50,85
	Thr112Lys	50	Like WT	Class V	50

Continued

Table 5.2 Functional analyses for the nonsynonymous variants in the *MC4R*—cont'd

Position within the MC4R[#]	Effect on amino acid sequence	Mutation identification	Function	Classification according to Tao 2005	Functional analyses
Transmembrane domain 3	Ile121Thr	34	RF, low cell surface expression, low αMSH binding	Class II, III	34,37
	Ile125Lys	28	LOF, no cell surface expression	Class II	42
	Asp126Tyr	18	LOF, no αMSH binding	Class III	18,19
	Ser127Leu	34 (35)	Like WT	Class V	30,31
			RF, low cell surface expression	Class II	24,34,42, 45,50,51, 55,67,84
			LOF	Class IV	33
	Ser136Phe	86	RF, impaired cAMP signaling	Class IV	65
			LOF, Loss of signal transduction activity, binding disrupted	Class IV	51,86,87
	Ser136Pro	65	RF, less cell surface trafficking, less binding, less cAMP production	Class II, III, IV	65
	Ile137Thr	82	RF, reduced potency for one or more endogenous agonists	Class IV	42
			LOF	Class III, IV	82
	Asp146Asn	18	Like WT	Class V	18
			Gain of function, increased basal activity, increased cAMP plateau		68

			LOF, increased αMSH response, no cell surface expression	Class I	19
	Thr150Ile	21	RF, reduced potency for all endogenous agonists, reduced binding and activation	Class II	20–22, 42,45
Intracellular loop 2	Ala154Asp	54 (56,80)	Like WT	Class V	54
			RF	Class II	22,37
	Gln156Pro	30			
	Gln156Arg	50	Like WT	Class V	50
	Tyr157Ser	32	RF, low cell surface expression, low αMSH binding	Class II, III	32,37
	His158Arg	50	Like WT	Class V	33,50
	Met161Thr	65	RF, less cell surface trafficking, less binding, less cAMP production	Class II, III, IV	65
	Thr162Arg	56	Like WT	Class V	56
	Thr162Ile	65	RF, less cell surface trafficking, less binding, less cAMP production	Class II, III, IV	65
Transmembrane domain 4	Arg165Trp	40	RF, reduced potency for all endogenous agonists	Class IV	21,42,45
			LOF, no cell surface expression	Class II	69,85
	Arg165Gln	38 (53,64)	RF, reduced potency for all endogenous agonists	Class IV	28,42,43, 45,50
			LOF	Class III	38,85
	Arg165Gly	43 (26,41)	RF	Class IV	43

Continued

Table 5.2 Functional analyses for the nonsynonymous variants in the *MC4R*—cont'd

Position within the MC4R[#]	Effect on amino acid sequence	Mutation identification	Function	Classification according to Tao 2005	Functional analyses
	Val166Ile	50	Like WT	Class V	50
	Ile169Ser	29	Like WT	Class V	42
	Ile170Val	21 (36,44)	RF	Class II	21,22
	Cys172Arg	66 (88)			
	Trp174Cys	23 (36,42,63)	RF, low cell surface expression	Class II	51
			LOF, no expression on cell surface (trapped in ER), no αMSH binding	Class II, III	25,37
	Ala175Thr	28	Like WT	Class V	33,42
			RF, dominant-negative effect	Class III	25,28,41
	Thr178Met	34 (23,83)	Like WT	Class V	34,42
	Gly181Asp	89	LOF, low cell surface expression, no αMSH binding	Class II, III	24,34,37, 41,50
	Ile186Val	19 (83)	Like WT	Class V	19
			LOF, no αMSH stimulation	Class IV	47
	Ile194Thr	19	LOF, no aMSH binding, no cell surface expression	Class II, III	19

Transmembrane domain 5	Ile195Val	33	Like WT	Class V	33
	Ile195Ser	20	LOF, no cell surface expression	Class II	19
	Met200Val	53	Like WT	Class V	50,54
	Phe201Leu	18	RF	Class IV	18,19
	Phe202Leu	27 (47,64)	Like WT	Class V	27,33,37, 54,66
	Met208Val	66 (36)			
	Met215Leu	67	Like WT	Class V	67
Intracellular loop 3	Met218Thr	60			
	Ala219Val	41	RF, low αMSH binding	Class III	37,41,51
	Ile226Thr	84	Like WT	Class V	37,50,84
	Pro230Leu	34	Like WT	Class V	42
			Gain of function, increased basal activity		90
			RF, low cell surface expression	Class II	34,51
	Gly231Ser	26	Like WT	Class V	33,37
			RF	Class II	22,43
	Gly231Val	18 (36)	Like WT	Class V	18,19
	Arg236Cys	30	Like WT	Class V	33
Intracellular loop 3	Arg236His	50	Like WT	Class V	50
	Gly238Asp	29	RF	Class III	37

Continued

Table 5.2 Functional analyses for the nonsynonymous variants in the *MC4R*—cont'd

Position within the MC4R[#]	Effect on amino acid sequence	Mutation identification	Function	Classification according to Tao 2005	Functional analyses
Transmembrane domain 6	Asn240Ser	27 (20)	Like WT	Class V	33,37
			RF, reduced binding and signaling	Class III, IV	54
			LOF		27
	Ala244Glu	30	RF, reduced potency for one or more endogenous agonists	Class IV	34,35,42
	Ala244Val	48	Like WT	Class V	48
	Leu250Gln	21 (56)	Gain of function, increased basal activity, increased cAMP plateau, constitutively active, AGRP as inverse agonist		21,68
			RF	Class III	22,42,45
			LOF, constitutively active	Class II	91
	Ile251Leu	40 (36,42, 47,80)	Like WT	Class V	43,67
			Gain of function		66,81
	Gly252Ser	40	Like WT	Class V	50
			RF	Class II	22,42,45,55
	Val253Ile	38	Like WT	Class V	42
			Partial acitivity	Class IV	38
	Pro260Gln	18 (20,35,39)	RF	Class IV	18

			LOF, reduced cAMP accumulation, no cell surface expression	Class II, IV	19,47
	Phe261Ser	92	RF, low cell surface expression	Class II	51,92
	Ile266Ala	93	Like WT	Class V	93
	Phe267Leu	93	Like WT	Class V	93
	Tyr268Ile	93	Gain of function (increased αMSH binding)		93
	Ile269Asn	30	RF, less cell surface trafficking, less binding, less cAMP production	Class II, III, IV	33,43,65
	Ile269Val	93	Gain of function (increased αMSH binding)		93
	Ser270Leu	93 (66)	Like WT	Class V	93
Extracellular loop 3	Cys271Arg	29	RF, low cell surface expression, low αMSH binding	Class II, III	37,94
	Cys271Phe	33	RF, reduced E_{max}	Class IV	33
	Cys271Tyr	38	RF, reduced potency for all endogenous agonists	Class IV	42,45
			LOF	Class II	28,38
	Asn274Ser	95	Like WT	Class V	42
	Pro275Ser	50	Like WT	Class V	43,50
	Phe280Leu	96	Gain of function, increased αMSH response		19
			LOF	Class II	47
	Met281Val	48	RF, low cell surface expression, low cAMP response	Class II, IV	48
Transmembrane domain 7	Ile289Leu	18	Like WT	Class V	19
			LOF	Class IV	18

Continued

Table 5.2 Functional analyses for the nonsynonymous variants in the *MC4R*—cont'd

Position within the MC4R[#]	Effect on amino acid sequence	Mutation identification	Function	Classification according to Tao 2005	Functional analyses
	Ser295Pro	54	Like WT	Class V	37,54
			RF	Class III	22
	Asp298Ala	72	RF, reduced interaction with AGRP	Class III	72
	Pro299His	30	RF, reduced E_{max}	Class IV	33
			LOF, low basal activity, no activation by αMSH, no cell surface expression	Class II, IV	31,42
	Pro299Leu	84	LOF, low cell surface expression, no αMSH binding	Class II, IV	37
	Pro299Ser	30	RF, reduced E_{max}	Class IV	33
	Leu300Pro	19	LOF, no cell surface expression	Class II	19
	Ile301Thr	21	RF, reduced potency for all endogenous agonists	Class IV	21,42,45
			LOF	Class I	22
	Tyr302Phe	88	RF, low cell surface expression, low cAMP response	Class II, IV	88
Intracellular C-terminus	Ala303Pro	43	RF		43
	Ala303Thr	31	RF, reduced E50, impaired cAMP signaling	Class IV	33,65
			LOF, low basal activity, low activation by αMSH	Class IV	31
	Leu304Phe	43	Like WT	Class V	43
	Arg305Trp	22	RF	Class IIc	22

	Arg305Gln	33	RF (E_{max})	Class VI	33
	Arg305Ser	18 (36)	RF	Class VI	18,19
	Glu308Lys	57	Like WT	Class V	37
			RF	Class IV	57
	Arg310Lys	44			
	Ile316Ser	28	RF, reduced potency for one or more endogenous agonists	Class IV	28,42
			LOF, no cell surface expression	Class II	69
	Ile317Val	23	Like WT	Class V	51
			RF, low cell surface expression	Class II	37
	Ile317Thr	40 (42)	LOF, no cell surface expression	Class II	69
			Like WT	Class V	50
	Leu325Phe	41	Like WT	Class V	51
	Leu325Phe		RF, low cell surface expression	Class II	37
			LOF		41
	Cys326Arg	31	Like WT	Class V	31,33
	Tyr332Cys	18	Like WT	Class V	18,19
	Tyr332His	18	Like WT	Class V	18,19,46

All detected MC4R variants in humans sorted by amino acid position and domain, including functional data if available (state of 02/2012). If a mutation was detected by several groups, additional references are given in brackets in the column "Mutation identification." When functional data were conflicting between different studies, all different classifications are given. Molecular classification of variants according to Tao 2005 (Class I: Defective Protein Production, Class II: Intracellular retained protein, Class III: Defective ligand binding, Class IV: Defective signaling, Class V: No known defect). Abbreviations are as follows: WT, wild type; RF, reduced function; LOF, loss of function.

carriers of haplo-insufficiency mutations were heavier than male carriers within the respective families. This finding was similar to the first established *MC4r*-knockout mice.[11] The gender effect was further followed up in a total of 181 phenotyped individuals from 25 *MC4R* mutation carrier families. The quantitative analyses revealed that females gained approximately 15–30 kg of body weight (9.5 BMI units), whereas males just gained approximately 15 kg (4.0 BMI units) compared to their family members without *MC4R* mutations.[97]

The general phenotypic implications were analyzed among 500 European mutation-screened individuals with early-onset extreme obesity, of which 29 had mutations in *MC4R* (23 heterozygous, 6 homozygous). Apart from their severe obesity, mutation carriers showed increased lean mass, increased linear growth, hyperphagia as well as extreme hyperinsulinemia, and homozygous carriers showed more severe phenotypes than heterozygotes.[29] A recent analysis of 153 carriers of *MC4R* deficiency mutations and 1392 age- and BMI-matched controls revealed increased linear growth in childhood that leads to increased final height in adults.[98] The increase in body height was greater than predicted by obesity alone. The pulsatility of growth hormone was maintained in individuals deficient for *MC4R*. As this is consistent with animal studies, it might be assumed that MC4R controls the hypothalamic somatostatinergic tone. In children harboring *MC4R* mutations, fasting insulin levels were increased. These factors might be relevant to the increased body weight described in some,[98] but not all studies.[11]

5. *MC4R* MUTATIONS IN POPULATION-BASED STUDY GROUPS

Initial *MC4R* mutation screens comprised (extremely) obese individuals and a usually smaller group of normal weight to lean controls. *MC4R* mutation rates in epidemiological cohorts covering the whole BMI range were not published until 2006. The first population-based study included 4068 individuals of a German epidemiologically ascertained study group (Cooperative Health Research in the Region of Augsburg: KORA-S4). The results were unexpected. Sixteen nonsynonymous mutations were detected in total (six previously unknown) heterozygously in 27 individuals of KORA-S4. Impaired *in vitro* receptor function was described for four of the mutation alleles. Contrary to the expectation, obesity (BMI ≥ 30 kg/m^2) was not found in any of these six heterozygous mutation carriers.[50] These results were compared with other epidemiological genetic analyses. In breast cancer, for example, the effect of mutations in the breast cancer 1 gene (*BRCA1*) was most severe in families with more than four affected members. In case–control studies, in contrast, the penetrance rates were lower and lowest estimates were obtained in a population-based study. Hence, the *MC4R* results in KORA-S4 are, although unexpected, in concordance with other genetic analyses.[50] Another population-based study screened for *MC4R* variability in 1100 individuals. Apart from the two polymorphisms (Val103Ile and Ile251Leu) and functionally silent *MC4R* variants, a novel mutation (Ala87Asp) was detected in an obese individual (BMI of 31.5 kg/m^2). This mutation showed *in vitro* a substantial loss of function of the mutated receptor.[71] Thus, this smaller sized study did not find normal weight individuals with functionally relevant *MC4R* mutations.

In summary, *MC4R* mutations leading to a reduced function are more consistent with a major gene (or oligenic effect) form of obesity than a monogenic form.[34,50,97] Hence, harboring such a mutation does not necessarily lead to obesity, but the odds are largely increased.[99]

6. FUNCTIONAL ANALYSES

6.1. Evolutionary perspective

A worldwide sample of 1015 humans from 51 populations and 8 chimpanzees was analyzed to determine evolutionary history at the *MC4R* locus. Diversity at the locus was reduced in humans but not in chimpanzees. The mutation spectrum in humans, combined with the low overall level of diversity, implied that most of the observed nonsynonymous variants are likely to be deleterious mutations.[100] The coding region of the *MC4R* was also analyzed in 12 primate species and in an additional 29 vertebrates. These data were included in molecular evolutionary analyses. It turned out that MC4R is highly conserved throughout vertebrate evolution. It has apparently been subject to high levels of continuous purifying selection. Nonsynonymous mutations, with relevance for MC4R function, occur by trend preferentially at sites that are subject to very strong purifying selection.[100] More than 90% of all mutations leading to a reduced function are located at highly conserved amino acid positions considering 450 million years of MC4R evolution in vertebrates.[101]

Altogether, currently 166 *MC4R* mutations (nonsynonymous, frameshift/ deletions, and nonsense) mostly resulting in an altered MC4R protein have been reported. They are located in all parts of the receptor (intracellular, extracellular, and transmembrane regions) (see Fig. 5.1, Tables 5.1 and 5.2).[103]

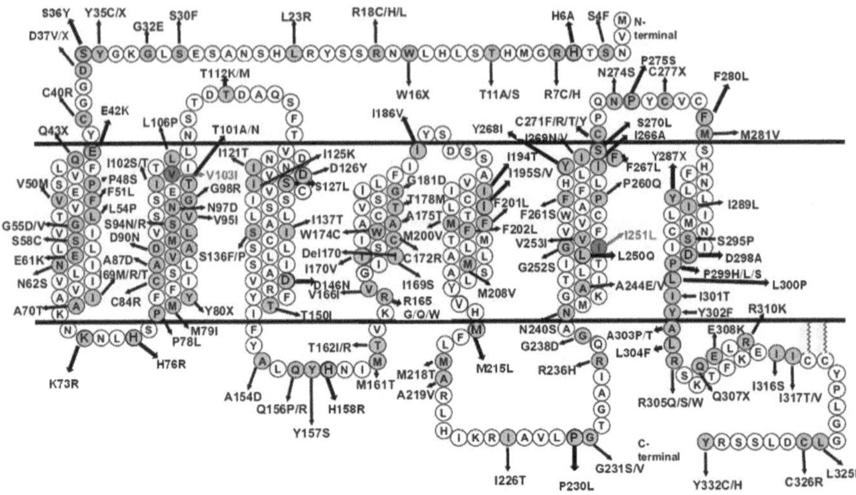

Figure 5.1 Nonsynonymous and nonsense mutations at the MC4R identified in different study groups. The two polymorphisms (V103I, I251) that confer protection from obesity are indicated with dark grey filling and letters. References describing these mutations are listed in the Tables 5.1 and 5.2. *Reprinted, with minor modification, with permission from Y.-X. Tao*[102] *©Elsevier 2009.* (See Color Insert.)

Each single *MC4R* mutation is infrequent to rare; most have only been detected in a single individual. For the analysis of relevance for obesity, the functional effects of the mutations have to be analyzed. As already implied from the mouse models, mutations leading to a reduced receptor function (from slightly reduced to complete loss of function) are associated with obesity. By contrast, mutations leading to an increased function (Val103Ile and Ile251Leu) have a slight weight-lowering effect.[81,104]

A substantial number of functional *in vitro* studies have been performed to elucidate how each mutation affects receptor function. The MC4R responds to agonist stimulation by increased intracellular cAMP levels. Mutations in *MC4R* can affect all functions of the receptor (e.g., cell surface expression, intracellular transport, agonist/antagonist binding). Up to now, functional data for 142 nonsynonymous, 13 frameshift, 3 deletions, and 8 nonsense mutations were reported (Table 5.1). The mutations result in (i) complete loss of receptor function, (ii) partial loss of function, or (iii) a function inseparable from the wild-type receptor. Two classification systems have been proposed:

1. The first classification system was described in 2005 by Tao proposing five mutation classes (e.g., Refs. 54,103):

   a. Class I mutations (null mutations) result in decreased/absent MC4R levels either because of defects in protein synthesis or/and increased protein degradation (all nonsense mutations, Table 5.1).
   b. Class II mutations (reduced cell surface expression) lead to the synthesis of mutant protein, which is not expressed at the cell surface. Most of the mutated proteins belong to this class.
   c. Class III mutations (reduced binding) lead to an MC4R that is expressed at the cell surface, but has a lowered ligand-binding capacity.
   d. Class IV mutated (reduced signaling) receptors show cell surface expression and ligand binding, whereas cell signaling via cAMP is impaired.
   e. Class V mutations (like wild type) have normal cell surface expression, ligand binding, and signaling. How they interfere with MC4R function has not been elicited so far with current state-of-the-art experimental assays. However, as appropriate human hypothalamic cell lines are not available, functional yet undetectable *in vivo* effects can be envisaged.[10]
2. The group of Vaisse[22,52] classified MC4R mutations in three groups:
   a. MC4R mutations leading to an altered cell surface expression
   b. MC4R mutations leading to altered basal activity
   c. MC4R mutations leading to an altered α-MSH response

There is no final consensus about the nomenclature. However, the categories help to classify the different functional receptor classes.[10] For statistical analyses, the mutations are grouped according to their functional effects; usually rather broad classifications are used (e.g., Tao 2005 class I–IV will be grouped together and contrasted to mutations with seemingly wild-type effect—class V; see e.g., Ref. 50).

6.2. Haploinsufficiency or dominant-negative effects?

Most *MC4R* mutations lead to a haplo-insufficiency of the receptor.[4,73] G-protein-coupled receptors (GPCRs) are known to di- or oligomerize. For MC4R, for instance, homo- and hetero-dimerization with other GPCRs involved in weight regulation was demonstrated (see Ref. 10). Only for a few mutations, dominant-negative effects have been described. The heterozygous Asp90Asn[73] as well the Ser136Phe mutations[87] cause dominant-negative effects, most probably by formation of heterodimers between the mutant receptor and the wild-type receptor. Thus, the mutant receptor impairs the function of the wild-type receptor copy. The dominant-negative effect implies a more severe functional outcome than a haplo-insufficiency.[73,87]

6.3. Constitutive activity?

In 2008, Tao elegantly summarized findings pertaining to the constitutive activity of the MC4R.[105] In early studies, basal cAMP levels of the mutants were not studied (see Ref. 106). Vaisse and coworkers suggested that loss of constitutive activity in the mutant MC4Rs is one cause of obesity. Several mutations in the N-terminal extracellular domain only resulted in loss of basal activity compared to wild-type MC4R. Hence, constitutive activity of the MC4R presumably transmits a tonic satiety signal. If this tonic signaling is disrupted, obesity can ensue.[52] Until now, only few studies identified mutations leading to a reduced basal activity.[51,54]

The existence of an endogenous inverse agonist (AgRP) in the melanocortinergic system further supports the constitutive activity at the MC4R (see Ref. 105). If for a mutated MC4R an increased affinity for AgRP is measured, reduced receptor function would presumably ensue. Some studies found decreased AgRP response for different mutations.[28] The group around Haskell-Luevano identified *MC4R* mutations with both increased and decreased AgRP potency.[42] Unexpectedly, increased constitutive activity was found for some mutations solely detected in (extremely) obese individuals.[42] Later, other mutations leading to high constitutive activity were identified in obese or overweight individuals.[34,50,84] However, whereas increased constitutive activity is expected to result in a lean phenotype, it is still a matter of debate if it can also lead to obesity.[105]

6.4. Nonsense mutations

Some of the described mutations lead to a premature stop codon (nonsense mutations and frameshift mutations; see Table 5.1). Interestingly, the phenomenon of nonsense-mediated mRNA decay (NMD) is not observed for the *MC4R*. The fidelity of gene expression is usually ensured by the phylogenetically conserved mechanism of NMD, which inhibits the accumulation of nonsense- or frameshift-mutated mRNA. Hence, synthesis of truncated proteins with potential dominant-negative effects would be reduced. However, Brocke *et al.*[107] showed for *MC4R* that nonsense-mutated variants of *MC4R* transcripts are stable and express truncated proteins.[107]

In the absence of NMD, the *MC4R* nonsense mutations lead to truncated and dysfunctional proteins. Aminoglycosides can bypass stop mutations during translation so that the protein expression can be rescued. Brumm *et al.*[108] showed that functional rescue of stop mutations in the *MC4R* is dependent on different parameters: (i) the sequence of the stop codon triplet, (ii) surrounding sequence, (iii) location within the receptor, and (iv) applied ligand and aminoglycoside. Functional rescue was possible for the two N-terminal nonsense mutations Trp16Stop and Tyr35Stop, but it was less successful for the C-terminus (Gln307Stop) and nearly impossible for Glu61Stop within the first transmembrane domain. Cell culture assays revealed that an aminoglycoside (G-418) restored Mc4r (X16; Mc4r termination at amino acid position 16; reflecting the human nonsense mutation Trp16Stop, see above) function *in vitro*. Mc4r(X16) knock-in mice are extremely obese, hyperphagic, have impaired glucose tolerance, and an accelerated body length so that this mouse model expectedly resembles typical characteristics of Mc4r deficiency. Disappointingly, in a first therapeutic trial, the aminoglycosides gentamicin and amikaçin did not induce reduction of obesity. Additional experiments with these mice are warranted to evaluate if nonsense suppression is possible at the *MC4R in vivo*.[109] Pharmaceutical applications will need to consider the potency of aminoglycosides to restore receptor function as well as the ability to pass the blood–brain barrier.[108]

7. NONSYNONYMOUS *MC4R* POLYMORPHISMS THAT CONFER PROTECTION FROM OBESITY

As delineated above, *MC4R* mutations leading to a reduced receptor function are associated with obesity. Surprisingly, two nonsynonymous SNPs (Val103Ile, Ile251Leu) in *MC4R* are associated with a slightly reduced body weight. About 2–9% of the general population is heterozygous for the 103Ile variant of the Val103Ile polymorphisms (rs2229616) of the *MC4R*.[110] Analysis (transmission disequilibrium test; TDT[111]) of 520 obesity trios (comprising extremely obese children and adolescents and both parents) revealed reduced transmission of the Ile103 variant to obese offspring. A meta-analysis on a total of 7713 individuals (3631 obese cases and 4082 controls) confirmed the negative association of the MC4R-Ile103 variant with obesity (odds ratio: 0.69; 95% confidence interval: 0.59–0.99). For carriers of the Ile103 variant, an effect estimate of -0.48 kg/m^2 was calculated, which is approximately equivalent to a reduction of 1.5 kg of body weight in a 1.8-m-tall individual.[110] Subsequently, the negative association of *MC4R*-103Ile with obesity was confirmed in (i) a large epidemiological study group of approximately 8000 individuals[112] and (ii) two additional meta-analyses encompassing a total of 55,195 individuals (19,822 obese cases and 35,373 nonobese controls[78,104]). The negative association with obesity for the 103Ile variant was again supported in a genome-wide association study (GWAS) with focus on early-onset extreme obesity when candidate genes for obesity had specifically been tested.[113]

As the effect of the infrequent SNP allele is contrary to the obesogenic effects of mutations leading to a reduced receptor function, it was speculated that the polymorphism confers an improved receptor function. In 2006, Xiang *et al.* published functional data confirming this expectation.[42] They showed that the MC4R-103Ile reveals a modest, but significant decrease in inverse agonist potency. Additionally, the effect of potent endogenous POMC-derived agonists at the MC4R seemed to be increased for that variant.[42] Hence, both the decreased inverse agonistic and the increased agonistic potencies are compatible with an increased MC4R function, which could explain the weight-reducing effect of the 103Ile variant.

The Val103Ile SNP was also analyzed in 1173 consecutive patients undergoing cardiac catheterization. Heterozygous carriers of the 103Ile allele had significantly lower triglyceride levels than individuals homozygous for the wild-type allele. The data suggest an influence of MC4R activity on triglyceride levels in cardiovascular patients.[114] Consistent with this finding, *Mc4r* knockout mice have increased triglyceride level in plasma, liver, and intestine.[115] However, central mechanisms were unlikely to be involved in the observed regulation of intestinal triglycerides. Instead, enterocytes were found to express *MC4R*. Hence it was proposed that MC4R is involved in a peripheral, local gut signaling mechanism that regulates intestinal triglycerides and controls intestinal lipid absorption.[115]

The *MC4R* Val103Ile polymorphism was analyzed in 7888 adults of a population-based cross-sectional study. The 3.7% who were carriers of the *MC4R* 103Ile variant exhibited a significantly decreased waist circumference, decreased glycosylated hemoglobin, and increased HDL-cholesterol, but no change in blood pressure. To exhibit features of the metabolic syndrome was more unlikely for carriers of *MC4R* 103Ile than for noncarriers. A borderline association of *MC4R* 103Ile with carbohydrate intake possibly mediates the association with leanness. The data supported the idea that this polymorphism plays a role in appetite regulation that affects not only BMI but also other parameters of the metabolic syndrome.[116] Surprisingly, another study in 2008[117] showed that carriers of the 103Ile variant had significantly higher daily energy and carbohydrate intakes than noncarriers. They found no association with BMI by analysis of the *MC4R* Val103Ile polymorphism in 1029 extremely obese individuals.[117]

The *Mc4r* in mice was shown to be relevant in cancer cachexia (e.g., Ref. 118). Pathogenic mechanisms of cancer cachexia are still not well understood. It was hypothesized that human carriers of the 103Ile variant should be more often or more severely affected by cancer cachexia. Analysis of 509 patients (295 males) with malignant neoplasms revealed that carriers of the 103Ile variant were not more prone to develop cancer cachexia than patients without this allele. Further studies of the melanocortinergic system in cachexia are warranted.[119]

The frequency of the *MC4R* Ile251Leu variant (rs52820871) is lower compared to the Val103Ile SNP. Only 0.41–1.21% of individuals is heterozygous for the *MC4R*-251Leu variant. Due to its low frequency, a large study group was necessary to detect a weight-lowering effect similar to the *MC4R*-103Ile variant. The contribution to body weight was analyzed in 16,797 individuals of European origin. In eight of nine studies, a consistent negative association of the *MC4R*-251Leu variant with extreme obesity of an early-onset and in adults (odds ratios 0.25–0.76) was detected. In population-based samples, the variant was also associated with reduced BMI. Strong evidence of the obesity protective effect of *MC4R*-251Leu was provided by a meta-analysis (odds ratio = 0.52,[37]). Again, as for the *MC4R*-103Ile variant, an improved MC4R function for the variant allele was speculated. In fact, an increased basal activity was described for the Leu251 variant.[42]

8. GENOME-WIDE ASSOCIATION STUDIES: SNPs DOWNSTREAM OF THE *MC4R* GENE

In 2008, a meta-analysis of GWAS within a large-scale international cooperation analyzed more than 90,000 individuals; they detected an SNP in the vicinity of the *MC4R*.[120] GWAS data from seven studies (a total of 16,876 Europeans) had initially been jointly analyzed. The second strongest association signal, after *FTO* (fat mass and obesity associated gene[121]), mapped 188 kb downstream of the *MC4R* (rs17782313). An effect of rs17782313 on weight regulation may be mediated through effects on *MC4R* expression as suggested by its location.[120] The association was confirmed in 60,352 independent adults and 5988 children and 660 nuclear German families encompassing one or more obese offspring and both biological parents. Each copy of the rs17782313 obesity risk allele (C) was associated with an increase in BMI of approximately 0.22 kg/m^2 in adults. A copy of the risk allele resulted in an 8% (12%) increased risk for overweight (obesity). Interestingly, a copy of the C-allele also resulted in a higher mean height (0.21 cm), suggesting that this SNP (or the functionally relevant variation in linkage disequilibrium) influences overall adult size.[120] The small effect size for BMI was confirmed by subsequent GWAS (meta-) analyses and studies analyzing specifically the previously described obesity risk alleles of human obesity genes.[113,122–130] The GWASs, which included in total more than 320,000 individuals, all picked up an SNP downstream of *MC4R* as a polygenic variant for obesity (Refs. 70,113,123,124; for review, see Refs. 99,131).

Genotyping of nearly 6000 children of the Avon Longitudinal Study of Children and Parents revealed that the effect of the C-allele was not detectable in children prior to age 7. However, in children aged 7–11, the effect size of a copy of the C-allele was twice the amount observed in adults, without observation of an effect for body height. The effect on weight was mainly due to fat mass.[120] In this and other large-scale studies, the *MC4R* risk alleles showed associations with childhood BMI, but little (no) association with birth weight.[132,133] There was a larger effect on weight gain in early infancy than on weight gain in childhood.[132] The *MC4R* SNP(s) are also relevant in early-onset extreme obesity.[134,135]

8.1. The effect of the SNPs downstream of the *MC4R* was also confirmed for other ethnicities

In 4992 individuals from six African-American study groups, significant association with BMI for the *MC4R* downstream SNP was shown,[136] as well as in European- and African-American youths.[137] In contrast, association to obesity was not detected in 1008 African-American obese children and 2715 controls.[130]

A GWAS in 2684 Indian Asians (including subsequent analyses in 11,955 individuals of Indian Asian or European ancestry) for insulin resistance and related phenotypes identified association of rs12970134, located 154 kb downstream of *MC4R*, with waist circumference. As the effect was also present in normal weight individuals, the authors suggested that the association is partially independent from obesity. Association of rs12970134 with BMI was also found in the whole study group ($N=11{,}955$) with a similar effect size determined in the 4561 Europeans alone.[70] A weak obesity association but a stronger association with height was described in South Asian Indians.[138] Genotyping of 1808 Indian patients with type 2 diabetes mellitus or related traits and 1549 controls revealed association of *MC4R* variants with BMI, fat mass per cent and other obesity-related traits (e.g., waist circumference and hip circumference), as well as insulin resistance.[139]

Initially, association of the *MC4R* SNP to obesity in Chinese was conflicting.[59,140] Recently, two independent large-scale GWAS meta-analyses for BMI in a total of 145,293 Chinese individuals revealed genome-wide significant findings for obesity for SNPs near *MC4R*.[141,142] Huang *et al.*[143] found association to obesity and coronary artery disease in 930 Han Chinese subjects; association to obesity was also described in 5000 Chinese women.[144] Among five GWAS ($N=10,482$) from Chinese (three cohorts, including one with type 2 diabetes and another one of children), Malay, and Indian ethnic groups from Singapore and in 3503 independent Chinese children, association to obesity was detected.[145,146] Association was also described in 18,264 participants from two general Japanese populations.[147] In 8842 individuals from South Korea, an *MC4R* SNP (rs571312) was associated with obesity.[148] Another study showed no significant association with obesity or BMI for rs17782313 in a Han Chinese population ($n=4638$ in total), but association with low-density lipoprotein cholesterol and total cholesterol.[149]

8.2. Association of *MC4R* SNPs to related phenotypes

Association with obesity and increased insulin levels have been reported for two variants (rs17782313 and rs12970134) located downstream of the *MC4R*. Investigation of these SNPs in 889 overweight and obese children and adolescents supported the associations with early-onset obesity and increased insulin levels. Exploratory evidence for sex-specific effects of the risk alleles was observed for BMI-SDS reduction and diastolic blood pressure.[150]

There is no indication that two *MC4R* SNPs (rs2229616 [Val103Ile] and rs17782313 in 3′ vicinity) are associated with stroke.[151] Among 598 French-Canadian adolescents, association for an *MC4R* SNP was detected for obesity but not for blood pressure.[152]

Polycystic ovary syndrome (PCOS) is the main reason for anovulatory infertility in women. Rates of obesity are highly increased in PCOS. The *MC4R* SNP was associated with obesity and contributed to increased BMI in PCOS, but it did not appear to play a BMI-independent role in PCOS.[153]

Atypical antipsychotics are frequently used for treatment of psychotic disorders. Substantial weight gain is a major side effect of this treatment (e.g., Ref. 58). A GWAS analysis and independent confirmation recently revealed that an SNP downstream of the *MC4R* (rs489693) is associated with antipsychotics-induced weight gain and related metabolic disturbances. It was suggested that the identification of high-risk subjects could be used to subject risk allele carriers to alternative treatment strategies.[154]

9. SYNTHETIC ASSOCIATION AT THE *MC4R* LOCUS?

It is still unclear how and if association signals of the two common noncoding GWAS-derived SNPs rs17782313 and rs12970134 located 154 and 188 kb of the 3′ end of the *MC4R*[70,120] are correlated to receptor function. As delineated above, more than 150 different mutations, mostly leading to a reduced receptor function, have been detected in obesity (Fig. 5.1). Additionally, the minor alleles of two *MC4R* nonsynonymous polymorphisms, Val103Ile and Ile251Leu, are protective against obesity (see above).[78,81,104,110,112]

In the GWAS that detected the association of obesity with a downstream SNP,[120] the respective finding could not be explained by the two nonsynonymous SNPs Val103Ile and Ile251Leu, analyzed by both imputed data (with a high confidence score for Val103Ile) and direct genotypes for both polymorphisms on a subsample of approximately 5000 individuals for each SNP.[120]

The hypothesis of synthetic association implies that GWAS signals of common nonfunctional SNPs outside of coding regions may be the result of a combination of rare coding/functional variants given that these rare variants arose on a haplotype that is tagged by the common SNP.[155,156] This contrasts the more widespread idea that association signals of common nonfunctional SNPs outside of coding regions point to a common causal variant. The distant GWAS-derived SNPs and the variation in the *MC4R*-coding region can be regarded as an excellent example for the proposed synthetic association. Conditional as well as haplotype analyses for a genomic region of approximately 310 kb downstream to 53 kb upstream of the *MC4R* coding region, which is between two recombination hotspots, were performed.

Haplotype analyses are known to be more powerful than single-marker analyses for the detection of genomic regions enriched for mutation(s)/causal variant(s) that are relevant for the phenotype. Haplotypes flanking the *MC4R* were analyzed. A relatively common risk haplotype was identified that covers the *MC4R* coding sequence and flanking regions up- and downstream. This haplotype was used to analyze if the *MC4R* region shows indications of synthetic association.[155,156] The influence of coding variants (i.e., two coding polymorphisms and several functionally relevant rare mutations) in *MC4R* on the transmission of the obesity effect haplotype was analyzed. The genetic variation within the coding region cannot explain the obesity association effect of the downstream SNPs or of the detected haplotype. Hence, the analyses did not find evidence for synthetic association for the *MC4R*. It could not be excluded that other functionally relevant variants (e.g., Refs. 72,157) outside the coding region of the *MC4R* contribute to the GWAS obesity association signal.[158]

A similar haplotype approach was recently taken to find if SNPs or haplotypes in the vicinity of *MC4R* could be used to track functionally relevant *MC4R* mutations, without the need to screen the gene. Although the initial analyses showed trends in this direction, a more thorough analysis demonstrated that a cryptic relatedness between carriers of the same functionally relevant *MC4R* mutation most likely founded the observed effect. Therefore, a thorough *MC4R* mutation screen cannot be substituted by SNP genotyping.[159]

10. RELEVANCE OF *MC4R* MUTATIONS FOR PHENOTYPES RELATED TO BODY WEIGHT REGULATION

Mutations in the *MC4R* might be relevant for disorders that are associated with an increased body weight or obesity. Numerous studies have analyzed various related disorders.

10.1. Blood pressure and sympathetic function

Weight gain and weight loss are associated with changes in blood pressure through unknown mechanisms. Central melanocortinergic signaling is implicated in the control of energy balance and blood pressure in rodents, but there is no information regarding such an association with blood pressure in humans. Analysis of individuals heterozygous for loss-of-function *MC4R* mutations implicated melanocortinergic signaling in the control of human blood pressure through an insulin-independent mechanism.[160] Melanocortinergic pathways are involved in obesity-associated sympathetic over-activity and its thermoregulatory and hemodynamic consequences. In eight heterozygous carriers of *MC4R* mutations leading to a reduced function and controls, vasoconstrictive muscle sympathetic nerve activity (MSNA) was measured. Obese subjects with *MC4R* mutations had an inverse relationship between obesity and MSNA. This implied that central sympathetic outflow to the vasculature could depend on melanocortinergic pathways. Their dysfunction might explain reduced sympathoexcitability and lower sympathetic nerve-induced lipolysis and the mostly normal blood pressure in these obese individuals.[161]

10.2. Eating disorders

Genes relevant for weight regulation could also be regarded as candidate genes for eating disorders. Hence mutations in the *MC4R* were analyzed in anorexia nervosa and bulimia nervosa (BN; for review, see Refs. 99,162). Obesity is described as a risk factor for the development of the eating disorder BN.[163] Thus, variants predisposing to obesity might be more frequent in patients with BN than in normal weight controls. Thus, it was not surprising that a mutation screen of the *MC4R* revealed a single extremely obese patient with BN (among 87 patients with BN) who is heterozygous for a haplo-insufficient mutation in the *MC4R*. This was the first time that a mutation predisposing to obesity was detected in a BN patient.[12]

Two studies reported that criteria for binge-eating disorder (BED) were fulfilled in all obese carriers of various *MC4R* variants.[53,64] However, there are some inconsistencies with other data in the initial study.[53] For instance, among other variants, the Ile103 variant of the Val103Ile SNP was regarded as an obesity risk variant. This variant was in large study groups and in meta-analyses shown to be associated with a slightly reduced BMI (e.g., Refs. 104,110 for details see above) and is thus protective of obesity (negatively associated with obesity).

Furthermore, subsequent studies in independent samples did not reveal any *MC4R* mutations among extremely obese adults with BED.[89] In obesity families, elevated rates of binge-eating episodes were not detected among 43 *MC4R* mutation carriers.[164] A rather small study (48 obese and 10 nonobese female patients with BED and 40 nonobese, healthy controls) identified, apart from the SNPs, a functionally relevant *MC4R* mutation in one obese patient with BED and none in the controls.[165] However, the study was too small for meaningful statistical analyses. Analysis of 769 extremely obese adults revealed 2.6% carriers of functionally relevant *MC4R* mutations, who did not have BED.[22] Taken together, the lack of confirmation led to the conclusion that it is unlikely that *MC4R* mutations have a strong effect on the development of BED.

10.3. Attention deficit hyperactivity disorder

Attention deficit hyperactivity disorder (ADHD) is a heritable psychiatric disorder in which rates of obesity are increased. We described a 2-year-old boy with early-onset extreme obesity who was heterozygous for a functionally relevant *MC4R* mutation (Glu308Lys). The boy also showed severe symptoms of ADHD. Treatment with the stimulant methylphenidate led to a sharp decrease of BMI. Potential mechanisms for this unexpectedly large weight loss suggest a link between the melanocortinergic and the dopaminergic systems, and the sympathetic nervous system. The potential benefit of methylphenidate in obese *MC4R* mutation carriers with and without comorbid ADHD warrants further studies.[166] A previous study already showed a family with obese carriers of an obesogenic *MC4R* mutation who were also diagnosed with ADHD.[167]

11. THERAPEUTIC IMPLICATIONS FOR *MC4R* MUTATION CARRIERS

This chapter focuses on molecular genetic findings pertaining to the *MC4R* in obesity and related phenotypes. Unfortunately, it was, within the scope of this chapter, impossible to describe the pharmacological properties and potential therapeutic implications pertaining to the MC4R. However, the respective data were recently elegantly described and in depth evaluated.[103] The interested reader is highly recommended to read this elegant state-of-the-art paper. The therapeutic options relevant for carriers of variation at the *MC4R* gene are delineated below.

11.1. Bariatric surgery

One of the most effective long-term therapies for the treatment of severe obesity is the Roux-en-Y gastric bypass (RYGB). Factors that influence long-term weight loss after RYGB surgery are poorly defined. Homozygous, but not heterozygous, *Mc4r* knockout mice lose less weight after surgery than wild-type animals so that MC4R signaling seems to be necessary for the weight-reducing effects of RYGB. Among 972 patients undergoing RYGB, patients heterozygous for *MC4R* mutations leading to a loss of function had the same weight loss as patients without such mutations. Hence, a single normal copy of the *MC4R* gene might be sufficient to mediate the weight loss effects of RYGB.[168] In another study,[169] 92 patients who had RYGB surgery with heterozygous *MC4R* mutations also benefited from the surgery. Weight loss seemed to be independent of the presence of such mutations.[169] In a case report, the outcome of bariatric surgery in an adolescent with compound heterozygosity and complete functional loss of both alleles of *MC4R* was described.[170] The data suggested that complete MC4R deficiency impairs response to gastric banding and results in poor weight loss after surgery. We had previously also analyzed a girl who was compound heterozygous for two different mutations in the *MC4R*, both leading to a reduced receptor function; she also did not profit from bariatric surgery.[11]

Another study focused on the influence of the two weight-lowering SNPs Val103Ile and Ile251Leu and the success of RYGB. Among 1433 extremely obese patients who underwent RYGB, the *MC4R* was screened for mutations: 36 and 26 patients carried one copy of the *MC4R*-103Ile or *MC4R*-251Leu variants, respectively. The study showed that individuals who carry the 251Leu allele are predisposed to better clinical outcome, reduced risk of type 2 diabetes mellitus, and better weight loss during diet and surgical interventions. The diabetic status has only a small, short-term effect on weight loss after RYGB.[171]

11.2. Conventional therapy

There are just few studies linking genetic variation in the *MC4R* with therapeutic options in obesity. The above-mentioned case report of an extremely obese boy with ADHD who is heterozygous for an *MC4R* mutation leading to a reduced receptor function implicated that treatment with methylphenidate (most commonly used drug for ADHD treatment) can have profound weight-reducing effects for *MC4R* mutation carriers.[166] A biological link between the melanocortinergic and the dopaminergic systems was proposed. A recent study underpinned this hypothesis with an animal model.[172] Further studies are warranted. Weight changes after a 1-year lifestyle intervention (OBELDICKS) were analyzed between 514 overweight children with and without *MC4R* variations. Fortunately, children with *MC4R* mutations leading to reduced receptor function could lose weight in a lifestyle intervention. However, they had much greater difficulties to maintain their weight loss. This finding again underscored the impact of these mutations on body weight.[48]

Hilbert *et al.*[173] analyzed the general psychological implications of an *MC4R* mutation screen for obese individuals. Four hundred and twenty-one obese individuals were psychologically tested in parallel to a mutation screen of the *MC4R*. It turned out that causal attributions of obesity to genetic/biological factors were neither cross-sectionally nor longitudinally related to weight regulation and longer-term weight outcome. However, those individuals who had a familial obesity risk attributed their obesity more often to genetic/biological factors than the others.[173] Analysis of obesity risk alleles in a number of obesity polygenes, including the SNPs in the vicinity of the *MC4R*, in a population-based study group ($n = 12{,}462$ subjects from the population-based MONICA/KORA Augsburg study) revealed that the genetic risk variants were not associated with lifestyle factors related to nutritional intake or energy expenditure.[174]

12. CONCLUSIONS AND FUTURE DIRECTIONS

Molecular genetic obesity research pertains to a large extent to the *MC4R*. Mutations leading to a reduced receptor function (currently 166 nonsynonymous, nonsense, deletions and frameshift *MC4R* mutations are known) lead to a major genetic form of obesity. The frequency of these mutations is up to 6% among (extremely) obese individuals. Two nonsynonymous mutations in the gene are associated with a slightly reduced body weight. Recently, GWAS for body weight led to the identification of SNPs downstream of the *MC4R*; an effect on the gene expression being most likely. Next-generation sequencing will presumably allow detection of more genetic variation at and near the *MC4R* in near future. These analyses will potentially unravel genetic variants that are functionally relevant for the GWAS signal.

The variants (mutations and SNPs) in the coding region and at the *MC4R* locus have been intensively analyzed for obesity-related traits; first, therapeutic options emerged at the horizon. However, the MC4R is not an easy target for pharmacological interventions, as the receptor is involved in a large number of physiological processes. Hence, any intervention is hampered by a number of unwanted effects (or side effects). As the focus of our chapter was the molecular elucidation of the *MC4R* gene in obesity and related phenotypes, it was impossible to also describe the pharmacological properties and potential therapeutic implications pertaining to the MC4R. The reader interested in this topic is recommended to read an elegant state-of-the-art paper (e.g., Ref. 103).

ACKNOWLEDGMENTS

Our work on the MC4R was supported by the German Research Foundation (DFG: Deutsche Forschungsgemeinschaft), the Federal Ministry of Education and Research (BMBF: Bundesministerium für Bildung und Forschung; 01KU0903; NGFNplus: 01GS0820, 01GS0830), and the IFORES program of the University of Duisburg-Essen. We thank Prof. Johannes Hebebrand for encouraging and supporting our work at the MC4R, for all fruitful discussions, and his inspiring support.

REFERENCES

1. Gantz I, Miwa H, Konda Y, Shimoto Y, Tashiro T, Watson SJ, et al. Molecular cloning, expression, and gene localization of a fourth melanocortin receptor. *J Biol Chem* 1993;**268**:15174–9.
2. Van der Ploeg LH, Martin WJ, Howard AD, Nargund RP, Austin CP, Guan X, et al. A role for the melanocortin 4 receptor in sexual function. *Proc Natl Acad Sci USA* 2002;**99**:11381–6.
3. Flier JS. Obesity wars: molecular progress confronts an expanding epidemic. *Cell* 2004;**116**:337–50.

4. Huszar D, Lynch CA, Fairchild-Huntress V, Dunmore JH, Fang Q, Berkemeier LR, et al. Targeted disruption of the melanocortin-4 receptor results in obesity in mice. *Cell* 1997;**88**:131–41.
5. Ste Marie L, Miura GI, Marsh DJ, Yagaloff K, Palmiter RD. A metabolic defect promotes obesity in mice lacking melanocortin-4 receptors. *Proc Natl Acad Sci USA* 2000;**97**:12339–44.
6. Weide K, Christ N, Moar KM, Arens J, Hinney A, Mercer JG, et al. Hyperphagia, not hypometabolism, causes early onset obesity in melanocortin-4 receptor knockout mice. *Physiol Genomics* 2003;**13**:47–56.
7. Balthasar N, Dalgaard LT, Lee CE, Yu J, Funahashi H, Williams T, et al. Divergence of melanocortin pathways in the control of food intake and energy expenditure. *Cell* 2005;**123**:493–505.
8. Garza JC, Kim CS, Liu J, Zhang W, Lu XY. Adeno-associated virus-mediated knockdown of melanocortin-4 receptor in the paraventricular nucleus of the hypothalamus promotes high-fat diet-induced hyperphagia and obesity. *J Endocrinol* 2008;**197**: 471–82.
9. Irani BG, Xiang Z, Moore MC, Mandel RJ, Haskell-Luevano C. Voluntary exercise delays monogenetic obesity and overcomes reproductive dysfunction of the melanocortin-4 receptor knockout mouse. *Biochem Biophys Res Commun* 2005;**326**: 638–44.
10. Biebermann H, Kühnen P, Kleinau G, Krude H. The neuroendocrine circuitry controlled by POMC, MSH, and AGRP. *Handb Exp Pharmacol* 2012;**209**:47–75.
11. Sina M, Hinney A, Ziegler A, Neupert T, Mayer H, Siegfried W, et al. Phenotypes in three pedigrees with autosomal dominant obesity caused by haploinsufficiency mutations in the melanocortin-4 receptor gene. *Am J Hum Genet* 1999;**65**:1501–7.
12. Hebebrand J, Fichter M, Gerber G, Görg T, Hermann H, Geller F, et al. Genetic predisposition to obesity in bulimia nervosa: a mutation screen of the melanocortin-4 receptor gene. *Mol Psychiatry* 2002;**7**:647–51.
13. Rossi J, Balthasar N, Olson D, Scott M, Berglund E, Lee CE, et al. Melanocortin-4 receptors expressed by cholinergic neurons regulate energy balance and glucose homeostasis. *Cell Metab* 2011;**13**:195–204.
14. Chen AS, Metzger JM, Trumbauer ME, Guan XM, Yu H, Frazier EG, et al. Role of the melanocortin-4 receptor in metabolic rate and food intake in mice. *Transgenic Res* 2000;**9**:145–54.
15. Yeo GS, Farooqi IS, Aminian S, Halsall DJ, Stanhope RG, O'Rahilly S. A frameshift mutation in MC4R associated with dominantly inherited human obesity. *Nat Genet* 1998;**20**:111–2.
16. Vaisse C, Clement K, Guy-Grand B, Froguel P. A frameshift mutation in human MC4R is associated with a dominant form of obesity. *Nat Genet* 1998;**20**:113–4.
17. (a) Antonarakis SE. Recommendations for a nomenclature system for human gene mutations. Nomenclature Working Group. *Hum Mutat* 1998;**11**:1–3. (b) Donohoue PA, Tao YX, Collins M, Yeo GS, O'Rahilly S, Segaloff DL. Deletion of codons 88–92 of the melanocortin-4 receptor gene: a novel deleterious mutation in an obese female. *J Clin Endocrinol Metab* 2003;**88**:5841–5.
18. Stutzmann F, Tan K, Vatin V, Dina C, Jouret B, Tichet J, et al. Prevalence of melanocortin-4 receptor deficiency in Europeans and their age-dependent penetrance in multigenerational pedigrees. *Diabetes* 2008;**57**:2511–8.
19. Wang ZQ, Tao YX. Functional studies on twenty novel naturally occurring melanocortin-4 receptor mutations. *Biochim Biophys Acta* 2011;**1812**:1190–9.
20. Wangensteen T, Kolsgaard ML, Mattingsdal M, Joner G, Tonstad S, Undlien D, et al. Mutations in the melanocortin 4 receptor (MC4R) gene in obese patients in Norway. *Exp Clin Endocrinol Diabetes* 2009;**117**:266–73.

21. Vaisse C, Clement K, Durand E, Hercberg S, Guy-Grand B, Froguel P. Melanocortin-4 receptor mutations are a frequent and heterogeneous cause of morbid obesity. *J Clin Invest* 2000;**106**:253–62.
22. Lubrano-Berthelier C, Dubern B, Lacorte JM, Picard F, Shapiro A, Zhang S, et al. Melanocortin 4 receptor mutations in a large cohort of severely obese adults: prevalence, functional classification, genotype-phenotype relationship, and lack of association with binge eating. *J Clin Endocrinol Metab* 2006;**91**:1811–8.
23. Buono P, Pasanisi F, Nardelli C, Ieno L, Capone S, Liguori R, et al. Six novel mutations in the proopiomelanocortin and melanocortin receptor 4 genes in severely obese adults living in southern Italy. *Clin Chem* 2005;**51**:1358–64.
24. Hainerová I, Larsen LH, Holst B, Finková M, Hainer V, Lebl J, et al. Melanocortin 4 receptor mutations in obese Czech children: studies of prevalence, phenotype development, weight reduction response, and functional analysis. *J Clin Endocrinol Metab* 2007;**92**:3689–96.
25. Alfieri A, Pasanisi F, Salzano S, Esposito L, Martone D, Tafuri D, et al. Functional analysis of melanocortin-4-receptor mutants identified in severely obese subjects living in Southern Italy. *Gene* 2010;**457**:35–41.
26. Ma L, Tataranni PA, Bogardus C, Baier LJ. Melanocortin 4 receptor gene variation is associated with severe obesity in Pima Indians. *Diabetes* 2004;**53**:2696–9.
27. Jacobson P, Ukkola O, Rankinen T, Snyder EE, Leon AS, Rao DC, et al. Melanocortin 4 receptor sequence variations are seldom a cause of human obesity: the Swedish Obese Subjects, the HERITAGE Family Study, and a Memphis cohort. *J Clin Endocrinol Metab* 2002;**87**:4442–6.
28. Yeo GS, Lank EJ, Farooqi IS, Keogh J, Challis BG, O'Rahilly S. Mutations in the human melanocortin-4 receptor gene associated with severe familial obesity disrupts receptor function through multiple molecular mechanisms. *Hum Mol Genet* 2003;**12**:561–74.
29. Farooqi IS, Keogh JM, Yeo GS, Lank EJ, Cheetham T, O'Rahilly S. Clinical spectrum of obesity and mutations in the melanocortin 4 receptor gene. *N Engl J Med* 2003;**348**:1085–95.
30. Lubrano-Berthelier C, Durand E, Dubern B, Shapiro A, Dazin P, Weill J, et al. Intracellular retention is a common characteristic of childhood obesity-associated MC4R mutations. *Hum Mol Genet* 2003;**12**:145–53.
31. Ahituv N, Kavaslar N, Schackwitz W, Ustaszewska A, Martin J, Hebert S, et al. Medical sequencing at the extremes of human body mass. *Am J Hum Genet* 2007;**80**:779–91.
32. Lee YS, Poh LK, Kek BL, Loke KY. Novel melanocortin 4 receptor gene mutations in severely obese children. *Clin Endocrinol (Oxf)* 2008;**68**:529–35.
33. Calton MA, Ersoy BA, Zhang S, Kane JP, Malloy MJ, Pullinger CR, et al. Association of functionally significant Melanocortin-4 but not Melanocortin-3 receptor mutations with severe adult obesity in a large North American case–control study. *Hum Mol Genet* 2009;**18**:1140–7.
34. Hinney A, Hohmann S, Geller F, Vogel C, Hess C, Wermter AK, et al. Melanocortin-4 receptor gene: case–control study and transmission disequilibrium test confirm that functionally relevant mutations are compatible with a major gene effect for extreme obesity. *J Clin Endocrinol Metab* 2003;**88**:4258–67.
35. Lubrano-Berthelier C, Le Stunff C, Bougnères P, Vaisse C. A homozygous null mutation delineates the role of the melanocortin-4 receptor in humans. *J Clin Endocrinol Metab* 2004;**89**:2028–32.
36. van den Berg L, van Beekum O, Heutink P, Felius BA, van de Heijning MP, Strijbis S, et al. Melanocortin-4 receptor gene mutations in a Dutch cohort of obese children. *Obesity (Silver Spring)* 2011;**19**:604–11.
37. Xiang Z, Proneth B, Dirain ML, Litherland SA, Haskell-Luevano C. Pharmacological characterization of 30 human melanocortin-4 receptor polymorphisms with the

endogenous proopiomelanocortin-derived agonists, synthetic agonists, and the endogenous agouti-related protein antagonist. *Biochemistry* 2010;**49**:4583–600.
38. Farooqi IS, Yeo GS, Keogh JM, Aminian S, Jebb SA, Butler G, et al. Dominant and recessive inheritance of morbid obesity associated with melanocortin 4 receptor deficiency. *J Clin Invest* 2000;**106**:271–9.
39. Marti A, Corbalán MS, Forga L, Martinez JA, Hinney A, Hebebrand J. A novel nonsense mutation in the melanocortin-4 receptor associated with obesity in a Spanish population. *Int J Obes Relat Metab Disord* 2003;**27**:385–8.
40. Hinney A, Schmidt A, Nottebom K, Heibült O, Becker I, Ziegler A, et al. Several mutations in the melanocortin-4 receptor gene including a nonsense and a frameshift mutation associated with dominantly inherited obesity in humans. *J Clin Endocrinol Metab* 1999;**84**:1483–6.
41. Larsen LH, Echwald SM, Sørensen TI, Andersen T, Wulff BS, Pedersen O. Prevalence of mutations and functional analyses of melanocortin 4 receptor variants identified among 750 men with juvenile-onset obesity. *J Clin Endocrinol Metab* 2005;**90**:219–24.
42. Xiang Z, Litherland SA, Sorensen NB, Proneth B, Wood MS, Shaw AM, et al. Pharmacological characterization of 40 human melanocortin-4 receptor polymorphisms with the endogenous proopiomelanocortin-derived agonists and the agouti-related protein (AGRP) antagonist. *Biochemistry* 2006;**45**:7277–88.
43. Thearle MS, Muller YL, Hanson RL, Mullins M, Abdussamad M, Tran J, et al. Greater impact of melanocortin-4 receptor deficiency on rates of growth and risk of type 2 diabetes during childhood compared with adulthood in Pima Indians. *Diabetes* 2012;**61**:250–7.
44. Wang CL, Liang L, Wang HJ, Fu JF, Hebebrand J, Hinney A. Several mutations in the melanocortin 4 receptor gene are associated with obesity in Chinese children and adolescents. *J Endocrinol Invest* 2006;**29**:894–8.
45. Xiang Z, Pogozheva ID, Sorenson NB, Wilczynski AM, Holder JR, Litherland SA, et al. Peptide and small molecules rescue the functional activity and agonist potency of dysfunctional human melanocortin-4 receptor polymorphisms. *Biochemistry* 2007;**46**:8273–87.
46. Santoro N, Cirillo G, Xiang Z, Tanas R, Greggio N, Morino G, et al. Prevalence of pathogenetic MC4R mutations in Italian children with early onset obesity, tall stature and familial history of obesity. *BMC Med Genet* 2009;**10**:25.
47. Beckers S, Zegers D, de Freitas F, Peeters AV, Verhulst SL, Massa G, et al. Identification and functional characterization of novel mutations in the melanocortin-4 receptor. *Obes Facts* 2010;**3**:304–11.
48. Reinehr T, Hebebrand J, Friedel S, Toschke AM, Brumm H, Biebermann H, et al. Lifestyle intervention in obese children with variations in the melanocortin 4 receptor gene. *Obesity (Silver Spring)* 2009;**17**:382–9.
49. Fleck BA, Ling N, Chen C. Substituted NDP-MSH peptides paired with mutant melanocortin-4 receptors demonstrate the role of transmembrane 6 in receptor activation. *Biochemistry* 2007;**46**:10473–83.
50. Hinney A, Bettecken T, Tarnow P, Brumm H, Reichwald K, Lichtner P, et al. Prevalence, spectrum, and functional characterization of melanocortin-4 receptor gene mutations in a representative population-based sample and obese adults from Germany. *J Clin Endocrinol Metab* 2006;**91**:1761–9.
51. Fan ZC, Sartin JL, Tao YX. Pharmacological analyses of two naturally occurring porcine melanocortin-4 receptor mutations in domestic pigs. *Domest Anim Endocrinol* 2008;**34**:383–90.
52. Srinivasan S, Lubrano-Berthelier C, Govaerts C, Picard F, Santiago P, Conklin BR, et al. Constitutive activity of the melanocortin-4 receptor is maintained by its

N-terminal domain and plays a role in energy homeostasis in humans. *J Clin Invest* 2004;**114**:1158–64.

53. Branson R, Potoczna N, Kral JG, Lentes KU, Hoehe MR, Horber FF. Binge eating as a major phenotype of melanocortin 4 receptor gene mutations. *N Engl J Med* 2003;**348**:1096–103.
54. Tao YX, Segaloff DL. Functional analyses of melanocortin-4 receptor mutations identified from patients with binge eating disorder and nonobese or obese subjects. *J Clin Endocrinol Metab* 2005;**90**:5632–8.
55. Roubert P, Dubern B, Plas P, Lubrano-Berthelier C, Alihi R, Auger F, et al. Novel pharmacological MC4R agonists can efficiently activate mutated MC4R from obese patient with impaired endogenous agonist response. *J Endocrinol* 2010;**207**:177–83.
56. Ochoa MC, Azcona C, Biebermann H, Brumm H, Razquin C, Wermter AK, et al. A novel mutation Thr162Arg of the melanocortin 4 receptor gene in a Spanish children and adolescent population. *Clin Endocrinol (Oxf)* 2007;**66**:652–8.
57. Santini F, Maffei M, Ceccarini G, Pelosini C, Scartabelli G, Rosellini V, et al. Genetic screening for melanocortin-4 receptor mutations in a cohort of Italian obese patients: description and functional characterization of a novel mutation. *J Clin Endocrinol Metab* 2004;**89**:904–8.
58. Theisen FM, Cichon S, Linden A, Martin M, Remschmidt H, Hebebrand J. Clozapine and weight gain. *Am J Psychiatry* 2001;**158**:816.
59. Ng MC, Tam CH, So WY, Ho JS, Chan AW, Lee HM, et al. Implication of genetic variants near NEGR1, SEC16B, TMEM18, ETV5/DGKG, GNPDA2, LIN7C/BDNF, MTCH2, BCDIN3D/FAIM2, SH2B1, FTO, MC4R, and KCTD15 with obesity and type 2 diabetes in 7705 Chinese. *J Clin Endocrinol Metab* 2010;**95**:2418–25.
60. Rong R, Tao YX, Cheung BM, Xu A, Cheung GC, Lam KS. Identification and functional characterization of three novel human melanocortin-4 receptor gene variants in an obese Chinese population. *Clin Endocrinol (Oxf)* 2006;**65**:198–205.
61. Demiralp DO, Berberoglu M, Akar N. Melanocortin-4 receptor polymorphisms in Turkish pediatric obese patients. *Clin Appl Thromb Hemost* 2011;**17**:70–4.
62. Miraglia Del Giudice E, Cirillo G, Nigro V, Santoro N, D'Urso L, Raimondo P, et al. Low frequency of melanocortin-4 receptor (MC4R) mutations in a Mediterranean population with early-onset obesity. *Int J Obes Relat Metab Disord* 2002;**26**:647–51.
63. Dubern B, Lubrano-Berthelier C, Mencarelli M, Ersoy B, Frelut ML, Bouglé D, et al. Mutational analysis of the pro-opiomelanocortin gene in French obese children led to the identification of a novel deleterious heterozygous mutation located in the alpha-melanocyte stimulating hormone domain. *Pediatr Res* 2008;**63**:211–6.
64. Potoczna N, Branson R, Kral JG, Piec G, Steffen R, Ricklin T, et al. Gene variants and binge eating as predictors of comorbidity and outcome of treatment in severe obesity. *J Gastrointest Surg* 2004;**8**:971–81 discussion 981–982.
65. Tan K, Pogozheva ID, Yeo GS, Hadaschik D, Keogh JM, Haskell-Leuvano C, et al. Functional characterization and structural modeling of obesity associated mutations in the melanocortin 4 receptor. *Endocrinology* 2009;**150**:114–25.
66. Cole SA, Butte NF, Voruganti VS, Cai G, Haack K, Kent Jr JW, et al. Evidence that multiple genetic variants of MC4R play a functional role in the regulation of energy expenditure and appetite in Hispanic children. *Am J Clin Nutr* 2009;**91**:191–9.
67. Nowacka-Woszuk J, Cieslak J, Skowronska B, Majewska KA, Stankiewicz W, Fichna P, et al. Missense mutations and polymorphisms of the MC4R gene in Polish obese children and adolescents in relation to the relative body mass index. *J Appl Genet* 2011;**52**:319–23.
68. Tao YX, Huang H, Wang ZQ, Yang F, Williams JN, Nikiforovich GV. Constitutive activity of neural melanocortin receptors. *Methods Enzymol* 2010;**484**:267–79.

69. Granell S, Mohammad S, Ramanagoudr-Bhojappa R, Baldini G. Obesity-linked variants of melanocortin-4 receptor are misfolded in the endoplasmic reticulum and can be rescued to the cell surface by a chemical chaperone. *Mol Endocrinol* 2010;**24**:1805–21.
70. Chambers JC, Elliott P, Zabaneh D, Zhang W, Li Y, Froguel P, et al. Common genetic variation near MC4R is associated with waist circumference and insulin resistance. *Nat Genet* 2008;**40**:716–8.
71. Alharbi KK, Spanakis E, Tan K, Smith MJ, Aldahmesh MA, O'Dell SD, et al. Prevalence and functionality of paucimorphic and private MC4R mutations in a large, unselected European British population, scanned by meltMADGE. *Hum Mutat* 2007;**28**:294–302.
72. Chen M, Celik A, Georgeson KE, Harmon CM, Yang Y. Molecular basis of melanocortin-4 receptor for AGRP inverse agonism. *Regul Pept* 2006;**136**:40–9.
73. Biebermann H, Krude H, Elsner A, Chubanov V, Gudermann T, Grüters A. Autosomal-dominant mode of inheritance of a melanocortin-4 receptor mutation in a patient with severe early-onset obesity is due to a dominant-negative effect caused by receptor dimerization. *Diabetes* 2003;**52**:2984–8.
74. Kobayashi H, Ogawa Y, Shintani M, Ebihara K, Shimodahira M, Iwakura T, et al. A Novel homozygous missense mutation of melanocortin-4 receptor (MC4R) in a Japanese woman with severe obesity. *Diabetes* 2002;**51**:243–6.
75. Buchbinder S, Bärtsch U, Müller M, Zorn M, Nawroth PP, Schilling T. A novel missense mutation T101N in the melanocortin-4 receptor gene associated with obesity. *Genet Mol Res* 2011;**10**:1042–9.
76. Gotoda T, Scott J, Aitman TJ. Molecular screening of the human melanocortin-4 receptor gene: identification of a missense variant showing no association with obesity, plasma glucose, or insulin. *Diabetologia* 1997;**40**:976–9.
77. Hart Sailors ML, Folsom AR, Ballantyne CM, Hoelscher DM, Jackson AS, Linda Kao WH, et al. Genetic variation and decreased risk for obesity in the Atherosclerosis Risk in Communities Study. *Diabetes Obes Metab* 2007;**9**:548–57.
78. Young EH, Wareham NJ, Farooqi S, Hinney A, Hebebrand J, Scherag A, et al. The V103I polymorphism of the MC4R gene and obesity: population based studies and meta-analysis of 29 563 individuals. *Int J Obes (Lond)* 2007;**31**:1437–41.
79. Zhang R, Jia WP, Hu C, Wang CR, Fang QC, Bao YQ, et al. Association between melanocortin 4 receptor gene polymorphisms and obesity. *Zhonghua Yi Xue Za Zhi* 2007;**87**:96–9.
80. Valladares M, Domínguez-Vásquez P, Obregón AM, Weisstaub G, Burrows R, Maiz A, et al. Melanocortin-4 receptor gene variants in Chilean families: association with childhood obesity and eating behavior. *Nutr Neurosci* 2010;**13**:71–8.
81. Stutzmann F, Vatin V, Cauchi S, Morandi A, Jouret B, Landt O, et al. Non-synonymous polymorphisms in melanocortin-4 receptor protect against obesity: the two facets of a Janus obesity gene. *Hum Mol Genet* 2007;**16**:1837–44.
82. Gu W, Tu Z, Kleyn PW, Kissebah A, Duprat L, Lee J, et al. Identification and functional analysis of novel human melanocortin-4 receptor variants. *Diabetes* 1999;**48**:635–9.
83. Zakel UA, Wudy SA, Heinzel-Gutenbrunner M, Görg T, Schäfer H, Gortner L, et al. Prevalence of melanocortin 4 receptor (MC4R) mutations and polymorphismsin consecutively ascertained obese children and adolescents from a pediatric health care utilization population. *Klin Padiatr* 2005;**217**:244–9.
84. Valli-Jaakola K, Lipsanen-Nyman M, Oksanen L, Hollenberg AN, Kontula K, Bjørbaek C, et al. Identification and characterization of melanocortin-4 receptor gene mutations in morbidly obese finnish children and adults. *J Clin Endocrinol Metab* 2004;**89**:940–5.

85. Nijenhuis WA, Garner KM, van Rozen RJ, Adan RA. Poor cell surface expression of human melanocortin-4 receptor mutations associated with obesity. *J Biol Chem* 2003;**278**:22939–45.
86. Rettenbacher E, Tarnow P, Brumm H, Prayer D, Wermter AK, Hebebrand J, et al. A novel non-synonymous mutation in the melanocortin-4 receptor gene (MC4R) in a 2-year-old Austrian girl with extreme obesity. *Exp Clin Endocrinol Diabetes* 2007; **115**:7–12.
87. Tarnow P, Rediger A, Brumm H, Ambrugger P, Rettenbacher E, Widhalm K, et al. A heterozygous mutation in the third transmembrane domain causes a dominant-negative effect on signalling capability of the MC4R. *Obes Facts* 2008;**1**:155–62.
88. Dubern B, Clément K, Pelloux V, Froguel P, Girardet JP, Guy-Grand B, et al. Mutational analysis of melanocortin-4 receptor, agouti-related protein, and alpha-melanocyte-stimulating hormone genes in severely obese children. *J Pediatr* 2001;**139**:204–9.
89. Herpertz S, Siffert W, Hebebrand J. Binge eating as a phenotype of melanocortin 4 receptor gene mutations. *N Engl J Med* 2003;**349**:606–9.
90. Kim DH, Shin SW, Baik JH. Role of third intracellular loop of the melanocortin 4 receptor in the regulation of constitutive activity. *Biochem Biophys Res Commun* 2008;**365**:439–45.
91. Proneth B, Xiang Z, Pogozheva ID, Litherland SA, Gorbatyuk OS, Shaw AM, et al. Molecular mechanism of the constitutive activation of the L250Q human melanocortin-4 receptor polymorphism. *Chem Biol Drug Des* 2006;**67**:215–29.
92. Fang QC, Jia WP, Cai SB, Shao XY, Zhang R, Wang CR, et al. Intracellular retention of human melanocortin-4 receptor: a molecular mechanism underlying early-onset obesity in F261S pedigree of Chinese. *Biomed Environ Sci* 2008;**21**:280–5.
93. Chen M, Cai M, McPherson D, Hruby V, Harmon CM, Yang Y. Contribution of the transmembrane domain 6 of melanocortin-4 receptor to peptide [Pro5, DNal (2′)8]-gamma-MSH selectivity. *Biochem Pharmacol* 2009;**77**:114–24.
94. Tarnow P, Schoneberg T, Krude H, Gruters A, Biebermann H. Mutationally induced disulfide bond formation within the third extracellular loop causes melanocortin 4 receptor inactivation in patients with obesity. *J Biol Chem* 2003;**278**:48666–73.
95. Mergen M, Mergen H, Ozata M, Oner R, Oner C. A novel melanocortin 4 receptor (MC4R) gene mutation associated with morbid obesity. *J Clin Endocrinol Metab* 2001;**86**:3448.
96. Farooqi IS, O'Rahilly S. Monogenic obesity in humans. *Annu Rev Med* 2005;**56**: 443–58.
97. Dempfle A, Hinney A, Heinzel-Gutenbrunner M, Raab M, Geller F, Gudermann T, et al. Large quantitative effect of melanocortin-4 receptor gene mutations on body mass index. *J Med Genet* 2004;**41**:795–800.
98. Martinelli CE, Keogh JM, Greenfield JR, Henning E, van der Klaauw AA, Blackwood A, et al. Obesity due to melanocortin 4 receptor (MC4R) deficiency is associated with increased linear growth and final height, fasting hyperinsulinemia, and incompletely suppressed growth hormone secretion. *J Clin Endocrinol Metab* 2011;**96**:E181–E188.
99. Hinney A, Scherag S, Hebebrand J. Genetic findings in anorexia and bulimia nervosa. *Prog Mol Biol Transl Sci* 2010;**94**:241–70 Review.
100. Hughes DA, Hinney A, Brumm H, Wermter AK, Biebermann H, Hebebrand J, et al. Increased constraints on MC4R during primate and human evolution. *Hum Genet* 2009;**124**:633–47.
101. Stäubert C, Tarnow P, Brumm H, Pitra C, Gudermann T, Grüters A, et al. Evolutionary aspects in evaluating mutations in the melanocortin 4 receptor. *Endocrinology* 2007;**148**:4642–8.

102. Tao YX. Mutations in melanocortin-4 receptor and human obesity. *Prog Mol Biol Transl Sci* 2009;**88**:173–204.
103. Tao YX. The melanocortin-4 receptor: physiology, pharmacology, and pathophysiology. *Endocr Rev* 2010;**31**:506–43.
104. Wang D, Ma J, Zhang S, Hinney A, Hebebrand J, Wang Y, et al. Association of the MC4R V103I polymorphism with obesity: a Chinese case–control study and meta-analysis in 55,195 individuals. *Obesity (Silver Spring)* 2010;**18**:573–9.
105. Tao YX. Constitutive activation of G protein-coupled receptors and diseases: insights into mechanisms of activation and therapeutics. *Pharmacol Ther* 2008;**120**:129–48.
106. Tao YX. Inactivating mutations of G protein-coupled receptors and diseases: structure-function insights and therapeutic implications. *Pharmacol Ther* 2006;**111**:949–73.
107. Brocke KS, Neu-Yilik G, Gehring NH, Hentze MW, Kulozik AE. The human intronless melanocortin 4-receptor gene is NMD insensitive. *Hum Mol Genet* 2002;**11**:331–5.
108. Brumm H, Mühlhaus J, Bolze F, Scherag S, Hinney A, Hebebrand J, et al. Rescue of melanocortin 4 receptor (MC4R) nonsense mutations by aminoglycoside-mediated read-through. *Obesity (Silver Spring)* 2012;**20**:1074–81.
109. Bolze F, Rink N, Brumm H, Kühn R, Mocek S, Schwarz AE, et al. Characterization of the melanocortin-4-receptor nonsense mutation W16X in vitro and in vivo. *Pharmacogenomics J* 2011;http://dx.doi.org/10.1038/tpj.2011.43.
110. Geller F, Reichwald K, Dempfle A, Illig T, Vollmert C, Herpertz S, et al. Melanocortin-4 receptor gene variant I103 is negatively associated with obesity. *Am J Hum Genet* 2004;**74**:572–81.
111. Spielman RS, McGinnis RE, Ewens WJ. Transmission test for linkage disequilibrium: the insulin gene region and insulin-dependent diabetes mellitus (IDDM). *Am J Hum Genet* 1993;**52**:506–16.
112. Heid IM, Vollmert C, Hinney A, Döring A, Geller F, Löwel H, et al. Association of the 103I MC4R allele with decreased body mass in 7937 participants of two population based surveys. *J Med Genet* 2005;**42**:e21.
113. Meyre D, Delplanque J, Chèvre JC, Lecoeur C, Lobbens S, Gallina S, et al. Genome-wide association study for early-onset and morbid adult obesity identifies three new risk loci in European populations. *Nat Genet* 2009;**41**:157–9.
114. Brönner G, Sattler AM, Hinney A, Soufi M, Geller F, Schäfer H, et al. The 103I variant of the melanocortin 4 receptor is associated with low serum triglyceride levels. *J Clin Endocrinol Metab* 2006;**91**:535–8.
115. Iqbal J, Li X, Chang BH, Chan L, Schwartz GJ, Chua Jr SC, et al. An intrinsic gut leptin-melanocortin pathway modulates intestinal microsomal triglyceride transfer protein and lipid absorption. *J Lipid Res* 2010;**51**:1929–42.
116. Heid IM, Vollmert C, Kronenberg F, Huth C, Ankerst DP, Luchner A, et al. Association of the MC4R V103I polymorphism with the metabolic syndrome: the KORA Study. *Obesity (Silver Spring)* 2008;**16**:369–76.
117. Pichler M, Kollerits B, Heid IM, Hunt SC, Adams TD, Hopkins PN, et al. Association of the melanocortin-4 receptor V103I polymorphism with dietary intake in severely obese persons. *Am J Clin Nutr* 2008;**88**:797–800.
118. Marks DL, Ling N, Cone RD. Role of the central melanocortin system in cachexia. *Cancer Res* 2001;**61**:1432–8.
119. Knoll S, Zimmer S, Hinney A, Scherag A, Neubauer A, Hebebrand J. Val103Ile polymorphism of the melanocortin-4 receptor gene (MC4R) in cancer cachexia. *BMC Cancer* 2008;**8**:85.
120. Loos RJ, Lindgren CM, Li S, Wheeler E, Zhao JH, Prokopenko I, et al. Common variants near MC4R are associated with fat mass, weight and risk of obesity. *Nat Genet* 2008;**40**:768–75.

121. Frayling TM, Timpson NJ, Weedon MN, Zeggini E, Freathy RM, Lindgren CM, et al. A common variant in the FTO gene is associated with body mass index and predisposes to childhood and adult obesity. *Science* 2007;**316**:889–94.
122. Willer CJ, Speliotes EK, Loos RJ, Li S, Lindgren CM, Heid IM, et al. Six new loci associated with body mass index highlight a neuronal influence on body weight regulation. *Nat Genet* 2009;**41**:25–34.
123. Thorleifsson G, Walters GB, Gudbjartsson DF, Steinthorsdottir V, Sulem P, Helgadottir A, et al. Genome-wide association yields new sequence variants at seven loci that associate with measures of obesity. *Nat Genet* 2009;**41**:18–24.
124. Speliotes EK, Willer CJ, Berndt SI, Monda KL, Thorleifsson G, Jackson AU, et al. Association analyses of 249,796 individuals reveal 18 new loci associated with body mass index. *Nat Genet* 2010;**42**:937–48.
125. Paternoster L, Evans DM, Nohr EA, Holst C, Gaborieau V, Brennan P, et al. Genome-wide population-based association study of extremely overweight young adults—the GOYA study. *PLoS One* 2011;**6**:e24303.
126. Rouskas K, Kouvatsi A, Paletas K, Papazoglou D, Tsapas A, Lobbens S, et al. Common variants in FTO, MC4R, TMEM18, PRL, AIF1, and PCSK1 show evidence of association with adult obesity in the Greek population. *Obesity (Silver Spring)* 2012;**20**:389–95.
127. Lurie G, Gaudet MM, Spurdle AB, Carney ME, Wilkens LR, Yang HP, et al. The obesity-associated polymorphisms FTO rs9939609 and MC4R rs17782313 and endometrial cancer risk in non-Hispanic white women. *PLoS One* 2011;**6**:e16756.
128. Beckers S, Zegers D, de Freitas F, Mertens IL, Van Gaal LF, Van Hul W. Association study of MC4R with complex obesity and replication of the rs17782313 association signal. *Mol Genet Metab* 2011;**103**:71–5.
129. Renström F, Payne F, Nordström A, Brito EC, Rolandsson O, Hallmans G, et al. Replication and extension of genome-wide association study results for obesity in 4923 adults from northern Sweden. *Hum Mol Genet* 2009;**18**:1489–96.
130. Grant SF, Bradfield JP, Zhang H, Wang K, Kim CE, Annaiah K, et al. Investigation of the locus near MC4R with childhood obesity in Americans of European and African ancestry. *Obesity (Silver Spring)* 2009;**17**:1461–5.
131. Hinney A, Hebebrand J. Three at one swoop!. *Obes Facts* 2009;**2**:3–8.
132. Elks CE, Loos RJ, Sharp SJ, Langenberg C, Ring SM, Timpson NJ, et al. Genetic markers of adult obesity risk are associated with greater early infancy weight gain and growth. *PLoS Med* 2010;**7**:e1000284.
133. Kilpeläinen TO, den Hoed M, Ong KK, Grøntved A, Brage S, Early Growth Genetics Consortium , et al. Obesity-susceptibility loci have a limited influence on birth weight: a meta-analysis of up to 28,219 individuals. *Am J Clin Nutr* 2011;**93**:851–60.
134. Scherag A, Dina C, Hinney A, Vatin V, Scherag S, Vogel CI, et al. Two new Loci for body-weight regulation identified in a joint analysis of genome-wide association studies for early-onset extreme obesity in French and german study groups. *PLoS Genet* 2010;**6**:e1000916.
135. Zhao J, Bradfield JP, Zhang H, Sleiman PM, Kim CE, Glessner JT, et al. Role of BMI-associated loci identified in GWAS meta-analyses in the context of common childhood obesity in European Americans. *Obesity (Silver Spring)* 2011;**19**:2436–9.
136. Hester JM, Wing MR, Li J, Palmer ND, Xu J, Hicks PJ, et al. Implication of European-derived adiposity loci in African Americans. *Int J Obes (Lond)* 2012;**36**:465–73.
137. Liu G, Zhu H, Dong Y, Podolsky RH, Treiber FA, Snieder H. Influence of common variants in FTO and near INSIG2 and MC4R on growth curves for adiposity in African- and European-American youth. *Eur J Epidemiol* 2011;**26**:463–73.
138. Vasan SK, Fall T, Neville MJ, Antonisamy B, Fall CH, Geethanjali FS, et al. Associations of Variants in FTO and Near MC4R With Obesity Traits in South Asian Indians. *Obesity (Silver Spring)* 2012;**20**:2268–77. http://dx.doi.org/10.1038/oby.2012.64.

139. Janipalli CS, Kumar MV, Vinay DG, Sandeep MN, Bhaskar S, Kulkarni SR, et al. Analysis of 32 common susceptibility genetic variants and their combined effect in predicting risk of Type 2 diabetes and related traits in Indians. *Diabet Med* 2012;**29**:121–7.
140. Cheung CY, Tso AW, Cheung BM, Xu A, Ong KL, Fong CH, et al. Obesity susceptibility genetic variants identified from recent genome-wide association studies: implications in a chinese population. *J Clin Endocrinol Metab* 2010;**95**:1395–13403.
141. Okada Y, Kubo M, Ohmiya H, Takahashi A, Kumasaka N, Hosono N, et al. Common variants at CDKAL1 and KLF9 are associated with body mass index in east Asian populations. *Nat Genet* 2012;**44**:302–6.
142. Wen W, Cho YS, Zheng W, Dorajoo R, Kato N, Qi L, et al. Meta-analysis identifies common variants associated with body mass index in east Asians. *Nat Genet* 2012;**44**:307–11.
143. Huang H, Zeng Z, Zhang L, Liu R, Li X, Qiang O, et al. Implication of genetic variants near TMEM18, BCDIN3D/FAIM2, and MC4R with coronary artery disease and obesity in Chinese: a angiography-based study. *Mol Biol Rep* 2012;**39**:1739–44.
144. Shi J, Long J, Gao YT, Lu W, Cai Q, Wen W, et al. Evaluation of genetic susceptibility loci for obesity in Chinese women. *Am J Epidemiol* 2010;**172**:244–54.
145. Dorajoo R, Blakemore AI, Sim X, Ong RT, Ng DP, Seielstad M, et al. Replication of 13 obesity loci among Singaporean Chinese, Malay and Asian-Indian populations. *Int J Obes (Lond)* 2012;**36**:159–63.
146. Wu L, Xi B, Zhang M, Shen Y, Zhao X, Cheng H, et al. Associations of six single nucleotide polymorphisms in obesity-related genes with BMI and risk of obesity in Chinese children. *Diabetes* 2010;**59**:3085–9.
147. Takeuchi F, Yamamoto K, Katsuya T, Nabika T, Sugiyama T, Fujioka A, et al. Association of genetic variants for susceptibility to obesity with type 2 diabetes in Japanese individuals. *Diabetologia* 2011;**54**:1350–9.
148. Hong KW, Oh B. Recapitulation of genome-wide association studies on body mass index in the Korean population. *Int J Obes (Lond)* 2012;**36**:1127–30. http://dx.doi.org/10.1038/ijo.2011.202.
149. Tao L, Zhang Z, Chen Z, Zhou D, Li W, Kan M, et al. A Common variant near the melanocortin 4 receptor is associated with low-density lipoprotein cholesterol and total cholesterol in the Chinese Han population. *Mol Biol Rep* 2012;**39**:6487–93.
150. Vogel CI, Boes T, Reinehr T, Roth CL, Scherag S, Scherag A, et al. Common variants near MC4R: exploring gender effects in overweight and obese children and adolescents participating in a lifestyle intervention. *Obes Facts* 2011;**4**:67–75.
151. Winter Y, Back T, Scherag A, Linseisen J, Rohrmann S, Lanczik O, et al. Evaluation of the obesity genes FTO and MC4R and the type 2 diabetes mellitus gene TCF7L2 for contribution to stroke risk: the Mannheim-Heidelberg Stroke Study. *Obes Facts* 2011;**4**:290–6.
152. Melka MG, Bernard M, Mahboubi A, Abrahamowicz M, Paterson AD, Syme C, et al. Genome-wide scan for loci of adolescent obesity and their relationship with blood pressure. *J Clin Endocrinol Metab* 2012;**97**:E145–E150.
153. Ewens KG, Jones MR, Ankener W, Stewart DR, Urbanek M, Dunaif A, et al. FTO and MC4R gene variants are associated with obesity in polycystic ovary syndrome. *PLoS One* 2011;**6**:e16390.
154. Malhotra AK, Correll CU, Chowdhury NI, Müller DJ, Gregersen PK, Lee AT, et al. Association between common variants near the melanocortin 4 receptor gene and severe antipsychotic drug-induced weight gain. *Arch Gen Psychiatry* 2012;**69**:904–12.
155. Dickson SP, Wang K, Krantz I, Hakonarson H, Goldstein DB. Rare variants create synthetic genome-wide associations. *PLoS Biol* 2010;**8**:e1000294.
156. Cirulli ET, Goldstein DB. Uncovering the roles of rare variants in common disease through whole-genome sequencing. *Nat Rev Genet* 2010;**11**:415–25.

157. Kleinjan DA, van Heyningen V. Long-range control of gene expression: emerging mechanisms and disruption in disease. *Am J Hum Genet* 2005;**76**:8–32.
158. Scherag A, Jarick I, Grothe J, Biebermann H, Scherag S, Volckmar AL, et al. Investigation of a genome wide association signal for obesity: synthetic association and haplotype analyses at the melanocortin 4 receptor gene locus. *PLoS One* 2010;**5**:e13967.
159. Mühlhaus J, Pütter C, Brumm H, Grallert H, Illig T, Scherag S, et al. Do common variants separate between obese melanocortin-4 receptor gene mutation carriers and non-carriers? The impact of cryptic relatedness. *Horm Res Paediatr* 2012;**77**:358–68.
160. Greenfield JR, Miller JW, Keogh JM, Henning E, Satterwhite JH, Cameron GS, et al. Modulation of blood pressure by central melanocortinergic pathways. *N Engl J Med* 2009;**360**:44–52.
161. Sayk F, Heutling D, Dodt C, Iwen KA, Wellhoner JP, Scherag S, et al. Sympathetic function in human carriers of melanocortin-4 receptor gene mutations. *J Clin Endocrinol Metab* 2010;**95**:1998–2002.
162. Scherag S, Hebebrand J, Hinney A. Eating disorders: the current status of molecular genetic research. *Eur Child Adolesc Psychiatry* 2010;**19**:211–26.
163. Fairburn CG, Welch SL, Doll HA, Davies BA, O'Connor ME. Risk factors for bulimia nervosa. A community-based case–control study. *Arch Gen Psychiatry* 1997;**54**:509–17.
164. Hebebrand J, Geller F, Dempfle A, Heinzel-Gutenbrunner M, Raab M, Gerber G, et al. Binge-eating episodes are not characteristic of carriers of melanocortin-4 receptor gene mutations. *Mol Psychiatry* 2004;**9**:796–800.
165. Tortorella A, Monteleone P, del Giudice EM, Cirillo G, Perrone L, Castaldo E, et al. Melanocortin-4 receptor molecular scanning and pro-opiomelanocortin R236G variant screening in binge eating disorder. *Psychiatr Genet* 2005;**15**:161.
166. Albayrak O, Albrecht B, Scherag S, Barth N, Hinney A, Hebebrand J. Successful methylphenidate treatment of early onset extreme obesity in a child with a melanocortin-4 receptor gene mutation and attention deficit/hyperactivity disorder. *Eur J Pharmacol* 2011;**660**:165–70.
167. Agranat-Meged A, Ghanadri Y, Eisenberg I, Ben Neriah Z, Kieselstein-Gross E, Mitrani-Rosenbaum S. Attention deficit hyperactivity disorder in obese melanocortin-4-receptor (MC4R) deficient subjects: a newly described expression of MC4R deficiency. *Am J Med Genet B Neuropsychiatr Genet* 2008;**147B**:1547–53.
168. Hatoum IJ, Stylopoulos N, Vanhoose AM, Boyd KL, Yin DP, Ellacott KL, et al. Melanocortin-4 receptor signaling is required for weight loss after gastric bypass surgery. *J Clin Endocrinol Metab* 2012;**97**:E1023–E1031.
169. Aslan IR, Campos GM, Calton MA, Evans DS, Merriman RB, Vaisse C. Weight loss after Roux-en-Y gastric bypass in obese patients heterozygous for MC4R mutations. *Obes Surg* 2011;**21**:930–4.
170. Aslan IR, Ranadive SA, Ersoy BA, Rogers SJ, Lustig RH, Vaisse C. Bariatric surgery in a patient with complete MC4R deficiency. *Int J Obes (Lond)* 2011;**35**:457–61.
171. Mirshahi UL, Stillm CD, Masker KK, Gerhard GS, Carey DJ, Mirshahi T. The MC4R (I251L) allele is associated with better metabolic status and more weight loss after gastric bypass surgery. *J Clin Endocrinol Metab* 2011;**96**:E2088–E2096.
172. Lim BK, Huang KW, Grueter BA, Rothwell PE, Malenka RC. Anhedonia requires MC4R-mediated synaptic adaptations in nucleus accumbens. *Nature* 2012;**487**:183–9.
173. Hilbert A, Dierk JM, Conradt M, Schlumberger P, Hinney A, Hebebrand J, et al. Causal attributions of obese men and women in genetic testing: implications of genetic/biological attributions. *Psychol Health* 2009;**24**:749–61.
174. Holzapfel C, Grallert H, Huth C, Wahl S, Fischer B, Döring A, et al. Genes and lifestyle factors in obesity: results from 12,462 subjects from MONICA/KORA. *Int J Obes (Lond)* 2010;**34**:1538–45.

CHAPTER SIX

G Protein-Coupled Estrogen Receptor in Energy Homeostasis and Obesity Pathogenesis

Haifei Shi*,†, Shiva Priya Dharshan Senthil Kumar†, Xian Liu†

*Department of Biology, Center for Physiology and Neuroscience, Miami University, Oxford, Ohio, USA
†Cellular, Molecular and Structural Biology, Miami University, Oxford, Ohio, USA

Contents

Progress in Molecular Biology and Translational Science, Volume 114
ISSN 1877-1173
http://dx.doi.org/10.1016/B978-0-12-386933-3.00006-6

Abstract

Obesity and its related metabolic diseases have reached a pandemic level worldwide. There are sex differences in the prevalence of obesity and its related metabolic diseases, with men being more vulnerable than women; however, the prevalence of these disorders increases dramatically in women after menopause, suggesting that sex steroid hormone estrogens play key protective roles against development of obesity and metabolic diseases. Estrogens are important regulators of several aspects of metabolism, including body weight and body fat, caloric intake and energy expenditure, and glucose and lipid metabolism in both males and females. Estrogens act in complex ways on their nuclear estrogen receptors (ERs) ERα and ERβ and transmembrane ERs such as G protein-coupled estrogen receptor. Genetic tools, such as different lines of knockout mouse models, and pharmacological agents, such as selective agonists and antagonists, are available to study function and signaling mechanisms of ERs. We provide an overview of the evidence for the physiological and cellular actions of ERs in estrogen-dependent processes in the context of energy homeostasis and body fat regulation and discuss its pathology that leads to obesity and related metabolic states.

ABBREVIATIONS

α-MSH α-melanocyte stimulating hormone
AgRP agouti-related protein
ARC arcuate nucleus
cAMP cyclic AMP
CNS central nervous system
E2 17β-estradiol
EGFRs epidermal growth factor receptors
ERs estrogen receptors
ERα estrogen receptor alpha
ERβ estrogen receptor beta
EREs estrogen response elements
GLUT4 glucose transporter 4
GPER G protein-coupled estrogen receptor
HSL hormone-sensitive lipase
LH lateral hypothalamus
LPL lipoprotein lipase
mER membrane subpopulation estrogen receptors
NPY neuropeptide Y
PI3K phosphoinositide-3 kinases
PKA protein kinase A
PKB protein kinase B (Akt)
POMC pro-opiomelanocortin
PTX pertussis toxin
PVN paraventricular nucleus
SERM selective estrogen receptor modulators
STZ streptozotocin
VMH ventromedial hypothalamus

1. INTRODUCTION

1.1. Obesity is a risk factor for metabolic diseases

Despite attempts to increase awareness of the deleterious health consequences associated with obesity, there continues to be a significant increase in the obese population worldwide.[1] Obesity has been recognized as a major and an independent risk factor for metabolic syndrome, a group of interrelated metabolic abnormalities characterized by increased visceral fat accumulation, disturbed glucose homeostasis with hyperglycemia and insulin resistance, increased triglycerides and decreased high-density lipoprotein cholesterol, and hypertension.[2] The economic costs of caring for patients with obesity and its metabolic complications are enormously high.[3] Although the pathophysiologic basis of obesity and metabolic syndrome is largely unknown, evidence suggests that modern diets with high fat content and sedentary lifestyle progressively lead to the accumulation of saturated fatty acids in nonadipose tissue. The subsequent metabolism of this ectopic fatty acid deposition leads to decreased insulin-stimulated glucose uptake in skeletal muscle, unsuppressed hepatic glucose production, and altered glucose-stimulated insulin release from pancreatic β-cells.[4] Once diabetes is declared, hyperglycemia and elevated free fatty acids cooperate to generate major oxidative stress in tissues, further aggravating insulin resistance and deficiency.

Development of adverse metabolic complications has been attributed to increased body fat, not just body weight, and the fat distributed particularly in the abdominal visceral compartment, a source of bioactive mediators that directly contribute to metabolic disorders.[5] Intra-abdominal visceral adipose tissue carries a greater risk for the development of metabolic disorders than subcutaneous adipose tissue does. Therefore, body composition and body fat distribution are closely associated with metabolic state.

1.2. Sex differences in the development of obesity and metabolic diseases

There are important sex differences in the prevalence of obesity and metabolic diseases. The prevalence of obesity and its related metabolic diseases is particularly high among men and middle-aged women. Prior to menopause, women have much less obesity-related metabolic disorders than obese men. For example, hypercholesterolemia is more common in males, whose lipid profile is characterized by lower HDL, higher LDL and triglyceride levels, and larger VLDL particles. The incidence of liver steatosis is also higher in men.[6] After menopause, however, women are at an increased risk of developing visceral obesity and metabolic syndrome[7] due to the loss of - endogenous ovarian hormone production.

Risk for metabolic syndrome incidence generally relates more to the central distribution of fat with abdominal, omental, and visceral adiposity than to the total amount of fat, with individuals with large amounts of abdominal visceral fat having particularly high risk.[8,9] Adipose tissue accumulation is sexually dimorphic.[10] The amount and regional distribution of body fat is different between males and females, as males and females differ in terms of how and where they store body fat. Premenopausal women usually have a higher amount of body fat and more subcutaneous gluteal/femoral pattern of fat distribution, whereas men have a leaner body composition and have more fat is stored in the central intra-abdominal visceral depot.[11,12] Because visceral fat is more likely to cause metabolic syndromes, incidence of obesity-related metabolic disorders is much lower in premenopausal women than men and women experiencing menopause.

1.3. Sex steroid estrogens play key roles in energy homeostasis and glucose regulation

1.3.1 Estrogens regulate energy homeostasis and body fat

The epidemiological and clinical evidence strongly suggests a major role for estrogens in the regulation of body weight and body fat. It is now evident that estrogens account for sexual dimorphism in body fat distribution and accretion. The differences in the fat distribution between premenopausal women and age-matched men are abolished between postmenopausal women and age-matched men.[13] Estrogens exert important antiobesity effects in females. Postmenopausal women with decreased levels of estrogens are associated with an increased risk for developing obesity.[14] Additionally, estrogens promote the accumulation of subcutaneous fat[15] while visceral fat varies inversely with levels of estrogens.[16] Consequently, the loss of endogenous estrogen production with menopause is associated with an increased risk to develop visceral obesity and symptoms of the metabolic syndrome.[9] The increased visceral obesity in menopause women can be altered by exogenous hormone replacement therapy.[17,18]

Studies using experimental animal models have provided important insights into the pathogenesis of human diseases and new therapeutic approaches. While metabolic diseases and obesity have been shown in a number of animal species, in the laboratory they are most commonly investigated in two species of rodents, rats (*Rattus norvegicus*) and mice (*Mus musculus/Mus domesticus*).

Reductions in circulating estrogens and estrogen signaling, which occurs following removal of ovaries (i.e., ovariectomy), transiently increase caloric intake, and result in increased adiposity,[19] specifically visceral fat with no change of subcutaneous fat.[20] Interestingly, the transient hyperphagia induced by ovariectomy does not account for the development of obesity,[21] as ovariectomized rats gain weight to a similar extent even when they are pair-fed to 17β–estradiol (E2)-treated rats,[22] suggesting that endogenous estrogens regulate body weight homeostasis primarily by modulating energy expenditure.

E2 administration decreases energy intake and increases energy expenditure,[23] indicating that exogenous estrogens may promote a negative energy balance by influencing both energy intake and energy expenditure. The ovariectomy-induced increase in adiposity can be ameliorated by exogenous E2 administration.[20] Furthermore, peripheral or central administration of E2 to ovariectomized rats changes their body fat distribution to mirror that of intact females. Additionally, altering the sex hormone milieu in males with E2 administration increases subcutaneous fat deposition.[20] An important implication from these findings is that estrogens are critical determinates of body fat distribution and energy balance.

1.3.2 Estrogens regulate lipid metabolism

Epidemiological and prospective studies associate estrogens to several aspects of the metabolic syndrome (Fig. 6.1). There is increasing evidence both in humans and laboratory rodents suggesting that estrogens are not only involved in regulating adiposity and body fat distribution, but also directly involved in regulation of glucose and lipid metabolism. Thus, estrogen action seems to be implicated in adipose tissue regulation and prevention of obesity.

Estrogens suppress lipogenesis in adipose tissue, skeletal muscle, and liver.[24,25] In order to understand the role of estrogens in liver, studies evaluated the consequences of hormone replacement therapy to postmenopausal women and the effects of estrogen-suppression therapy in breast cancer patients. Estrogens increase HDL and triglycerides, decrease LDL and total cholesterol, and decrease fasting insulin and glucose levels, suggesting beneficial effects of estrogen on lipid and glucose metabolism.[26] Recent studies demonstrate that the use of antiestrogen therapy, such as tamoxifen and ovariectomy, leads to abnormal lipid profile and liver steatosis.[27] Furthermore, studies reported that islet activation of estrogen receptors (ERs), estrogen receptor alpha (ERα), estrogen receptor beta (ERβ), and

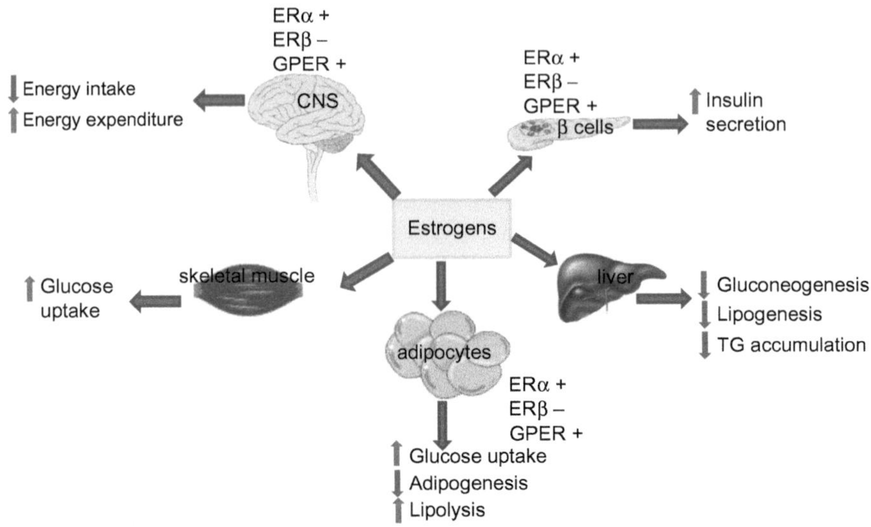

Figure 6.1 Overview of estrogenic control of energy homeostasis, glucose, and lipid metabolism by regulatory actions in the CNS, adipose tissue, pancreatic β cells, skeletal muscles, and liver. Deficiency in estrogenic action leads to opposite events and subsequent disturbances in metabolic tissues. ERα, ERβ, and GPER play roles to regulate energy balance and glucose homeostasis. ERα and ERβ generally mediate opposing effects, while GPER generally has similar roles as ERα in the brain, adipose tissues, and pancreas. (See Color Insert.)

G protein-coupled ER (GPER) by estrogens suppresses fatty acid synthase expression and enzymatic activity.[28] This effect reduces islet lipid accumulation in rodent and human β-cells and prevents β-cell failure in rodent models of type 2 diabetes.

Another essential argument for estrogens' antilipogenic roles on lipid metabolism comes from the study of men with estrogen deficiency. In both sexes, estrogens are biosynthesized by the cytochrome P450 enzyme complex called aromatase.[29] In men, estrogen secretion comes principally from the conversion of androgens in adipose tissue by aromatase. The aromatase-deficient patients, who cannot synthesize estrogens thus lack endogenous estrogen secretion secondary to missense mutations in the aromatase gene, show impaired liver functions, hepatic steatosis, and altered lipid profile,[30] as well as develop hypertriglyceridemia and insulin resistance.[31,32] These metabolic abnormalities are improved or corrected by estrogen replacement therapy.[30,32,33] In the E2-deficient aromatase knockout (ArKO) mice, insulin resistance develops due to hepatic steatosis.[34]

1.4. Focus of this chapter

In both humans and rodent models, estrogens are important sex hormones central to health in both genders that have protective effects in many physiological processes in mammals, including but not limited to reproduction, cardiovascular health, metabolic systems, bone integrity, cognition, and behavior.[35] Given these widespread roles for estrogens, it is not surprising that estrogens are implicated in the development or progression of numerous diseases, which include but are not limited to obesity, insulin resistance, osteoporosis, cardiovascular disease, various types of cancer of reproductive tissues, and neurodegenerative diseases. Pertinent to this chapter is estrogens' regulation in energy homeostasis, glucose and lipid metabolism, and insulin sensitivity (Fig. 6.1). Estrogens are protective against the development of visceral obesity and glucose intolerance, and thus contribute to the maintenance of insulin sensitivity. Deficiency in estrogen leads to development of visceral obesity and insulin resistance.

We discuss in this chapter the mechanisms of action by estrogens in metabolic processes, focusing on the central nervous system (CNS) and adipose tissues. We first introduce the estrogenic system, including the estrogens and their receptors. We then discuss the current knowledge and questions about estrogenic action in the brain and adipose tissue, and how these processes are involved in the regulation of energy homeostasis and lipid metabolism. We review studies that have investigated the roles of estrogens and their receptors in the regulation of such metabolism, focusing on GPER-dependent signaling and function. Controversies that have complicated our understanding of GPER, including interactions with nuclear and transmembrane ERs, will also be discussed.

2. ESTROGEN SIGNALING: ESTROGENS AND ERs

2.1. Estrogenic action through genomic and nongenomic signaling mechanisms

Estrogens are the best-characterized member of the family of steroid hormones that includes progesterone, testosterone, glucocorticoids, and mineralocorticoids. Human estrogens comprise a group of structurally related steroid molecules, namely 17β-estradiol (E2), estrone, and estriol, which are the most important regulators of the female and male reproductive systems. Estrogens are synthesized from cholesterol by aromatase via a series of chemical reactions that are part of the steroidogenic pathways predominantly in the ovaries. Estrogens are important hormones in mammalian physiology exerting important functions on the regulation of female secondary sexual characteristics. In males, relatively small amount of estrogens are produced by Leydig and germ cells of the testis. Additionally, estrogens are synthesized at multiple discrete sites throughout the body from the liver, adrenal glands, and adipocytes,[36] where they have highly localized effects on various unrelated processes in many tissues.[37] Estrogens also interact with and exert regulatory effects in a variety of tissues in the body that do not secrete estrogen, including the CNS, cardiovascular system, immune system, breast, uterus, and bone.[38–42]

Estrogenic actions are mediated by a number of different ERs (Fig. 6.2). The highly hydrophobic nature of estrogens allows them to pass through cellular membranes by passive diffusion[43] and also allows its concentration in cellular membranes.[44] Cellular responses to estrogens are mediated by receptors that initiate a complex array of cellular events upon ligand binding. These responses are divided into two broad categories. One is genomic

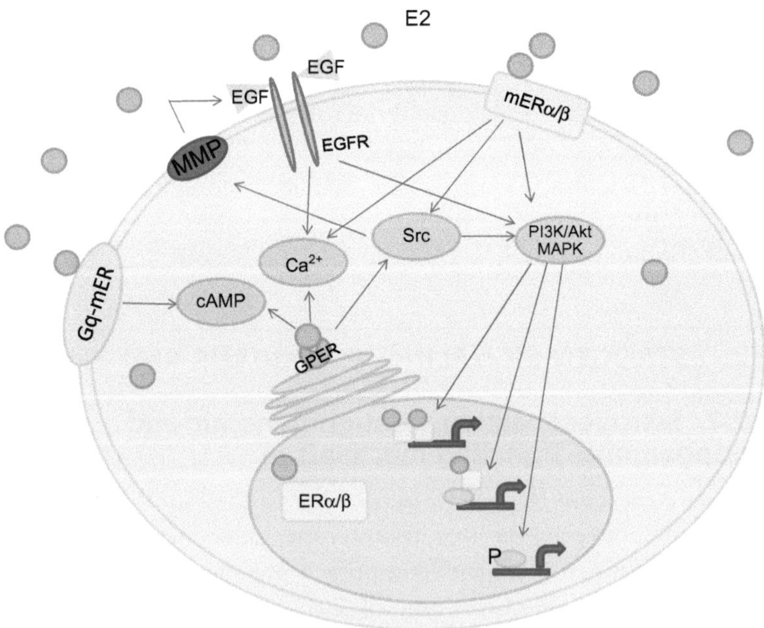

Figure 6.2 Schematic overview of estradiol-mediated genomic signaling pathway via ERα/β and rapid nongenomic signaling pathways via GPER, Gq-mER, and membrane subpopulation mERα/β. (See Color Insert.)

responses that are characterized by changes in gene transcription and occur on the time frame of hours to days after estrogens bind to nuclear ERS, ERα and ERβ. The other one is nongenomic rapid signaling events that occur within seconds to minutes of cell stimulation after estrogens bind to novel GPER[45] which is localized to the endoplasmic reticulum, membrane subpopulation ERs (mERα/β), or G protein-coupled receptor membrane ERs (Gq-mER), and mediates nongenomic estrogen signaling via cytosolic pathways involving second messengers.[45,46] Thus, an organism's responses to estrogens are generally the result of a complex interaction between genomic and nongenomic signaling (Fig. 6.2). The genomic mechanisms are relatively well characterized, while the nongenomic estrogenic signaling is less well understood and is beginning to be explored.

2.2. Estrogenic genomic actions via nuclear ERs

Estrogens activate their nuclear ERs in target cells, acting as transcription factors to regulate the expression of target genes, ultimately controlling cell growth, differentiation, and homeostasis.[47] Two nuclear ERs located on distinct chromosomes have been identified.[48–51] The first described classical ER, today known as ERα, was discovered and characterized in 1973, on the basis of specific binding activity in rat uterus/vagina extracts.[52] The classical nuclear ER was cloned in 1985,[51] its DNA sequence was determined in 1986,[50] and the first crystal structure of an ER ligand-binding domain was described in 1997.[53] ER was renamed ERα when a second nuclear ER, ERβ, was identified 10 years later.[49]

The major physiological estrogen is E2 that has a similar affinity for both ERs. Pharmacological reagents have been developed that permit the investigation of the specific role of ERs. ERs are activated by a range of ligands, including selective ER modulators (SERM) such as raloxifene and tamoxifen; the ERα-selective agonist 4,4′,4″-(4-propyl-[1*H*]-pyrazole-1,3,5-triyl) *tris*phenol (PPT), which has 410 times greater affinity for ERα than ERβ[54]; and two commonly used ERβ agonists, 2,3-bis(4-Hydroxyphenyl)-propionitrile (DPN) and 7-Bromo-2-(4-hydroxyphenyl)-1,3-benzoxazol-5-ol (WAY200070), which show similar receptor specificity and neither is as selective for ERβ as PPT is for ERα. DPN is 70-fold more selective for ERβ than ERα,[55] while the affinity of WAY200070 (compound 92) for ERβ in mice is 68 times that for ERα.[56]

Estrogens' genomic action on cell functions is via the activation of ERα and ERβ, both of which belong to the nuclear hormone receptor family and are transcription factors. They have classically been described as ligand-activated transcription factors mediating long-term genomic effects in central and peripheral estrogen-regulated tissues. ERα and ERβ exist primarily in the nucleus, where they are complexed with chaperones. Briefly, binding of estrogens to ERα and ERβ leads to conformational changes, chaperone dissociation from their inhibitory protein complexes, and dimerization of the receptor, followed by receptor translocation to the nucleus to recognize and binding by receptor dimers to *cis*-acting estrogen response elements (EREs) on the promoters of hormonally regulated genes, where the recruitment of coactivators and/or corepressors leads to transcription of important estrogen responsive genes and alterations in the rate of gene expression.[57–59] Estrogen also modulates gene expression by a non-ERE-mediated mechanism in which ERα and ERβ dimers bind to other non-ERE promoter sites and interact with other transcription factors through protein–protein interactions, such as activating protein-1 and stimulating protein-1, Fos-Jun, and a number of coactivators to regulate estrogenic genomic actions, a process referred to as transcription factor cross talk. Furthermore, the ERs have some ligand-independent transcriptional responses and can also interact with other transcription factors in regulating gene expression.[60]

Although ERα and ERβ are highly homologous in their DNA- and ligand-binding domains, ERα and ERβ lack homology in their transcriptional activation domains and their splice variants differ in their affinities for E2.[60,61] Additionally, ERα and ERβ differ in the tissue distribution.[62] ERα is mainly expressed in reproductive tissues, kidney, bone, white adipose tissue, and liver, whereas ERβ is expressed in the ovary, prostate, lung, gastrointestinal tract, bladder, hematopoietic cells, and the CNS.[63] These differences suggest that functional differences exist between the two subtypes ERα and ERβ, which is supported by characterization of ERα and ERβ KO mice.[64]

Male and female ERα KO mice are diabetogenic and obese with severe hepatic insulin resistance induced by targeted deletion in the ERα gene.[65] Thus, ERα is believed to mediate most estrogenic effects on energy homeostasis and glucose regulation. The cause of obesity in ERα KO mice is unclear; however, overeating does not appear to be essential,[65–68] and the increased adiposity in ERα KO mice is attributed primarily to decreased energy expenditure.[65,67] While E2 replacement prevents ovariectomized wild type mice from developing hyperphagia and obesity, such protective effects are blocked in ovariectomized ERα KO mice.[66] Therefore, although no hyperphagia is observed in ERα KO mice,[65–67] ERα is clearly required to mediate normal satiation process because E2-induced hypophagia and CCK-induced satiation in wild type mice are blocked in ERα KO mice.[66]

The role of ERβ in the central regulation of feeding is less known. In ovariectomized rats, there is increased food intake, body weight, and abdominal fat accumulation, which can be reversed by E2. However, intracerebroventricular administration of E2 together with ERβ antisense oligodeoxynucleotides abrogates the anorexic effects of E2. Ovariectomy of ERβ KO mice leads to normalized homeostasis of circulating glucose and insulin levels and reverses the obese phenotype, suggesting that ERβ activity may result in a diabetogenic and adipogenic phenotype, which is opposite to the effects of ERα. Additionally, ERβ KO mice display improved insulin sensitivity and glucose tolerance without increased body fat content, suggesting that ERβ also plays an important role in maintaining metabolic control.[69] These studies suggest opposite effects of E2 via ERα and ERβ.[70]

2.3. Estrogenic nongenomic actions via GPER

Estrogens are best known for their long-term transcriptional effects involving the genomic mechanism of action that typically takes several hours for the effect to be manifested due to the time needed for transcription and translation of estrogen-regulated genes.[71] However, a growing body of evidence has emerged to describe rapid estrogen signal transduction events in various cellular systems,[72,73] which occur within minutes and thus cannot be attributed to a genomic mechanism of action. These rapid effects have thus been designated as nongenomic effects of estrogen.[38,74–78] The earliest reports for nongenomic effects of estrogens date back to 1967.[79] In 1977, Pietras and Szego discovered that estrogens bind to a cell surface receptor,[80] introducing the possibility that estrogenic signaling may involve nongenomic nontranscriptional mechanisms or indirect actions to influence gene expression.

Membrane-bound ERs have been described recently. GPER, previously known as G protein-coupled receptor 30 (GPR30), has been identified as an estrogen-binding intracellular membrane receptor.[45,81–84] GPER expression is required for rapid cellular responses to E2.[85,86] In addition, another estradiol-responsive, G protein-coupled membrane ER has been described (Gq-mER).[87] Even though its distribution is currently unknown, Gq-mER appears to be involved in maintaining homeostatic functions by reducing food intake and fat accumulation, as well as other functions such as increasing bone density.[88–90] Finally, in 2002, the existence of an additional mER, named ER-X, was hypothesized.[91] However, its gene sequence and amino acid structure are currently unknown.

It is noteworthy that membrane ERs have been the focus of a surge of studies over the past decade.[92] A search in PubMed in May 2012 with the keywords "estrogen and nongenomic" yields 576 published papers since1984, with 480 (83%) of these papers being published after 2000. This area has seen an explosion of interest in the past decade, and represents one of the fastest emerging areas in the field of estrogen research.

2.3.1 GPER distribution and expression in tissues

GPER is a seven transmembrane-domain G protein-coupled receptor. GPER was first identified and cloned by Carmechi *et al.* in 1997.[81] Using a differential cDNA library screening technique, the investigators identified GPER as a differentially expressed gene in a screen of genes from an ER-positive breast cancer cell line (MCF-7) versus an ER-negative breast cancer cell line (MDA-MB-361).

Antibodies have been raised against GPER, facilitating a detailed analysis of its tissue distribution. Expression of GPER protein is not restricted to traditionally estrogen responsive tissues. Indeed, characterization of GPER using immunohistochemical analysis revealed ubiquitous expression of the receptor in normal and malignant human tissues. GPER expression is tissue and cell type specific, with high levels of expression found in numerous human organs, including male and female reproductive systems, heart, intestine, ovary, CNS, pancreatic islets, adipose tissue, neurons, and inflammatory cells. GPER is moderately positive at skeletal/cardiac muscle, but is negative at most smooth muscle.[83,93–105] GPER is expressed in many regions of the brain with high expression noted in the hypothalamic–pituitary axis, hippocampus, brainstem autonomic nuclei, cortex, and striatum.[106] Several primary breast cancers[81] and lymphomas[107] also expressed GPER transcripts, although many others were negative. There are discrepancies in the reported expression levels in some tissues, such as the rodent embryonic and fetal tissues, placenta, lung, and liver.[81,107,108] Therefore a thorough analysis of the regulation of GPER expression remains to be completed.

GPER expression is developmentally regulated during the development of murine female reproductive tissues. For example, in the mammary gland, GPER expression in elongating ducts is low during puberty but rises once sexual maturity is reached. GPER expression in the mammary ductal epithelia exhibits estrous-cycle-dependent changes and differentiation-dependent changes during pregnancy, lactation, and involution. Similarly, GPER expression in both luminal and glandular uterine epithelia is regulated during the estrous cycle. Finally, Roy and colleagues recently demonstrated that the distribution and level of expression of GPER in the hamster ovary also varies with the estrous cycle; both granulosa and theca cell layers are rich in GPER, and expression is reduced at estrous.[109]

Interestingly, sexual dimorphisms for GPER expression and/or function have been described in the brain[110] and in the pancreatic islets, where it is expressed at a much higher level in women than in men.[95] Accordingly, a role for GPER in the regulation of obesity-associated metabolic functions has been proposed.

These studies have provided a framework in which to interpret many of the reported functions of GPER, and suggest that GPER has been implicated to be involved in estrogen-dependent physiology of immune function and cardiovascular systems.[47,93] For example, GPER is associated with diseases such as obesity, insulin resistance, and hormone-sensitive cancers,[93,98,99] and GPER plays a role in estrogen-sensitive tumor growth in the absence of ERα/β.[81,111]

2.3.2 GPER location in cells

GPER is expected to localize in the plasma membrane and its plasma membrane localization is reported by Thomas *et al.*[84] However, other groups reported GPER locations in the membrane of intracellular compartments. Revankar *et al.* used novel fluorescent estrogen derivatives (E2-Alexas) to directly visualize the cellular and subcellular binding properties of GPER. Confocal microscopy revealed that, whereas E2-Alexas detected ERα and ERβ in the nucleus of cells, E2-Alexa binding to GPER occurred predominantly in the endoplasmic reticulum, with no detectable binding at the plasma membrane.[45] In addition, binding of the fluorescent E2-Alexas was directly proportional to expression of endogenous GPER on a cell-by-cell basis, strongly supporting the idea that expression of GPER protein directly generates estrogen-binding sites. In all cell lines where GPER expression has been observed, the E2-Alexa or antibody staining is colocalized in the endoplasmic reticulum, with staining also present in the Golgi apparatus and nuclear membrane. These results, although they are in contrast to the expected localization of GPER in the plasma membrane, clearly demonstrate that GPER is localized in the membrane of intracellular compartments, such as the endoplasmic reticulum and the Golgi complex.[45,81,83–85,93]

2.3.3 GPER intracellular signaling

GPER binds E2 with high affinity to mediate subsequent nongenomic and rapid estrogen signaling via multiple pathways (Fig. 6.2).[45,83–85] As a G protein-coupled receptor, GPER couples to G proteins and modulates second messenger pathways. Signaling mechanisms of this action include rapid increases in Ca^{2+},[112] cAMP[79,113] and phosphoinositide-3 kinase (PI3K)/Akt and mitogen-activated protein kinase (MAPK) activation[47] (Fig. 6.2). cAMP and Ca^{2+} have emerged as the second messengers most frequently associated with GPER signaling. GPER is involved in the activation of downstream signaling cascades such as protein kinase A (PKA), protein kinase C, and MAPK via membrane-localized ERs[82] and intracellular Ca^{2+} regulation.[106] The GPER couples to both a Gαs protein yielding stimulation of cyclic AMP (cAMP) production and a pertussis toxin (PTX)-sensitive Gαi/o protein in part yielding attenuation of cAMP production. Ding *et al.*[114] have reported that E2-promoted vascular smooth muscle cell apoptosis is dependent on GPER-mediated PKA inhibition and PI3K activation, both of which are PTX sensitive. Furthermore, Maggiolini *et al.*[86] demonstrate GPER-mediated and PTX-sensitive c-fos upregulation in SkBr3 breast cancer cells, a cell line that has been shown to have GPER-dependent stimulation of cAMP production.[82]

The rapid effects from stimulation of GPER also include release of intracellular Ca^{2+}. This signal has been described as phospholipase C-independent,[45] PTX-sensitive,[45,105] and dependent on epidermal growth factor receptor (EGFR) activation.[45] Increased cytosolic Ca^{2+} subsequently activates Ca^{2+}–calmodulin-dependent kinases or activation of MAPK and PI3K pathways.

Filardo *et al.* demonstrated that in MCF-7 cells, which express ERα, ERβ, and GPER, estrogen mediated a rapid increase in MAPK phosphorylation but that in MDA-MB-231 breast cancer cells, which express ERβ but not ERα or GPER, this response was absent, suggesting a possible role for ERα or GPER. When MDA-MB-231 cells were transfected to express GPER, E2 stimulated both cAMP production and PTX-sensitive MAPK activation in the cells.[85] Additionally, they identified that GPER is involved in the estrogen-mediated activation of MAPK[82,85,115] and PI3K[45] in SKBr3 breast cancer-derived cells that express only GPER but lack classical nuclear ERα and ERβ. Together, these results strongly implicated GPER is essential in estrogen-mediated cellular responses.

Stimulation of GPER also leads to PI3K and PTX-sensitive Src activation, which in turn yields phosphorylation of Akt and SHC, respectively.

While Akt phosphorylation has been linked to antiapoptotic and cardioprotective effects,[116] SHC phosphorylation triggers cleavage of membrane-tethered heparin-bound EGF, presumably through activation of a metalloproteinase, resulting in transactivation of the EGF receptor.[85] However, there is a debate on whether the PI3K/Akt pathway lies downstream of Gαi/o or not. While several groups have shown that the GPER-coupled Src/SHC/EGF/MAPK pathway is PTX sensitive,[85,114] and thus dependent on Gαi/o signaling, Revankar *et al.*[45] reported that PI3K/Akt activation is downstream of the Src/SHC/EGF/MAPK pathway but insensitive to PTX.

2.3.4 Pharmacologic tools—GPER agonists and antagonists

Elucidation of the biological roles of GPER has been facilitated by the identification of specific pharmacological reagents. GPER agonists, including G-1, (±)-1-[(3a*R**,4*S**,9b*S**)-4-(6-Bromo-1,3-benzodioxol-5-yl)-3a,4,5,9b-tetrahydro-3H-cyclopenta[c]quinolin-8-yl]-ethanone,[117] and 5408–0877, which behaved similarly to G-1,[118] and an antagonist G-15, (3a*S**,4*R**,9b*R**)-4-(6-Bromo-1,3-benzodioxol-5-yl)-3a,4,5,9b-3H-cyclopenta[c] quinoline,[119] exhibit high affinity and high selectivity for GPER over classical ERα and ERβ, with none of them showing affinity for ERα or ERβ.[117,119] These tools have revealed that GPER activation may have several beneficial effects in the metabolic system including stimulation of insulin release and protection against pancreatic β-cell apoptosis. Thus, GPER is emerging as a candidate therapeutic target in metabolic disease.

To demonstrate the selectivity of G-1 toward GPER as compared with classical ERs, Ca^{2+} mobilization in COS7 cells expressing either ERα or ERβ revealed no response to G-1 compared with estrogen, which yielded a rapid response with both ERα and ERβ. G-1 also activates transfected and endogenous GPER, resulting in PI3K activation. With the identification of the first GPER-selective agent, it is now possible to investigate the specific functions of GPER compared with those of ERs in complex systems expressing both receptor types.

The pharmacological agents are currently being widely employed to investigate GPER function and physiology in various systems.[117,119] GPER agonists including G-1 and 5408-0877 and antagonist G-15 remain to be fully evaluated for unequivocal specificity for this receptor.

GPER can also be activated by tamoxifen and raloxifene, SERMs[45,55,120,121] traditionally thought only to modulate the function of ERα and ERβ.[45,55,85] These studies provide experimental evidence that support the concept that

GPER activation has inhibitory effects on food intake, body weight, and fat mass.[122,123]

2.3.5 Genetic tools—GPER knockout mice

The availability of genetic tools (i.e., different mouse models lacking the GPER gene) and pharmacological agents (i.e., GPER-selective agonists and antagonists) has advanced our understanding on GPER. Four mouse models lacking the GPER gene (GPER knockout, GPER KO) have been used to study the pathophysiological role of this receptor.[98,124–126]

GPER KO mice are found to associate with increased visceral adiposity[100,127] whereas Mårtensson *et al.*[98] using a different GPER KO strategy found changes in body weight that were limited to GPER KO females. The same investigators found no effect of GPER deficiency on the antiobesity effects of E2.[128] By contrast, Isensee *et al.* found no effect of GPER deficiency on body weight in animals on either a normal or high-fat diet in another model, LacZ-GPER reporter mice, which represent a partial GPER deletion.[125,129,130] Data obtained from different GPER KO models generated by different transgenic techniques do not show a homogenous phenotype under physiological conditions.[126] One group was unable to demonstrate an E2-dependent phenotype in GPER KO mice indicating that GPER might not be relevant for estrogen signaling,[130,131] whereas other groups did.[98,101,124] Such discrepancy would not exclude the possibility that E2-mediated effects might involve overlapping signaling of various ERs, and might be due to cross talk effects between GPER and other ERs[132] as described below.

2.4. Cross talk between GPER and other ERs

2.4.1 Cross talk between membrane subpopulation ERs (mERα/β) and GPER

Rapid estrogen effects are mediated by ERs localized to the plasma membrane, comprising subpopulations of the classical ERs, namely membrane subpopulations of ERα and ERβ (mERα/β).[47,57,133] Activation of mERα/β induces a variety of intracellular signaling cascades and mediates rapid estrogenic actions that involve cell-signaling mechanisms, mostly the phosphorylation of various kinases[134,135] and the regulation of intracellular Ca^{2+} levels.[136,137]

An mERα/β selective ligand, STX, [2-(4-hydroxyphenyl)-3-phenylpent-2-enoic acid [4-(2-dimethyl amino ethoxy)phenyl] amide, E-enantiomer], has been developed and used in studies on homeostasis.[138] Most behavioral investigations with this drug have utilized systemic administrations.[139,140] Less commonly, direct brain infusions have attempted to identify its specific site of action.[141]

Membrane-associated full-length ERα and two truncated isoforms, ERα 46 kDa and ERα 36 kDa, exist that are at least in part responsible for nongenomic estrogenic signaling.[142–144] Wang and colleagues have identified ERα 36 as a novel splice variant of the human ER.[145] In breast cancer cell lines lacking ERα, expression of ERα 36 is associated with improved survival.[146] It has been recently proposed that ERα 36 is the main effector and mediator of GPER-dependent effects mediated by estradiol and that its induction is dependent on GPER.[147] However, others found that regulation of ERα 36 or ERα 36 is not detectable despite effects of G-1 being observed.[148] A recent report demonstrating protective effects of G-1 in hypertensive rats in the kidney did not find any expression of ERα 36 while ERα 46 and ERα 66 were abundantly expressed.[148] At present, data—particularly *in vivo* data—are still scarce. A recent report found effects of GPER agonists tamoxifen and ICI 182,780 are dependent on ERα 36 in cancer cells.[149] Levin has proposed that GPER might act as "collaborator" of ERα—which it might be under certain conditions, as cross talk between both receptors has been described.[150,151] Recently, it has been suggested that GPER signaling requires and might even solely occur through ERα 36.[143]

Different experimental settings from *in vitro* observations, immortalized cells vs. experimental animals, cancer cell vs. normal kidney, may be applicable to certain but not all conditions. In addition, results from *in vitro* findings cannot be extrapolated to the *in vivo* functions of receptors and thus are only of limited relevance. Another aspect to consider is that, unlike GPER, rapid ERα 36-dependent signaling is also activated by estrone, estradiol, estriol, and estetrol, and that E2-induced MAPK activation is further potentiated by GPER agonists such as ICI or tamoxifen.[143] By contrast, the ERα 36 agonist estriol is an antagonist for GPER.[152] Importantly, *in vivo* evidence for a role of ERα 36 in mediating the E2-mediated effects of GPER under physiological or disease conditions is still lacking.

2.4.2 Cross talk between EGFRs and GPER

Cellular signaling by GPER activates EGFRs via a G protein-dependent pathway, which involves Src and metalloproteinase (MMP) cleavage of proheparin-binding epidermal growth factor-like growth factor. Filardo and colleagues described GPER-mediated elevation of cAMP by estrogen as a mechanism to restore EGFR-activated MAPK to basal levels through

PKA-dependent inhibition of Raf-1 activity.[82] Two questions arose from this observation: (a) Do different ligands specifically activate the two receptors? (b) Do the two receptors utilize identical signaling pathways? To answer the first question, the same cells are treated with 4-hydroxytamoxifen, a well-described ER partial antagonist previously shown to activate MAPK via GPER. Stimulation with 4-hydroxytamoxifen resulted in no activation of PI3K in ERα-expressing cells but did result in PI3K activation in GPER-expressing cells, demonstrating for the first time in a side-by-side comparison the differential effects of tamoxifen on ERα and GPER. To answer the second question, the EGFR inhibitor AG1478 is employed. Two groups have shown independently that both ERα and GPER mediated EGFR transactivation.[85,153] When EGFR phosphorylation is blocked, GPER-, but not ERα-mediated PI3K activation is inhibited. These studies demonstrated the close link in cellular signaling between GPER and EGFR.

2.4.3 Cross talk between classical ER and GPER

There appears to be complex interplay between ERα and GPER although such interplay has not yet been fully defined.[93,150,154] Classical ERs may couple to G protein signaling via an unknown mechanism to mediate rapid estrogen signaling. The coexpression of ERα and ERβ with GPER in breast cancer cells has suggested a possible role for interactions between these two receptors.[132,155,156]

Indeed, functional cross talk has been demonstrated in cancer cells and uterine epithelial cells,[150] where GPER functions as inhibitor of ERα-dependent, estradiol-mediated cell growth.[151] Functional cross talk between ERs is also evident from isolated vascular preparations, where the E2-induced acute vasodilator response is only seen with ERα selective agonists but completely abrogated when ERα, ERβ, and GPER are activated simultaneously by E2.[157] Navarro *et al.* demonstrated by coimmunoprecipitation studies that ERα interacts with Gαi3 in immortalized GnRH neurons, and that estrogen-induced inhibition of cAMP can be blocked by pertussis toxin, a G protein signaling inhibitor.[158] Additionally, Kelly and coworkers[87,159] have provided evidence of a Gαq-coupled ER that regulates the PKC and PKA signaling pathway in hypothalamic neurons and is involved in hypothalamic control of energy homeostasis.

Mitochondrial protection in response to E2 unrelated to GPER[160] has been described for ERβ.[161] E2 effect on insulin is absent in GPER KO mice,[28,98,101] and protective effects on pancreatic islets are comparable independent of whether selective activation of ERα, ERβ, or GPER occurs.[28]

Thus, possible cross talk between GPER and ERs (also possibly other steroid receptors) as well as redundancy of ER function must be considered when assessing possible pathogenic and therapeutic questions related to GPER activation or inhibition.[132] This should also be considered when determining the specific activity of compounds or drugs targeting either one[117,119] or several membrane ERs.[162]

3. ESTROGENIC ACTION IN THE BRAIN: CENTRAL REGULATION OF ENERGY HOMEOSTASIS

3.1. Estrogenic action in the CNS

3.1.1 Estrogens directly act on hypothalamic areas involved in energy balance

A series of complex systems regulate energy homeostasis in order to keep energy storage levels and body weight stable. Central brain circuits receive peripheral signals indicating satiety, energy levels, and energy stores. The brain is made aware of the metabolic state of the body through signals released from adipose tissue, skeletal muscle, liver, pancreas, and gut. There is a wide variety of these signals, including hormones (insulin, leptin, gut hormones), cytokines (IL-6, TNF-α), and nutrients (glucose, free fatty acids, lipids). The hypothalamus is a key regulator of food intake and maintenance of energy homeostasis.[163] The hypothalamus processes afferent signals from the gut and brain stem and efferent signals that modulate food intake and energy expenditure. Anorexigenic and orexigenic signals from the CNS then induce behavioral and metabolic changes targeting energy homeostasis and survival.

The main hypothalamic areas involved in food intake and satiety are the arcuate nucleus (ARC), the lateral hypothalamus (LH), the paraventricular nucleus (PVN), and the ventromedial hypothalamus (VMH).[134] Several melanocortin receptors composing the melanocortin system are expressed throughout these nuclei and are one of the main regulators of hunger and satiety.[164]

Estrogens are major effectors for the regulation of energy balance and body fat. Energy homeostasis and feeding behavior in the hypothalamus follows the menstrual cycle, and food intake in women varies across the cycle with lowest daily food intake during the periovulatory period when estrogen levels are maximum.[165] In several mammals and in women, higher levels of E2 during the estrus of menstrual cycles and pregnancy lead to decreased food intake and fat accumulation. In addition to its central actions, E2 also regulates peripheral feedback hormones that influence feeding and meal size through their actions in the CNS.

In contrast, ovariectomy, antiestrogen treatment, and menopause lead to an increase in food consumption and meal size, a decrease in running wheel activities, and increased fat mass, which can be reversed by E2 replacement.[166] The importance of the brain for the observed phenotypes has been demonstrated by studies showing that direct intracranial injections of E2 into the PVN leads to anorexia, reduced food intake, and body weight and increased running activities.[167,168]. These observations indicate that E2 has direct anorexigenic action in the hypothalamus.[134]

3.1.2 Estrogens act on central melanocortin system

Estrogens act on central melanocortin system to regulate energy balance and body fat distribution. The melanocortin system is responsible for the opposing actions of these neurons; however, so far a clear relation between E2, the melanocortin system, and energy balance has not been established.[169,170] In the ARC, several types of neurons and receptors of the melanocortin system (MC3R and MC4R) are involved in the regulation of appetite and respond to variations in glucose, lipids, leptin, and insulin. The proopiomelanocortin (POMC) neurons have an anorexigenic action and, when activated, reduce food intake through the release of two peptides, α-melanocyte stimulating hormone (α-MSH) and cocaine-and-amphetamine-regulated transcripts. The neuropeptide Y (NPY) neurons, on the other hand, release NPY hormone and agouti gene-related protein (AgRP), which prevent the binding of α-MSH to MC3R and MC4R, increasing food intake.[164] Whole animal studies in ovariectomized rats have reported that a significant reduction in food intake by E2 treatment compared to oil treatment was not detected after administration of the melanocortin receptor antagonist SHU919.

A recent study demonstrated that NPY/AgRP neurons are required to mediate the anorexigenic effects of estrogens. In this study, Xu and colleagues showed that hypothalamic expression of NPY and AgRP in wild type mice is tightly regulated across the estrus cycle, with the lowest levels during the estrus, which coincides with the plasma estrogen peak and feeding nadir.[171] They further showed that central E2 administration suppresses fasting-induced c-Fos activation in NPY/AgRP neurons and blunts the refeeding response.[171] Importantly, the cyclic changes in food intake and E2-induced anorexia are blunted in mice with degenerated NPY/AgRP neurons. This study indicates that NPY/AgRP neurons are functionally required for the cyclic changes in feeding across estrous cycles. Surprisingly, these authors also found that ERα is not expressed in NPY/AgRP neurons,[171] suggesting that estrogen may regulate these neurons indirectly via presynaptic neurons that express ERα (e.g., POMC neurons). Indeed, POMC neurons coexpress ERα.[172–174] In addition, estrogens regulate excitability of POMC neurons. Using electron microcopy, Horvath and colleagues have reported that the number of excitatory synaptic inputs to ARC POMC neurons rises as mice enter proestrus when estrogen levels are high.[23]

Further, central E2 administration rapidly increases the excitatory synapses on POMC neurons, an effect that is also reflected by increased miniature excitatory postsynaptic current recorded from POMC neurons.[23] These synaptological rearrangements in POMC neurons are tightly paralleled with the effects of E2 on food intake and body weight.[23] Similarly, Ronnekleiv and colleagues reported that E2 administration in hypothalamic slices activates POMC neurons by rapidly uncoupling $GABA_B$ receptors from the G protein—gated inwardly rectifying K+ channels.[175] Importantly, female mice lacking ERα in POMC neurons only develop hyperphagia.[172] These observations, together with the findings from the Horvath and Ronnekleiv groups, indicate that estrogen signals in POMC neurons are physiologically relevant in the regulation of food intake.

3.1.3 Estrogens act on nuclear and membrane ER in the CNS

Classical ERα and ERβ and GPER are all highly expressed in the brain, their distribution being both overlapping and unique.[61] The critical intracellular signals that mediate the anorexigenic effects of estrogens have been explored. Stimulation of nuclear and membrane ERs engages genomic and nongenomic events that may mediate distinct functions of estrogens, as detailed below.

3.2. Estrogen acts on nuclear ERs in the CNS

Genomic activation of ERs results from estrogens binding to EREs in promoter regions of target genes.[176,177] ERα is ubiquitously expressed in rodent brains. Areas of predominant ERα expression are the VMH and ARC.[178–181] ERβ is found in the same hypothalamic nuclei as ERα, but ERβ expression is significantly reduced relative to ERα. ERβ predominates in the suprachiasmatic, supraoptic, and PVN of the hypothalamus, the cerebellum, dorsal raphe, hippocampus, and cerebral cortex.[68,179,182–184] ERα and ERβ expression overlaps in the ARC, LH, medial amygdala, preoptic area, bed nucleus of the stria terminalis, nucleus of the solitary tract and periaqueductal gray.

Pertinent to the estrogenic action on melanocortin system is that both ERα and ERβ have been identified in the ARC. Specifically, ERα is predominantly expressed in the POMC neurons, while both ERα and ERβ are present in the NPY neurons.[134,173] The PVN and VMH are also mediators of energy ingestion and expenditure and are interconnected with the ARC. The PVN is highly responsive to glucose, lipids, NPY, and AgRP, and its activation leads to increased food intake.[134,185] The PVN is the hypothalamic region with the highest expression of ERβ and is weakly ERα positive.[182] Despite this apparent low level of ERα, in rats and mice, the ERα-selective agonist, PPT, rapidly results in a decrease in food intake and increase in expression of c-fos in the corticotrophin-releasing hormone-expressing cells in the PVN.[186] The VMH is ERα regulated and has an important function in the inhibition of food intake through anorexigenic signals from the ARC. Silencing of ERα in the VMH by RNA interference leads to hyperphagia, obesity, decreased glucose tolerance, and reduced energy expenditure.[187]

The role of estrogens and ERα in the VMH in the regulation of energy balance has been reexamined using the ERα silencing approach.[187] In this study, ERα in the VMH is knocked down with an AAV-shRNA.[187] Animals with impaired ERα signaling in the VMH are less sensitive to E2-induced weight loss and develop obesity characteristic of increased visceral fat. The obesity syndrome is likely caused by decreased physical activity and impaired diet-induced thermogenesis, whereas food intake of these animals is not directly affected.

Xu *et al.* generated mice with ERα deleted in VMH SF1 neurons.[172] They found that deletion of ERα in VMH SF1 neurons in female mice, while not affecting food intake, significantly reduces basal metabolic rate and diet-induced thermogenesis, which consequently results in increased body weight and adiposity. Interestingly, a significant increase in visceral fat deposition but not subcutaneous fat deposition is observed in these mutant females.[172] Finally, they showed that the decreased energy expenditure and increased visceral fat distribution in mice lacking ERα in SF1 neurons presumably results from decreased sympathetic tone, as demonstrated by decreased plasma norepinephrine levels.[172] These findings are largely consistent with those obtained from the VMH-specific ERα knockdown model. Collectively, these results support the hypothesis that ERα signaling in VMH neurons plays an important role in regulating energy expenditure and fat distribution.

To summarize, ERα and ERβ are expressed in brain regions implicated in the regulation of appetite, satiety, and energy expenditure,[49,68,178–180,183] which can explain the involvement of these receptors in changes of feeding behavior and energy expenditure and thus regulation of energy homeostasis. The differential distribution further suggests that, like for other behaviors, the respective involvement of ERα and ERβ in regulation of energy homeostasis may be different.

3.3. Estrogen acts on membrane ERs in the CNS

3.3.1 Identification of three membrane ERs in the CNS

Recent evidence suggests that estrogen has at least three different transmembrane receptors in the brain.[115,188]

The first membrane ER identified and localized in the brain is the estrogen-binding G protein-coupled receptor GPER. Brain GPER expression in adult male and female rats was thoroughly characterized by Brailoiu *et al.* using immunohistochemical analysis.[106] GPER in the hypothalamus is localized primarily to the PVN and the supraoptic nucleus but is also found in ARC neurons.[106] GPER is also expressed in nonhypothalamic regions, including the anterior and posterior pituitary, striatum, hippocampus, and the autonomic nuclei of the brainstem including substantia nigra, area postrema, nucleus of the solitary tract (NTS), and dorsal motor nucleus of the vagus.[106] These hypothalamic and brainstem areas are typically associated with energy homeostasis function. GPER immunostaining exhibits a similar pattern in male and female mice,[189] suggesting that GPER expression in these autonomic areas is not regulated by estrogen status. In the NTS, GPER immunostaining is predominantly localized to neurons that coexpress ERα, whereas in the PVN, GPER displays moderate colocalization with nitric oxide synthase.[190] Consistent with this study, Sakamoto and coworkers have also shown expression of GPER mRNA and protein in the oxytocin neurons of the rat PVN and supraoptic nuclei.[191]

The second membrane ER is the putative G protein-coupled receptor membrane estrogen receptor (Gq-mER) functionally characterized by Kelly and coworkers.[87,88] Although the Gq-mER has not been cloned, it has been functionally identified in at least three types of ARC neurons, including POMC, dopamine, and γ-aminobutyric acid (GABA) neurons.

The third membrane ER is membrane subpopulation of ERα and ERβ has been identified. Ultrastructural electron microscopic studies have provided direct evidence of extranuclear localization of ERs in the brain.[192] Milner *et al.*[192] performed ultrastructural studies of ERα localization in the rat hippocampal formation and demonstrated that ERα, in addition to nuclear localization, is found in cytoplasm and plasma membrane of neurons, and in unmyelinated axons and axon terminals. Furthermore, approximately 25% of ERα is localized in dendritic spines, often associated with spine apparati and/or polyribosomes, suggesting it may act to regulate Ca^{2+} availability, local protein translation, and synaptic growth. Using Western blot and estrogen-horseradish protein-FITC or estrogen-BSA-FITC conjugates, Marin *et al.* showed plasma membrane localization of ERα in SN56 septal-derived cholinergic neurons, and that the estrogen conjugates protected the cells against A-beta toxicity as effectively as estrogen.[193,194] Xu *et al.*[195] transfected GFP-tagged human ERα in rat cortical neurons *in vitro* and determined its subcellular localization. The studies revealed that there was localization of the GFP-tagged human ERα in neurites of the cortical neurons and that the ERα appears to be directed to the neurites directly from its site of translation and not from nuclear stores. Furthermore, biotinylated membrane protein studies also support localization of ERα at the plasma membrane of rat cortical neurons. Finally, ultrastructural studies have also demonstrated that ERβ is also localized at extranuclear sites such as cytoplasm, plasma membrane, dendritic spines, axons, and axon terminals in the rat hippocampus.[196] Thus, ERβ could also have a role in mediating nongenomic actions of estrogen in the brain. We should also mention that a novel plasma membrane-associated ER called ER-X has been proposed as well in neurons of the brain, but attempts to clone it have so far been unsuccessful.[91]

3.3.2 Estrogenic nongenomic signaling in the CNS

Estrogens can initiate rapid signaling cascade events within minutes to affect cell physiology and activate gene transcription through membrane-initiated steroid signaling via nongenomic mechanisms at non-ERE transcription factor promoter sites. Instead, estrogens activate transcription via complexes with other transcription factors through protein–protein interactions, including CREB, STATs, Elk-1-SRF, ATF-2-Jun, and NFκB-inducing transcription via their respective promoter sites.[197]

Reports of rapid E2 effects in the brain first began to appear in the literature over 40 years ago, when it was observed that estrogen rapidly modulated hypothalamic electrical activity.[198–200] The latency of the estrogen effects was within minutes, and typically two types of effects were observed on hypothalamic neurons—some units had a decreased firing rate that lasted up to 150 min, whereas other units responded with a transitory increase in firing rates. The remaining units were unresponsive to E2.[200] Kelly *et al.*[201–204] showed that neurons from several hypothalamic regions displayed rapid electrophysiological responses to estrogen, including preoptic, ARC, and VMH neurons. These early observations set the stage for subsequent studies on the mechanisms underlying the rapid effects of estrogen in the brain. ERs in the plasma membrane are able to activate the MAPK cascade and PI3K pathway, causing a rise in intracellular Ca^{2+}.[205,206] ERs also activate PKB or Akt in neurons.[207–209] The PI3K/Akt cascade mediates a variety of estrogens' central actions, including neuronal excitability, neuroprotection, reductions in inflammation, neurite outgrowth, and potentially body weight regulation.[210]

The rapid effects of estrogen on Ca^{2+} and ion channel signaling and kinase activation suggest a potential membrane receptor-mediated effect of estrogen. One piece of evidence supporting a membrane receptor-mediated mechanism is the use of membrane impermeable conjugates of estrogen, such as estrogen-BSA and estrogen dendrimers. Similar to estrogen, estrogen-BSA was shown to rapidly enhance phosphorylation of MAPK in the human neuroblastoma SK-N-SHR cells,[211] in rat hypothalamic neurons *in vitro*,[212] and in the rat hippocampus *in vivo*.[213] Estrogen-BSA conjugates have also been shown to rapidly enhance PKA activation in hippocampal neurons,[214] increase intracellular Ca^{2+} levels within minutes in both astrocytes and neurons,[215,216] and to rapidly modulate potassium channel activation in hypothalamic neurons.[87] Nevertheless, it should be pointed out that there has been some criticism of estrogen-BSA conjugates, in that some estrogen (3–5%) can be found in the free unconjugated form in commercially available estrogen-BSA conjugates.[217] For this reason, it is advisable that the estrogen-BSA be subjected to filtration in order to remove free estrogen, and that studies that fail to perform the filtration should be interpreted cautiously. Stevis *et al.*[217] reported that estrogen-BSA did not bind ERα or ERβ in binding studies; however, a subsequent study demonstrated that estrogen-BSA indeed binds to purified ERs *in vitro* and to membrane ERs in intact cells, albeit at a slower rate than estrogen.[218] Taken as a whole, the estrogen conjugate data suggest that rapid signaling effects of estrogen involve mediation via a membrane ER.

3.3.2.1 Rapid estrogen signaling in ER KO mice

One approach to determine the role of ERs in mediating rapid signaling effects of estrogen in the brain is to utilize mutant mice with one or both classical ERs knocked out. Using this approach, Chaban and Micevych demonstrated that rapid estrogen inhibition of ATP-induced Ca^{2+} signaling in dorsal root ganglion neurons is blocked by an ER antagonist, and that the effect of estrogen on Ca^{2+} signaling is lost in ERα KO mice.[219] Likewise, Herbison and colleagues demonstrated that rapid estrogen-induced phospho-CREB responses in hypothalamic neurons, where ERβ is densely localized, is preserved in ERα KO mice, but lost in ERβ KO mice.[220]

With regard to MAPK regulation, there are contradicting reports on the estrogen regulation of phospho-MAPK in ER KO mice. Singh *et al.*[221] reported that estrogen-induced phosphorylation of MAPK was not lost in cortical neurons from ERα knockout mice, suggesting that ERα may not mediate the rapid phospho-MAPK-enhancing effects of estrogen. In contrast, Abraham *et al.*[220] reported that rapid estrogen induction of phosphorylation of MAPK in the medial preoptic nucleus was lost in both ERα and ERβ knockout mice, suggesting that both ERα and ERβ may mediate the MAPK regulatory effect of estrogen. Finally, estrogen can also rapidly induce c-Fos protein in the brain, an effect which has been shown to be lost in double ERα/β KO mice in several brain areas including the VMH.[222]

As a whole, results with the ER KO mice shed some light on involvement of ERα or ERβ in rapid signaling effects of estrogen in the brain. Additionally, they suggest that the role of ERα or ERβ in rapid estrogen signaling effects may differ from one brain region to another, and results in one region cannot be immediately extrapolated to another region or another kinase.

3.3.2.2 Rapid estrogen signaling in the brain via G protein-coupled receptors

G protein-coupled receptors are vital membrane mediators for a host of central and peripheral signals that control energy homeostasis in the hypothalamus.[223] Because G protein-coupled receptors target potassium channels to control neuronal excitability, they may account for one of the rapid mechanisms estrogen utilizes to control energy homeostasis. A candidate for an estrogenic mechanism recently identified is the control of neuronal excitability through a G protein-coupled receptor membrane ER (Gq-mER).[87,88] The Gq-mER affects neuronal activity, alters metabolic processes such as body weight, and regulates energy balance.

The Gq-mER was initially characterized in female guinea pigs, but has also been functionally examined in both male and female ERα KO mice and ERβ KO mice, and in male ERα/β double KO mice.[88] Estrogen through Gq-mER activates multiple signaling pathways including PI3K, phospholipase C (PLC), MAPK and PKA and PKC pathways.[210,224,225] Briefly, the

Gq-mER activates a Gq-PLC-PKA pathway that inhibits the activation of G protein-coupled inwardly rectifying potassium channels. The inhibition of the potassium channel activity will depolarize the POMC neuron and increase neuronal activity and potentially add another mechanism for estrogen to control energy homeostasis.

The Gq-mER involved in estrogenic membrane-initiated steroid signaling is thought to be ERα,[225] although recent data suggest that, in a transfected neuronal cell line, ERβ is also associated with the membrane and is translocated to the membrane by estrogen to activate MAPK signaling.[226] In the hypothalamus, estrogen membrane-initiated steroid signaling has been identified in the VMH to potentiate the effects of estrogenic nuclear-initiated steroid signaling during particular reproductive behaviors such as lordosis[227] and may also affect estrogen's control of homeostasis via VMH neurons.

There is *in vivo* evidence suggesting a role for the activation of membrane-initiated estrogenic signaling events that affect energy homeostasis. E2 can attenuate food intake within 6 h of administration into the third ventricle via cannula after an overnight fast compared to saline in mice[228] and between 4 and 14 h in fed rats.[23] Estrogen can activate signaling cascades initiated through Gq-mER that would rapidly alter neuronal activity and initiate changes in feeding behavior.[87,88] Indeed, acute administration of estrogen to hypothalamic slices will alter neuronal excitability of many different types of relevant neurons through the modulation of various ion channels.[229,230]

To elucidate this estrogen pathway, STX, a selective agonist for the Gq-mER, has been developed that has no binding capacity to classical ER.[138] STX is more potent than estrogen in attenuating the activation of K^+ channels.[87,88] STX is an excellent tool for the elucidation of the role of the Gq-mER in energy homeostasis and gene regulation.[89] The ability of STX to alter POMC neuronal excitability leads to the hypothesis that the putative Gq-mER has a role in energy homeostasis. Whole animal studies in which STX was systemically injected in ovariectomized female guinea pigs over a period of 4 weeks support this hypothesis. This study demonstrated a dose-dependent effect on the reduction in body weight gain by systemic STX treatment. Furthermore, the administration of systemic STX-generated new transcription in the ARC of these STX-treated female guinea pigs.[89] The genes regulated include NPY gene involved in the control of energy homeostasis and Cav3.1 gene involved in neuronal activity. Therefore, Gq-mER may function in the estrogenic control of energy homeostasis, presumably through activation of POMC neurons in the ARC, although other hypothalamic nuclei may be involved.[88]

The molecular characterization of this Gq-mER, including cloning, localization, and regulation, is crucial for a full understanding of the role of the rapid, membrane-mediated effects of estrogen on energy homeostasis. Additionally, STX can be utilized to target the Gq-mER to determine whether Gq-mER actually contributes to the regulation of energy homeostasis in ERKO mice.

4. ESTROGEN ACTION IN ADIPOSE TISSUE: REGULATION OF LIPOGENESIS/LIPOLYSIS AND ADIPOSITY

Organisms store energy for later use during times of nutrient scarcity. Excess energy is stored as triacylglycerol (TG) in lipid droplets produced by lipogenesis. When energy is required, TG is catabolized into free fatty acids via lipolytic pathways. Obesity results from excess white adipose tissue, which is considered to be an endocrine organ and it is able to store and metabolize steroid hormones. ERs are expressed in adipose tissue, where they may influence adiposity and inflammation. Estrogens regulate both the metabolism and the location of white adipose tissue and play a role in adipose deposition, lipogenesis, lipolysis, and adipocyte proliferation.[231]

4.1. Estrogenic action in adipose tissue prevents obesity

Obesity, high body fat percentage, and increased hip circumference are all strongly correlated to increased E2 levels.[232] Sex steroids are known to regulate lipid metabolism and are responsible for accumulation, distribution, and development and function of adipose tissues. The differences in body fat distribution suggest that the sex steroid hormones could have an effect on adipose tissues. Direct effects seem plausible, as sex steroid hormone receptors have now been found in adipose tissues. The amount of the receptor varies with the fat depot, with subcutaneous adipose tissue having a higher concentration of ERs than intra-abdominal adipose.

Estrogen activates lipogenesis under normal physiological conditions in lean animals but suppresses lipogenesis under pathologic conditions. In normal lean animals, increased intracellular E2 activates ERα and stimulates lipogenesis.[233] Consequently, female mice with higher E2 level have greater lipogenic capacities in adipocytes compared with male mice.[234] Specifically, E2 increase the deposition of subcutaneous fat on the hips, upper thighs, and lower abdomen, ectopic fat in the liver and muscle, and visceral fat in and about abdominal organs.[235] As the enlarging fat cells produce more E2, they increase their leptin output,[236] to cyclically upregulate 11β-hydroxysteroid dehydrogenase,[237] increase intracellular E2 production.[238] This stimulation of lipogenesis by E2 is ERα-dependent, as revealed by global gene expression analysis from ERα KO and wild type mice that demonstrated ERα-dependent increased expression of lipogenic genes and decreased expression of genes regulating lipid transport.[239] Interestingly, despite the increased lipogenesis, adipocytes from females are smaller than adipocytes from males.

It is noteworthy that androgen receptors (AR) are also expressed in adipose tissues.[240] Subcutaneous adipose tissue has higher concentrations of ERs whereas visceral adipose tissue has higher concentrations of AR.[241] Additionally, subcutaneous adipose tissue has few AR, and estrogens down-regulate AR expression in subcutaneous fat.[12] Adipose tissue-specific AR knockout mice have increased intra-adipose estrogens, which leads to increased subcutaneous obesity and hyperleptinemia.[242] The sexual dimorphism in adipose tissue distribution may partially explain the greater risk for the metabolic syndrome in men compared with premenopausal women.

Estrogens suppress lipogenesis under certain conditions leading to obesity. E2 suppresses the expression of lipogenic genes, i.e. fatty acid synthase, stearoyl-coenzyme A desaturase 1, and glycerol-3-phosphate acyltransferase, and thus suppresses lipogenesis and TG accumulation in adipose tissue and liver in high-fat diet fed, leptin-deficient *ob/ob* mice.[243] When circulating levels of estrogen are raised above the physiological range due to enlarged adipocytes, adipose tissue metabolism is altered resulting in reduced lipogenic rates and fat depot size to prevent further expansion of adipose tissue. E2-treated ovariectomized mice display reduced lipogenesis in adipocytes.[24] This is further supported by epidemiological observations that serum triglyceride levels increase in postmenopausal women with increased adiposity and that estrogen hormone replacement therapy reduces the activity of lipoprotein lipase (LPL),[244] the enzyme that catalyzes the conversion of triglycerides into free fatty acids and glycerol.[245] Therefore, normal amount of estrogen favors homeostasis in adipose tissues and estrogenic action in adipose tissue prevents the development of obesity by decreasing the lipogenic activity in adipose tissue. However, decreases in the amount of estrogen, as occur with menopause in women and ageing or gonadectomy in rodents, usually result in central obesity with increased lipid accumulation and decreased lipid utilization at visceral fat.[246]

Genetic mouse models of either E2 deficiency by knockout of the aromatase gene (ArKO) or E2 resistance in ERKO have provided important information on the role of E2 signaling in adipose tissue physiology and support roles of E2 and ER in adipose tissue.

ArKO, ERα KO, and mice with double knockout of both ERα and ERβ exhibit an increased white adipose tissue mass but not brown adipose tissue mass,[65,67,247] whereas knockout of the ERβ gene alone does not develop obesity,[248] suggesting that ERα plays an important role in adipogenesis. In addition, ArKO mice and ERα KO mice show increase in adipocyte size (hypertrophy) and number (hyperplasia), display dyslipidemia including an increase in serum cholesterol and a change in the lipoprotein profile, and insulin resistance, implicating a role for ERα in adipocyte growth and proliferation. By studying estrogen replacement in ovariectomized ERα and/or ERβ knockout mice, Lindberg *et al.*[249] demonstrated a reduction of fat in these animals with estrogen replacement mediated mainly by ERα. These studies demonstrate the importance of E2 signaling through the ERα in adipocyte differentiation and metabolism.[65,247,250] This is further supported by studies on 3T3-L1 preadipocytes with stably transfected ERα, which show decreased triglyceride accumulation and reduced expression of LPL. The observation from ArKO mice is consistent with men with aromatase deficiency who have central obesity, elevated blood lipids, and severe insulin resistance.[251]

ERβ KO female mice have a higher body weight under HFD feeding than wild type littermates.[252] This could be a result from increased adipogenesis with subsequent increased mass of adipose tissue and improved insulin sensitivity. Furthermore, the key adipogenic and lipogenic factor PPARγ is negatively regulated by ERβ, suggesting that PPARγ could be a mediator of the metabolic effects observed in ERβ KO mice.[252] Female ERβ KO mice on HFD displayed impaired food efficiency and increased respiratory quotient, which is an indication of disturbed fatty acid oxidation. PPARγ DNA-binding properties and target gene activation are markedly induced in the gonadal fat of the ERβ KO mice and inhibition of adipose PPARγ signaling reversed this metabolic phenotype.[252] In summary, both ER isoforms participate in the antilipogenic actions of estrogens.

Estrogens carry out their action directly on adipose tissues by a combination of genomic (receptor DNA interactions) and nongenomic (membrane receptors) mechanisms in estrogen regulation of adipose accretion and metabolism.[253,254] In the genomic mechanism, estrogens bind to their receptors and the steroid–receptor complex regulates the transcription of given genes. LPL is a key adipose protein regulated by estrogens in a classical genomic mechanism. Second messenger pathways via the cAMP and phosphoinositide cascades are modulated by membrane receptors for estrogens in a nongenomic mechanism. See below for details.

4.2. Estrogenic action on adipogenesis via genomic mechanism

4.2.1 Classical nuclear receptors ERα and ERβ in adipose tissue

The mechanism for the regulation of the amount and distribution of adipose tissues by estrogens is not completely clear. One possible mechanism would be the regulation of key proteins in adipose tissues at the genomic level by transcriptional means. This would require ERs to be present in adipose tissues.

Many groups have demonstrated that ERα and ERβ mRNA and protein are expressed in human adipose tissue.[240,255–260] In initial experiments, it was not possible to detect ER in subcutaneous or omental fat from normal women.[261,262] Later, low levels of ER have been found in human adipose tissues with increased sensitivity of molecular techniques.

Price & O'Brien[258] demonstrated by PCR the presence of ER mRNA in human subcutaneous abdominal tissue. Mizutani *et al.*[260] identified ER protein and ER mRNA in subcutaneous abdominal tissue. The ER protein had a molecular weight of 65 kDa, which is similar to the ER from other tissues including reproductive tissues. Human subcutaneous and visceral adipose tissues express both ERα and ERβ. Expression of ERα is similar in subcutaneous gluteal and subcutaneous abdominal adipose tissues, and is similar between men and women.[256] The ER is higher in concentration in subcutaneous fat than in visceral fat in both males and females. These differences in concentrations of ER in visceral adipose tissue and subcutaneous adipose tissue may offer a possible explanation why men have a more central accumulation of fat, whereas women have a more gluteal/femoral accumulation.

Pedersen *et al.* demonstrated by ligand binding and RT-PCR the presence of ER in human adipose tissue and mature adipocytes.[263] ERα and ERβ are present in cytosols of adipose tissues from both men and women. Cytosols from visceral adipose tissue and subcutaneous abdominal tissue of women showed about the same amount of ER binding, whereas cytosol of subcutaneous abdominal tissue from men showed a higher amount of binding than those from visceral adipose tissue. By immunohistochemistry and RT-PCR analyses, ERα and ERβ are detected in both stromal cells and adipocytes from subcutaneous and omental adipose tissues.[257] A higher amount of ER mRNA was found in adipocytes than in adipose stromal cells.

In mature human adipocytes, ERα is the more highly expressed isoform. ERβ is expressed in breast and subcutaneous abdominal tissues,[240] and it appears to be more highly expressed in women than in men.[259] Interestingly, only ERα, not ERβ, is expressed in human preadipocytes by RT-PCR, although the expression of ERα in preadipocytes is very low compared with that in mature adipocytes.[264]

E2 treatment regulates ERα and ERβ in stromal cells from subcutaneous and omental adipose tissues. In adipocytes from subcutaneous adipose tissue, E2 treatment decreases the expression of ERα, while increases ERβ expression. In omental adipocytes, the decrease in the expression of ERα with E2 treatment is not observed.

Laboratory rodents have been used to study the relationship of estrogens to obesity. Obesity in laboratory rodents is easily induced by overeating and reduced energy expenditure, and the animals develop many of the same deleterious effects, such as insulin resistance and hyperinsulinemia, as seen in human obesity. The major advantage of working with mice is the development of techniques that allow for the manipulation of their genome to generate transgenic mice that lack functional ER.[265] Both ERs are expressed in rodent white adipose tissue[256,266–269] with a predominance of ERα[270] and colocalization of ERα and ERβ in the same nuclei of some cells,[271] whereas only ERα but not ERβ mRNA has been identified in brown adipose tissue.[272] Additionally, brown adipocytes from male rats have higher ERα expression compared to those of female rats, suggesting that this receptor could have a role in thermogenesis.[272]

One of the first indication that ERs exists in adipose tissues was reported by Wade & Gray.[266] They showed estrogen-specific, high affinity [^{3}H]-E2 binding in cytosolic extracts of ovariectomized rat adipose tissues with similar properties to ER in reproductive tissue, but at a much lower concentration. Binding was highest in parametrial fat and lowest in inguinal fat. In a subsequent paper, Gray & Wade[273] showed estrogen-binding sites in male rat adipose tissues. Gray *et al.*[274] also demonstrated binding of injected [^{3}H]-E2 in the nuclei in adipose tissues. Using ovariectomized and adrenalectomized rats for conditions of low endogenous sex steroid production, Rebuffe-Scrive[275] demonstrated cytoplasmic E2 binding in adipose tissues from these animals that decreased dramatically after estrogen administration. Pedersen *et al.*[276] showed nuclear binding of [^{3}H]-E2 in isolated rat adipocytes. The binding is specific for E2 and showed regional variation. Using immunogold electron microscopic localization, Echeverria *et al.*[277] demonstrated significant labeling of ER in adipose cells from various locations of both male and female rats. A higher density of label is found in the nucleus, but some label was observed in the cytoplasm of the adipose cells. All these results support the notion that estrogens may have a direct effect on adipose tissues via its binding to their receptors.

ER knockout mice have abnormalities in their adipose tissues.[278] Indeed, ERα is now considered a candidate gene for obesity.[279] In male mice, deletion of ERα causes progressive increase in epididymal, perirenal, and inguinal adipose tissue with advancing age. Epididymal and perirenal adipocyte size is increased and adipocyte number is greater in the fat pads. Female ERα KO mice have increased fat, adipocyte size, number, and leptin and cholesterol levels and smaller LDL particles. Both males and females are insulin resistant with impaired glucose tolerance.[65] E2 administered to ovariectomized mice fed a high-fat diet preserved glucose tolerance and insulin sensitivity and improved glucose tolerance in wild type but not in ERα KO mice, suggesting that targeting of the ERα could represent a strategy to reduce the impact of high-fat diet-induced insulin resistance.[280]

ERs play roles on GLUT4 function in adipose tissue. The administration of tamoxifen to ERα KO or ERβ KO mice resulted in different responses on GLUT4 expression. In ERα KO mice, tamoxifen did not interfere with GLUT4 expression, whereas treatment of ERβ KO mice with tamoxifen substantially decreased it, suggesting that ERα is the main regulator of GLUT4 expression in adipose tissue.[271] Ovariectomy of ERβ KO mice prevented accumulation of triglycerides and reduced body fat and adipocyte size, and preserved regular insulin signaling in liver and skeletal muscle, and improved whole body insulin sensitivity and glucose tolerance, suggesting that ERα and ERβ have opposite functions on lipid and glucose metabolism, with ERβ involved in adipogenesis.[252,281] An imbalance between ERα and ERβ in adipose tissue may have important implications for the development of metabolic diseases.

4.2.2 Estrogenic genomic action on LPL in adipose tissue

In the classical mechanism of action of sex steroid hormones, the steroid enters the cell and binds with high affinity to a specific receptor. The steroid–receptor complex then undergoes a conformation change and binds to a specific response element on the DNA. Once bound, it regulates the transcription of a given gene. Transcription may be up- or downregulated depending upon the gene, the steroid, and the coregulatory proteins. Many genes in adipose tissues are transcriptionally regulated by sex steroid hormones. An ERE is present in the LPL genome and thus *LPL* gene is a direct transcriptional target of estrogen. LPL is the key protein involved in lipid deposition.

LPL is the key enzyme for the hydrolysis of circulating triglycerides into free fatty acids and glycerol. Its highest activity is located in adipose tissues, heart, and skeletal muscle, and it plays a major role in adipocyte lipid storage and muscle fuel supply and hence the regulation of lean muscle mass and obesity. It is synthesized and secreted by the adipocyte and then transported to the capillary endothelium, where it hydrolyses the circulating triglycerides into free fatty acids, which in turn are stored in the adipose tissues. LPL in adipocytes is activated by insulin. During feasting, increased insulin activates LPL to promote adipogenesis.

Although the mechanisms responsible for estrogen's regulation of adipogenesis remain unclear,[255] they may involve estrogens' suppression of LPL, a lipogenic enzyme that regulates the metabolism of plasma triglycerides to free fatty acids and increases lipid storage by adipocytes. Specifically, the effect of estrogens on LPL activity may be concentration dependent, as Palin *et al.* found that a high concentration (10^{-7} mol L^{-1}) of estrogen reduced LPL expression in isolated adipocytes from subcutaneous abdominal fat of women, while lower estrogen concentrations increased LPL expression.[282] Consistent with this notion, ovariectomy of female rats results in increased adipose tissue LPL while estrogen replacement decreases the LPL activity.[283–288] E2 administration to male rats also significantly reduces adipocyte LPL activity[289] to suppress adipogenesis.

In obese women, plasma LPL activity has an inverse correlation with plasma E2 levels.[244] The administration of estrogens alone or estrogen–progestin in women decreases LPL activity and increases serum triglyceride concentrations.[290–292] A transdermal E2 patch placed in the gluteal region of women decrease LPL in adipose tissue from beneath the patch as compared to tissue beneath the placebo patch.[293] However, there is no difference in LPL mRNA levels between these two groups, suggesting that a possible posttranscriptional modification of LPL led to the difference. In women, LPL levels increase after menopause.[294] In addition to inhibiting fatty acid uptake (lipogenesis), estrogen also increases fatty acid and glycerol release (lipolysis) by increasing expression of hormone-sensitive lipase (HSL), which catalyzes the breakdown of stored TG and the release of fatty acids.[282,295] However, estrogen has also been shown to inhibit lipolysis. Estrogen treatment of postmenopausal women increases expression of the α2A-adrenergic receptor in subcutaneous adipose tissue. Activation of this receptor inhibits lipolysis; therefore, increased expression due to estrogen would inhibit lipolysis.

This increase was recapitulated *in vitro* using cultured human adipose tissue fragments incubated with E2 and was correlated with inhibition of lipolysis,[296] suggesting a direct role for ER in regulating lipolysis. The complex relationship between the ERs and adipogenesis remains to be elucidated and may provide critical information for the prevention and treatment of obesity.

4.3. Estrogenic nongenomic action via second messengers in adipose tissue

Another possible mechanism would be the regulation of second messengers at the cell membrane by nongenomic effects. Although transcriptional regulation by steroid hormones has been the most studied mechanism of steroid action, recent studies have shown that a nongenomic mechanism also exists for steroid hormone action.[297] This would require a sex steroid receptor to be present in the membrane. There is also evidence for rapid nongenomic estrogen actions via a membrane-bound ER in adipocytes.[112,298] Nongenomic actions of steroids are generally rapid responses occurring in less than 10 min, while transcriptional regulation requires several hours to days.[197,299,300]

In the nongenomic mechanism, estrogens bind to their receptors in the plasma membrane, and second messengers are formed to carry out their action. Estrogens appear to regulate various components of the membrane-signaling systems, including both the cAMP cascade and the phosphoinositide cascade.[301,302]

Activation of the cAMP cascade by sex steroid hormones would activate HSL leading to lipolysis in adipose tissues. In the phosphoinositide cascade, diacylglycerol and inositol 1,4,5-trisphosphate are formed as second messengers ultimately causing the activation of protein kinase C. Their activation appears to be involved in the control of preadipocyte proliferation and differentiation.

Sex steroid membrane receptors have been characterized in many different tissues,[112] but little is known about the receptors in adipose tissues. Anwar *et al.*[257] showed immunostaining for both ERα and ERβ in the cellular membranes of subcutaneous abdominal and omental human adipose tissues. Dos Santos *et al.*[298] found immunoreactive ERα in the membrane fraction from cultured rat adipocytes, but not in the membrane fraction from preadipocytes.

These plasma membrane ERs are associated with caveolae endocytic vesicles. Caveolin-1 coimmunoprecipitate with ERα and bind directly to ERα.[303] Caveolin-1 has also been reported to bind to endoplasmic reticulum membranes.[304]

4.3.1 cAMP cascade

Early studies showed an increase in cAMP and lipolysis within 5 min when isolated rat adipocytes were treated with E2[305] and adenylate cyclase activity in fat cell mass increased by E2 treatment of ovariectomized rats.[306] This response to estrogen appears to be depot specific as ovariectomy caused a decrease in cAMP production in rat parametrial adipocytes, but not in femoral subcutaneous adipocytes.[307] These studies with rats would suggest that estrogen increased cAMP in adipose tissues, resulting in stimulation of HSL and increased lipolysis.

In humans, estrogens also appear to decrease lipolysis in adipose tissues. Lindberg *et al.* found that women who were given oral ethinyl E2 had a decrease in noradrenaline-stimulated lipolysis in subcutaneous abdominal adipocytes.[308] To study the effect of estrogens on lipolysis in women, Pedersen *et al.* used adipocytes that were isolated from adipose tissue fragments, which were incubated *in vitro* with or without estradiol.[296] A decrease in adrenaline-stimulated lipolysis was observed in adipocytes from subcutaneous E2-treated adipose tissue fragments. E2 also increased the number of antilipolytic α2A-adrenergic receptors in subcutaneous adipocytes. These results suggested that estrogen lowered lipolysis by increasing the number of antilipolytic α2A-adrenergic receptors. In contrast, no effect of estrogen was observed in adipocytes from intra-abdominal fat. This difference in the E2 effect on α2A-adrenergic receptors between subcutaneous fat and intra-abdominal fat may be one factor to explain how estrogen maintains the gynoid fat pattern in women. This increase in α2A-adrenergic receptors with estrogen in subcutaneous fat from women may also explain the difference in response to estrogen between humans and rats as adipose tissue from rats do not have α2A-adrenergic receptors.[308]

4.3.2 Phosphoinositide cascade

The phosphoinositide cascade also appears to be regulated by sex steroid hormones. Protein kinase C content of rat subcutaneous and parametrial fat cells are reduced by ovariectomy and restored to normal by E2 and progesterone treatment.[309] Lacasa *et al.*[310] demonstrated that ovariectomy increased proliferation, MAPK activity, and c-fos protein in preadipocytes from rat perirenal fat depots. Treatment of the ovariectomized animals with E2 and progesterone reversed the increased proliferation and c-fos protein induction, but not the MAPK activation. The ovarian status had no effect on these parameters in subcutaneous preadipocytes, indicating that this regulation is depot specific. Recently, Dos Santos *et al.*[298] found that estrogen activated p42/p44 MAPK by phosphorylation in rat white adipocytes through a rapid nongenomic pathway. This pathway was postulated to involve membrane ERα, G protein, protein kinase C, raf, and MAPK kinase (MEK). The phosphoinositide cascade appears to be involved in adipose proliferation and differentiation.

5. ESTROGENS INTERACT WITH LEPTIN

5.1. Estrogen interact with leptin in the CNS

Leptin, the product of the obese gene, is a key adipocytokine hormone that is normally released by adipose tissue in direct proportion to fat content and is related to the amount and distribution of body fat.[311] It is a potent anorexigenic and catabolic hormone in central regulation of metabolism by crossing the blood–brain barrier to interact with leptin receptors (lepr) in the hypothalamus. Although there are several isoforms of the lepr, the b form (leprb) is the critical variant for regulating energy balance.[312] Leprb has been identified in the ARC, PVN, and VMH, where they colocalize with ERα.[313] Leptin transfers a catabolic signal to the brain to inhibit food intake and increase energy expenditure.[314–316] Thus, leptin plays a key role in the regulation of body fat and energy homeostasis. Estrogen not only modulates leptin receptor leprb mRNA in the ARC and VMH via an ERE in the lepr gene, but also increases hypothalamic sensitivity to leptin, altering peripheral fat distribution.[317] In addition, raised levels of estrogens have been associated with increased leptin sensitivity in the brain,[318] although ovariectomy reduces the sensitivity to leptin when compared with intact females, an effect that can be restored by E2 replacement. Furthermore, administration of E2 in male rats increases the sensitivity to leptin.[20] Thus, interactions between E2 and leptin exist for the regulation of energy homeostasis.[313,319]

5.2. Estrogen interact with leptin in adipose tissue

Leptin levels are higher in females compared with males.[320] After puberty, estrogen modulates leptin synthesis and secretion via ER-dependent transcriptional mechanisms in adipose tissue.[321] The presence of rising intracellular bioactive E2 stimulates leptin production by fat cells.[236] In support of this, Hardie *et al.*[322] reported a correlation between serum leptin and estrogen levels. As estrogen levels decrease with menopause, a decrease in leptin levels might be expected with menopause.[320,323,324] Additionally, increase in leptin with estrogen replacement therapy has been reported.[324,325] Thomas *et al.*[326] found that bioavailable estrogen levels in postmenopausal women who were not on hormone replacement therapy are significantly and directly related to serum leptin levels, but in premenopausal women or in women on hormone replacement therapy, variations in estrogen level are not associated with the higher leptin levels, suggesting a possible threshold effect. The most convincing evidence that the sex steroid hormones affect blood leptin levels comes from studies on cross-sex administration of sex steroid hormones to transsexuals. Elbers *et al.*[327] were able to show that estrogen and antiandrogen administration to male-to-female transsexuals greatly increased the serum leptin levels and testosterone administration to female-to-male transsexuals decreased the serum leptin levels. They concluded that the sex steroid milieu plays an important role in the regulation of blood leptin levels.

Studying human omental adipose tissue *in vitro*, Casabiell *et al.*[328] demonstrated that leptin secretion was higher in tissue from women than in tissue from men. Dexamethasone and E2 stimulated leptin secretion in fat from women, but not in fat from men. The increased E2[237,329] and the increased leptin[330] both upregulate aromatase to further increase extra-glandular E2 production,[237] to increase subcutaneous fat deposition[331] and to potentiate the action of E2 on ERα[332] and transmembrane GPER.[333]

As in humans, female rats have higher circulating levels of leptin than do males.[334–336] Ovariectomy of female rats has been shown by some investigators to lower circulating leptin levels, and estrogen replacement in these animals has been found to return the levels back to normal.[320,337] Other investigators, however, found no effect on circulating leptin levels after ovariectomy and estrogen replacement.[321,334–336,338,339] These discrepancies could be caused by the time of the blood samples, as Chu *et al.*[340] showed that leptin levels declined during the first 7 weeks after ovariectomy. Subsequently, its concentration increased, and by the end of 13 weeks the value was higher than the level before ovariectomy. The fluctuation of serum leptin levels with ovariectomy was eliminated by estrogen replacement. These discrepancies could also be caused by weight gain in the ovariectomized animals, as leptin levels are related to the amount of fat, which was not always controlled in these experiments.

Ovariectomy has also been reported by some authors to decrease leptin gene expression in white adipose tissues of rats, and this decrease can be reversed by estrogen replacement.[320,321,337,339] In contrast, other authors have reported no effect of estrogen replacement on leptin gene expression in white adipose tissues.[336,338] These discrepancies could be caused by differences in leptin secretion and regulation in various adipose sites, as Shimizu *et al.*[320] showed that although leptin gene expression decreased in subcutaneous and retroperitoneal white adipose tissues of ovariectomized rats as compared to controls, leptin gene expression actually increased in mesenteric white adipose tissues of the ovariectomized rats.

Machinal *et al.*[321] also observed that ovariectomy resulted in a 25% decrease in leptin gene expression in isolated perirenal adipocytes and a small but insignificant decrease in subcutaneous adipocytes. Therefore, the regional distribution of adipose tissues may greatly influence leptin levels and its regulation by sex steroid hormones. Basal leptin production is equal in epididymal fat fragments from males and parametrial fat fragments from females. E2 increased leptin secretion in both male and female adipose fragments, but the effect was greater in female adipose tissue. Parametrial fat from ovariectomized rats had lower leptin secretion compared to sham-operated controls, and E2 treatment of ovariectomized rats maintained a normal leptin secretion in fat fragments.

Sex steroid hormone regulation of leptin production has also been studied in normal and mutant mice. Nedvidkova *et al.* showed a decrease of serum leptin levels in estrogen-treated male mice.[341] However, in female mice, Pelleymounter *et al.* reported that ovariectomy or E2 administration did not alter leptin levels.[342] Kronfeld-Schor *et al.* demonstrated that subcutaneous implants of E2 or corticosterone into lactating mice stimulated leptin secretion rates of omental adipose tissue *in vitro*, and corticosterone, but not estradiol, stimulated leptin secretion rates when added to isolated omental adipose tissue.[343] From this study, the authors suggest that E2 could be acting indirectly on the omental adipose tissue possibly through an increase in plasma corticosterone levels.

In ERα KO male mice, Ohlsson *et al.* demonstrated an enhanced serum leptin level as compared to controls that was not seen in ERβ KO mice.[67] Jones *et al.* showed an elevation of leptin levels in ArKO male and female mice.[247] However, both the ERα KO mice and ArKO mice had an increase in fat mass, which could account for the increased leptin levels.

In summary, estrogens appear to increase leptin production. However, many discrepancies exist in the data. If sex steroid hormones are going to directly regulate the production of leptin, an ERE should be present in the leptin gene. After reviewing the cDNA sequence of the leptin gene, Shimizu *et al.*[320] concluded that this gene had a consensus sequence of the ERE in its promoter region. These results suggest that the leptin gene has an ERE; that leptin promoter activation may depend upon coactivators present in leptin-producing cells; and that different effects of estrogen may be due to the type of ER expressed in target tissue.

6. CONCLUSIONS

Obesity and its associated health-related problems have become so prevalent in developed countries that many programs and strategies have been developed to prevent and/or treat this affliction, but without much success.[344,345] Although sex steroid hormones have a major role in lipid metabolism and deposition in adipose tissues, they or their derivatives have not been extensively used to prevent or treat obesity.

There is abundant evidence that estrogen can utilize both genomic and nongenomic signaling mechanisms to regulate energy homeostasis, body fat, and glucose regulation. Estrogens exert rapid nongenomic effects in the brain and adipose tissues to regulate second messengers and kinase signaling pathways, and play physiologically important roles in estrogen-induced regulation of homeostasis.

One area where sex steroid hormones appear to have a positive effect on obesity is in hormone replacement therapy in older women. Menopause in women causes a large decrease in endogenous estrogens, which leads to central obesity and an increased risk in metabolic disorders. A relationship of central obesity and its complications to the decrease in sex steroid hormones is unclear. If it is related, hormone replacement therapy in postmenopausal women should reduce the occurrence of these problems.[346] Several studies were able to show that hormone replacement therapy did decrease central obesity in postmenopausal women and hence prevent at least some of the problems of menopause.[18,347–354] In contrast, other studies have not been able to confirm the decrease in central obesity with hormone treatment.[17,355,356] In contrast, a recent report showed that hormone replacement therapy in postmenopausal women results in an increased risk for coronary heart disease, stroke, thromboembolic events, and breast cancer,[357] suggesting caution when using sex steroid treatment. For the amelioration of obesity by sex steroid hormones, it would be helpful to develop compounds that would be specific for adipose tissues and that would not have the adverse side effects of estrogen. Considerable time was devoted to the need for SERMs that are adipose tissue specific and have pathways related to lipid metabolism for safe intervention in a number of endocrine-regulated conditions.[358] SERMs, such as estren[359] and CP-336,156,[360] that were shown to prevent bone loss in animal models without serious side effects, and CP-336,156 that was shown to prevent the increase in body fat in ovariectomized rats, may be an option for obesity regulation in the future.

Regulation of some key proteins in adipose tissues by estrogens, such as downregulation of LPL in visceral adipose tissue, may also be a mechanism for the treatment and/or prevention of obesity.[361] Estrogens are known to decrease LPL activity at least in gluteal adipose tissue, which may be important in body fat distribution.[292]

Leptin is another adipose protein that could be a target for regulation by estrogens to control central obesity. The extension of leptin therapy in experimental animals to humans has been only partially successful.[362,363] In patients with a genetic defect of leptin production, leptin therapy greatly reduces the obesity in these individuals.[364] However, these individuals only represent an extremely small percentage of the obese population, and in the common diet-induced obesity, leptin therapy remains questionable because leptin levels are very high in obese individuals. The high leptin levels in obese individuals might suggest that exogenous leptin therapy would be ineffective in decreasing fat stores. It is possible that regulation of endogenous leptin by sex steroid hormones would be more effective than exogenous leptin.

By gaining a more thorough understanding of estrogen's effects on the CNS and peripheral adipose tissues, and a better grasp of the signaling cascade involved, we will gain new insights that lead to better diagnostic tools and improved therapeutics in humans.

ACKNOWLEDGMENTS

This work was supported by the National Institutes of Health National Institute of Diabetes and Digestive and Kidney Diseases (DK090823) and the American Heart Association (SDG4520028) to H. S., and doctoral fellowship from the Cellular, Molecular and Structural Biology program to S. P. D. S. K. and X. L.

REFERENCES

1. James WPT. The epidemiology of obesity: the size of the problem. *J Intern Med* 2008;**263**(4):336–52.
2. Cummings DE, Schwartz MW. Genetics and Pathophysiology of Human Obesity. *Annu Rev Med* 2003;**54**(1):453–71.

3. Alberti K, Zimmet P, Shaw J. IDF epidemiology task force consensus group: the metabolic syndrome—a worldwide definition. *Lancet* 2005;**366**:1059–62.
4. McGarry JD. Banting lecture 2001: dysregulation of fatty acid metabolism in the etiology of type 2 diabetes. *Diabetes* 2002;**51**(1):7–18.
5. Xu H, et al. Chronic inflammation in fat plays a crucial role in the development of obesity-related insulin resistance. *J Clin Invest* 2003;**112**(12):1821–30.
6. Kolovou G, et al. Apolipoprotein E gene polymorphism and gender. *Ann Clin Lab Sci* 2009;**39**(2):120–33.
7. Ford ES. Prevalence of the metabolic syndrome defined by the International Diabetes Federation among adults in the U.S. *Diabetes Care* 2005;**28**(11):2745–9.
8. Kannel WB, et al. Regional obesity and risk of cardiovascular disease; the Framingham study. *J Clin Epidemiol* 1991;**44**(2):183–90.
9. Lee CG, et al. Adipokines, inflammation, and visceral adiposity across the menopausal transition: a prospective study. *J Clin Endocrinol Metab* 2009;**94**(4):1104–10.
10. Regitz-Zagrosek V, Lehmkuhl E, Weickert M. Gender differences in the metabolic syndrome and their role for cardiovascular disease. *Clin Res Cardiol* 2006;**95**(3):136–47.
11. Björntorp P. Body fat distribution, insulin resistance, and metabolic diseases. *Nutrition* 1997;**13**(9):795–803.
12. Björntorp P. Hormonal control of regional fat distribution. *Hum Reprod* 1997;**12** (Suppl. 1):21–5.
13. Kotani K, et al. Sexual dimorphism of age-related changes in whole-body fat distribution in the obese. *Int J Obes Relat Metab Disord* 1994;**18**(4):207–12.
14. Carr MC. The emergence of the metabolic syndrome with menopause. *J Clin Endocrinol Metab* 2003;**88**(6):2404–11.
15. Krotkiewski M, et al. Impact of obesity on metabolism in men and women. Importance of regional adipose tissue distribution. *J Clin Invest* 1983;**72**(3):1150–62.
16. Bouchard C, Despres J-P, Mauriege P. Genetic and nongenetic determinants of regional fat distribution. *Endocr Rev* 1993;**14**(1):72–93.
17. Ryan AS, Nicklas BJ, Berman DM. Hormone replacement therapy, insulin sensitivity, and abdominal obesity in postmenopausal women. *Diabetes Care* 2002;**25**(1):127–33.
18. Samaras K, et al. Effects of postmenopausal hormone replacement therapy on central abdominal fat, glycemic control, lipid metabolism, and vascular factors in type 2 diabetes: a prospective study. *Diabetes Care* 1999;**22**(9):1401–7.
19. Blaustein JD, Wade GN. Ovarian influences on the meal patterns of female rats. *Physiol Behav* 1976;**17**(2):201–8.
20. Clegg DJ, et al. Gonadal hormones determine sensitivity to central leptin and insulin. *Diabetes* 2006;**55**(4):978–87.
21. Wallen WJ, Belanger MP, Wittnich C. Sex hormones and the selective estrogen receptor modulator tamoxifen modulate weekly body weights and food intakes in adolescent and adult rats. *J Nutr* 2001;**131**(9):2351–7.
22. Roy EJ, Wade GN. Role of food intake in estradiol-induced body weight changes in female rats. *Horm Behav* 1977;**8**(3):265–74.
23. Gao Q, et al. Anorectic estrogen mimics leptin's effect on the rewiring of melanocortin cells and Stat3 signaling in obese animals. *Nat Med* 2007;**13**(1):89–94.
24. D'Eon TM, et al. Estrogen regulation of adiposity and fuel partitioning Evidence of genomic and non-genomic regulation of lipogenic and oxidative pathways. *J Biol Chem* 2005;**280**(43):35983–91.
25. Gao H, et al. Long-term administration of estradiol decreases expression of hepatic lipogenic genes and improves insulin sensitivity in ob/ob Mice: a possible mechanism is through direct regulation of signal transducer and activator of transcription 3. *Mol Endocrinol* 2006;**20**(6):1287–99.

26. Espeland MA, et al. Effect of postmenopausal hormone therapy on lipoprotein(a) concentration. PEPI Investigators. Postmenopausal estrogen/progestin interventions. *Circulation* 1998;**97**(10):979–86.
27. Redig AJ, Munshi HG. Care of the cancer survivor: metabolic syndrome after hormone-modifying therapy. *Am J Med* 2010;**123**(1):87.e1–87.e6.
28. Tiano JP, et al. Estrogen receptor activation reduces lipid synthesis in pancreatic islets and prevents β cell failure in rodent models of type 2 diabetes. *J Clin Invest* 2011;**121**(8):3331–42.
29. Simpson ER, et al. Estrogen—the good, the bad, and the unexpected. *Endocr Rev* 2005;**26**(3):322–30.
30. Maffei L, et al. Dysmetabolic syndrome in a man with a novel mutation of the aromatase gene: effects of testosterone, alendronate, and estradiol treatment. *J Clin Endocrinol Metab* 2004;**89**(1):61–70.
31. Morishima A, et al. Aromatase deficiency in male and female siblings caused by a novel mutation and the physiological role of estrogens. *J Clin Endocrinol Metab* 1995;**80**(12):3689–98.
32. Bilezikian JP, et al. Increased bone mass as a result of estrogen therapy in a man with aromatase deficiency. *N Engl J Med* 1998;**339**(9):599–603.
33. Rochira V, et al. Oestradiol replacement treatment and glucose homeostasis in two men with congenital aromatase deficiency: evidence for a role of oestradiol and sex steroids imbalance on insulin sensitivity in men. *Diabet Med* 2007;**24**(12):1491–5.
34. Takeda K, et al. Progressive development of insulin resistance phenotype in male mice with complete aromatase (CYP19) deficiency. *J Endocrinol* 2003;**176**(2):237–46.
35. Deroo BJ, Korach KS. Estrogen receptors and human disease. *J Clin Invest* 2006;**116**(3):561–70.
36. Hess R. Estrogen in the adult male reproductive tract: a review. *Reprod Biol Endocrinol* 2003;**1**(1):52.
37. Baquedano MS, et al. Identification and developmental changes of aromatase and estrogen receptor expression in prepubertal and pubertal human adrenal tissues. *J Clin Endocrinol Metab* 2007;**92**(6):2215–22.
38. Kelly MJ, Qiu J, Rønnekleiv OK. Estrogen signaling in the hypothalamus. In: Gerald L, editor. *Vitamins & hormones*. Academic Press; 2005. p. 123–45.
39. McEwen BS, Alves SE. Estrogen actions in the central nervous system. *Endocr Rev* 1999;**20**(3):279–307.
40. Teede HJ. Sex hormones and the cardiovascular system: effects on arterial function in women. *Clin Exp Pharmacol Physiol* 2007;**34**(7):672–6.
41. Turner RT, Riggs BL, Spelsberg TC. Skeletal effects of estrogen. *Endocr Rev* 1994;**15**(3):275–300.
42. Gruber CJ, et al. Production and actions of estrogens. *N Engl J Med* 2002;**346**(5):340–52.
43. Muller RE, et al. Studies on the mechanism of estradiol uptake by rat uterine cells and on estradiol binding to uterine plasma membranes. *Adv Exp Med Biol* 1979;**117**:401–21.
44. Jackson V, Chalkley R. The binding of estradiol-17 beta to the bovine endometrial nuclear membrane. *J Biol Chem* 1974;**249**(5):1615–26.
45. Revankar CM, et al. A transmembrane intracellular estrogen receptor mediates rapid cell signaling. *Science* 2005;**307**(5715):1625–30.
46. Nadal A, et al. Nongenomic actions of estrogens and xenoestrogens by binding at a plasma membrane receptor unrelated to estrogen receptor α and estrogen receptor β. *Proc Natl Acad Sci U S A* 2000;**97**(21):11603–8.
47. Meyer MR, et al. Non-genomic regulation of vascular cell function and growth by estrogen. *Mol Cell Endocrinol* 2009;**308**(1–2):9–16.

48. Walter P, et al. Cloning of the human estrogen receptor cDNA. *Proc Natl Acad Sci U S A* 1985;**82**(23):7889–93.
49. Kuiper GG, et al. Cloning of a novel receptor expressed in rat prostate and ovary. *Proc Natl Acad Sci U S A* 1996;**93**(12):5925–30.
50. Greene G, et al. Sequence and expression of human estrogen receptor complementary DNA. *Science* 1986;**231**(4742):1150–4.
51. Green S, et al. Human oestrogen receptor cDNA: sequence, expression and homology to v-erb-A. *Nature* 1986;**320**(6058):134–9.
52. Jensen EV, DeSombre ER. Estrogen–receptor interaction. *Science* 1973;**182**(4108): 126–34.
53. Brzozowski AM, et al. Molecular basis of agonism and antagonism in the oestrogen receptor. *Nature* 1997;**389**(6652):753–8.
54. Stauffer SR, et al. Pyrazole ligands: structure–affinity/activity relationships and estrogen receptor-alpha-selective agonists. *J Med Chem* 2000;**43**(26):4934–47.
55. Meyer MR, et al. Dilation of epicardial coronary arteries by the G protein-coupled estrogen receptor agonists G-1 and ICI 182,780. *Pharmacology* 2010;**86**(1):58–64.
56. Malamas MS, et al. Design and synthesis of aryl diphenolic azoles as potent and selective estrogen receptor-beta ligands. *J Med Chem* 2004;**47**(21):5021–40.
57. Hammes SR, Levin ER. Extranuclear steroid receptors: nature and actions. *Endocr Rev* 2007;**28**(7):726–41.
58. Hall JM, Couse JF, Korach KS. The multifaceted mechanisms of estradiol and estrogen receptor signaling. *J Biol Chem* 2001;**276**(40):36869–72.
59. Nilsson S, et al. Mechanisms of estrogen action. *Physiol Rev* 2001;**81**(4):1535–65.
60. Thomas C, Gustafsson J-Å. The different roles of ER subtypes in cancer biology and therapy. *Nat Rev Cancer* 2011;**11**(8):597–608.
61. Handa RJ, et al. Roles for oestrogen receptor β in adult brain function. *J Neuroendocrinol* 2012;**24**(1):160–73.
62. Kuiper GGJM, et al. Comparison of the ligand binding specificity and transcript tissue distribution of estrogen receptors alpha and beta. *Endocrinology* 1997;**138**(3):863–70.
63. Matthews J, Gustafsson J-Å. Estrogen signaling: a subtle balance between ER alpha and ER beta. *Mol Interv* 2003;**3**(5):281–92.
64. Hewitt S, Korach K. Oestrogen receptor knockout mice: roles for oestrogen receptors alpha and beta in reproductive tissues. *Reproduction* 2003;**125**(2):143–9.
65. Heine PA, et al. Increased adipose tissue in male and female estrogen receptor-α knockout mice. *P Natl Acad Sci USA* 2000;**97**(23):12729–34.
66. Geary N, et al. Deficits in E2-dependent control of feeding, weight gain, and cholecystokinin satiation in ER-α null mice. *Endocrinology* 2001;**142**(11):4751–7.
67. Ohlsson C, et al. Obesity and disturbed lipoprotein profile in estrogen receptor-α-deficient male mice. *Biochem Biophys Res Commun* 2000;**278**(3):640–5.
68. Shughrue PJ, Lane MV, Merchenthaler I. Comparative distribution of estrogen receptor-alpha and -beta mRNA in the rat central nervous system. *J Comp Neurol* 1997;**388**(4):507–25.
69. Nilsson S, Gustafsson JA. Estrogen receptors: therapies targeted to receptor subtypes. *Clin Pharmacol Ther* 2011;**89**(1):44–55.
70. Liang YQ, et al. Estrogen receptor beta is involved in the anorectic action of estrogen. *Int J Obes Relat Metab Disord* 2002;**26**(8):1103–9.
71. O'Lone R, et al. Genomic targets of nuclear estrogen receptors. *Mol Endocrinol* 2004;**18** (8):1859–75.
72. Edwards DP. Regulation of signal transduction pathways by estrogen and progesterone. *Annu Rev Physiol* 2005;**67**(1):335–76.
73. Losel RM, et al. Nongenomic steroid action: controversies, questions, and answers. *Physiol Rev* 2003;**83**(3):965–1016.

74. Brann DW, Hendry LB, Mahesh VB. Emerging diversities in the mechanism of action of steroid hormones. *J Steroid Biochem Mol Biol* 1995;**52**(2):113–33.
75. Manavathi B, Kumar R. Steering estrogen signals from the plasma membrane to the nucleus: two sides of the coin. *J Cell Physiol* 2006;**207**(3):594–604.
76. Moriarty K, Kim KH, Bender JR. Minireview: estrogen receptor-mediated rapid signaling. *Endocrinology* 2006;**147**(12):5557–63.
77. Mhyre AJ, Dorsa DM. Estrogen activates rapid signaling in the brain: role of estrogen receptor α and estrogen receptor β in neurons and glia. *Neuroscience* 2006;**138**(3):851–8.
78. Boonyaratanakornkit V, Edwards DP. Receptor mechanisms mediating non-genomic actions of sex steroids. *Semin Reprod Med* 2007;**25**(03):139,153.
79. Szego CM, Davis JS. Adenosine 3',5'-monophosphate in rat uterus: acute elevation by estrogen. *Proc Natl Acad Sci U S A* 1967;**58**(4):1711–8.
80. Pietras RJ, Szego CM. Specific binding sites for oestrogen at the outer surfaces of isolated endometrial cells. *Nature* 1977;**265**(5589):69–72.
81. Carmeci C, et al. Identification of a gene (GPR30) with homology to the G-protein-coupled receptor superfamily associated with estrogen receptor expression in breast cancer. *Genomics* 1997;**45**(3):607–17.
82. Filardo EJ, et al. Estrogen action via the G protein-coupled receptor, GPR30: stimulation of adenylyl cyclase and cAMP-mediated attenuation of the epidermal growth factor receptor-to-MAPK signaling axis. *Mol Endocrinol* 2002;**16**(1):70–84.
83. Prossnitz ER, et al. Estrogen signaling through the transmembrane G protein-coupled receptor GPR30. *Annu Rev Physiol* 2008;**70**(1):165–90.
84. Thomas P, et al. Identity of an estrogen membrane receptor coupled to a G protein in human breast cancer cells. *Endocrinology* 2005;**146**(2):624–32.
85. Filardo EJ, et al. Estrogen-induced activation of Erk-1 and Erk-2 requires the G protein-coupled receptor homolog, GPR30, and occurs via trans-activation of the epidermal growth factor receptor through release of HB-EGF. *Mol Endocrinol* 2000;**14**(10):1649–60.
86. Maggiolini M, et al. The G protein-coupled receptor GPR30 mediates c-fos up-regulation by 17beta-estradiol and phytoestrogens in breast cancer cells. *J Biol Chem* 2004;**279**(26):27008–16.
87. Qiu J, et al. Rapid signaling of estrogen in hypothalamic neurons involves a novel G-protein-coupled estrogen receptor that activates protein kinase C. *J Neurosci* 2003;**23**(29):9529–40.
88. Qiu J, et al. A G-protein-coupled estrogen receptor is involved in hypothalamic control of energy homeostasis. *J Neurosci* 2006;**26**(21):5649–55.
89. Roepke TA, et al. Genes associated with membrane-initiated signaling of estrogen and energy homeostasis. *Endocrinology* 2008;**149**(12):6113–24.
90. Roepke TA, et al. Contribution of a membrane estrogen receptor to the estrogenic regulation of body temperature and energy homeostasis. *Endocrinology* 2010;**151**(10):4926–37.
91. Toran-Allerand CD, et al. ER-X: a novel, plasma membrane-associated, putative estrogen receptor that is regulated during development and after ischemic brain injury. *J Neurosci* 2002;**22**(19):8391–401.
92. Wang Z, et al. Identification, cloning, and expression of human estrogen receptor-α36, a novel variant of human estrogen receptor-α66. *Biochem Biophys Res Commun* 2005;**336**(4):1023–7.
93. Prossnitz ER, Barton M. Signaling, physiological functions and clinical relevance of the G protein-coupled estrogen receptor GPER. *Prostaglandins Other Lipid Mediat* 2009;**89**(3–4):89–97.
94. O'Dowd BF, et al. Discovery of three novel G-protein-coupled receptor genes. *Genomics* 1998;**47**(2):310–3.

95. Balhuizen A, et al. Activation of G protein-coupled receptor 30 modulates hormone secretion and counteracts cytokine-induced apoptosis in pancreatic islets of female mice. *Mol Cell Endocrinol* 2010;**320**(1–2):16–24.
96. Hugo ER, et al. Bisphenol A at environmentally relevant doses inhibits adiponectin release from human adipose tissue explants and adipocytes. *Environ Health Perspect* 2008;**116**(12):1642–7.
97. Kanda N, Watanabe S. 17[beta]-estradiol inhibits oxidative stress-induced apoptosis in keratinocytes by promoting Bcl-2 expression. *J Invest Dermatol* 2003;**121**(6):1500–9.
98. Mårtensson UEA, et al. Deletion of the G protein-coupled receptor 30 impairs glucose tolerance, reduces bone growth, increases blood pressure, and eliminates estradiol-stimulated insulin release in female mice. *Endocrinology* 2009;**150**(2):687–98.
99. Nadal A, et al. The role of oestrogens in the adaptation of islets to insulin resistance. *J Physiol* 2009;**587**(21):5031–7.
100. Haas E, et al. Regulatory role of G protein-coupled estrogen receptor for vascular function and obesity. *Circ Res* 2009;**104**(3):288–91.
101. Liu S, et al. Importance of extranuclear estrogen receptor-alpha and membrane G protein-coupled estrogen receptor in pancreatic islet survival. *Diabetes* 2009;**58** (10):2292–302.
102. Blasko E, et al. Beneficial role of the GPR30 agonist G-1 in an animal model of multiple sclerosis. *J Neuroimmunol* 2009;**214**(1–2):67–77.
103. Noel SD, et al. Involvement of G protein-coupled receptor 30 (GPR30) in rapid action of estrogen in primate LHRH neurons. *Mol Endocrinol* 2009;**23**(3):349–59.
104. Rettew JA, McCall SH, Marriott I. GPR30/GPER-1 mediates rapid decreases in TLR4 expression on murine macrophages. *Mol Cell Endocrinol* 2010;**328**(1–2):87–92.
105. Terasawa E, Noel SD, Keen KL. Rapid action of oestrogen in luteinising hormone-releasing hormone neurones: the role of GPR30. *J Neuroendocrinol* 2009;**21**(4):316–21.
106. Brailoiu E, et al. Distribution and characterization of estrogen receptor G protein-coupled receptor 30 in the rat central nervous system. *J Endocrinol* 2007;**193**(2):311–21.
107. Owman C, et al. Cloning of human cDNA encoding a novel heptahelix receptor expressed in Burkitt's lymphoma and widely distributed in brain and peripheral tissues. *Biochem Biophys Res Commun* 1996;**228**(2):285–92.
108. Takada Y, et al. Cloning of cDNAs encoding G protein-coupled receptor expressed in human endothelial cells exposed to fluid shear stress. *Biochem Biophys Res Commun* 1997;**240**(3):737–41.
109. Wang C, Prossnitz ER, Roy SK. Expression of G protein-coupled receptor 30 in the hamster ovary: differential regulation by gonadotropins and steroid hormones. *Endocrinology* 2007;**148**(10):4853–64.
110. Canonaco M, et al. A sexually dimorphic distribution pattern of the novel estrogen receptor G-protein-coupled receptor 30 in some brain areas of the hamster. *J Endocrinol* 2008;**196**(1):131–8.
111. Wang D, et al. G protein-coupled receptor 30 in tumor development. *Endocrine* 2010;**38**(1):29–37.
112. Pietras R, Nemere I, Szego C. Steroid hormone receptors in target cell membranes. *Endocrine* 2001;**14**(3):417–27.
113. Aronica SM, Kraus WL, Katzenellenbogen BS. Estrogen action via the cAMP signaling pathway: stimulation of adenylate cyclase and cAMP-regulated gene transcription. *Proc Natl Acad Sci U S A* 1994;**91**(18):8517–21.
114. Ding Q, et al. Estradiol-mediated ERK phosphorylation and apoptosis in vascular smooth muscle cells requires GPR 30. *Am J Physiol Cell Physiol* 2009;**297**(5): C1178–C1187.
115. Filardo E, et al. Activation of the novel estrogen receptor G protein-coupled receptor 30 (GPR30) at the plasma membrane. *Endocrinology* 2007;**148**(7):3236–45.

116. Weil BR, et al. Signaling via GPR30 protects the myocardium from ischemia/reperfusion injury. *Surgery* 2010;**148**(2):436–43.
117. Bologa CG, et al. Virtual and biomolecular screening converge on a selective agonist for GPR30. *Nat Chem Biol* 2006;**2**(4):207–12.
118. Broughton BRS, Miller AA, Sobey CG. Endothelium-dependent relaxation by G protein-coupled receptor 30 agonists in rat carotid arteries. *Am J Physiol Heart Circ Physiol* 2010;**298**(3):H1055–H1061.
119. Dennis MK, et al. In vivo effects of a GPR30 antagonist. *Nat Chem Biol* 2009;**5**(6):421–7.
120. Chow RWY, Handelsman DJ, Ng MKC. Minireview: rapid actions of sex steroids in the endothelium. *Endocrinology* 2010;**151**(6):2411–22.
121. Lin BC, et al. Stimulating the GPR30 estrogen receptor with a novel tamoxifen analogue activates SF-1 and promotes endometrial cell proliferation. *Cancer Res* 2009;**69**(13):5415–23.
122. Meli R, et al. Estrogen and raloxifene modulate leptin and its receptor in hypothalamus and adipose tissue from ovariectomized rats. *Endocrinology* 2004;**145**(7):3115–21.
123. Baptista T, et al. Tamoxifen prevents sulpiride-induced weight gain in female rats. *Pharmacol Biochem Behav* 1997;**57**(1–2):215–22.
124. Wang C, et al. GPR30 contributes to estrogen-induced thymic atrophy. *Mol Endocrinol* 2008;**22**(3):636–48.
125. Isensee J, et al. Expression pattern of G protein-coupled receptor 30 in LacZ reporter mice. *Endocrinology* 2009;**150**(4):1722–30.
126. Otto C, et al. GPR30 does not mediate estrogenic responses in reproductive organs in mice. *Biol Reprod* 2009;**80**(1):34–41.
127. Ford J, et al. GPR30 deficiency causes increased bone mass, mineralization, and growth plate proliferative activity in male mice. *J Bone Miner Res* 2011;**26**(2):298–307.
128. Windahl SH, et al. The role of the G protein-coupled receptor GPR30 in the effects of estrogen in ovariectomized mice. *Am J Physiol Endocrinol Metab* 2009;**296**(3):E490–E496.
129. Olde B, Leeb-Lundberg LMF. GPR30/GPER1: searching for a role in estrogen physiology. *Trends Endocrinol Metab* 2009;**20**(8):409–16.
130. Langer G, et al. A critical review of fundamental controversies in the field of GPR30 research. *Steroids* 2010;**75**(8–9):603–10.
131. Ohlsson C, Vandenput L. The role of estrogens for male bone health. *Eur J Endocrinol* 2009;**160**(6):883–9.
132. Prossnitz ER, Barton M. The G-protein-coupled estrogen receptor GPER in health and disease. *Nat Rev Endocrinol* 2011;**7**(12):715–26.
133. Pedram A, et al. Functional estrogen receptors in the mitochondria of breast cancer cells. *Mol Biol Cell* 2006;**17**(5):2125–37.
134. Roepke TA. Oestrogen modulates hypothalamic control of energy homeostasis through multiple mechanisms. *J Neuroendocrinol* 2009;**21**(2):141–50.
135. Zhang C, Kelly MJ, Rønnekleiv OK. 17β-Estradiol rapidly increases K(ATP) activity in GnRH via a protein kinase signaling pathway. *Endocrinology* 2010;**151**(9):4477–84.
136. Kenealy BP, et al. STX, a novel nonsteroidal estrogenic compound, induces rapid action in primate GnRH neuronal calcium dynamics and peptide release. *Endocrinology* 2011;**152**(8):3182–91.
137. Kuo J, et al. Membrane estrogen receptors stimulate intracellular calcium release and progesterone synthesis in hypothalamic astrocytes. *J Neurosci* 2010;**30**(39):12950–7.
138. Tobias SC, et al. Synthesis and biological evaluation of SERMs with potent nongenomic estrogenic activity. *ChemMedChem* 2006;**1**(5):565–71.
139. Clipperton Allen AE, et al. Agonistic behavior in males and females: effects of an estrogen receptor beta agonist in gonadectomized and gonadally intact mice. *Psychoneuroendocrinology* 2010;**35**(7):1008–22.

140. Clipperton-Allen AE, et al. Effects of an estrogen receptor alpha agonist on agonistic behaviour in intact and gonadectomized male and female mice. *Psychoneuroendocrinology* 2011;**36**(7):981–95.
141. Chang Y-J, et al. Estrogen modulates sexually dimorphic contextual fear extinction in rats through estrogen receptor β. *Hippocampus* 2009;**19**(11):1142–50.
142. Chen Z, et al. Estrogen receptor α mediates the nongenomic activation of endothelial nitric oxide synthase by estrogen. *J Clin Invest* 1999;**103**(3):401–6.
143. Kang L, et al. Involvement of estrogen receptor variant ER-alpha36, not GPR30, in nongenomic estrogen signaling. *Mol Endocrinol* 2010;**24**(4):709–21.
144. Chambliss KL, et al. Non-nuclear estrogen receptor α signaling promotes cardiovascular protection but not uterine or breast cancer growth in mice. *J Clin Invest* 2010;**120** (7):2319–30.
145. Wang Z, et al. A variant of estrogen receptor-α, hER-α36: transduction of estrogen- and antiestrogen-dependent membrane-initiated mitogenic signaling. *Proc Natl Acad Sci U S A* 2006;**103**(24):9063–8.
146. Pelekanou V, et al. ERα36, a new variant of the ERα is expressed in triple negative breast carcinomas and has a specific transcriptomic signature in breast cancer cell lines. *Steroids* 2012;**77**:928–34.
147. Wallacides A, et al. Estrogens promote proliferation of the seminoma-like TCam-2 cell line through a GPER-dependent ERα36 induction. *Mol Cell Endocrinol* 2012;**350** (1):61–71.
148. Lindsey SH, et al. Estrogen receptor GPR30 reduces oxidative stress and proteinuria in the salt-sensitive female mRen2. Lewis rat. *Hypertension* 2011;**58**(4):665–71.
149. Zhang X, et al. Estrogen receptor-alpha 36 mediates mitogenic antiestrogen signaling in ER-negative breast cancer cells. *PLoS One* 2012;**7**(1):e30174.
150. Albanito L, et al. G protein-coupled receptor 30 (GPR30) mediates gene expression changes and growth response to 17beta-estradiol and selective GPR30 ligand G-1 in ovarian cancer cells. *Cancer Res* 2007;**67**(4):1859–66.
151. Gao F, et al. GPR30 activation opposes estrogen-dependent uterine growth via inhibition of stromal ERK1/2 and estrogen receptor alpha (ERα) phosphorylation signals. *Endocrinology* 2011;**152**(4):1434–47.
152. Lappano R, et al. Estriol acts as a GPR30 antagonist in estrogen receptor-negative breast cancer cells. *Mol Cell Endocrinol* 2010;**320**(1–2):162–70.
153. Razandi M, et al. Proximal events in signaling by plasma membrane estrogen receptors. *J Biol Chem* 2003;**278**(4):2701–12.
154. Vivacqua A, et al. G protein-coupled receptor 30 expression is up-regulated by EGF and TGF alpha in estrogen receptor alpha-positive cancer cells. *Mol Endocrinol* 2009;**23** (11):1815–26.
155. Huang GS, et al. Co-expression of GPR30 and ERβ and their association with disease progression in uterine carcinosarcoma. *Am J Obstet Gynecol* 2010;**203**(3):242.e1–242.e5.
156. Filardo EJ, et al. Distribution of GPR30, a seven membrane-spanning estrogen receptor, in primary breast cancer and its association with clinicopathologic determinants of tumor progression. *Clin Cancer Res* 2006;**12**(21):6359–66.
157. Tica AA, et al. G protein-coupled estrogen receptor 1-mediated effects in the rat myometrium. *Am J Physiol Cell Physiol* 2011;**301**(5):C1262–C1269.
158. Navarro CE, et al. Regulation of cyclic adenosine 3',5'-monophosphate signaling and pulsatile neurosecretion by Gi-coupled plasma membrane estrogen receptors in immortalized gonadotropin-releasing hormone neurons. *Mol Endocrinol* 2003;**17** (9):1792–804.
159. Rønnekleiv OK, Kelly MJ. Diversity of ovarian steroid signaling in the hypothalamus. *Front Neuroendocrinol* 2005;**26**(2):65–84.

160. Bopassa JC, et al. A novel estrogen receptor GPER inhibits mitochondria permeability transition pore opening and protects the heart against ischemia-reperfusion injury. *Am J Physiol Heart Circ Physiol* 2010;**298**(1):H16–H23.
161. Jayachandran M, et al. Loss of estrogen receptor β decreases mitochondrial energetic potential and increases thrombogenicity of platelets in aged female mice. *Age (Dordr)* 2010;**32**(1):109–21.
162. Lappano R, et al. MIBE acts as antagonist ligand of both estrogen receptor alpha and GPER in breast cancer cells. *Breast Cancer Res* 2012;**14**(1):R12.
163. Morton GJ, et al. Central nervous system control of food intake and body weight. *Nature* 2006;**443**(7109):289–95.
164. Pillot B, et al. Role of hypothalamic melanocortin system in adaptation of food intake to food protein increase in mice. *PLoS One* 2011;**6**(4):e19107.
165. Asarian L, Geary N. Modulation of appetite by gonadal steroid hormones. *Philos T Roy Soc B* 2006;**361**(1471):1251–63.
166. Laudenslager ML, et al. Energy balance in ovariectomized rats with and without estrogen replacement. *Am J Physiol* 1980;**238**(5):R400–R405.
167. Ahdieh HB, Wade GN. Effects of hysterectomy on sexual receptivity, food intake, running wheel activity, and hypothalamic estrogen and progestin receptors in rats. *J Comp Physiol Psychol* 1982;**96**(6):886–92.
168. Colvin GB, Sawyer CH. Induction of running activity by intracerebral implants of estrogen in overiectomized rats. *Neuroendocrinology* 1969;**4**(4–5):309–20.
169. Polidori C, Geary N. Estradiol treatment fails to affect the feeding responses to melanocortin-3/4 receptor agonism or antagonism in ovariectomized rats. *Peptides* 2002;**23**(9):1697–700.
170. Rocha M, et al. Physiologic estradiol levels enhance hypothalamic expression of the long form of the leptin receptor in intact rats. *J Nutr Biochem* 2004;**15**(6):328–34.
171. Olofsson LE, Pierce AA, Xu AW. Functional requirement of AgRP and NPY neurons in ovarian cycle-dependent regulation of food intake. *Proc Natl Acad Sci U S A* 2009;**106**(37):15932–7.
172. Xu Y, et al. Distinct hypothalamic neurons mediate estrogenic effects on energy homeostasis and reproduction. *Cell Metab* 2011;**14**(4):453–65.
173. de Souza FSJ, et al. The estrogen receptor α colocalizes with proopiomelanocortin in hypothalamic neurons and binds to a conserved motif present in the neuron-specific enhancer nPE2. *Eur J Pharmacol* 2011;**660**(1):181–7.
174. Miller MM, et al. Effects of age and long-term ovariectomy on the estrogen–receptor containing subpopulations of beta-endorphin-immunoreactive neurons in the arcuate nucleus of female C57BL/6J mice. *Neuroendocrinology* 1995;**61**(5):542–51.
175. Malyala A, et al. PI3K signaling effects in hypothalamic neurons mediated by estrogen. *J Comp Neurol* 2008;**506**(6):895–911.
176. Pappas T, Gametchu B, Watson C. Membrane estrogen receptors identified by multiple antibody labeling and impeded-ligand binding. *FASEB J* 1995;**9**(5):404–10.
177. Razandi M, et al. Cell membrane and nuclear estrogen receptors (ERs) originate from a single transcript: studies of ERα and ERβ expressed in Chinese hamster ovary cells. *Mol Endocrinol* 1999;**13**(2):307–19.
178. Merchenthaler I, et al. Distribution of estrogen receptor α and β in the mouse central nervous system: in vivo autoradiographic and immunocytochemical analyses. *J Comp Neurol* 2004;**473**(2):270–91.
179. Österlund M, et al. Differential distribution and regulation of estrogen receptor-α and -β mRNA within the female rat brain. *Mol Brain Res* 1998;**54**(1):175–80.
180. Simerly RB, et al. Distribution of androgen and estrogen receptor mRNA-containing cells in the rat brain: an in situ hybridization study. *J Comp Neurol* 1990;**294**(1):76–95.

181. Shima N, Yamaguchi Y, Yuri K. Distribution of estrogen receptor β mRNA-containing cells in ovariectomized and estrogen-treated female rat brain. *Anat Sci Int* 2003;**78**(2):85–97.
182. Laflamme N, et al. Expression and neuropeptidergic characterization of estrogen receptors (ERα and ERβ) throughout the rat brain: anatomical evidence of distinct roles of each subtype. *J Neurobiol* 1998;**36**(3):357–78.
183. Mitra SW, et al. Immunolocalization of estrogen receptor β in the mouse brain: comparison with estrogen receptor α. *Endocrinology* 2003;**144**(5):2055–67.
184. Shughrue PJ, et al. Comparative distribution of estrogen receptor-α (ER-α) and β (ER-β) mRNA in the rat pituitary, gonad, and reproductive tract. *Steroids* 1998;**63**(10):498–504.
185. Horvath TL. Synaptic plasticity in energy balance regulation. *Obesity (Silver Spring)* 2006;**14**(S8):228S–233S.
186. Thammacharoen S, et al. Divergent effects of estradiol and the estrogen receptor-[alpha] agonist PPT on eating and activation of PVN CRH neurons in ovariectomized rats and mice. *Brain Res* 2009;**1268**:88–96.
187. Musatov S, et al. Silencing of estrogen receptor α in the ventromedial nucleus of hypothalamus leads to metabolic syndrome. *Proc Natl Acad Sci U S A* 2007;**104**(7):2501–6.
188. Kelly MJ, et al. Estrogen modulation of G-protein-coupled receptor activation of potassium channels in the central nervous system. *Ann N Y Acad Sci* 2003;**1007**(1):6–9.
189. Hazell GGJ, et al. Localisation of GPR30, a novel G protein-coupled oestrogen receptor, suggests multiple functions in rodent brain and peripheral tissues. *J Endocrinol* 2009;**202**(2):223–36.
190. Spary EJ, Maqbool A, Batten TFC. Oestrogen receptors in the central nervous system and evidence for their role in the control of cardiovascular function. *J Chem Neuroanat* 2009;**38**(3):185–96.
191. Sakamoto H, et al. Expression of G protein-coupled receptor-30, a G protein-coupled membrane estrogen receptor, in oxytocin neurons of the rat paraventricular and supraoptic nuclei. *Endocrinology* 2007;**148**(12):5842–50.
192. Milner TA, et al. Ultrastructural evidence that hippocampal alpha estrogen receptors are located at extranuclear sites. *J Comp Neurol* 2001;**429**(3):355–71.
193. Marin R, et al. An oestrogen membrane receptor participates in estradiol actions for the prevention of amyloid-β peptide1−40-induced toxicity in septal-derived cholinergic SN56 cells. *J Neurochem* 2003;**85**(5):1180–9.
194. Marin R, et al. An ICI 182,780-sensitive, membrane-related estrogen receptor contributes to estrogenic neuroprotective actions against amyloid-beta toxicity. *Ann N Y Acad Sci* 2003;**1007**(1):108–16.
195. Xu Y, et al. Neurite-localized estrogen receptor-α mediates rapid signaling by estrogen. *J Neurosci Res* 2003;**74**(1):1–11.
196. Milner TA, et al. Ultrastructural localization of estrogen receptor β immunoreactivity in the rat hippocampal formation. *J Comp Neurol* 2005;**491**(2):81–95.
197. Kelly MJ, Levin ER. Rapid actions of plasma membrane estrogen receptors. *Trends Endocrinol Metab* 2001;**12**(4):152–6.
198. Hukshorn CJ, et al. The effect of pegylated recombinant human leptin (PEG-OB) on weight loss and inflammatory status in obese subjects. *Int J Obes Relat Metab Disord* 2002;**26**(4):504–9.
199. Bueno J, Pfaff DW. Single unit recording in hypothalamus and preoptic area of estrogen-treated and untreated ovariectomized female rats. *Brain Res* 1976;**101**(1):67–78.
200. Yagi K. Changes in firing rates of single preoptic and hypothalamic units following an intravenous administration of estrogen in the castrated female rat. *Brain Res* 1973;**53**(2):343–52.

201. Kelly MJ, Moss RL, Dudley CA. The effect of ovariectomy on the responsiveness of preoptic-septal neurons to microelectrophoresed estrogen. *Neuroendocrinology* 1978;**25**(4):204–11.
202. Kelly MJ, Moss RL, Dudley CA. Differential sensitivity of preoptic-septal neurons to microelectrophoresed estrogen during the estrous cycle. *Brain Res* 1976;**114**(1):152–7.
203. Kelly MJ, Moss RL, Dudley CA. The effects of microelectrophoretically applied estrogen, cortisol and acetylcholine on medial preoptic-septal unit activity throughout the estrous cycle of the female rat. *Exp Brain Res* 1977;**30**(1):53–64.
204. Kelly MJ, et al. The specificity of the response of preoptic-septal area neurons to estrogen: 17α-estradiol versus 17β-estradiol and the response of extrahypothalamic neurons. *Exp Brain Res* 1977;**30**(1):43–52.
205. Balthazart J, Baillien M, Ball GF. Phosphorylation processes mediate rapid changes of brain aromatase activity. *J Steroid Biochem Mol Biol* 2001;**79**(1–5):261–77.
206. Sutter-Dub M-T. Rapid non-genomic and genomic responses to progestogens, estrogens, and glucocorticoids in the endocrine pancreatic B cell, the adipocyte and other cell types. *Steroids* 2002;**67**(2):77–93.
207. Ivanova T, et al. Rapid stimulation of the PI3-kinase/Akt signalling pathway in developing midbrain neurones by oestrogen. *J Neuroendocrinol* 2002;**14**(1):73–9.
208. Singh M. Ovarian hormones elicit phosphorylation of akt and extracellular-signal regulated kinase in explants of the cerebral cortex. *Endocrine* 2001;**14**(3):407–15.
209. Wilson ME, Liu Y, Wise PM. Estradiol enhances Akt activation in cortical explant cultures following neuronal injury. *Mol Brain Res* 2002;**102**(1–2):48–54.
210. Vasudevan N, Pfaff DW. Non-genomic actions of estrogens and their interaction with genomic actions in the brain. *Front Neuroendocrinol* 2008;**29**(2):238–57.
211. Watters JJ, et al. Rapid membrane effects of steroids in neuroblastoma cells: effects of estrogen on mitogen activated protein kinase signalling cascade and c-fos immediate early gene transcription. *Endocrinology* 1997;**138**(9):4030–3.
212. Carrer HF, Cambiasso MJ, Gorosito S. Effects of estrogen on neuronal growth and differentiation. *J Steroid Biochem Mol Biol* 2005;**93**(2–5):319–23.
213. Kuroki Y, et al. Putative membrane-bound estrogen receptors possibly stimulate mitogen-activated protein kinase in the rat hippocampus. *Eur J Pharmacol* 2000;**400**(2–3):205–9.
214. Shingo AS, Kito S. Estradiol induces PKA activation through the putative membrane receptor in the living hippocampal neuron. *J Neural Transm* 2005;**112**(11):1469–73.
215. Beyer C, Raab H. Nongenomic effects of oestrogen: embryonic mouse midbrain neurones respond with a rapid release of calcium from intracellular stores. *Eur J Neurosci* 1998;**10**(1):255–62.
216. Micevych PE, et al. Estradiol stimulates progesterone synthesis in hypothalamic astrocyte cultures. *Endocrinology* 2007;**148**(2):782–9.
217. Stevis PE, et al. Differential effects of estradiol and estradiol-BSA conjugates. *Endocrinology* 1999;**140**(11):5455–8.
218. Taguchi Y, Koslowski M, Bodenner D. Binding of estrogen receptor with estrogen conjugated to bovine serum albumin (BSA). *Nuclear Receptor* 2004;**2**(1):5.
219. Chaban VV, Micevych PE. Estrogen receptor-α mediates estradiol attenuation of ATP-induced Ca2+ signaling in mouse dorsal root ganglion neurons. *J Neurosci Res* 2005;**81**(1):31–7.
220. Ábrahám IM, et al. Critical in vivo roles for classical estrogen receptors in rapid estrogen actions on intracellular signaling in mouse brain. *Endocrinology* 2004;**145**(7):3055–61.
221. Singh M, et al. Estrogen-induced activation of mitogen-activated protein kinase in cerebral cortical explants: convergence of estrogen and neurotrophin signaling pathways. *J Neurosci* 1999;**19**(4):1179–88.

222. Dominguez-Salazar E, Shetty S, Rissman EF. Rapid neural Fos responses to oestradiol in oestrogen receptor alphabeta double knockout mice. *J Neuroendocrinol* 2006;**18**(3):195–202.
223. Schioth HB. G protein-coupled receptors in regulation of body weight. *CNS Neurol Disord Drug Targets* 2006;**5**(3):241–9.
224. Björnström L, Sjöberg M. Mechanisms of estrogen receptor signaling: convergence of genomic and nongenomic actions on target genes. *Mol Endocrinol* 2005;**19**(4):833–42.
225. Levin ER. Integration of the extranuclear and nuclear actions of estrogen. *Mol Endocrinol* 2005;**19**(8):1951–9.
226. Sheldahl LC, et al. Estrogen induces rapid translocation of estrogen receptor β, but not estrogen receptor α, to the neuronal plasma membrane. *Neuroscience* 2008;**153**(3):751–61.
227. Vasudevan N, Kow LM, Pfaff D. Integration of steroid hormone initiated membrane action to genomic function in the brain. *Steroids* 2005;**70**(5–7):388–96.
228. Qiu J, et al. Serotonin 5-hydroxytryptamine2C receptor signaling in hypothalamic proopiomelanocortin neurons: role in energy homeostasis in females. *Mol Pharmacol* 2007;**72**(4):885–96.
229. Kelly MJ, et al. Rapid effects of estrogen on G protein-coupled receptor activation of potassium channels in the central nervous system (CNS). *J Steroid Biochem Mol Biol* 2002;**83**(1–5):187–93.
230. Kow LM, et al. Acute estradiol application increases inward and decreases outward whole-cell currents of neurons in rat hypothalamic ventromedial nucleus. *Brain Res* 2006;**1116**(1):1–11.
231. Cooke PS, Naaz A. Role of estrogens in adipocyte development and function. *Exp Biol Med* 2004;**229**(11):1127–35.
232. Group, E.H.B.C.C. Body Mass Index, Serum Sex Hormones, and Breast Cancer Risk in Postmenopausal Women. *J Natl Cancer Inst* 2003;**95**(16):1218–26.
233. Iguchi T, et al. Developmental effects: oestrogen-induced vaginal changes and organotin-induced adipogenesis. *Int J Androl* 2008;**31**(2):263–8.
234. Macotela Y, et al. Sex and depot differences in adipocyte insulin sensitivity and glucose metabolism. *Diabetes* 2009;**58**(4):803–12.
235. Dyck DJ, Heigenhauser GJF, Bruce CR. The role of adipokines as regulators of skeletal muscle fatty acid metabolism and insulin sensitivity. *Acta Physiol (Oxf)* 2006;**186**(1):5–16.
236. Tanaka M, et al. Effects of estrogen on serum leptin levels and leptin mRNA expression in adipose tissue in rats. *Horm Res* 2001;**56**(3–4):98–104.
237. Dieudonné M-N, et al. Sex steroids and leptin regulate 11β-hydroxysteroid dehydrogenase I and P450 aromatase expressions in human preadipocytes: sex specificities. *J Steroid Biochem Mol Biol* 2006;**99**(4–5):189–96.
238. Thornton MJ, et al. The modulation of aromatase and estrogen receptor alpha in cultured human dermal papilla cells by dexamethasone: a novel mechanism for selective action of estrogen via estrogen receptor beta? *J Invest Dermatol* 2006;**126**(9):2010–8.
239. Bryzgalova G, et al. Evidence that oestrogen receptor-α plays an important role in the regulation of glucose homeostasis in mice: insulin sensitivity in the liver. *Diabetologia* 2006;**49**(3):588–97.
240. Crandall DL, et al. Identification of estrogen receptor β RNA in human breast and abdominal subcutaneous adipose tissue. *Biochem Biophys Res Commun* 1998;**248**(3):523–6.
241. Lu SF, et al. Androgen receptor in mouse brain sex: differences and similarities in autoregulation. *Endocrinology* 1998;**139**(4):1594–601.
242. Yu I-C, et al. Hyperleptinemia without obesity in male mice lacking androgen receptor in adipose tissue. *Endocrinology* 2008;**149**(5):2361–8.

243. Bryzgalova G, et al. Mechanisms of antidiabetogenic and body weight-lowering effects of estrogen in high-fat diet-fed mice. *Am J Physiol Endocrinol Metab* 2008;**295**(4):E904–E912.
244. Iverius PH, Brunzell JD. Relationship between lipoprotein lipase activity and plasma sex steroid level in obese women. *J Clin Invest* 1988;**82**(3):1106–12.
245. Homma H, et al. Estrogen suppresses transcription of lipoprotein lipase gene. Existence of a unique estrogen response element on the lipoprotein lipase promoter. *J Biol Chem* 2000;**275**(15):11404–11.
246. Wohlers LM, Spangenburg EE. 17β-estradiol supplementation attenuates ovariectomy-induced increases in ATGL signaling and reduced perilipin expression in visceral adipose tissue. *J Cell Biochem* 2010;**110**(2):420–7.
247. Jones MEE, et al. Aromatase-deficient (ArKO) mice have a phenotype of increased adiposity. *Proc Natl Acad Sci U S A* 2000;**97**(23):12735–40.
248. Couse JF, Korach KS. Estrogen receptor null mice: what have we learned and where will they lead us? *Endocr Rev* 1999;**20**(3):358–417.
249. Lindberg M, et al. Estrogen receptor specificity for the effects of estrogen in ovariectomized mice. *J Endocrinol* 2002;**174**(2):167–78.
250. Misso ML, et al. Cellular and molecular characterization of the adipose phenotype of the aromatase-deficient mouse. *Endocrinology* 2003;**144**(4):1474–80.
251. Jones MEE, et al. Of mice and men: the evolving phenotype of aromatase deficiency. *Trends Endocrinol Metab* 2006;**17**(2):55–64.
252. Foryst-Ludwig A, et al. Metabolic actions of estrogen receptor beta (ERβ) are mediated by a negative cross-talk with PPARγ. *PLoS Genet* 2008;**4**(6):e1000108.
253. Luconi M, Forti G, Baldi E. Genomic and nongenomic effects of estrogens: molecular mechanisms of action and clinical implications for male reproduction. *J Steroid Biochem Mol Biol* 2002;**80**(4–5):369–81.
254. Pedram A, et al. Integration of the non-genomic and genomic actions of estrogen Membrane-initiated signaling by steroid to transcription and cell biology. *J Biol Chem* 2002;**277**(52):50768–75.
255. Mayes JS, Watson GH. Direct effects of sex steroid hormones on adipose tissues and obesity. *Obes Rev* 2004;**5**(4):197–216.
256. Pedersen SB, et al. Demonstration of estrogen receptor subtypes α and β in human adipose tissue: influences of adipose cell differentiation and fat depot localization. *Mol Cell Endocrinol* 2001;**182**(1):27–37.
257. Anwar A, et al. Site-specific regulation of oestrogen receptor-α and -β by oestradiol in human adipose tissue. *Diabetes Obes Metab* 2001;**3**(5):338–49.
258. Price TM, O'Brien SN. Determination of estrogen receptor messenger ribonucleic acid (mRNA) and cytochrome P450 aromatase mRNA levels in adipocytes and adipose stromal cells by competitive polymerase chain reaction amplification. *J Clin Endocrinol Metab* 1993;**77**(4):1041–5.
259. Dieudonné MN, et al. Evidence for functional estrogen receptors α and β in human adipose cells: regional specificities and regulation by estrogens. *Am J Physiol Cell Physiol* 2004;**286**(3):C655–C661.
260. Mizutani T, et al. Identification of estrogen receptor in human adipose tissue and adipocytes. *J Clin Endocrinol Metab* 1994;**78**(4):950–4.
261. Rebuffé-Scrive M, et al. Steroid hormone receptors in human adipose tissues. *J Clin Endocrinol Metab* 1990;**71**(5):1215–9.
262. Brönnegård M, et al. Lack of evidence for estrogen and progesterone receptors in human adipose tissue. *J Steroid Biochem Mol Biol* 1994;**51**(5–6):275–81.
263. Pedersen SB, et al. Identification of oestrogen receptors and oestrogen receptor mRNA in human adipose tissue. *Eur J Clin Invest* 1996;**26**(4):262–9.
264. Joyner J, Hutley L, Cameron D. Estrogen receptors in human preadipocytes. *Endocrine* 2001;**15**(2):225–30.

265. Lubahn DB, et al. Alteration of reproductive function but not prenatal sexual development after insertional disruption of the mouse estrogen receptor gene. *Proc Natl Acad Sci U S A* 1993;**90**(23):11162–6.
266. Wade GN, Gray JM. Cytoplasmic 17beta-[3H] estradiol binding in rat adipose tissues. *Endocrinology* 1978;**103**(5):1695–701.
267. Watson GH, et al. Biochemical and immunological characterization of oestrogen receptor in the cytosolic fraction of gluteal, omental and perirenal adipose tissues from sheep. *J Endocrinol* 1993;**139**(1):107–15.
268. Mayes JS, et al. Regional differences and up-regulation of progesterone receptors in adipose tissues from oestrogen-treated sheep. *J Endocrinol* 1996;**148**(1):19–25.
269. McCann J, et al. Subcellular distribution and glycosylation pattern of androgen receptor from sheep omental adipose tissue. *J Endocrinol* 2001;**169**(3):587–93.
270. Rodriguez-Cuenca S, et al. Depot differences in steroid receptor expression in adipose tissue: possible role of the local steroid milieu. *Am J Physiol Endocrinol Metab* 2005;**288**(1):E200–E207.
271. Barros RPA, et al. Participation of ERα and ERβ in glucose homeostasis in skeletal muscle and white adipose tissue. *Am J Physiol Endocrinol Metab* 2009;**297**(1):E124–E133.
272. Rodriguez-Cuenca S, et al. Sex steroid receptor expression profile in brown adipose tissue Effects of hormonal status. *Cell Physiol Biochem* 2007;**20**(6):877–86.
273. Gray JM, Wade GN. Cytoplasmic estrogen, but not progestin, binding sites in male rat adipose tissues. *Am J Physiol* 1980;**239**(4):E237.
274. Gray JM, Dudley SD, Wade GN. In vivo cell nuclear binding of 17 beta-[3H]estradiol in rat adipose tissues. *Am J Physiol* 1981;**240**(1):E43–E46.
275. Rebuffe-Scrive M. Sex steroid hormones and adipose tissue metabolism in ovariectomized and adrenalectomized rats. *Acta Physiol Scand* 1987;**129**(4):471–7.
276. Pedersen SB, et al. Nuclear estradiol binding in rat adipocytes Regional variations and regulatory influences of hormones. *Biochim Biophys Acta* 1991;**1093**(1):80–6.
277. Echeverría OM, et al. Immuno-electron microscopic localization of estradiol receptor in cells of male and female reproductive and non-reproductive organs. *Biol Cell* 1994;**81**(3):257–65.
278. Cooke PS, et al. The role of estrogen and estrogen receptor-α in male adipose tissue. *Mol Cell Endocrinol* 2001;**178**(1–2):147–54.
279. Casazza K, Page GP, Fernandez JR. The association between the rs2234693 and rs9340799 estrogen receptor alpha gene polymorphisms and risk factors for cardiovascular disease: a review. *Biol Res Nurs* 2010;**12**(1):84–97.
280. Riant E, et al. Estrogens protect against high-fat diet-induced insulin resistance and glucose intolerance in mice. *Endocrinology* 2009;**150**(5):2109–17.
281. Naaz A, et al. Effect of ovariectomy on adipose tissue of mice in the absence of estrogen receptor alpha (ERα): a potential role for estrogen receptor beta (ERβ). *Horm Metab Res* 2002;**34**(11/12):758,763.
282. Palin SL, et al. 17Beta-estradiol and anti-estrogen ICI:Compound 182,780 regulate expression of lipoprotein lipase and hormone-sensitive lipase in isolated subcutaneous abdominal adipocytes. *Metabolism* 2003;**52**(4):383–8.
283. Pedersen SB, et al. Effects of in vivo estrogen treatment on adipose tissue metabolism and nuclear estrogen receptor binding in isolated rat adipocytes. *Mol Cell Endocrinol* 1992;**85**(1–2):13–9.
284. Hamosh M, Hamosh P. The effect of estrogen on the lipoprotein lipase activity of rat adipose tissue. *J Clin Invest* 1975;**55**(5):1132–5.
285. Kim HJ, Kalkhoff RK. Sex steroid influence on triglyceride metabolism. *J Clin Invest* 1975;**56**(4):888–96.
286. Wilson DE, et al. Estrogen treatment and gonadal function in the regulation of lipoprotein lipase. *Atherosclerosis* 1976;**24**(3):491–9.

287. Steingrimsdottir L, Brasel J, Greenwood MR. Hormonal modulation of adipose tissue lipoprotein lipase may alter food intake in rats. *Am J Physiol* 1980;**239**(2): E162–E167.
288. Ramirez I. Estradiol-induced changes in lipoprotein lipase, eating, and body weight in rats. *Am J Physiol* 1981;**240**(5):E533–E538.
289. Ellis GS, et al. Effects of estradiol on lipoprotein lipase activity and lipid availability in exercised male rats. *J Appl Physiol* 1994;**77**(1):209–15.
290. Hazzard WR, et al. Studies on the mechanism of increased plasma triglyceride levels induced by oral contraceptives. *N Engl J Med* 1969;**280**(9):471–4.
291. Fabian E, et al. Lipoprotein lipase and postheparin esterase activity and postheparin increase in free fatty acids after the short-term administration of a sequential contraceptive. *Am J Obstet Gynecol* 1971;**109**(8):1212–3.
292. Ence TJ, et al. Heparin metabolism and heparin-released lipase activity during long-term estrogen-progestin treatment. *Metabolism* 1976;**25**(2):139–45.
293. Price TM, et al. Estrogen regulation of adipose tissue lipoprotein lipase—possible mechanism of body fat distribution. *Am J Obstet Gynecol* 1998;**178** (1,part 1):101–7.
294. Urabe M, et al. Effect of Estrogen Replacement Therapy on Hepatic Triglyceride Lipase, Lipoprotein Liase and lipids Including Apolipoprotein E in Climacteric and Elderly Women. *Endocr J* 1996;**43**(6):737–42.
295. Benz, V., et al., Sexual dimorphic regulation of body weight dynamics and adipose tissue lipolysis. *PLoS ONE* 2012;**7**(5):p. e37794.
296. Pedersen SB, et al. Estrogen controls lipolysis by up-regulating α2A-adrenergic receptors directly in human adipose tissue through the estrogen receptor α Implications for the female fat distribution. *J Clin Endocrinol Metab* 2004;**89**(4):1869–78.
297. Simoncini T, Genazzani A. Non-genomic actions of sex steroid hormones. *Eur J Endocrinol* 2003;**148**(3):281–92.
298. Dos Santos EG, et al. Rapid nongenomic E2 effects on p42/p44 MAPK, activator protein-1, and cAMP response element binding protein in rat white adipocytes. *Endocrinology* 2002;**143**(3):930–40.
299. Falkenstein E, et al. Multiple actions of steroid hormones—a focus on rapid, nongenomic effects. *Pharmacol Rev* 2000;**52**(4):513–56.
300. Schmidt BMW, et al. Rapid, nongenomic steroid actions: a new age? *Front Neuroendocrinol* 2000;**21**(1):57–94.
301. Segars JH, Driggers PH. Estrogen action and cytoplasmic signaling cascades Part I: membrane-associated signaling complexes. *Trends Endocrinol Metab* 2002;**13** (8):349–54.
302. Driggers PH, Segars JH. Estrogen action and cytoplasmic signaling pathways Part II: the role of growth factors and phosphorylation in estrogen signaling. *Trends Endocrinol Metab* 2002;**13**(10):422–7.
303. Schlegel A, et al. Caveolin-1 Potentiates Estrogen Receptor α (ERα) Signaling. *J Biol Chem* 1999;**274**(47):33551–6.
304. Schlegel A, Arvan P, Lisanti MP. Caveolin-1 binding to endoplasmic reticulum membranes and entry into the regulated secretory pathway are regulated by serine phosphorylation. Protein sorting at the level of the endoplasmic reticulum. *J Biol Chem* 2001; **276**(6):4398–408.
305. Mitznegg P, Estler CJ, Schubert E. Effect of estradiol on lipolysis and adenosine 3',5'-monophosphate accumulation in isolated rat adipocytes. *Biochem Pharmacol* 1974;**23** (16):2337–9.
306. Pasquier Y-N, Pecquery R, Giudicelli Y. Increased adenylate cyclase catalytic activity explains how estrogens "in vivo" promote lipolytic activity in rat white fat cells. *Biochem Biophys Res Commun* 1988;**154**(3):1151–9.

307. Lacasa D, et al. Influence of ovariectomy and regional fat distribution on the membranous transducing system controlling lipolysis in rat fat cells. *Endocrinology* 1991;**128**(2):747–53.
308. Lindberg U-B, et al. Regional adipose tissue metabolism in postmenopausal women after treatment with exogenous sex steroids. *Horm Metab Res* 1990;**22**(06):345,351.
309. Lacasa D, et al. Influence of ovarian status and regional fat distribution on protein kinase C in rat fat cells. *J Endocrinol* 1994;**140**(2):275–82.
310. Lacasa D, et al. Control of rat preadipocyte adipose conversion by ovarian status: regional specificity and possible involvement of the mitogen-activated protein kinase-dependent and c-fos signaling pathways. *Endocrinology* 1997;**138**(7):2729–34.
311. Friedman JM, Halaas JL. Leptin and the regulation of body weight in mammals. *Nature* 1998;**395**(6704):763–70.
312. Chen S-J, et al. Estrogen reduces myointimal proliferation after balloon injury of rat carotid artery. *Circulation* 1996;**93**(3):577–84.
313. Diano S, et al. Leptin receptors in estrogen receptor-containing neurons of the female rat hypothalamus. *Brain Res* 1998;**812**(1–2):256–9.
314. Ahima RS, et al. Regulation of neuronal and glial proteins by leptin: implications for brain development. *Endocrinology* 1999;**140**(6):2755–62.
315. Elias CF, et al. Leptin differentially regulates NPY and POMC neurons projecting to the lateral hypothalamic area. *Neuron* 1999;**23**(4):775–86.
316. Elmquist JK, Elias CF, Saper CB. From lesions to leptin: hypothalamic control of food intake and body weight. *Neuron* 1999;**22**(2):221–32.
317. Chakraborty S, et al. Stereological analysis of estrogen receptor expression in the hypothalamic arcuate nucleus of ob/ob and agouti mice. *Brain Res* 2008;**1217**:86–95.
318. Ainslie DA, et al. Estrogen deficiency causes central leptin insensitivity and increased hypothalamic neuropeptide Y. *Int J Obes Relat Metab Disord* 2001;**25**(11):1680–8.
319. Lindell K, et al. Leptin receptor 5′untranslated regions in the rat: relative abundance, genomic organization and relation to putative response elements. *Mol Cell Endocrinol* 2001;**172**(1–2):37–45.
320. Shimizu H, et al. Estrogen increases in vivo leptin production in rats and human subjects. *J Endocrinol* 1997;**154**(2):285–92.
321. Machinal F, et al. In vivo and in vitro ob gene expression and leptin secretion in rat adipocytes: evidence for a regional specific regulation by sex steroid hormones. *Endocrinology* 1999;**140**(4):1567–74.
322. Hardie L, et al. Circulating leptin in women: a longitudinal study in the menstrual cycle and during pregnancy. *Clin Endocrinol* 1998;**53**(2):87–9.
323. Rosenbaum M, et al. Effects of gender, body composition, and menopause on plasma concentrations of leptin. *J Clin Endocrinol Metab* 1996;**81**(9):3424–7.
324. Konukoglu D, Serin Ö, Ercan M. Plasma leptin levels in obese and non-obese postmenopausal women before and after hormone replacement therapy. *Maturitas* 2000;**36**(3):203–7.
325. Elbers JMH, et al. Effects of administration of 17β-oestradiol on serum leptin levels in healthy postmenopausal women. *Clin Endocrinol* 1999;**51**(4):449–54.
326. Thomas T, et al. Relationship of serum leptin levels with body composition and sex steroid and insulin levels in men and women. *Metabolism* 2000;**49**(10):1278–84.
327. Elbers JMH, et al. Effects of sex steroid hormones on regional fat depots as assessed by magnetic resonance imaging in transsexuals. *Am J Physiol Endocrinol Metab* 1999;**276**(2):E317–E325.
328. Casabiell X, et al. Gender differences in both spontaneous and stimulated leptin secretion by human omental adipose tissue in vitro: dexamethasone and estradiol stimulate leptin release in women, but not in men. *J Clin Endocrinol Metab* 1998;**83**(6):2149–55.

329. Santner SJ, et al. Aromatase activity and expression in breast cancer and benign breast tissue stromal cells. *J Clin Endocrinol Metab* 1997;**82**(1):200–8.
330. Geisler J, et al. Total body aromatization in postmenopausal breast cancer patients is strongly correlated to plasma leptin levels. *J Steroid Biochem Mol Biol* 2007;**104**(1–2):27–34.
331. Shin J-H, et al. The ratio of estrogen receptor α to estrogen receptor β in adipose tissue is associated with leptin production and obesity. *Steroids* 2007;**72**(6–7):592–9.
332. Catalano S, et al. Leptin induces, via ERK1/ERK2 signal, functional activation of estrogen receptor α in MCF-7 cells. *J Biol Chem* 2004;**279**(19):19908–15.
333. Vecchione C, et al. Leptin effect on endothelial nitric oxide is mediated through Akt–endothelial nitric oxide synthase phosphorylation pathway. *Diabetes* 2002;**51**(1):168–73.
334. Pinilla L, et al. Regulation of serum leptin levels by gonadal function in rats. *Eur J Endocrinol* 1999;**140**(5):468–73.
335. Watanobe H, Suda T. A detailed study on the role of sex steroid milieu in determining plasma leptin concentrations in adult male and female rats. *Biochem Biophys Res Commun* 1999;**259**(1):56–9.
336. Wu-Peng S, et al. Effects of exogenous gonadal steroids on leptin homeostasis in rats. *Obes Res* 1999;**7**(6):586–92.
337. Brann DW, et al. Regulation of leptin gene expression and secretion by steroid hormones. *Steroids* 1999;**64**(9):659–63.
338. Luukkaa V, et al. Effects of estrous cycle and steroid replacement on the expression of leptin and uncoupling proteins in adipose tissue in the rat. *Gynecol Endocrinol* 2001;**15**(2):103–12.
339. Yoneda N, et al. The influence of ovariectomy on ob gene expression in rats. *Horm Metab Res* 1998;**30**(05):263–5.
340. Chu S-C, et al. Fluctuation of serum leptin level in rats after ovariectomy and the influence of estrogen supplement. *Life Sci* 1999;**64**(24):2299–306.
341. Nedvídková J, Haluzík M, Schreiber V. The decrease of serum leptin levels in oestrogen-treated male mice. *Physiol Res* 1997;**46**(4):291–4.
342. Pelleymounter MA, Baker MB, McCaleb M. Does estradiol mediate leptin's effects on adiposity and body weight? *Am J Physiol Endocrinol Metab* 1999;**276**(5):E955–E963.
343. Kronfeld-Schor N, et al. Steroid-dependent up-regulation of adipose leptin secretion in vitro during pregnancy in mice. *Biol Reprod* 2000;**63**(1):274–80.
344. Halpern A, Mancini MC. Treatment of obesity: an update on anti-obesity medications. *Obes Rev* 2003;**4**(1):25–42.
345. Swinburn B, Egger G. Preventive strategies against weight gain and obesity. *Obes Rev* 2002;**3**(4):289–301.
346. Tchernof A, et al. Menopause, central body fatness, and insulin resistance: effects of hormone-replacement therapy. *Coron Artery Dis* 1998;**9**(8):503–12.
347. Gambacciani M, et al. Body weight, body fat distribution, and hormonal replacement therapy in early postmenopausal women. *J Clin Endocrinol Metab* 1997;**82**(2):414–7.
348. Haarbo J, et al. Postmenopausal hormone replacement therapy prevents central distribution of body fat after menopause. *Metabolism* 1991;**40**(12):1323–6.
349. Espeland MA, et al. Effect of postmenopausal hormone therapy on body weight and waist and hip girths. Postmenopausal Estrogen-Progestin Interventions Study Investigators. *J Clin Endocrinol Metab* 1997;**82**(5):1549–56.
350. Samaras K, et al. Tobacco smoking and oestrogen replacement are associated with lower total and central fat in monozygotic twins. *Int J Obes* 1998;**22**(2):149–56.
351. Hänggi W, et al. Differential impact of conventional oral or transdermal hormone replacement therapy or tibolone on body composition in postmenopausal women. *Clin Endocrinol* 1998;**48**(6):691–9.

352. Gambacciani M, et al. Prospective evaluation of body weight and body fat distribution in early postmenopausal women with and without hormonal replacement therapy. *Maturitas* 2001;**39**(2):125–32.
353. Sites CK, et al. Relationship between hormone replacement therapy use with body fat distribution and insulin sensitivity in obese postmenopausal women. *Metabolism* 2001;**50** (7):835–40.
354. Chmouliovsky L, et al. Beneficial effect of hormone replacement therapy on weight loss in obese menopausal women. *Maturitas* 1999;**32**(3):147–53.
355. Kritz-Silverstein D, Barrett-Connor E. Long-term postmenopausal hormone use, obesity, and fat distribution in older women. *JAMA* 1996;**275**(1):46–9.
356. Kanaley JA, et al. Abdominal fat distribution in pre- and postmenopausal women: the impact of physical activity, age, and menopausal status. *Metabolism* 2001;**50**(8):976–82.
357. Nelson HD, et al. Postmenopausal hormone replacement therapy: scientific review. *JAMA* 2002;**288**(7):872–81.
358. Smith CL, O'Malley BW. Coregulator function: a key to understanding tissue specificity of selective receptor modulators. *Endocr Rev* 2004;**25**(1):45–71.
359. Kousteni S, et al. Reversal of bone loss in mice by nongenotropic signaling of sex steroids. *Science* 2002;**298**(5594):843–6.
360. Ke HZ, et al. Effects of CP-336,156, a new, nonsteroidal estrogen agonist/antagonist, on bone, serum cholesterol, uterus and body composition in rat models. *Endocrinology* 1998;**139**(4):2068–76.
361. McCarty MF. Modulation of adipocyte lipoprotein lipase expression as a strategy for preventing or treating visceral obesity. *Medical Hypotheses* 2001;**57**(2):192–200.
362. Mantzoros CS, Flier JS. Leptin as a therapeutic agent—trials and tribulations. *J Clin Endocrinol Metab* 2000;**85**(11):4000–2.
363. Lee DW, et al. Leptin and the treatment of obesity: its current status. *Eur J Pharmacol* 2002;**440**(2–3):129–39.
364. Farooqi IS, et al. Effects of recombinant leptin therapy in a child with congenital leptin deficiency. *N Engl J Med* 1999;**341**(12):879–84.

CHAPTER SEVEN

Free Fatty Acid Receptor GPR120 and Pathogenesis of Obesity and Type 2 Diabetes Mellitus

Xiu-Lei Mo[*,1], **Hong-Kui Wei**[*,†,1], **Jian Peng**[†], **Ya-Xiong Tao**[*]
[*]Department of Anatomy, Physiology and Pharmacology, College of Veterinary Medicine, Auburn University, Auburn, Alabama, USA
[†]Department of Animal Nutrition and Feed Science, College of Animal Science and Technology, Huazhong Agricultural University, Wuhan, People's Republic of China
[1]Both authors contributed equally

Contents

Abstract

G protein-coupled receptor 120 (GPR120) was initially identified as an orphan receptor through mining the human genome databases. In 2005, GPR120 was deorphanized and shown to be a receptor for long-chain free fatty acids. GPR120 regulates various physiological processes, including gut hormone secretion, islet function, food preference, osteoclastogenesis, anti-inflammation, adipogenesis, and appetite control. Recently, a

Progress in Molecular Biology and Translational Science, Volume 114
ISSN 1877-1173
http://dx.doi.org/10.1016/B978-0-12-386933-3.00007-8

human genetic study conducted in European populations identified a loss-of-function *GPR120* mutation associated with obesity and insulin resistance. Therefore, GPR120, the sensing receptor for long-chain free fatty acids, represents a novel drug target for the treatment of obesity and diabetes.

1. INTRODUCTION

Obesity has reached epidemic proportions worldwide with monumental health, societal, and economic problems. In the United States, more than 35% of the adult population is considered as obese (defined by a body mass index of 30 or more).[1] Obesity is closely associated with type II diabetes mellitus, dyslipidemia, hypertension, cardiovascular disease, and certain types of cancers. The elevated level of plasma free fatty acids (FFAs) in most obese individuals has been demonstrated to be a physiological link between obesity and associated comorbidities.[2]

Fatty acids have conventionally been considered as an essential source of energy and important metabolic substrate.[3] According to the length of aliphatic tails, FFAs are categorized into short-chain fatty acids with less than six carbons, medium-chain fatty acids with 6–12 carbons, and long-chain fatty acids with 12 or more carbons. Recently, multiple lines of investigations demonstrated that FFAs also act as signaling molecules involved in various physiological mechanisms.[4] Nuclear receptors, including peroxisome proliferator-activated receptors (PPARs), were shown to act as FFA sensors and to regulate expression of genes that maintain homeostasis in various physiological conditions.[5] However, some FFA-mediated physiological effects were shown to be PPAR independent and might be mediated by other mechanisms[6,7] such as signaling through G protein-coupled receptors (GPCRs).

In the last decade, a series of rhodopsin-like GPCRs, including GPR40, GPR41, GPR43, GPR84, and GPR120, have been identified as FFA receptors (FFARs) through the GPCR deorphanization strategy.[8] GPR120 and GPR40 are receptors for long-chain FFAs. GPR84 is activated by medium-chain FFAs, while GPR43 and GPR41 are activated by short-chain FFAs.[9–11] GPCRs represent the most successful drug target that ~30% of prescription drugs target GPCRs with proven efficacy in the treatment of a broad range of human diseases.[12] Therefore, FFARs remain to be explored and have attracted considerable attention as potential drug targets.

This chapter summarizes the studies on one of the FFARs, GPR120, from its cloning and tissue distribution to its physiological roles in regulating gut hormone secretion, islet function and glucose homeostasis, food preference, osteoclastogenesis, anti-inflammation, adipogenesis, and appetite and energy homeostasis. We will then review the studies on the pharmacology of the receptor, including ligand binding, receptor activation, and signaling pathways. Recent genetic study of *GPR120* variants and obesity will be highlighted.

2. MOLECULAR CLONING AND LOCALIZATION OF GPR120

2.1. Molecular cloning

With the goal of disclosing the identities of new orphan GPCRs, Fredriksson *et al.* revealed seven new GPCR sequences through mining the human NCBI and Celera genome databases. One of these GPCRs was named GPR120, with gene located on chromosome 10q23.33.[8] The human (h) GPR120 was first isolated from genomic DNA fragment by Hirasawa and colleagues.[13] The *GPR120* gene contains four exons separated by three introns, and has two splice variants. The variant without exon 3 (ENST00000371481) produces a short isoform of GPR120 (GPR120-S) containing 361 residues (ENSP00000360536). The long isoform of GPR120 (GPR120-L) contains 377 residues (ENSP00000360538), and the additional 16 residues of GPR120-L are located in the third intracellular loop (ICL) from Thr233 to Glu248.

Since the cloning of the hGPR120-L, the GPR120-L has been predicted to be specific to other primates including chimpanzee and rhesus monkey, as shown in Fig. 7.1A, using the NCBI gene prediction method GNOMON. Comparison of the putative sequences of chimpanzee and rhesus monkey GPR120-L with that of the hGPR120-L shows that hGPR120-L is 97.7 and 96.6% identical at amino acid level to chimpanzee and rhesus monkey, respectively (Fig. 7.1B). Phylogenetic analysis indicates that hGPR120-L clusters together with chimpanzee GPR120-L, but is a little more divergent from rhesus monkey GPR120-L at amino acid level (Fig. 7.1C).

GPR120-S has been cloned from human, cynomolgus monkey,[14] mouse,[13] and rat[15]. As shown in Fig. 7.2A, GPR120-S has also been predicted to exist in other mammalian species including cow, pig, dog, rabbit, and horse, as well as in other nonmammalian species including chicken, turkey, zebra finch, green anoles, frog, and tilapia, using the NCBI gene prediction method GNOMON. The homologies of GPR120-S across those species are summarized in Fig. 7.2B. The comparison shows that human GPR120-S is 97.0, 88.9, and 88.4% identical at amino acid level to that of cynomolgus monkey,

A

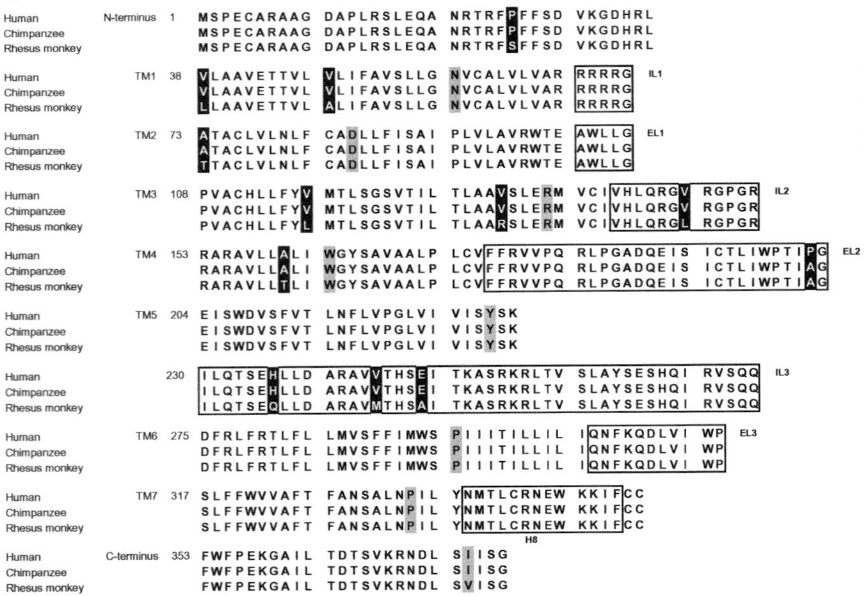

B

	Human-L	Chimpanzee	Rhesus monkey
Human-L	100.0	99.7	96.6
Chimpanzee		100.0	96.8
Rhesus monkey			100.0

C

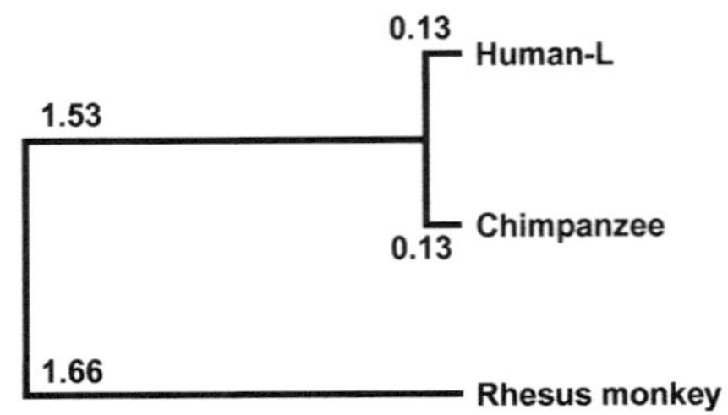

Figure 7.1 Comparison of GPR120-L amino acid sequences between hGPR120-L and GPR120-L predicted from other species. (A) Alignment of the amino acid sequences between hGPR120-L and GPR120-L predicted from other species. The ECLs and ICLs are *boxed*. The most conserved residues in each TM are shaded in *gray*. The divergent amino acids are in *white* and shaded in *black*. (B) Homology and (C) phylogenetic analysis of GPR120-L amino acid sequences from several species. NCBI reference numbers for the proteins are: human, NP_859529.2; chimpanzee, XP_001148255.1; rhesus monkey, XP_002808552.1.

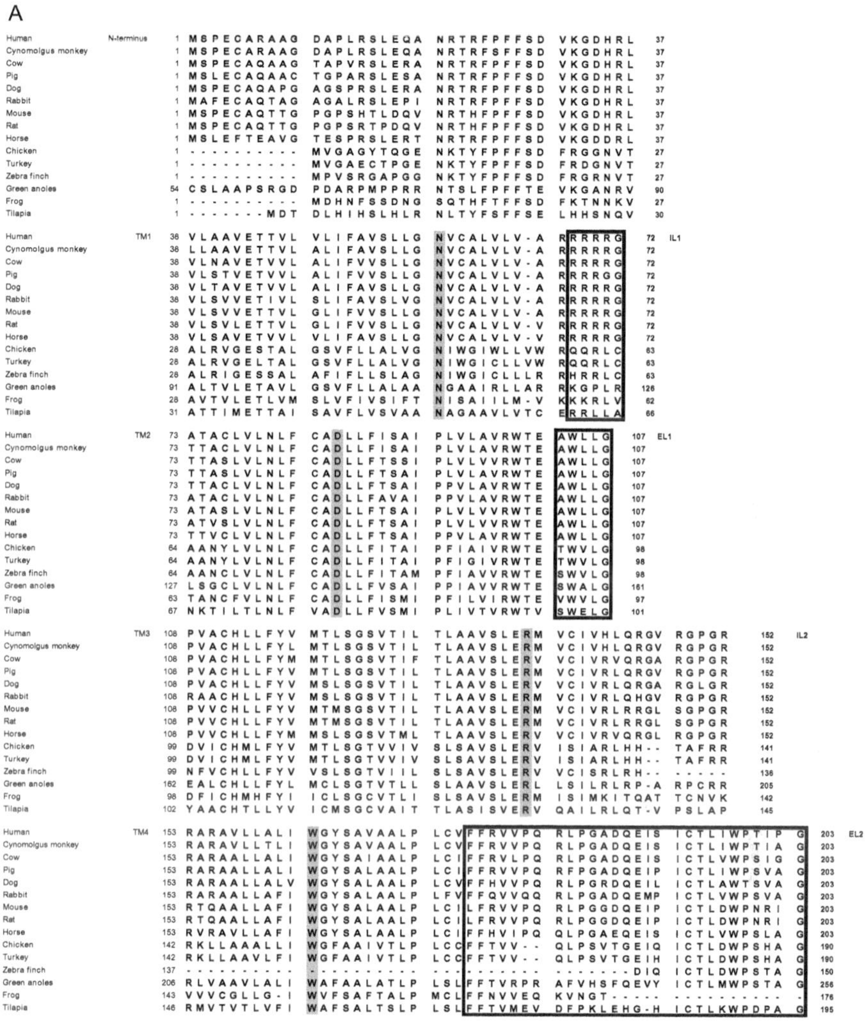

Continued

```
Human              TM5   204 E I SWDVSFVT  LNFLVPGLVI  VISYSK      229
Cynomolgus monkey        204 E I SWDVSFVT  LNFLVPGLVI  VISYSK      229
Cow                      204 E I TWDVSFAI  FNFLVPGLLI  VISYSK      229
Pig                      204 E I SWDVSFVT  LNFLVPGLLI  VISYSK      229
Dog                      204 E I SWDVSFVT  LNFLVPGLLI  VISYSK      229
Rabbit                   204 E I SWDVSFVT  LNFLVPGLVI  VISYSK      229
Mouse                    204 E I SWDVFFVT  LNFLVPGLVI  VISYSK      229
Rat                      204 E I SWDVFFVT  LNFLVPGLVI  VISYSK      229
Horse                    204 E I SWDVLFVT  LNFLVPGLLI  VISYSK      229
Chicken                  191 E I VWDVTYAV  AVFLIPGLIT  VISYSK      216
Turkey                   191 E I VWDVTFAI  VVFLIPGLII  VISYSK      216
Zebra finch              151 ETVWDTTYAI    VFFLIPGFII  VISYSK      176
Green anoles             257 E I SWDVCFAF  VVFLVPGLVI  VISYSK      282
Frog                     176 ----------    ----------  ------      176
Tilapia                  196 E I VWNVVFTA  LCFLFPGLII  LVSYSK      221

Human                    230 ILQITKASRK  RL---TVSLA  YSESHQIRVS  QQ   258  IL3
Cynomolgus monkey        230 ILQITKASRK  RL---TVSLA  YSESHQIRVS  QQ   258
Cow                      230 ILQITKASRK  RL---TMSLA  YSESHQLRVS  QQ   258
Pig                      230 ILQITKASRR  RL---TMSLA  YSESHQIRVS  QQ   258
Dog                      230 ILQITKASRK  RL---TVNLA  YSESHHIRVS  QQ   258
Rabbit                   230 ILQITKASRK  RL---TVSLA  YSESHQIRVS  QQ   258
Mouse                    230 ILQITKASRK  RL---TLSLA  YSESHQIRVS  QQ   258
Rat                      230 ILQITKASRK  RL---TLSLA  YSESHQIRVS  QQ   258
Horse                    230 ILQITKASRK  RL---TVSQA  YSESHQIRVS  RQ   258
Chicken                  217 ILQITKASRR  SL---NAGLA  YSENHQIRVS  QQ   245
Turkey                   217 ILQITKASRR  SL---NAGLA  YSENHQIRVS  QQ   245
Zebra finch              177 ILQITKASRR  SL---NAGLA  YPEHHQIHVS  QQ   205
Green anoles             283 ILQIAKASRR  RL---DAGMA  YSVSSHIRIS  RQ   311
Frog                     177 ---ITKSVRN  RL---ISCTT  YPENNQMKVS  HR   202
Tilapia                  222 ILQIAKNCRL  RLRPRDRQHP  NGDSPHYYVS  RQ   253

Human              TM6   259 DFRLFRTLFL  LMVSFFIMWS  PIIITILLIL  IQNFKQDLVI  WP  300  EL3
Cynomolgus monkey        259 DFRLFRTLFL  LMVSFFIMWS  PIIITILLIL  IQNFKQDLVI  WP  300
Cow                      259 DVRLFRTLFL  LMISFFIMWS  PIIITILLIL  IQNFKQNLVI  WP  300
Pig                      259 DVRLFRTLFL  LMISFFIMWS  PIIITILLIL  IQNFKQNLVI  WP  300
Dog                      259 DFRLFRTLFL  LMISFFIMWS  PIIITILLIL  IQNFKQDLVI  WP  300
Rabbit                   259 DFRLFRTLFL  LMVSFFIMWS  PIIITILLIL  IQNFQQDLVI  WP  300
Mouse                    259 DYRLFRTLFL  LMVSFFIMWS  PIIITILLIL  IQNFRQDLVI  WP  300
Rat                      259 DYRLFRTLFL  LMVSFFIMWS  PIIITILLIL  IQNFRQDLVI  WP  300
Horse                    259 DFRLFRTLFL  LMVSFFIMWS  PIIVTILLIL  IQSFKQDLVI  WP  300
Chicken                  246 DYKLFRALIV  LMISFFIMWS  PIIIIIXLIL  IRNYKQDLNI  LP  287
Turkey                   246 DYKLFRALIV  LMISFFIMWS  PIIIIIFLIL  VQNYKQDLNI  LP  287
Zebra finch              206 DYKLFRALFL  LMTSFFIMWT  PIIVIIFLIL  IQNFKKDLNI  LP  247
Green anoles             312 DFKLLRTLFV  LMISFFVMWS  PIIVTILLIL  VQELKPELNI  VP  353
Frog                     203 DYKLFRTLFI  LMISFFIMWT  PVAIIVLLLL  LQNLHKHVSI  PP  244
Tilapia                  254 DMKLFRTMLV  LVLSFLVMWS  PIFIITFVIL  AHNIQGHIYV  SS  295

Human              TM7   301 SLFFWVVAFT  FANSALNPIL  YNMTLCRNEW  KKIFC-C   336
Cynomolgus monkey        301 SLFFWVVAFT  FANSALNPIL  YNMTLCRNEW  KKIFC-C   336
Cow                      301 SLFFWVMAFT  FANAAVNPIL  YNMSLFRNEW  RRFFH-C   336
Pig                      301 SLFFWVMAFT  FANSAVNPIL  YNMSLFRNEW  RKFFH-C   336
Dog                      301 SLFFWVVAFT  FANSALNPIL  YNMSLFRNEW  RKIFH-C   336
Rabbit                   301 SLFFWVVAFT  FANSALNPIL  YNMSLFRNEW  RKTFC-C   336
Mouse                    301 SLFFWVVAFT  FANSALNPIL  YNMSLFRNEW  RKIFC-C   336
Rat                      301 SLFFWVVAFT  FANSALNPIL  YNMSLFRSEW  RKIFC-C   336
Horse                    301 TLFFWVVAFT  FANSAVNPIL  YNMSLFRNEW  RKIFC-C   336
Chicken                  288 SVFFWIMLFT  FANSAVNPVL  YNVAHFRRKC  QEILLCC   324
Turkey                   288 SVFFWIVLFT  FSNSAVNPVL  YNVAHFRRKC  QEIFLCC   324
Zebra finch              248 SVFFWIALFT  FANSVVNPIL  YNVAHFRRKC  QEILLCC   284
Green anoles             354 TFFFWVMAFT  FSNSVVNPVL  YNVAHFRHEW  RRIFF-C   389
Frog                     245 TVFFWITTLT  FSNSVLNPIL  YNINLFRQKW  VHIIL-C   280
Tilapia                  296 TMFFWVVTFT  LANSALNPIL  YSVCQFKNSW  RKRCCGS   332
                                                                H8

Human              C-terminus 337 FWFPEKGAIL  TDTSVKRND-  ----LSIISG  361
Cynomolgus monkey        337 FWFPEKGAIL  TDTSVKRND-  ----LSVISG  361
Cow                      337 CFFAEKGAML  TDTSVRRND-  ----LSVISS  361
Pig                      337 FFFPEKGAMF  TDTSVRRND-  ----LSVISS  361
Dog                      337 FFYPEKGAMF  TDTSVRRND-  ----LSIIYS  361
Rabbit                   337 FLFPEKGALF  TDTSARKND-  ----LSVISS  361
Mouse                    337 FFFPEKGAIF  TDTSVRRND-  ----LSVISS  361
Rat                      337 FFFPEKGAIF  TETSIRRND-  ----LSVIST  361
Horse                    337 FLFLEKGAMF  TDSSGRKND-  ----LSIISS  361
Chicken                  325 TGNPVRTRVG  SETSA-RSKR  EQPKLSVITR  353
Turkey                   325 TGNPVRTGAG  SETST-RSKR  EKPHLSVITR  353
Zebra finch              285 TGNPERHGPG  TETTARRSNH  EQPHLSFITR  314
Green anoles             390 CPGSSRRKTV  TDTTLRHHNN  GHFTVSVISR  419
Frog                     281 HSVEEIAD--  TETTTKRNEN  ANISHGTF--  306
Tilapia                  333 VVFPVRKRPT  SG--------  -----       344
```

B

	Human	Cynomolgus monkey	Cow	Pig	Dog	Rabbit	Mouse	Rat	Horse	Chicken	Turkey	Zebra finch	Green anoles	Frog	Tilapia
Human	100.0	97.0	87.7	89.4	89.6	88.9	88.4	87.0	86.1	57.8	57.8	54.2	45.8	46.0	49.5
Cynomolgus monkey		100.0	85.0	87.0	87.0	86.0	85.0	83.0	82.0	51.0	51.0	54.0	52.0	44.0	43.0
Cow			100.0	92.0	88.7	84.4	85.8	84.2	85.4	57.5	57.3	53.8	47.4	45.0	47.4
Pig				100.0	91.0	88.7	88.9	87.3	86.3	58.7	58.3	54.7	48.3	45.3	49.3
Dog					100.0	89.2	87.0	85.4	88.0	57.8	57.8	55.2	48.3	45.8	49.5
Rabbit						100.0	87.7	86.3	85.8	57.8	58.0	55.0	47.6	45.3	50.5
Mouse							100.0	97.9	84.2	58.5	59.0	55.2	47.2	46.0	49.5
Rat								100.0	83.7	58.7	59.2	55.7	47.4	46.7	48.8
Horse									100.0	55.4	55.9	53.5	48.1	45.3	49.3
Chicken										100.0	94.6	73.1	46.5	48.3	47.9
Turkey											100.0	73.3	47.2	49.1	48.3
Zebra finch												100.0	42.5	47.2	43.6
Green anoles													100.0	31.8	36.3
Frog														100.0	41.0
Tilapia															100.0

C

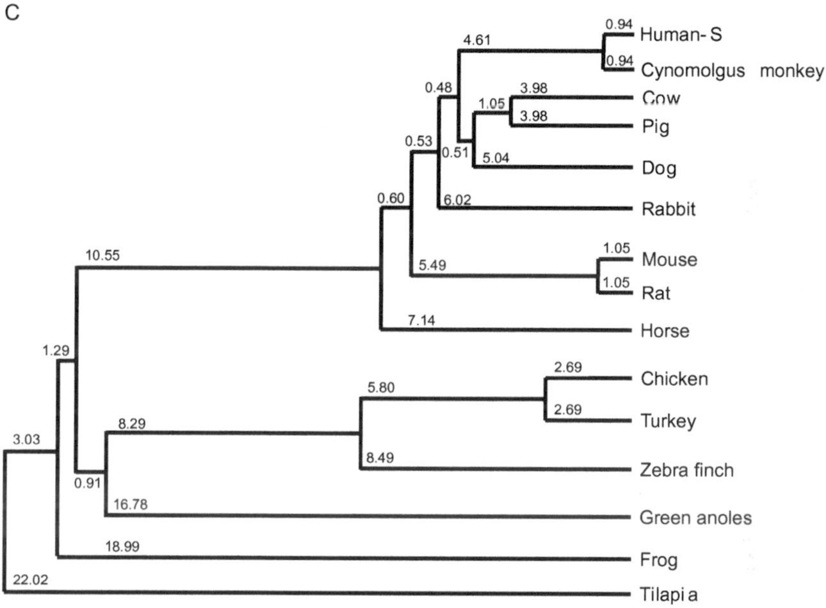

Figure 7.2 Comparison of GPR120-S amino acid sequences between hGPR120-S and GPR120-S predicted from other species. (A) Alignment of the amino acid sequences between hGPR120-S and GPR120-S cloned or predicted from other species. The ECLs and ICLs are *boxed*. The most conserved residues in each TM are shaded in *gray*. (B) Homology and (C) phylogenetic analysis of GPR120-S amino acid sequences from several species. NCBI reference numbers for the proteins are: human, NP_001182684.1; cynomolgus monkey, ACU82853.1; mouse, NP_861413.1; rat, NP_001040553.1; rabbit, XP_002718560.1; cow, XP_870359.2; pig; NP_001191695.1; horse, XP_001500917.1; dog, NP_001245268.1; chicken, XP_003641529.1; turkey, XP_003208098.1; zebra finch, XP_002189647.1; green anole, XP_003223819.1; frog, XP_002937240.1; tilapia, XP_003452042.1.

rat, and mouse, respectively. Phylogenetic analysis indicates that hGPR120-S clusters together with GPR120-S from other mammalian species but is significantly divergent from nonmammalian species at amino acid level (Fig. 7.2C). No orthologue was found in fugu or zebra fish.[8]

GPR120 is a member of the rhodopsin-like family A GPCRs with seven transmembrane domains (TM) connected by alternating extracellular loops (ECLs) and ICLs. GPR120 has two signature motifs, (D/E)R(Y/W) at the end of TM3 (ERM in GPR120) and NSxxNPxxY in TM7, that are highly conserved in this family.[8] There is one potential N-linked glycosylation site at Asn21. Whether the receptor is indeed glycosylated has not been reported yet. Two conserved Cys residues at the C terminus, Cys351 and Cys352, might serve as sites for palmitoylation anchoring the C terminus to the plasma membrane, forming a fourth ICL. It is not known whether the GPR120 is indeed palmitoylated or where this modification is located.

GPCRs, upon agonist stimulation, are typically phosphorylated on their cytoplasmic tail or ICLs by G protein-coupled receptor kinases (GRKs)[16] and second messenger-associated kinases, such as protein kinase A (PKA) and protein kinase C (PKC).[17] We identified several potential phosphorylation sites in GPR120-L using online prediction tool.[18] There are three potential PKA/PKC phosphorylation sites within the third ICL. These sites are Ser253, Thr258, and Ser272. In addition, there are six potential GRK phosphorylation sites. Three sites, Ser247, Thr250, and Ser264, are within the third ICL, and the other three sites, Ser366, Ser373, and Ser376, are within the C terminus tail. GPR120-S, lacking the additional 16 residues in the third ICL, was reported to exhibit more pronounced basal phosphorylation than GPR120-L in the absence of agonist.[19] There is no significant difference in the degree and rate of phosphorylation of both isoforms upon agonist stimulation. The precise mechanisms and site(s) of phosphorylation have not been reported.

2.2. Tissue distribution

Through quantitative PCR detection of human and mouse mRNA, it was shown that the GPR120 gene is primarily expressed in intestinal tract and lung.[13] A later study investigating tissue expression in cynomolgus monkey also revealed the highest GPR120 expression in lung and colon.[14] However, in the rat, GPR120 is only highly expressed in colon.[15] In mouse lung, the GPR120 protein is expressed in Clara cell, indicating that the GPR120 might be involved in the function of Clara cells, such as the production and/or secretion of surfactant.[20] In mouse intestine, GPR120 displays increasing expression along the longitudinal axis with the highest expression in the proximal colon.[21]

In the original cloning paper, Hirasawa *et al.* pointed out that GPR120 is expressed in mouse enteroendocrine cell line STC-1, but not in other intestinal cell lines, including NCI-H716, CaCo-2, HT29, IEC-6, IEC-18, and

Intestine 407.[13] Using double immunostaining for GPR120 and chromogranin A, a well-established marker of endocrine cells, Matsumura *et al.* reported that GPR120 immunoreactivities are co-localized with chromogranin A, suggesting that GPR120 is expressed in the enteroendocrine cells.[22] This correlates functionally with subsequent findings that GPR120 mediates fatty acid-induced glucagon-like peptide-1 (GLP-1) and cholecystokinin (CCK) secretion from intestine. It has been demonstrated that GPR120 is expressed in another enteroendocrine cell, K cell, that might contribute to fatty acid-induced glucose-dependent insulinotropic polypeptide secretion.[23]

GPR120 is also expressed in the mouse GI tract, including the endocrine cells and the brush cells at the limiting ridge of the stomach, as well as in the duodenum.[24] In the stomach, GPR120 expression does not colocalize with ghrelin expression; however, in the duodenum, there is high colocalization of GPR120 and ghrelin expression. The ghrelinoma cell line, MGN3-1,[25] also expresses GPR120.[24]

In addition to intestinal tract and lung, it was shown that GPR120 is also abundantly expressed in mouse adipose tissue using immunohistochemical staining.[20] Using RT-PCR technique, Gotoh *et al.* showed that GPR120 is present in four different adipose tissues, including subcutaneous, perirenal, mesenteric, and epididymal adipose tissues.[26] Developmentally, GPR120 expression is increased during adipogenesis. The high expression of GPR120 is only detected in differentiated adipocyte, but not in stromal-vascular fraction (SVF) and/or preadipocytes.[26] Moreover, the mRNA expression of GPR120 is increased during the differentiation of 3T3-L1 adipocytes.[26] Recently, macrophage has been reported to express GPR120 as observed by measuring GPR120 mRNA abundance in SVF (containing macrophages) of mouse adipose tissue.[27] Further observation confirmed that GPR120 is highly expressed in proinflammatory CD11c$^+$ macrophages, intraperitoneal macrophages, monocytic RAW 264.7 cells, and specialized macrophages in the liver.[27]

Gotoh *et al.* showed that the mRNA expression of GPR120 is increased in adipose tissues of mice fed high-fat diet (HFD).[26] Interestingly, HFD-induced GPR120 expression in adipose tissue might be caused by elevated GPR120 expression in adipose tissue macrophages, in consideration that elevated GPR120 expression is only detected in adipose tissue macrophages but not in adipocytes.[27] In addition, expression of GPR120 is increased in adipose tissue macrophages and specialized macrophages in the liver of mice fed HFD.[27]

GPR120 mRNA is also detected in the epithelium of circumvallate papillae containing taste buds but not in the nonsensory epithelium.[22] GPR120-immunoreactive cells were found in the taste cells (possibly gustatory cells) of the foliate and fungiform papillae. This result suggests that GPR120 might be involved in the reception of gustatory signals of dietary fat.

Although GPR120 has two splice variants in humans, Galindo *et al.* reported that GPR120-S is the predominant variant expressed in human colon.[28] In other tissues, such as islet,[29] taste cell-free lingual epithelium, fungiform papillae, and circumvallate papillae,[28] only the short form of GPR120 was found to be expressed. However, the expression patterns of these two splice variants in other tissues remain unknown.

At present, the expression of GPR120 in other tissues, such as pancreas, skeletal muscle, liver, and heart, is still controversial. Expression of GPR120 is not detected in rat pancreas[15] and pancreatic β-cell line MIN6.[13] However, in another study, it was shown that GPR120 is expressed in pancreatic β-cell line (including MIN6, IαTC1-6, NS-1 and BRIN-BD11cells), isolated mouse islets, and human islets.[30] Another group also reported that only the short form can be found in human islet.[29] Similarly, in skeletal muscle, heart, and liver, expression of GPR120 cannot be detected in the human and the mouse in the initial study.[13] In contrast, later work showed that GPR120 is present in these tissues in the rat.[31] HFD can increase the expression of GPR120 in rat cardiac tissue and extensor digitorum longus skeletal muscle.[31] Thus, expression of GPR120 protein in some tissues remains to be confirmed. Moreover, it has been shown that GPR120 is highly expressed in mouse pituitary[26] and human breast cancer cells.[32] The potential physiological functions of GPR120 in these tissues remain to be elucidated.

3. PHYSIOLOGY OF GPR120

3.1. Intestinal peptide hormone secretion

The GPR120 was first identified as a FFAR regulating GLP-1 secretion in mice.[13] GLP-1 production has been reported to be regulated by dietary fatty acids. Hirasawa *et al.* showed that unsaturated long-chain FFAs, especially α-linolenic acid (ALA), show a dose-dependent stimulation of GLP-1 secretion in STC-1 cells.[13] Oral administration of ALA significantly increases plasma concentrations of GLP-1 in both mice[13] and rats.[15] Transfection with GPR120-specific siRNA, but not GPR40-specific siRNA, significantly reduces ALA-induced GLP-1 secretion,[13] which indicate that the FFA-induced GLP-1 secretion in STC-1 cells is mediated through GPR120.

CCK is another intestinal peptide hormone of which the secretion is regulated by GPR120. CCK plays a key role in regulating a range of intestinal responses, including stimulation of pancreatic secretion, gall bladder emptying, and inhibition of gastric motility.[33] In humans, plasma CCK concentration is elevated by infusion of long-chain FFAs into the upper gut, but does not change when short- or medium-chain FFAs was infused.[34] In C57/BL6 mice, similar results are observed[35]. Intra-gastric administration of ALA significantly increases the plasma concentrations of CCK compared with mice administered octanoic acid or vehicle.[35] Moreover, *in vitro* study confirmed that regulation of CCK secretion by FFAs depends on chain length. In that study, 100 μM ALA but not octanoic acid increases CCK secretion in STC-1 cells.[35] Furthermore, using GPR120-specific or GPR40-specific short hairpin RNA, Tanaka *et al.* showed that ALA-induced CCK secretion in STC-1 cells is also mediated through GPR120.[35] ALA, through a G protein-dependent pathway, initiates membrane depolarization with cation influx via the monovalent cation channel, transient receptor potential channel type M5, which results in the opening of voltage-gated calcium channel, and the rise of intracellular calcium, leading to the release of satiety hormones such as CCK and GLP-1.[36] The molecular mechanism of regulating GLP-1 and CCK secretion through GPR120 is illustrated in Fig. 7.3, and will be elaborated in *Signaling Pathways and Regulation* below.

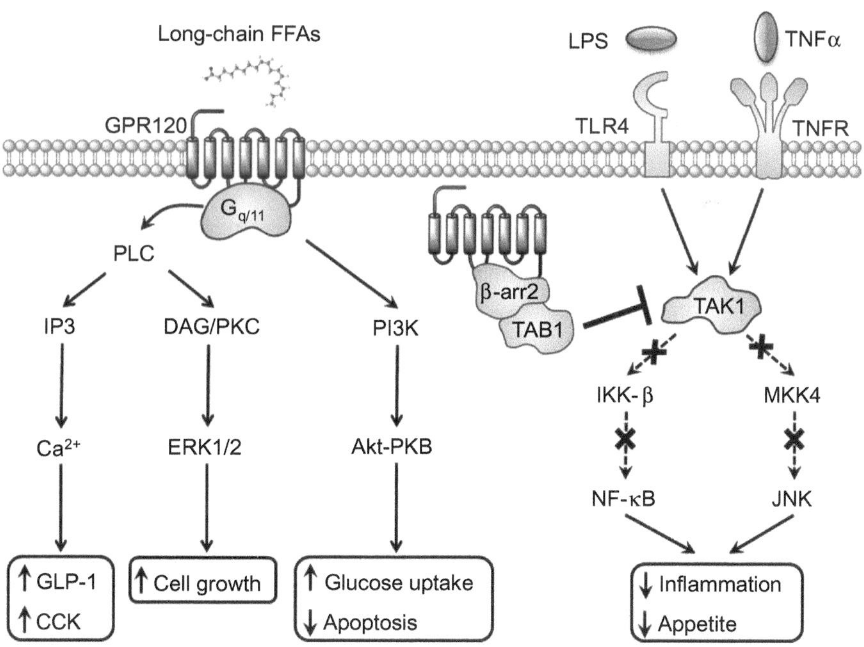

Figure 7.3 Molecular mechanisms of GPR120-associated physiological functions, including regulation of intestinal peptide secretion, anti-inflammation, anti-apoptosis, and energy homeostasis, upon GPR120 activation by long-chain FFAs.

GPR120 is also potentially involved in regulating ghrelin secretion. Oral gavage of grifolic acid, a specific GPR120 partial agonist,[37] increases plasma ghrelin levels in mice.[24] In the ghrelinoma cell line, MGN3-1, ALA decreases octanoyl ghrelin levels although partial knockdown of GPR120 does not affect this effect. Grifolic acid, although effective *in vivo*, does not affect ghrelin secretion in this cell line.[24] Therefore more studies need to be done to clarify the role of GPR120 in ghrelin secretion and octanoylation.

Recently, it was shown that in germ-free mice that lack intestinal microbiota, intestinal epithelial expression of GPR120, as well as several other fatty acid receptors, is significantly decreased compared to normal control mice, suggesting decreased intestinal signaling for fats.[38] The germ-free mice are leaner than normal control mice although they consume more calories, with increased oral detection for fats, and increased preference, and intake of fat in two-bottle test.[38]

3.2. Islet function and glucose homeostasis

Since GLP-1 has potent effects on glucose-mediated insulin secretion, proliferation of β-cells and β-cell mass, GPR120-mediated GLP-1 secretion in intestine may affect islet function. Hirasawa *et al.* showed that increased plasma concentration of GLP-1 is associated with elevated plasma insulin level in mice fed with ALA.[13] In a long-term feeding study, Tanaka *et al.* showed that, compared with vehicle or octanoic acid administration, administration of ALA for 4 weeks increases insulin secretion in rats fed a HFD.[15] Interestingly, this study also found that proliferation of pancreatic β-cell is increased as observed with increased number of KI67-positive cells in the insulin-positive area (pancreatic islet) in rats administered ALA.[15]

As mentioned above, expression of GPR120 in islet remains unclear. However, previous work suggested that GPR120 might not play a direct role in the regulation of insulin secretion from pancreatic β-cell. Acute insulin secretory response to fatty acids in mouse islets is impaired by deletion of GPR40.[39] Moreover, GPR40 knockdown in MIN6 cells also leads to loss of this response.[40] These results suggest that GPR40 might mediate the insulin secretory response to fatty acids.

It was shown that in *Gpr120* KO mice, compared to WT mice, high-fat feeding causes more severe insulin resistance, such as glucose intolerance, hyperinsulinemia, decreased hypoglycemic response to injected insulin, decreased insulin signaling in white adipose tissue and insulin, and downregulation of several insulin-signaling genes.[41] The results suggest that GPR120 signaling is important for maintaining normal insulin sensitivity.

A comprehensive systems genetics study just published showed that *GPR120* is expressed in human islets, with higher expression associated with increased insulin secretion and lower hemoglobin A_{1c} (HbA_{1c}).[42] Differential expression of *GPR120* is observed in islets from diabetic vs. nondiabetic donors, and normoglycemic ($HbA_{1c} < 6\%$) vs. hyperglycemic ($HbA_{1c} > 6\%$) donors.[42] Treatment of human islets with eicosapentanoic acid (EPA), an agonist of GPR120, prevents the apoptosis induced by palmitate and linoleic acid and increases cell viability. The effect of EPA is decreased by 50% when *GPR120* was knocked down using short hairpin RNA. These results suggest that GPR120 signaling protects islets from lipotoxicity.[42]

3.3. Taste and food preference

Traditionally, the oral perception of dietary fat was considered to rely predominantly on trigeminal,[43] olfactory,[44] and postingestive mechanisms.[45] There were circumstantial studies indicating that oral fat also activates the gustatory system[46–49], similar to other basic taste qualities, including sweet, bitter, sour, salty, and umami. Delayed-rectifying K^+ channels[50] and CD36 (a membrane fatty acid transporter)[48,51] have been demonstrated to be important in the fatty acid perception mechanisms by the gustatory system. For those basic taste qualities (except for salty and sour), the taste sensations are initiated by binding of tastants to their receptors, which are all GPCRs reported so far.[51–54]

In rodents, GPR120 has been shown to be activated by medium- and long-chain FFAs *in vitro* and later was found to be also primarily expressed in type II taste receptor cells of foliate and vallate papillae.[22,55] Recently, Cartoni *et al.* demonstrated that *Gpr120* knockout (KO) mice show a diminished preference to FFAs, including linoleic acid and oleic acid, using two-bottle preference tests with xanthan gum as the control.[55] In addition, the glossopharyngeal nerve responses of *Gpr120* KO mice to unsaturated long-chain FFAs are significantly weaker than that of wild-type (WT) mice.[55]

In humans, a recent study suggests that GPR120 might be also involved in human gustatory perception of dietary fatty acids.[28] Galindo *et al.* demonstrated that GPR120 mRNA and protein are expressed in human taste buds and surrounding epithelial cells. Moreover, the monitored fatty acid dose–response profile for GPR120 *in vitro* fits with the spectrum of human "fatty-scratchy" sensory analyses *in vivo*.[28]

3.4. Bone physiology

Since early 1990s, a number of groups carried out studies of the relationship between fat mass and bone density.[56–58] The clinical studies have shown that the total body fat mass is positively related to the bone density and fracture risk.[59] Several endocrine mechanisms, including adipokines, β-cell hormones, and gut hormones, have been demonstrated to be important in mediating the positive relationship between fat mass and bone density.[59] Additionally, lipids might also act directly on the bone.

Recently, Cornish *et al.* assessed the direct effect of FFAs on osteoclastogenesis in bone marrow cultures.[60] The results demonstrated that saturated FFAs with chain lengths of 14–18 have maximal inhibition of osteoclastogenesis, particularly palmitic acid (C16:0). The unsaturated FFAs eliminate or even reverse the anti-osteoclastogenic effect. Furthermore, they demonstrated that the effect of fatty acids is directly on osteoclastogenesis rather than on the activity of mature osteoclasts, stromal cells, or blockade of the mevalonate pathway. Most importantly, it was found that GPR120 is highly expressed in the osteoclast-like monocytic RAW264.7 cells, at least 100-fold than that of GPR40. GPR120 is also expressed in a number of osteoblastic cell lines, including ST2 and MC3T3-E1. GW9508, a synthetic GPR40/GPR120 agonist, inhibits osteoclastogenesis in murine bone marrow cultures to a similar extent as the long-chain FFAs.[60]

This study suggests that the osteoclastogenesis mediated by GPR120, a receptor of long-chain FFAs, might be a new mechanism underlying the positive relationship between total fat mass and bone density. GPR120 might provide a novel aspect for understanding human bone physiology and has potential therapeutic value in treating osteoporosis.

3.5. Anti-inflammation

Chronic low-grade inflammation has been shown to be associated with various diseases such as obesity,[61,62] type 2 diabetes,[62,63] cardiovascular disease,[62,64] cancer,[65] psoriasis,[66] periodontitis[67], and multiple sclerosis.[68] ω-3 fatty acids have been shown to have potential anti-inflammatory effect and have been used in multiple clinical trials to assess the therapeutic benefits in treatment and/or prevention of various inflammatory diseases.[69]

In 2010, a landmark study was published that established the critical importance of the GPR120 as an ω-3 fatty acid receptor in mediating potent anti-inflammatory effects both *in vitro* and *in vivo*.[27] Oh *et al.* showed that GPR120 is highly expressed in proinflammatory $CD11c^+$ macrophages and monocytic RAW 264.7 cells. In RAW 264.7 cells, they observed inhibition of lipopolysaccharide-stimulated inflammatory responses with the treatment of GW9508, docosahexaenoic acid (DHA), and ALA. More importantly, all the observed anti-inflammatory effects are completely abrogated by siRNA-mediated knockdown of GPR120.

Moreover, they performed *in vivo* studies with *Gpr120* KO mice and WT littermates fed standard chow diet, HFD, and ω-3 fatty acids supplemented HFD. For WT mice, they found that ω-3 fatty acids decrease proinflammatory M1 macrophages as well as the M1 proinflammatory gene expression, and increase anti-inflammatory M2 macrophages as well as the M2 anti-inflammatory gene expression. However, no anti-inflammatory effect of ω-3 fatty acids as described above is observed in *Gpr120* KO mice. The molecular mechanism of anti-inflammation through GPR120 activation by ω-3 fatty acid is illustrated in Fig. 7.3, and will be elaborated in *Signaling Pathways and Regulation* below.

The identification of GPR120 as the receptor for ω-3 fatty acids in mediating potent anti-inflammatory effects established a new route toward understanding the pathogenesis of multiple inflammatory-associated diseases. Hopefully, synthetic ω-3 fatty acids analogs targeting GPR120 could be developed as more efficient and potent anti-inflammatory drugs.

Im recently raised several concerns on whether GPR120 is the real mediator of the beneficial cardiovascular effects of ω-3 fatty acids such as DHA and EPA.[70] Firstly, ω-6 fatty acids and saturated fatty acids, which do not have beneficial cardiovascular effects, can also activate GPR120. Secondly, fatty acids have to be dissolved in DMSO or methanol to be able to activate GPR120 *in vitro*. Fatty acids conjugated with albumin, more relevant in blood and exudates, cannot activate GPR120 *in vitro*. Thirdly, the EC_{50} values obtained from the *in vitro* experiments, at 1–10 μM, are relatively high. Finally, prolonged stimulation of GPCRs is known to lead to desensitization. The beneficial cardiovascular effects of ω-3 fatty acids usually need long-term, high-dose intake.[70] Whether GPR120 is desensitized in this scenario has not been investigated. It should be pointed out that the Nomenclature Committee of the International Union of Basic and Clinical Pharmacology has not formally ratified GPR120 as the receptor for long-chain FFAs yet.[71]

3.6. Adipogenesis

In 2007, Gotoh *et al.* for the first time reported the physiological role of GPR120 in adipose tissue development and adipocyte differentiation *in vitro*.[26] They demonstrated that GPR120 is highly expressed in both rodent and human-differentiated adipocyte, but not SVF and preadipocytes of adipose tissues. Furthermore, they demonstrated that siRNA knockdown of GPR120 results in dramatic decrease of adipogenic gene expression such as PPAR-γ2 and aP2, as well as decrease in the number of 3T3-L1 adipocytes stained with oil red O. This study identified the important role of GPR120 in adipogenesis as a mediator facilitating adipocyte maturation and differentiation.

3.7. Appetite and energy homeostasis

Recently, Cintra *et al.* investigated the anorexigenic effect of ω3- and ω9-fatty acids in rodents.[72] They found that, in HFD-induced obese mouse model, partial substitution of fatty acids component of the diet by flax seed oil (rich in ω3-fatty acids) or olive oil (rich in ω9-fatty acids) results in decreased food intake and body weight gain. This effect has been shown not due to palatability since there is no difference in food preference between normal HFD and unsaturated FFA-substituted HFD. In addition, intracerebroventricular injection of ω3- and ω9-fatty acids results in significantly decreased food intake and increased leptin-induced anorexigenic effect, and thus results in significant reduction of adipocyte size and loss of body weight. Most importantly, they identified that GPR120 is co-expressed with neuropeptide Y in the arcuate nucleus in hypothalamus. Moreover, the intracerebroventricular injection of ω3- and ω9-fatty acids activates GPR120-mediated β-arrestin 2/TAK1 binding protein 1 (TAB1) signaling pathways[72] (Fig. 7.3) that were identified previously.[27] These results indicate that GPR120 might play a role in mediating unsaturated fatty acids-induced anorexigenic effect.[72] However, further detailed studies such as studies using *Gpr120* KO mice and GPR120 specific ligands are needed to confirm and elucidate the relationship between GPR120 and appetite control in rodent hypothalamus.

Recently, in 2012, Ichimura *et al.* published a systematic study providing convincing evidence that GPR120 is involved in diet-induced obesity in mice.[41] They showed that HFD-fed *Gpr120* KO mice develop more severe obesity than WT mice with lower energy expenditure and higher body weight gain. The *Gpr120* KO mice also exhibit more severe fatty liver with decreased adipocyte differentiation, lipogenesis and enhanced hepatic lipogenesis, and more severe obesity-associated insulin resistance with decreased insulin signaling and increased inflammation in adipose tissue. The results demonstrated that dysfunction of GPR120 could be a potential mechanism for HFD-induced obesity and obesity-associated metabolic syndrome in mice, suggesting that the activation of GPR120 has an anti-obesity effect.

4. PHARMACOLOGY OF THE GPR120

4.1. Ligand binding and receptor activation

Long-chain FFAs were first identified as endogenous ligands that led to internalization of fluorescently labeled GPR120.[13] Concentration-dependent $[Ca^{2+}]_i$ response in HEK293 cells stably expressing GPR120-Gα16 proteins are detected for saturated FFAs (with chain length from C14 to C18) and unsaturated FFAs (with chain length from C16 to C22), in particular for ω-3 FFAs such as ALA.[13] However, long-chain FFA, such as ALA, has low potency for the GPR120 with pEC_{50} value of 5.59, and poor selectivity between other receptors such as PPAR[73] and GPR40.[74] In addition, fatty acids acting as metabolism substrate rather than signaling molecules complicate the pharmacological characterization of GPR120. Therefore, the pharmacological study of GPR120 is currently hindered due to the lack of potent and selective ligands.

In the past few years, great efforts have been carried out to discover potent and selective GPR120 ligands. Briscoe *et al.* first described a potent GPR40 agonist, GW9508,[75] which also moderately activates GPR120 with pEC_{50} value of 5.46. Later, a PPARγ derivative, NCG21,[76] has been reported to be a potent and selective GPR120 agonist with ~16-fold higher selectivity over GPR40. Some natural compounds,[37] including grifolin derivatives, grifolic acid, and grifolic acid methyl ether, have been demonstrated to be partial agonists that activate $[Ca^{2+}]_i$ and ERK signaling and inhibit the ALA-induced GPR120 responses. Recently, Shimpukade *et al.* reported a much more potent and selective GPR120 agonist, compound 43 (3-(4-[(4-fluoro-4′-methyl-[1,1′-biphenyl]-2-yl)methoxy]phenyl)propanoic acid) with EC_{50} around 44 nM,[77] that has ~1500-fold higher selectivity over GPR40.

Computer modeling of GPR120, constructed from the crystal structure of bovine rhodopsin, docked with ALA, ALA methyl ester, or NCG21, reveals that there seem to be hydrogen bonds between the Arg99 residue of GPR120 and the carboxylic acid residue of the ligands.[78] A more precise modeling of GPR120, using the crystal structure of a nanobody-stabilized active β_2-adrenoceptor as template, docked with compound 43, show that the carboxyl group interacts with Arg99 through hydrogen bond and electrostatic interactions.[77] Indeed, the R99Q mutant of GPR120 totally abolishes the compound 43-induced β-arrestin recruitment.[77] It was also reported that R99A substitution in GPR120-S reduces cell surface receptor expression and abolishes agonist-induced calcium responses.[79] The modeling also reveals a potential lipophilic pocket (consisting of Met118, Thr119, Gly122, Phe211, Asn215, Ile280, Ile281, and Trp277) that might interact with the biphenyl moiety of compound 43.[77]

Currently, no ligand binding assay has been set up to directly measure the binding of endogenous or synthetic ligands to the receptor.[80] Rather, results from signaling assays were used to indirectly gauge the binding of the ligands to the receptor. The establishment of a binding assay would be an important contribution to understanding the molecular pharmacology of the receptor, especially in elucidating the molecular mechanism of the ligand interaction with the receptor. In the face of normal cell surface expression, lack of signaling can be due to either defect in ligand binding or G protein coupling/activation.[81,82]

4.2. Signaling pathways and regulation

The classical signaling pathway for the GPR120 is by coupling to the heterotrimeric Gαq/11 protein as shown in Fig. 7.3. Hirasawa *et al.* first demonstrated that GPR120 activation leads to dose-dependent increase of intracellular Ca^{2+} concentration ($[Ca^{2+}]_i$) in HEK293 cells stably expressing GPR120-Gα16, and neither stimulation nor inhibition of cAMP production is detected.[13] The Gαq/11-coupled signaling pathway is further shown to be also activated in other cell lines, such as enteroendocrine cell,[13] macrophages[27] and adipocyte,[27] stimulated by various natural or synthetic GPR120 ligands.

The physiological relevance of Ca^{2+} signaling pathway is still unclear. It was reported that ALA-induced GLP-1 secretion from STC-1 cell is inhibited by using Ca^{2+}-free medium, but is not affected by L-type calcium channel blocker, endoplasmic reticulum Ca^{2+} ATPase inhibitor, or phospholipase C inhibitor.[13] Some evidence shows that ALA-induced CCK secretion from STC-1 cells is abolished by removal of extracellular Ca^{2+} or addition of L-type Ca^{2+} channel blocker.[35] siRNA knockdown of Gαq/11 blocks the DHA- and GW9508-induced enhancement of glucose uptake in adipocytes.[27]

In addition to activation of Gq-coupled signaling pathway, GPR120 also activates the mitogen-activated protein kinases pathway (Fig. 7.3). Hirasawa *et al.* reported the increased phosphorylation of extracellular signal-regulated kinase (ERK) 1/2 upon stimulation with long-chain FFAs in HEK293 cells transiently transfected with GPR120.[13] The activation of ERK1/2 pathway is also observed in various cell lines induced by other GPR120 ligands.[27,32,35,83,84] Both $[Ca^{2+}]_i$ and ERK1/2 pathways are used to characterize the function of GPR120 and screening potential GPR120 ligands.

β-Arrestin2-dependent signaling has been demonstrated to be essential in ω-3 fatty acid-induced anti-inflammatory effect. Oh *et al.* reported that β-arrestin2 knockdown blocks the lipopolysaccharide-induced proinflammatory signaling.[27] They suggested that, in RAW 264.7 cells, DHA treatment leads to β-arrestin2 mediated GPR120 internalization, and then causes association of β-arrestin2 and TAB1, which largely blocks the association of TAB1 and transforming growth factor-β activated kinase 1 (TAK1), and thus results in inhibition of TAK1-induced inflammatory signaling (Fig. 7.3).[27]

It was reported that GPR120 activation also stimulates phosphorylation of protein kinase B (also known as Akt).[83] Inhibition of Akt pathway, as well as ERK1/2 pathway, results in reduced anti-apoptotic effect of ALA in STC-1 cell (Fig. 7.3).[83] In murine GLUTag cells stimulated with oleic acid, however, no activation of calcium, ERK1/2 or Akt pathway is observed.[84] Instead, it is suggested that the atypical PKC isoform, PKCζ, is involved in fatty acid-induced GLP-1 secretion.[84]

Recently, Watson *et al.* argued that GPR120-L and GPR120-S have different signaling profiles.[79] They demonstrated that GPR120-S isoform has normal agonist stimulated Gq-coupled signaling, but no specific $[Ca^{2+}]_i$ signaling or dynamic mass redistribution responses are observed in HEK293 cells transfected with GPR120-L isoform, despite similar levels of cell surface expression of GPR120-L compared to the GPR120-S. This finding is indirectly supported by the requirement of receptor fusion to Gα16,[13] viral receptor overexpression,[75] or overexpression of chimeric Gα subunit[28] for the normal function of GPR120-L reported by others. In another study characterizing naturally occurring *GPR120* variants, the WT GPR120-L was reported to signal in response to ALA stimulation with increased $[Ca^{2+}]_i$.[41]

Activated and phosphorylated GPCRs recruit arrestins and other adapter proteins to underdo internalization. It was shown that GPR120-S and GPR120-L isoforms of the human receptor recruit β-arrestin2 and undergo β-arrestin2-mediated internalization and lysosome degradation to a similar degree.[79] The internalized receptor is not recycled to the cell surface rapidly. Rather, it is more likely to be destined for lysosomal degradation.[79]

5. *GPR120* MUTATIONS IN HUMAN OBESITY

Ichimura *et al.* recently demonstrated that in humans, *GPR120* expression levels in both subcutaneous and omental adipose tissues are significantly higher in nondiabetic obese subjects than in lean subjects.[41] *GPR120* expression in gastric mucosa is also increased in morbidly obese patients.[85]

Human genetic studies identified six *GPR120* mutations, including two missense mutations (R67C and R270H) and four synonymous mutations (V38V, S192S, V243V, and S264S) (Fig. 7.4) [41]. In addition, through mining the EMBL database, we identified eight additional missense and three synonymous *GPR120* mutations (Fig. 7.4).

Although R270H is associated with obesity, R67C only shows a trend for association with obesity. *in vitro* functional studies showed that R67C has normal signaling properties in response to the endogenous agonist, ALA, stimulation, compared to the WT GPR120, whereas R270H signaling is severely diminished. In addition, R270H has dominant negative activity. Transfecting R270H GPR120 in cells stably expressing WT GPR120 results

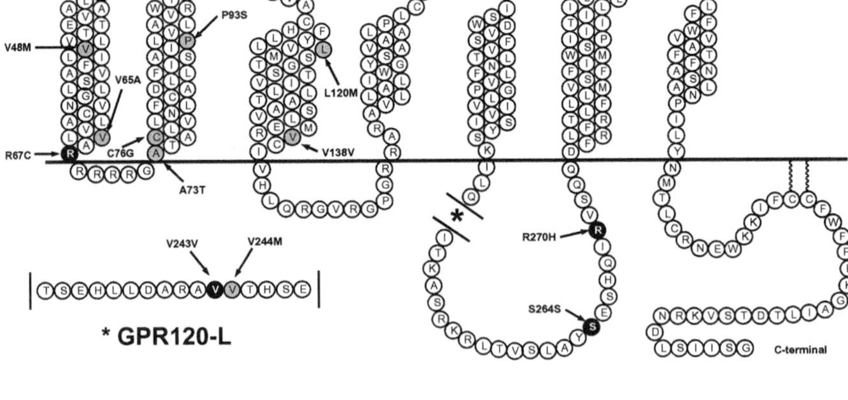

Figure 7.4 Naturally occurring mutations of hGPR120. The six polymorphisms that were identified from obese patients are indicated in *white* letter and shaded in *black*. The mutations identified from EMBL database are shaded in *gray*.

in decreased ALA-induced $[Ca^{2+}]_i$ signaling compared to transfecting empty vector or WT GPR120. Moreover, the $[Ca^{2+}]_i$ signaling is significantly impaired in cells stably expressing bicistronic WT/R270H or R270/WT receptors compared with that of WT/WT, despite that almost equal level of receptor expression is observed.[41]

Furthermore, in a cell line with regulated GLP-1 secretion that lacks endogenous *GPR120* expression, human intestinal NCI-H716 cells, transfection with WT or R67C GPR120 results in ALA-induced GLP-1 secretion, whereas ALA does not induce GLP-1 secretion in NCI-H716 cells transfected with R270H GPR120. In summary, the R270H mutation located at ICL3 is strongly associated with obesity and has a dominant negative effect on GPR120. The R67C variant that is not associated with obesity does not have signaling defect or dominant negative effect.[41]

From a structure–function point of view, as shown in Fig. 7.4, R270 is located in ICL3, a domain known to be critical for conformational changes and G protein coupling upon agonist binding.[86,87] The exact defect(s) that cause(s) R270H signaling defect, trafficking, ligand binding, G protein coupling and/or activation, is (are) not known.[82]

Synonymous mutations have been identified to contribute to disease severity and susceptibility by affecting splicing efficiency.[88] However, the functional studies of the other four synonymous *GPR120* mutations have not been reported.

6. CONCLUSIONS

Obesity is a complex and heritable disease resulting from intricate interactions of behavior, environmental factor, and genetic disorders. Although several genetic variants have been identified to be associated with human obesity,[89] the discovery of the physiological roles of GPR120, particularly in obesity, anti-inflammation, insulin resistance, and type 2 diabetes, provides a novel route for investigating human obesity. Detailed studies on the defects of the GPR120 mutants and identifying novel ligands represent important approaches toward treatment of obesity and diabetes.

REFERENCES

1. Flegal KM, Carroll MD, Kit BK, Ogden CL. Prevalence of obesity and trends in the distribution of body mass index among US adults, 1999–2010. *JAMA* 2012;**307**:491–7.
2. Boden G. Obesity and free fatty acids. *Endocrinol Metab Clin North Am* 2008;**37**:635–46.

3. Yaney GC, Corkey BE. Fatty acid metabolism and insulin secretion in pancreatic beta cells. *Diabetologia* 2003;**46**:1297–312.
4. Schmitz G, Ecker J. The opposing effects of n-3 and n-6 fatty acids. *Prog Lipid Res* 2008;**47**:147–55.
5. Chawla A, Repa JJ, Evans RM, Mangelsdorf DJ. Nuclear receptors and lipid physiology: opening the X-files. *Science* 2001;**294**:1866–70.
6. Sauer LA, Dauchy RT, Blask DE. Mechanism for the antitumor and anticachectic effects of n-3 fatty acids. *Cancer Res* 2000;**60**:5289–95.
7. Louet JF, Chatelain F, Decaux JF, Park EA, Kohl C, Pineau T, et al. Long-chain fatty acids regulate liver carnitine palmitoyltransferase I gene (L-CPT I) expression through a peroxisome-proliferator-activated receptor α (PPARα)-independent pathway. *Biochem J* 2001;**354**:189–97.
8. Fredriksson R, Hoglund PJ, Gloriam DEI, Lagerstrom MC, Schioth HB. Seven evolutionarily conserved human rhodopsin G protein-coupled receptors lacking close relatives. *FEBS Lett* 2003;**554**:381–8.
9. Hirasawa A, Hara T, Katsuma S, Adachi T, Tsujimoto G. Free fatty acid receptors and drug discovery. *Biol Pharm Bull* 2008;**31**:1847–51.
10. Miyauchi S, Hirasawa A, Ichimura A, Hara T, Tsujimoto G. New frontiers in gut nutrient sensor research: free fatty acid sensing in the gastrointestinal tract. *J Pharmacol Sci* 2010;**112**:19–24.
11. Talukdar S, Olefsky JM, Osborn O. Targeting GPR120 and other fatty acid-sensing GPCRs ameliorates insulin resistance and inflammatory diseases. *Trends Pharmacol Sci* 2011;**32**:543–50.
12. Overington JP, Al-Lazikani B, Hopkins AL. Opinion—how many drug targets are there? *Nat Rev Drug Discov* 2006;**5**:993–6.
13. Hirasawa A, Tsumaya K, Awaji T, Katsuma S, Adachi T, Yamada M, et al. Free fatty acids regulate gut incretin glucagon-like peptide-1 secretion through GPR120. *Nat Med* 2005;**11**:90–4.
14. Moore K, Zhang Q, Murgolo N, Hosted T, Duffy R. Cloning, expression, and pharmacological characterization of the GPR120 free fatty acid receptor from cynomolgus monkey: comparison with human GPR120 splice variants. *Comp Biochem Physiol B Biochem Mol Biol* 2009;**154**:419–26.
15. Tanaka T, Yano T, Adachi T, Koshimizu TA, Hirasawa A, Tsujimoto G. Cloning and characterization of the rat free fatty acid receptor GPR120: in vivo effect of the natural ligand on GLP-1 secretion and proliferation of pancreatic beta cells. *Naunyn Schmiedebergs Arch Pharmacol* 2008;**377**:515–22.
16. Reiter E, Lefkowitz RJ. GRKs and beta-arrestins: roles in receptor silencing, trafficking and signaling. *Trends Endocrinol Metab* 2006;**17**:159–65.
17. Tobin AB, Butcher AJ, Kong KC. Location, location, location . . . site-specific GPCR phosphorylation offers a mechanism for cell-type-specific signalling. *Trends Pharmacol Sci* 2008;**29**:413–20.
18. Xue Y, Ren J, Gao XJ, Jin CJ, Wen LP, Yao XB. GPS 2.0, a tool to predict kinase-specific phosphorylation sites in hierarchy. *Mol Cell Proteomics* 2008;**7**:1598–608.
19. Burns RN, Moniri NH. Agonism with the omega-3 fatty acids alpha-linolenic acid and docosahexaenoic acid mediates phosphorylation of both the short and long isoforms of the human GPR120 receptor. *Biochem Biophys Res Commun* 2010;**396**:1030–5.
20. Miyauchi S, Hirasawa A, Iga T, Liu N, Itsubo C, Sadakane K, et al. Distribution and regulation of protein expression of the free fatty acid receptor GPR120. *Naunyn Schmiedebergs Arch Pharmacol* 2009;**379**:427–34.
21. Ito J, Ito M, Nambu H, Fujikawa T, Tanaka K, Iwaasa H, et al. Anatomical and histological profiling of orphan G-protein-coupled receptor expression in gastrointestinal tract of C57BL/6J mice. *Cell Tissue Res* 2009;**338**:257–69.

22. Matsumura S, Mizushige T, Yoneda T, Iwanaga T, Tsuzuki S, Inoue K, et al. GPR expression in the rat taste bud relating to fatty acid sensing. *Biomed Res* 2007;**28**:49–55.
23. Parker HE, Habib AM, Rogers GJ, Gribble FM, Reimann F. Nutrient-dependent secretion of glucose-dependent insulinotropic polypeptide from primary murine K cells. *Diabetologia* 2009;**52**:289–98.
24. Janssen S, Laermans J, Iwakura H, Tack J, Depoortere I. Sensing of fatty acids for octanoylation of ghrelin involves a gustatory g-protein. *PLoS One* 2012;**7**:e40168.
25. Iwakura H, Li Y, Ariyasu H, Hosoda H, Kanamoto N, Bando M, et al. Establishment of a novel ghrelin-producing cell line. *Endocrinology* 2010;**151**:2940–5.
26. Gotoh C, Hong YH, Iga T, Hishikawa D, Suzuki Y, Song SH, et al. The regulation of adipogenesis through GPR120. *Biochem Biophys Res Commun* 2007;**354**:591–7.
27. Oh DY, Talukdar S, Bae EJ, Imamura T, Morinaga H, Fan WQ, et al. GPR120 is an omega-3 fatty acid receptor mediating potent anti-inflammatory and insulin-sensitizing effects. *Cell* 2010;**142**:687–98.
28. Galindo MM, Voigt N, Stein J, van Lengerich J, Raguse JD, Hofmann T, et al. G protein-coupled receptors in human fat taste perception. *Chem Senses* 2012;**37**:123–39.
29. Morgan NG, Dhayal S. G-protein coupled receptors mediating long chain fatty acid signalling in the pancreatic beta-cell. *Biochem Pharmacol* 2009;**78**:1419–27.
30. Whalley NM, Pritchard LE, Smith DM, White A. Processing of proglucagon to GLP-1 in pancreatic alpha-cells: is this a paracrine mechanism enabling GLP-1 to act on beta-cells? *J Endocrinol* 2011;**211**:99–106.
31. Cornall LM, Mathai ML, Hryciw DH, McAinch AJ. Diet-induced obesity up-regulates the abundance of GPR43 and GPR120 in a tissue specific manner. *Cell Physiol Biochem* 2011;**28**:949–58.
32. Soto-Guzman A, Robledo T, Lopez-Perez M, Salazar EP. Oleic acid induces ERK1/2 activation and AP-1 DNA binding activity through a mechanism involving Src kinase and EGFR transactivation in breast cancer cells. *Mol Cell Endocrinol* 2008;**294**:81–91.
33. Sayegh AI. The role of cholecystokinin receptors in the short-term control of food intake. *Prog Mol Biol Trans Sci* 2012;**114**:1–40.
34. McLaughlin J, Grazia Luca M, Jones MN, D'Amato M, Dockray GJ, Thompson DG. Fatty acid chain length determines cholecystokinin secretion and effect on human gastric motility. *Gastroenterology* 1999;**116**:46–53.
35. Tanaka T, Katsuma S, Adachi T, Koshimizu TA, Hirasawa A, Tsujimoto G. Free fatty acids induce cholecystokinin secretion through GPR120. *Naunyn Schmiedebergs Arch Pharmacol* 2008;**377**:523–7.
36. Shah BP, Liu P, Yu T, Hansen DR, Gilbertson TA. TRPM5 is critical for linoleic acid-induced CCK secretion from the enteroendocrine cell line, STC-1. *Am J Physiol Cell Physiol* 2012;**302**:C210–C219.
37. Hara T, Hirasawa A, Sun Q, Sadakane K, Itsubo C, Iga T, et al. Novel selective ligands for free fatty acid receptors GPR120 and GPR40. *Naunyn Schmiedebergs Arch Pharmacol* 2009;**380**:247–55.
38. Duca FA, Swartz TD, Sakar Y, Covasa M. Increased oral detection, but decreased intestinal signaling for fats in mice lacking gut microbiota. *PLoS One* 2012;**7**:e39748.
39. Latour MG, Alquier T, Oseid E, Tremblay C, Jetton TL, Luo J, et al. GPR40 is necessary but not sufficient for fatty acid stimulation of insulin secretion in vivo. *Diabetes* 2007;**56**:1087–94.
40. Itoh Y, Kawamata Y, Harada M, Kobayashi M, Fujii R, Fukusumi S, et al. Free fatty acids regulate insulin secretion from pancreatic beta cells through GPR40. *Nature* 2003;**422**:173–6.
41. Ichimura A, Hirasawa A, Poulain-Godefroy O, Bonnefond A, Hara T, Yengo L, et al. Dysfunction of lipid sensor GPR120 leads to obesity in both mouse and human. *Nature* 2012;**483**:350–4.

42. Taneera J, Lang S, Sharma A, Fadista J, Zhou Y, Ahlqvist E, et al. A systems genetics approach identifies genes and pathways for type 2 diabetes in human islets. *Cell Metab* 2012;**16**:122–34.
43. Tepper BJ, Nurse RJ. Fat perception is related to PROP taster status. *Physiol Behav* 1997;**61**:949–54.
44. Ramirez I. Role of olfaction in starch and oil preference. *Am J Physiol* 1993;**265**: R1404–R1409.
45. Greenberg D, Smith GP. The controls of fat intake. *Psychosom Med* 1996;**58**:559–69.
46. Fukuwatari T, Shibata K, Iguchi K, Saeki T, Iwata A, Tani K, et al. Role of gustation in the recognition of oleate and triolein in anosmic rats. *Physiol Behav* 2003;**78**: 579–83.
47. Hiraoka T, Fukuwatari T, Imaizumi M, Fushiki T. Effects of oral stimulation with fats on the cephalic phase of pancreatic enzyme secretion in esophagostomized rats. *Physiol Behav* 2003;**79**:713–7.
48. Laugerette F, Passilly-Degrace P, Patris B, Niot I, Febbraio M, Montmayeur JP, et al. CD36 involvement in orosensory detection of dietary lipids, spontaneous fat preference, and digestive secretions. *J Clin Invest* 2005;**115**:3177–84.
49. Chale-Rush A, Burgess JR, Mattes RD. Multiple routes of chemosensitivity to free fatty acids in humans. *Am J Physiol Gastrointest Liver Physiol* 2007;**292**:G1206–G1212.
50. Gilbertson TA, Liu L, Kim I, Burks CA, Hansen DR. Fatty acid responses in taste cells from obesity-prone and -resistant rats. *Physiol Behav* 2005;**86**:681–90.
51. Fukuwatari T, Kawada T, Tsuruta M, Hiraoka T, Iwanaga T, Sugimoto E, et al. Expression of the putative membrane fatty acid transporter (FAT) in taste buds of the circumvallate papillae in rats. *FEBS Lett* 1997;**414**:461–4.
52. Hoon MA, Adler E, Lindemeier J, Battey JF, Ryba NJ, Zuker CS. Putative mammalian taste receptors: a class of taste-specific GPCRs with distinct topographic selectivity. *Cell* 1999;**96**:541–51.
53. Nelson G, Hoon MA, Chandrashekar J, Zhang Y, Ryba NJ, Zuker CS. Mammalian sweet taste receptors. *Cell* 2001;**106**:381–90.
54. Nelson G, Chandrashekar J, Hoon MA, Feng L, Zhao G, Ryba NJ, et al. An amino-acid taste receptor. *Nature* 2002;**416**:199–202.
55. Cartoni C, Yasumatsu K, Ohkuri T, Shigemura N, Yoshida R, Godinot N, et al. Taste preference for fatty acids is mediated by GPR40 and GPR120. *J Neurosci* 2010;**30**: 8376–82.
56. Ribot C, Tremollieres F, Pouilles JM, Bonneu M, Germain F, Louvet JP. Obesity and postmenopausal bone loss - the influence of obesity on vertebral density and bone turnover in postmenopausal women. *Bone* 1987;**8**:327–31.
57. Reid IR, Plank LD, Evans MC. Fat mass is an important determinant of whole-body bone-density in premenopausal women but not in men. *J Clin Endocrinol Metab* 1992;**75**:779–82.
58. Reid IR, Ames R, Evans MC, Sharpe S, Gamble G, France JT, et al. Determinants of total-body and regional bone-mineral density in normal postmenopausal women—a key role for fat mass. *J Clin Endocrinol Metab* 1992;**75**:45–51.
59. Reid IR. Relationships among body mass, its components, and bone. *Bone* 2002;**31**:547–55.
60. Cornish J, MacGibbon A, Lin JM, Watson M, Callon KE, Tong PC, et al. Modulation of osteoclastogenesis by fatty acids. *Endocrinology* 2008;**149**:5688–95.
61. Gregor MF, Hotamisligil GS. Inflammatory mechanisms in obesity. *Annu Rev Immunol* 2011;**29**:415–45.
62. Hotamisligil GS. Inflammation and metabolic disorders. *Nature* 2006;**444**:860–7.
63. Shoelson SE, Lee J, Goldfine AB. Inflammation and insulin resistance. *J Clin Invest* 2006;**116**:1793–801.

64. Berg AH, Scherer PE. Adipose tissue, inflammation, and cardiovascular disease. *Circ Res* 2005;**96**:939–49.
65. Qian BZ, Deng Y, Im JH, Muschel RJ, Zou YY, Li JF, et al. A distinct macrophage population mediates metastatic breast cancer cell extravasation, establishment and growth. *PLoS One* 2009;**4**:e6562.
66. Ricketts JR, Rothe MJ, Grant-Kels JM. Nutrition and psoriasis. *Clin Dermatol* 2010;**28**:615–26.
67. Naqvi AZ, Buettner C, Phillips RS, Davis RB, Mukamal KJ. n-3 fatty acids and periodontitis in US adults. *J Am Diet Assoc* 2010;**110**:1669–75.
68. Shinto L, Marracci G, Baldauf-Wagner S, Strehlow A, Yadav V, Stuber L, et al. Omega-3 fatty acid supplementation decreases matrix metalloproteinase-9 production in relapsing-remitting multiple sclerosis. *Prostaglandins Leukot Essent Fatty Acids* 2009; **80**:131–6.
69. Fetterman JW, Zdanowicz MM. Therapeutic potential of n-3 polyunsaturated fatty acids in disease. *Am J Health Syst Pharm* 2009;**66**:1169–79.
70. Im DS. Omega-3 fatty acids in anti-inflammation (pro-resolution) and GPCRs. *Prog Lipid Res* 2012;**51**:232–7.
71. Smith NJ. Low affinity GPCRs for metabolic intermediates: challenges for pharmacologists. *Front Endocrinol (Lausanne)* 2012;**3**:1.
72. Cintra DE, Ropelle ER, Moraes JC, Pauli JR, Morari J, de Souza CT, et al. Unsaturated fatty acids revert diet-induced hypothalamic inflammation in obesity. *PLoS One* 2012;**7**: e30571.
73. Gottlicher M, Widmark E, Li Q, Gustafsson JA. Fatty-acids activate a chimera of the clofibric acid-activated receptor and the glucocorticoid receptor. *Proc Natl Acad Sci USA* 1992;**89**:4653–7.
74. Brown AJ, Jupe S, Briscoe CP. A family of fatty acid binding receptors. *DNA Cell Biol* 2005;**24**:54–61.
75. Briscoe CP, Peat AJ, McKeown SC, Corbett DF, Goetz AS, Littleton TR, et al. Pharmacological regulation of insulin secretion in MIN6 cells through the fatty acid receptor GPR40: identification of agonist and antagonist small molecules. *Br J Pharmacol* 2006;**148**:619–28.
76. Suzuki T, Igari SI, Hirasawa A, Hata M, Ishiguro M, Fujieda H, et al. Identification of G protein-coupled receptor 120-selective agonists derived from PPAR gamma agonists. *J Med Chem* 2008;**51**:7640–4.
77. Shimpukade B, Hudson BD, Hovgaard CK, Milligan G, Ulven T. Discovery of a potent and selective GPR120 agonist. *J Med Chem* 2012;**55**:4511–5.
78. Sun Q, Hirasawa A, Hara T, Kimura I, Adachi T, Awaji T, et al. Structure-activity relationships of GPR120 agonists based on a docking simulation. *Mol Pharmacol* 2010;**78**:804–10.
79. Watson SJ, Brown AJH, Holliday ND. Differential signaling by splice variants of the human free fatty acid receptor GPR120. *Mol Pharmacol* 2012;**81**:631–42.
80. Holliday ND, Watson SJ, Brown AJ. Drug discovery opportunities and challenges at G protein-coupled receptors for long chain free fatty acids. *Front Endocrinol (Lausanne)* 2011;**2**:112.
81. Tao YX, Segaloff DL. Functional characterization of melanocortin-3 receptor variants identify a loss-of-function mutation involving an amino acid critical for G protein-coupled receptor activation. *J Clin Endocrinol Metab* 2004;**89**:3936–42.
82. Tao YX. Inactivating mutations of G protein-coupled receptors and diseases: structure-function insights and therapeutic implications. *Pharmacol Ther* 2006;**111**:949–73.
83. Katsuma S, Hatae N, Yano T, Ruike Y, Kimura M, Hirasawa A, et al. Free fatty acids inhibit serum deprivation-induced apoptosis through GPR120 in a murine enteroendocrine cell line STC-1. *J Biol Chem* 2005;**280**:19507–15.

84. Iakoubov R, Izzo A, Yeung A, Whiteside CI, Brubaker PL. Protein kinase C zeta is required for oleic acid-induced secretion of glucagon-like peptide-1 by intestinal endocrine L cells. *Endocrinology* 2007;**148**:1089–98.
85. Widmayer P, Kuper M, Kramer M, Konigsrainer A, Breer H. Altered expression of gustatory-signaling elements in gastric tissue of morbidly obese patients. *Int J Obes (Lond)* 2012;**36**:1353–9. http://dx.doi.org/10.1038/ijo.2011.216.
86. Kobilka BK, Deupi X. Conformational complexity of G-protein-coupled receptors. *Trends Pharmacol Sci* 2007;**28**:397–406.
87. Rosenbaum DM, Cherezov V, Hanson MA, Rasmussen SGF, Thian FS, Kobilka TS, et al. GPCR engineering yields high-resolution structural insights into β_2-adrenergic receptor function. *Science* 2007;**318**:1266–73.
88. Wang GS, Cooper TA. Splicing in disease: disruption of the splicing code and the decoding machinery. *Nat Rev Genet* 2007;**8**:749–61.
89. Rankinen T, Zuberi A, Chagnon YC, Weisnagel SJ, Argyropoulos G, Walts B, et al. The human obesity gene map: the 2005 update. *Obesity* 2006;**14**:529–644.

CHAPTER EIGHT

The Role of Cholecystokinin Receptors in the Short-Term Control of Food Intake

Ayman I. Sayegh
Gastroenterology Laboratory, Department of Biomedical Sciences, College of Veterinary Medicine, Tuskegee University, Tuskegee, Alabama, USA

Contents

Abstract

Cholecystokinin (CCK) is a hormone secreted by the I-cells of the upper small intestine in response to fat, protein, and some nonnutrients, for example, camostat, and a peptide/neurotransmitter secreted by neurons of the central and peripheral nervous systems. There are multiple molecular forms of CCK, for example, CCK-8, CCK-33, and CCK-58, with an active site located within the first eight amino acids of the carboxyl terminus and with a sulfate group on the seventh tyrosine residue. Physiologically, CCK increases pancreatic secretions and gallbladder and smooth muscle contractions as well as inhibits gastric emptying and food intake. CCK evokes these responses by activating two G protein-coupled receptors: CCK_1 and CCK_2. CCK_1 receptors are located mainly in the alimentary tract and contain two affinity states, high and low, whereas CCK_2 receptors are found mainly in the brain. Although a CCK-mediated reduction in cumulative food intake occurs by the activation of low-affinity CCK_1 receptors located on vagal afferents, the vagus, and the splanchnic nerves are necessary for the reduction of meal size (MS) and the prolongation of the inter-meal interval (IMI) by CCK. Finally, the

Progress in Molecular Biology and Translational Science, Volume 114
ISSN 1877-1173
http://dx.doi.org/10.1016/B978-0-12-386933-3.00008-X

reduction of food intake by CCK occurs by three possible modes of action: paracrine, endocrine, and neurocrine; thus far, the data favor a paracrine mode. In addition, the gut, which is the main source of peripheral CCK, contains the first neuronal component that senses the presence of food, the enteric nervous system. The enteric nervous system may have a role in the reduction of MS and the prolongation of the IMI by CCK.

1. CONTROL OF FOOD INTAKE: A GENERAL OVERVIEW

Food is the main source of energy that is utilized to maintain the various physiological functions of an organism. One of these functions is the control of body weight, which is governed by a natural process known as energy homeostasis—a process that balances energy intake and energy expenditure. If an imbalance between these two entities takes place, an increase or a decrease in body weight will result. For example, when energy expenditure exceeds energy intake, the organism will suffer emaciation and weight loss. In contrast, when energy intake exceeds energy expenditure, the organism will become overweight or obese.[1,2]

It is well established that obesity is a worldwide epidemic.[3] Therefore, to establish rational therapeutic strategies in the fight against this disease, a clear understanding of the mechanisms that control food intake is essential. There are two major, closely interacting mechanisms that control food intake: a long-term mechanism and a short-term mechanism. The long-term mechanism controls food intake over long periods of time such as days or years, whereas the short-term mechanism controls food intake by regulating individual meal size (MS) and the time between two consecutive meals, also known as the inter-meal interval (IMI; Fig. 8.1).

The signals that control MS and IMI are primarily peptides and hormones secreted by the digestive system, which consists of the gastrointestinal (GI) tract and the accessory digestive glands, pancreas and liver. Such signals include orexigenic peptides and hormones that increase food intake and anorexigenic peptides and hormones, which decrease food intake. Orexigenic peptides include the stomach peptide ghrelin; anorexigenic peptides include the stomach peptide gastrin-releasing peptide (GRP), the upper small intestinal hormone cholecystokinin (CCK), the lower small intestinal peptides peptide tyrosine tyrosine (PYY) and glucagon-like peptide-1 (GLP-1), and the pancreatic peptides glucagon and amylin. The long-term control of food intake occurs through signals that monitor the metabolic status of the organism, for example, energy stores and availability. Such signals include the anorexigenic white adipose tissue hormone leptin and the pancreatic β cell hormone insulin, both of which reduce food intake.

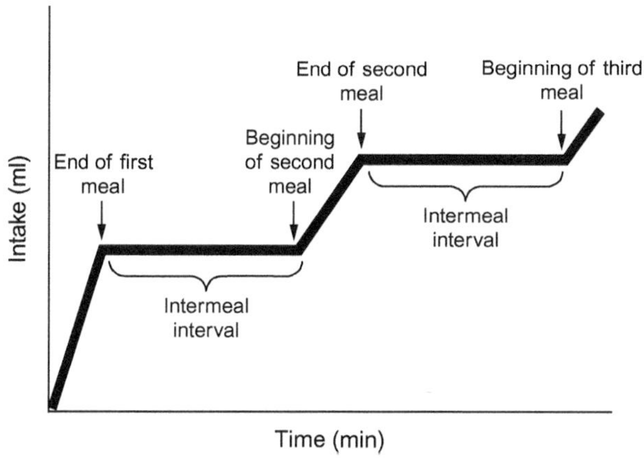

Figure 8.1 A diagram depicting normal feeding behavior and the components of the short-term control of food intake, meal size, and inter-meal interval (time between two consecutive meals). These components are generally controlled by gut–brain peptides, such as gastrin-releasing peptide from the stomach, cholecystokinin from the proximal small intestine and glucagon-like peptide-1 from the distal small intestine.

As stated previously, the control of food intake requires the integration of long-term signals, for example, leptin and insulin, and short-term signals, for example, ghrelin, GRP, CCK, PYY, GLP-1, glucagon and amylin. This interaction occurs at the hypothalamic level in the forebrain and at the dorsal vagal complex (DVC) level in the hindbrain.

The ability of the hypothalamus to regulate food intake results from two areas. First, the hypothalamus expresses receptors for leptin and insulin,[4,5] which monitor the systemic energy status of the body, that is, energy stores and availability. Second, the hypothalamus contains two populations of neurons that also control food intake by inhibiting or stimulating feeding: (1) neurons expressing propeptide proopiomelanocortin (POMC), a precursor for the anorexigenic peptide α-melanocyte-stimulating hormone (α-MSH), and (2) neurons expressing neuropeptide Y (NPY) and the endogenous melanocortin antagonist agouti-related protein (AgRP), both of which stimulate feeding (Fig. 8.2). Furthermore, POMC- and NPY/AgRP-expressing neurons project from the arcuate nucleus, the ventral-most part of the hypothalamus, to the paraventricular nucleus (PVN) of the hypothalamus,[6,7] which has a role in the reduction of food intake, and to the lateral hypothalamus,[5] which has a role in stimulating food intake (Fig. 8.2).

Leptin activates the POMC-expressing[5,8,9] neurons and inhibits the NPY/AgRP-expressing neurons.[9,10] Therefore, when there is an excess of food (Fig. 8.2), the circulating levels of leptin are increased. This increase

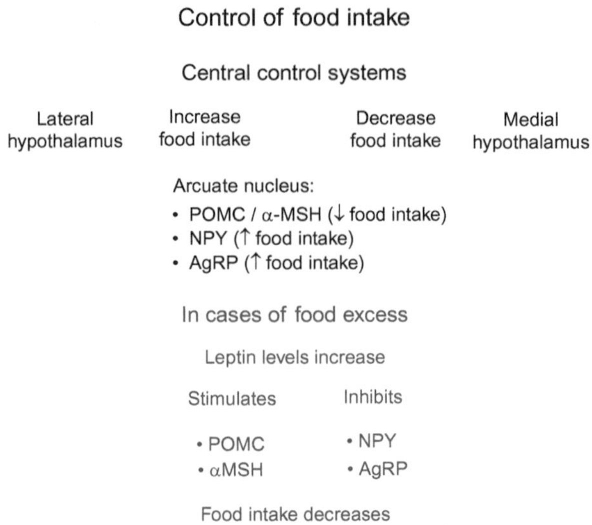

Figure 8.2 A diagram summarizing the control of the food intake process. The lateral hypothalamus controls the increase in food intake while the medial hypothalamus controls the reduction of food intake. These areas are connected to the arcuate nucleus, which contains neurons that secrete proopiomelanocortin (POMC) to stimulate neurons that secrete α-melanocyte-stimulating hormone (α-MSH) that in turn reduces food intake. In addition, the arcuate nucleus contains neurons that secrete neuropeptide Y (NPY) and the endogenous melanocortin receptor antagonist agouti-related protein (AgRP), both of which increase food intake. For example, in the case of excessive amounts of available food, leptin levels increase, resulting in a stimulation of the neurons that decrease food intake (α-MSH) and an inhibition of the peptides that increase food intake (NPY and AgRP). As a result, food intake decreases. (For color version of this figure, the reader is referred to the online version of this chapter.)

leads to the activation of POMC-expressing neurons that release α-MSH, causing a reduction in food intake, and the inhibition of NPY/AgRP-expressing neurons, which also results in a reduction in food intake. However, when there is a deficit of food, leptin levels decrease. As a result, POMC neurons will not be activated, and the NPY/AgRP neurons will be less inhibited. Therefore, the individual starts eating (Fig. 8.2).

Because feeding in most species occurs in distinct bouts or meals, the hindbrain, unlike the hypothalamus, processes peripheral signals that originate mainly from the digestive system and control MS and IMI (Fig. 8.1), for example, GRP and CCK.[11–13] This review will focus on the role of the gut–brain peptide CCK in the reduction of MS and the prolongation of the IMI. It will include a historical background about the discovery of CCK followed by listing the various forms, receptors, and physiological responses evoked by this peptide. Then, it will present evidence for the role of the CCK_1 receptor in the reduction of food intake by CCK and the mechanism of action by which this reduction takes place. In the last section, this review will provide data in support of a possible role for the GI tract and its nervous system, the enteric nervous system (ENS), in the reduction of food intake by CCK.

2. CCK: HISTORICAL PROSPECTIVE

In 1902, the British scientists Bayliss and Starling described a substance obtained from the dog intestinal extract that stimulated pancreatic secretion.[14] At that time, the extract was suspected to be secretin. However, in 1919, Braga and Campos found that an extract obtained by the same method as Bayliss and Starling also contracted the gallbladder. In 1928, Ivy and Oldberg[15] demonstrated that the substance that contracted the gallbladder could be separated from that which increased pancreatic secretion, and they called that substance CCK.[15] The word *cholecystokinin* originates from three Latin words: *chole,* which means bile, *cysto,* which means sac, and *kinin,* which means movement or contraction. Thus, the complete name cholecystokinin or CCK means the substance that contracts the gallbladder.

In 1943, Harper and Raper found that the intravenous administration of an upper intestinal mucosal extract stimulated pancreatic secretion.[16] They named this extract pancreozymin. Finally, in 1944, Greengard and colleagues confirmed that this substance was different from secretin.[17]

In 1968 and 1971, Mutt and Jorpes presented two publications describing the full amino acid sequence of porcine pancreozymin[18,19] to be a 33-amino acid peptide. They found that pancreozymin and CCK were one and the same. Therefore, because CCK was the original name given by Ivy and Oldberg in 1928, it replaced pancreozymin, the name that was given by Harper and Raper in 1943.

According to the modified classification of Wiesbaden, the main source of peripheral CCK is the endocrine I-cells of the upper small intestine.[20–22] In addition, CCK can be found in enteric neurons, both myenteric and to a much less extent submucosal,[23] and also in the pancreas.[24] The enteric neurons that contain CCK are located primarily in the colon, followed by the stomach, duodenum, jejunum, and ileum.[25] CCK-containing nerves are also located in the celiacomesenteric ganglia,[23] vagus, splanchnic, and sciatic nerves.[26–28] The central nervous system contains the majority of neuronal

CCK, especially colocalization with dopamine-containing neurons in the mesencephalon,[29,30] which are involved in the pathogenesis of schizophrenia.

3. MOLECULAR FORMS OF CCK

CCK is encoded by a gene located on chromosome 3p22–p21.3 and is synthesized by the multiple-step processing of large peptidic precursors with posttranslational modification that includes the sulfation of the seventh tyrosine residue of the COOH-terminus and the amidation of the same terminus.[31] The peptide then is stored and secreted from different tissues including the GI tract,[20–22] central and peripheral neurons,[23,32,33] the pituitary[34,35] and adrenal glands and also from various types of tumors including those of the pancreas, lung, stomach, colon, and breast (see reviews Refs. 36–38).

The work of Mutt and Jorpes[18,19] has reported that the COOH-terminal tyrosine residue-containing octapeptide of CCK-33 contains the complete, active portion of CCK responsible for contracting the gallbladder and increasing pancreatic secretion. Later, they reported another form of CCK, CCK-39, that contains an extension at the NH_2-terminus[39] of six additional amino acids. Eysselein and colleagues discovered CCK-58 in various species,[40,41] and Eberlein et al. found CCK-83 in the human intestine.[42] Shorter forms of CCK have also been isolated from the canine intestine. These forms include CCK-25, -22, -18, -8, -7, and -5.[43–45] CCK-8[46] and -58,[47–49] but not the remaining forms of CCK, have been isolated from the central nervous systems of various species.

4. CCK RECEPTORS

CCK evokes responses, for example, the reduction of food intake, the inhibition of gastric emptying and the contraction of the gallbladder, by activating two membrane G protein-coupled receptors, CCK_1 or CCK1R, formerly known as the CCK-A receptor (A stands for alimentary), and CCK_2 or CCK2R, formerly known as the CCK-B receptor (B stands for brain) (Fig. 8.3). The CCK_1 receptor was first identified on pancreatic acinar cells,[50–54] whereas CCK_2 receptors were first identified in the brain.[53,55] In addition, sulfated CCK binds to the CCK_1 receptor with a 500- to 1000-fold greater affinity than sulfated gastrin or nonsulfated CCK, both of which bind to the CCK_2 receptor with similar affinities.

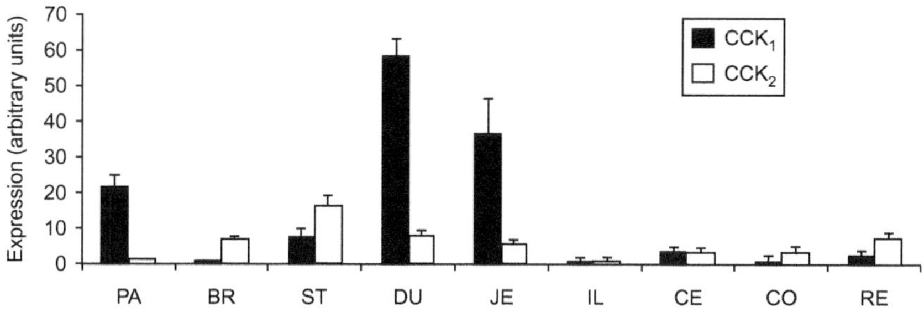

Figure 8.3 The distribution of mRNA for the two CCK receptors, CCK_1 and CCK_2, in whole tissue of the pancreas (PA), brain (BR), stomach (ST), duodenum (DU), jejunum (JE), ileum (IL), cecum (CE), colon (CO), and rectum (RE). Most CCK_1 receptor mRNA was found in the duodenum, jejunum, and pancreas, while CCK_2 receptor mRNA was found primarily in the stomach, duodenum, brain, jejunum, and rectum.

CCK receptors are distributed in many tissues including the digestive system, the GI tract—stomach and intestines—and the accessory digestive glands such as the liver and the pancreas (Fig. 8.3). In addition, the two CCK receptors are also found in the kidneys, adrenal glands, adipocytes, blood cells, and vagal afferents (see also Tables 8.1 and 8.2).

Table 8.1 Distribution of CCK_1 receptors in various tissues

CCK_1 receptor tissue	**Species**	**Reference**
Stomach	H, D, R	56–58
Exocrine pancreas	H, R, M, Bu, G, D, B, P	55,59–66
Endocrine pancreas	R, P, H	67–69
Liver	–	–
Gallbladder	Mo, B, G, H, Hm, Al	70–76
Smooth muscles of the stomach	G, H, R,	57,77–79
Smooth muscles of intestine	H, G	80–82
Adipocytes	–	–
Adrenal gland	H, R	83,84
Blood cells	H	56
Kidney	H, R, M	85,86
Vagal afferents	H, R, Ra	87–91

H, human; D, dog; R, rat; M, mouse; G, guinea pig; Ra, rabbit; B, bovine; Mo, monkey; Bu, bullfrog; P, pig; Ha, hamster; Al, alligator.

Table 8.2 Distribution of CCK_2 receptors in various tissues

CCK_2 receptor tissue	Species	Reference
Stomach	H, D, R, G	56–58,92,93
Exocrine pancreas	H, D, G, B, P	59,60,64–66,94–97
Endocrine pancreas	H, R, P, Ho, D	67,98,99
Liver	R	100
Gallbladder	–	–
Smooth muscles of the stomach	G, H	57,78
Smooth muscles of intestine	G	82
Adipocytes	R	101
Adrenal gland	H, R	83,84
Blood cells	H	56,102–104
Kidney	G, R, M	86,105–107
Vagal afferents	H, R, Ra	88–90

H, human; D, dog; R, rat; M, mouse; G, guinea pig; Ra, rabbit; B, bovine; P, pig; Ho, horse.

The CCK1R gene is located on human chromosome 4p15.1–p15.2 or mouse chromosome 5, whereas the CCK2R gene is located on human chromosome 11p15.4 or mouse chromosome 7[108,109]; each of these genes contains five exons and four introns. The cloning of pancreatic CCK_1 receptor cDNA, which encodes 444 amino acids, was described in rats and dogs by Wank and colleagues.[110,111] The cDNA encoding the CCK_2 receptor, a clone that encodes 453 amino acids, was first identified by Kopin and colleagues using a COS-7 cell line.[112] This finding was confirmed in brain and gastric tissues.[94,113,114]

There is approximately a 50% sequence homology between the CCK_1 and the CCK_2 receptors. Each receptor has seven hydrophobic transmembrane domains with an extracellular NH_2-terminus containing three to four N-linked glycosylation sites and an intracellular COOH-terminus, an E/DRY motif at the bottom of transmembrane domain III, and an NPXXY motif within transmembrane domain VII. There is little species variation in the cDNA sequence of CCK receptors.[115–119]

CCK binds to CCK1R and CCK2R with three different affinities: high, low, and very low. The K_d values for the binding of CCK to the high- and low-affinity sites of CCK1R are in the range of 50–300 pM and 50–200 nM, respectively.[52,54] The K_d values for the binding of CCK to the high- and low-affinity CCK2R sites are in the range of 100–300 pM and 2–5 nM, respectively.[120] The K_d value for very low-affinity binding to either receptor is 10 μM.[121] The different forms of sulfated CCK bind to CCK1R with similar affinities,[122,123] while sulfated CCK and sulfated gastrin-17 bind to CCK2R with similar affinities.[124]

A list of CCK receptor antagonists can be found in Table 8.3. There are three groups of antagonists for the CCK receptors: amino acid derivatives, for example, benotript and proglumide; peptide analogs of CCK and gastrin, such as JMV-180 and JMV-179[135,136] (both antagonize the CCK_1 receptor); and nonpeptide ligands, such as members of the benzodiazepine-derivative family, for example, L364,718 (MK-329), an antagonist for the CCK_1 receptor,[34,125] and L365,260, an antagonist for the CCK_2 receptor.[128]

The activation of CCK receptors involves several pathways. Both forms of CCK receptors activate phospholipase C β isoforms through heterotrimeric G-proteins of the G_q family, which produce inositol triphosphate and result in the mobilization of intracellular calcium and the activation of protein kinase C.[137–139] In addition, CCK receptors activate the mitogen-activated kinase (MAPK) cascade, which increases pancreatic secretion through the CCK_1 receptor[140] and cell proliferation through the CCK_2 receptor.[141,142] The two receptors also activate the phosphatidylinositol 3 kinase (PI3K) pathway to

Table 8.3 List of CCK antagonists and their selectivity

Name of antagonist	Receptor selectivity	Reference
L364,718 (devazepide)	$CCK_1 \ggg CCK_2$	125
CR2017 (dexloxiglumide)	$CCK_1 \gg CCK_2$	126
SR-27,897 (Iinitrip)	$CCK_1 \ggg CCK_2$	127
L365,260	$CCK_2 \gg CCK_1$	128
L369,466	$CCK_2 \ggg CCK_1$	129
YM-022	$CCK_2 \ggg CCK_1$	130
CR-2945	$CCK_2 \gg CCK_1$	131
YF-476	$CCK_2 \ggg CCK_1$	132
GV150013X	$CCK_2 \ggg CCK_1$	133
PD-1343308 (CI-988)	$CCK_2 \gg CCK_1$	134

increase pancreatic secretion via the CCK_1 receptor[143–145] and cell proliferation via the CCK_2 receptor.[146–149] Both CCK receptors also activate the focal adhesion kinase pathways, which mediate cell adhesion, morphology, and proliferation.[150–152] The CCK_1 receptor activates adenylyl cyclase, leading to increased cyclic AMP,[153,154] and also activates the nitric oxide/cGMP pathway, which leads to cell proliferation,[155,156] pancreatic secretion,[157,158] and gastroprotection.[159,160] Lastly, the CCK_2 receptor activates the JAK2/STAT3 pathway, which mediates growth effects.[148,161]

5. PHYSIOLOGICAL RESPONSES EVOKED BY CCK

CCK signaling evokes a number of physiological responses. As mentioned earlier, the earliest discovered response evoked by CCK was the contraction of the gallbladder[15,162] followed by an increase in pancreatic secretion.[16] More specifically, CCK increases pancreatic exocrine secretions, such as alkaline phosphatase,[163,164] disccharidase,[165] and enterokinase.[166] In addition, CCK stimulates the synthesis of pancreatic enzymes, such as amylase,[167–169] as well as fluids and bicarbonates.[170,171] Furthermore, CCK, especially CCK-8 and -33, increases pancreatic endocrine secretions such as insulin and glucagon in dogs and rats but not in humans or pigs.[172,173] As reported by Rothamn and Wells, CCK also increases pancreatic weight.[168] Due to the presence of CCK-secreting nerves that innervate the smooth muscles of the GI tract, primarily in the small and large intestines,[23,174] exogenous CCK increases intestinal motility[175–177] and intestinal blood flow.[178] However, the effect of CCK on gastric motility is species- and region-specific: in humans and dogs, CCK inhibits gastric emptying[179–182]; in guinea pigs, CCK stimulates gastric motility.[183] Finally in 1973, Gibbs, Young, and Smith demonstrated that exogenous CCK also inhibits food intake in rats.[184] This finding has also been extended to different species including rabbits,[185] humans,[186] monkeys,[187] pigs,[188] and sheep.[189] The satiety and the satiation effects evoked by CCK will be discussed in detail in the following sections.

6. THE ROLE OF CCK_1 RECEPTORS IN THE SHORT-TERM CONTROL OF FOOD INTAKE BY CCK

The presence of fat, partially digested proteins, and some nonnutrient trypsin inhibitors in the lumen of the gut stimulate the secretion of CCK from the I-cells of the upper small intestine. Triglycerides[190] and long-chain fatty acids that are greater than 12 carbons, for example, oleate,[191] and unhydrolyzed or partially hydrolyzed proteins, for example, casein and soybean trypsin inhibitors[192,193] are potent stimulators of endogenous CCK secretion. In addition, the nonnutrient trypsin inhibitor camostat has been shown to stimulate endogenous CCK secretion from those cells.[194]

The mechanism by which these substances stimulate the secretion of endogenous CCK is through the protection against digestion by intestinal proteases of three intestinal peptides: (1) monitor peptide from pancreatic juice[195]; (2) luminal CCK-releasing factor from intestinal secretions[196]; and porcine diazepam-binding inhibitor from porcine intestinal mucosa.[197] Once these three peptides are protected, they can safely reach the I-cells and cause the stimulation of CCK secretion.

Both exogenous and endogenous CCK reduce food intake. Gibbs and colleagues demonstrated that intraperitoneal injections of CCK-8 reduce cumulative food intake in rats.[184,198] This result has been confirmed in other species such as monkeys[187,199] and humans.[186,200] Furthermore, the reduction of food intake by CCK-8 occurs in rats as young as 5 days old.[201] The reduction of food intake by other forms of CCK, for example, CCK-33 and -58, has not been tested in this age group.

Similar to CCK-8, other forms of CCK have been shown to reduce food intake with different potencies. For example, CCK-33[202,203] and -58[204–206] reduced cumulative food intake (30 and 60 min following 3-h or overnight food deprivation, respectively) similar to CCK-8; however, they prolonged the IMI more than CCK-8 (Fig. 8.4). Recently, we have shown that CCK-33 and -58 reduce the size of the first meal (10% sucrose solution) similar to CCK-8, but only CCK-33 and -58 were able to prolong the IMI.[11,12] This prolongation by CCK-33 and -58 resulted in an increase in the satiety ratio, which is the amount of food consumed per given unit of time (i.e., the MS divided by the IMI). The greater this value, the more potent the peptide is in reducing food intake. This difference in the potency between CCK-33/58 and CCK-8 can be attributed to two factors. First, CCK-33 and -58 have a longer half-life, which also results in a delayed hepatic clearance time for the peptides compared with CCK-8.[207–209] This allows the two peptides to stay in the circulation longer and to act more potently than CCK-8. Second, the tertiary structure of the longer peptides is more complicated compared to CCK-8[210]; with an arm folding around the active site, the longer peptides become more resistant to digestion by plasma proteases. As a result, the longer peptides stay in the circulation longer and act more efficiently compared with CCK-8.

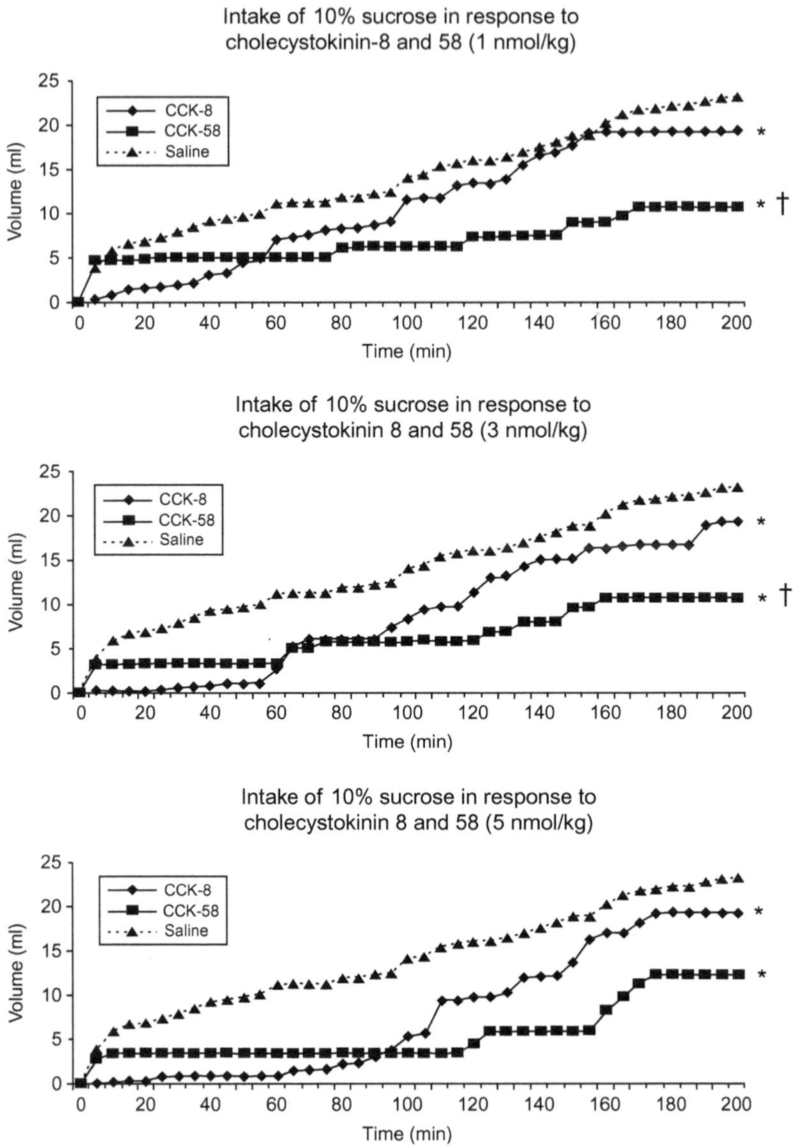

Figure 8.4 Cumulative food intake (10% sucrose, 200 min test) in rats deprived of food overnight in response to CCK-8, CCK-58 (1, 3, 5 nmol/kg) or saline administered intraperitoneally. Both peptides reduced food intake to a greater extent than saline, and CCK-58 reduced food intake more than CCK-8 and for longer durations. *Significant difference from saline, †Significant difference between CCK-8 and CCK-58.

Soybean,[211] oleate, and casein[212–215] reduce food intake via a CCK-dependent mechanism. In addition, oro-gastric gavage of the nonnutrient trypsin inhibitor camostat has also been shown to reduce food intake in rats.[216–218] Recently, we have shown that camostat, which stimulates the secretion of CCK-58, the only endocrine form of CCK,[219,220] reduces food intake by decreasing the size of the first meal and prolonging the IMI.[11,25]

The reduction of food intake by both exogenous and endogenous CCK is mediated by the CCK_1 receptor. Devazepide, a potent CCK_1 receptor antagonist, abolished the reduction of food intake evoked in response to CCK-8 in rats[221,222] and in monkeys.[223] The same antagonist also abolished the reduction of food intake caused by intestinal infusions of oleate and casein[213,215] and camostat.[217]

Furthermore, studies of a genetic rat model that lacks the CCK_1 receptor, the obese Otsuka Long Evans Tokushima Fatty (OLETF) rat, and its control, the Long Evans Tokushima Otsuka (LETO) rat,[224] confirmed the roles of CCK and the CCK_1 receptor in the short-term regulation of food intake and clarified many of the mechanisms by which CCK evokes this physiological response (see Ref. 225 for details). This rat model was developed by selective breeding that resulted in a 6847-base pair deletion in the CCK_1 receptor gene that spanned the promoter region and the first and second exons leading to a nonfunctional receptor.[226] The OLETF rat is hyperphagic due to eating large meals as a result of a loss in the feedback satiety signals and an increase in dorsomedial hypothalamic NPY expression.[227] In addition, these rats are diabetic and suffer from polydipsia and polyuria.[224] In OLETF rats, CCK-8 failed to reduce food intake,[228] and these rats showed a decreased response to the satiety effects of lipid infusion and were also less responsive to a CCK_1 receptor antagonist compared to LETO controls.[229,230] The reduction of food intake by CCK-33 and -58 has not been tested in this animal model.

Finally, it has been shown that the reduction of food intake by CCK-8 was attenuated in obese Zucker fatty rats,[231] which lack the leptin receptor, and in rats maintained on a high-fat diet.[232] This result suggests an interaction between CCK-8 and the fat hormone leptin. In addition, the adaptation to a high-fat diet causes many morphological changes in the intestine, for example, hypertrophy and hyperplasia,[233,234] and results in increased plasma CCK.[235,236] Furthermore, it causes an attenuation of enteric and hindbrain neuronal activation by CCK-8, as measured by the quantification of Fos-like immunoreactive (Fos-LI) neurons; Fos is marker for neuronal activation.[237] These results could be due to a downregulation of the CCK_1 receptor, which mediates the reduction of food intake evoked by CCK-8, as a result of the continuous secretion of CCK by fat, as demonstrated in the pancreas.[238,239] However, these results have not been tested using CCK-33 or the only endocrine form of CCK in the rat, CCK-58.[219]

7. CCK: MECHANISM OF ACTION

It is generally accepted that CCK is secreted by the I-cells of the small intestine in response to fat and protein and reduces food intake by activating CCK_1 receptors located on pyloric/duodenal vagal afferents that terminate in the hindbrain areas that regulate food intake including the area postrema (AP), the dorsal motor nucleus of the vagus (DMV), and the nucleus tractus solitarius (NTS). The following data support this mechanism: the secretion of CCK from the I-cells of the small intestine[20–22] depends primarily on digested fats and proteins,[192,193] and intra-intestinal infusions of oleate, maltotriose, and maltose in real[240] and sham-fed rats,[213] orogastric gavage of the nonnutrient trypsin inhibitor camostat,[217] and intraperitoneal injections of CCK-8[221] reduce cumulative food intake through a CCK_1 receptor-dependent mechanism, as pretreatment with devazepide, a specific CCK_1 receptor antagonist (Table 8.3), attenuates this reduction.

The CCK_1 receptor can be found on nerves,[87] smooth muscles, and mucosal cells of the pyloric sphincter, duodenum, jejunum, and ileum.[25] In addition, this receptor can be found on certain vagal afferents in the aforementioned areas, namely the interganglionic laminar endings (IGLE), which reside in close proximity to the I-cells that secret CCK and exhibit CCK receptor immunoreactivity,[77,241] implying that they can potentially be activated by CCK. In fact, it is known that CCK reduces food intake by activating CCK_1 receptors on vagal, capsaicin-sensitive, sensory afferents[114,242–244] that synapse within central hindbrain areas, for example, AP, NTS and DMV, that control food intake.[245] In addition, it has been documented that a selective branch or total subdiaphragmatic vagotomy attenuated the reduction in cumulative food intake evoked by CCK-8 and -33,[246–249] intra-intestinal infusions of nutrients,[250] and orogastric gavage of camostat.[217] Pretreatment with the neurotoxin capsaicin, which destroys vagal afferents, attenuates the reduction in cumulative food intake evoked by CCK-8[251–253] or the intra-intestinal infusion of nutrients,[212] which stimulates the release of endogenous CCK. Recently, we have shown that CCK-8, CCK-33 and camostat only reduce the MS via the vagus nerve,[254] whereas the prolongation of the IMI by CCK-33 and camostat

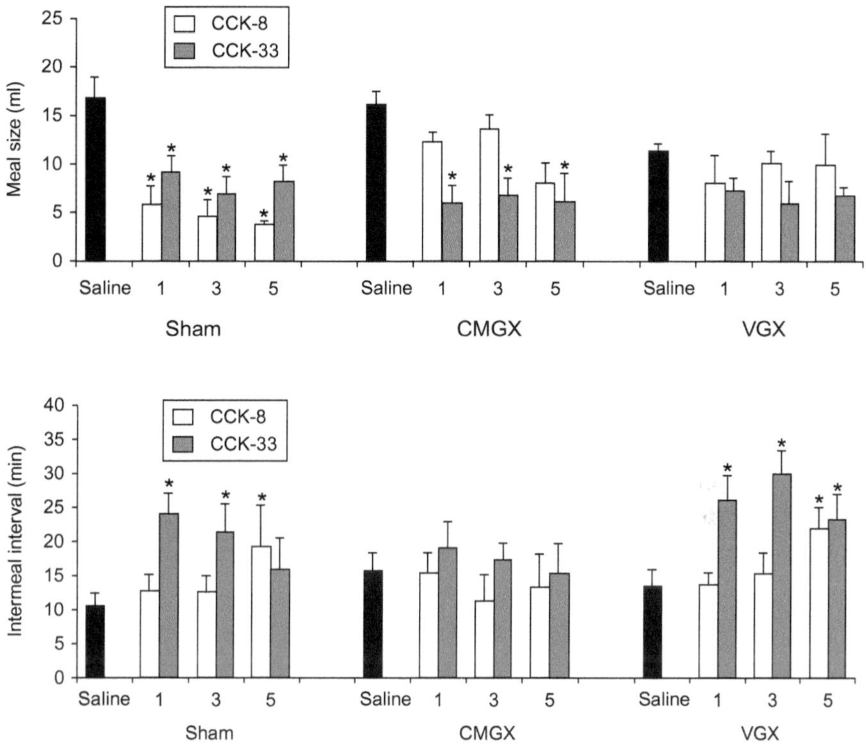

Figure 8.5 The effects of vagotomy (VGX) and celiacomesenteric ganglionectomy (CMGX) on the reduction of meal size following administration of CCK-8 or -33 (1, 3, and 5 nmol/kg, i.p., upper panel) and the prolongation of the inter-meal interval by CCK-33 (lower panel). The reduction of meal size by both peptides was attenuated by VGX (upper panel), while the prolongation of the inter-meal interval by CCK-33 was attenuated by CMGX. *Significant difference from saline.

occurred through the splanchnic nerve, which is the arm of the innervation of the gut that carries sympathetic and spinal afferents[254] (Fig. 8.5). The testing of changes in MS and IMI in response to CCK-8, CCK-33, and camostat, in capsaicin-pretreated rats requires further evaluation.

Three studies have been performed to localize the potential area of the gut in which CCK is utilized to transmit satiety and satiation signals to the DVC. Moran and colleagues have shown that the surgical resection of the pyloric sphincter reduces the satiety effect of CCK-8.[255] This result indicates that the CCK_1 receptor, which mediates the reduction of food intake evoked by CCK-8, is located in the pyloric sphincter. Cox et al. found that a continuous infusion of CCK-8 near the celiac artery, which supplies

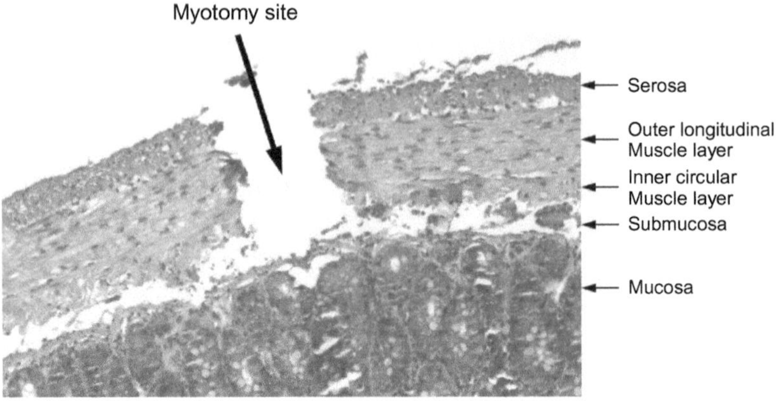

Figure 8.6 Photomicrograph of a myotomy site in the duodenum—the procedure severs the myenteric and submucosal plexuses of the enteric nervous system (ENS) that are located under the submucosa and between the two muscle layers, respectively. This procedure can be used as a valuable tool to study the role of the ENS in the various functions evoked by gut peptides, such as CCK and GRP.[13,25,258,259] (See Color Insert.)

the upper GI tract, produced a significant reduction in food intake compared with infusions given in the jugular vein.[256,257] Recently, Lateef et al.[25] demonstrated that duodenal myotomy (Fig. 8.6), a surgical procedure that damages the ENS of the duodenum and the vagal afferents in this area, attenuated the reduction in the MS and the prolongation of the IMI by camostat (Fig. 8.7). Therefore, these data support a duodenal site of action for both exogenous and endogenous CCK, which transmits satiety and satiation signals to the hindbrain.

The vagus nerve terminates in central hindbrain areas that control food intake such as the AP, NTS, and DMV. Significant evidence has shown a potential role for these areas in the reduction of food intake evoked by CCK. First, these nuclei contain the CCK_1 receptor, which mediates the reduction of food intake evoked by CCK,[245] and lesions of these nuclei attenuate this reduction of food intake.[260] Second, the AP, NTS, and DMV, can be activated, as measured by increased Fos-LI, by CCK-8,[261] CCK-33,[262] CCK-58,[263] intestinal nutrients,[264] and camostat[217,218] through a CCK_1 receptor-dependent mechanism. As such, these data demonstrate that CCK, released by the I-cells in response to nutrients, reduces MS by activating CCK_1 receptors located on duodenal vagal afferents and prolongs the IMI[11,254] by activating CCK_1 receptors on splanchnic afferents[265] that terminate in food control areas in the hindbrain.

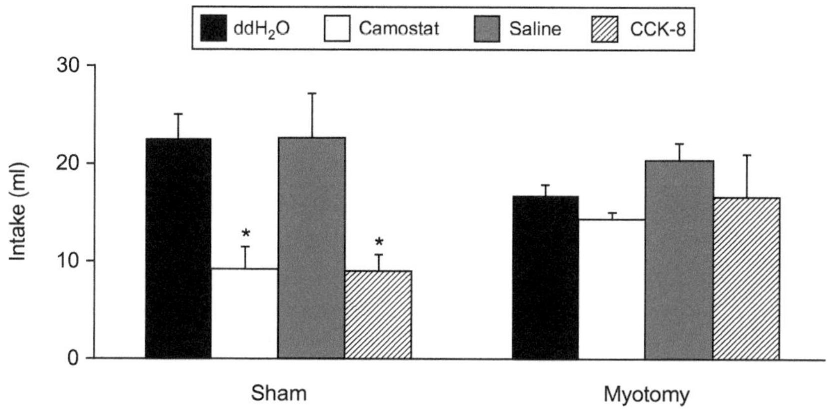

Figure 8.7 The effect of duodenal myotomy on the reduction of meal size (10% sucrose solution following overnight food deprivation) evoked by the administration of CCK-8 (5 nmol/kg intraperitoneally) or camostat (200 mg/kg orally). Both treatments reduced the size of the first meal, and myotomy attenuated this response, which indicates that the duodenum is the site of action for both CCK-8 and camostat. * Significant difference from ddH_2O or saline.

There are three possibilities that could explain the mode of action for CCK: endocrine, paracrine, and neurocrine. The endocrine route assumes that the I-cells secrete CCK into the blood stream, primarily the circulation of the upper GI tract (the celiac and cranial mesenteric arteries), and activates the DMV, NTS, and AP in the hindbrain through circumventricular organs that lack the blood–brain barrier such as the AP. The result of this activation is a reduction in food intake. Adopting the endocrine route requires an elevation of plasma CCK following a high-fat or high-protein meal, which are the main stimulators of endogenous CCK release. However, data from vagotomy/capsaicin, myotomy,[25,212,247,250] and injections of CCK-8 via the hepatic portal vein, argue against this mechanism. First, the activation of the DVC by CCK, which evokes a reduction in food intake, requires duodenal vagal afferents. Second, gut peptides may enter the hepatic circulation through the portal vein to be processed by the liver prior to entering the general circulation. Therefore, if CCK-8 acts through an endocrine route, then food intake should be reduced when it is infused into the hepatic portal vein; however, CCK-8 infused into this vein reduced food intake with only 25% of the potency of an intraperitoneal injection.[266,267] Therefore, CCK is not working through an endocrine route. Furthermore, while CCK has two affinity states, high- and low-affinity,[268,269] the peptide reduces food intake by activating the low-affinity CCK_1 receptor[270] located on vagal afferents.[271] This finding also does not support an endocrine route for CCK-8 to reduce food intake. Instead, all of the data described above support a paracrine route of action for this form of CCK.

The paracrine route suggests that CCK that is secreted by I-cells travels through the interstitial space and activates neighboring vagal afferents, which intensely supply the duodenum and terminate in the DVC to reduce food intake. All of the previous data support this route of action.[25,212,247,250] However, most of these data were generated using a synthetic form of CCK, CCK-8, which is not an endocrine form according to Reeve and colleagues,[219] who demonstrated that the only endocrine form of CCK is CCK-58.[219] Therefore, re-examining the previous data using this form of CCK is required to rule out or to confirm the paracrine route of action.

The neurocrine route suggests that CCK is secreted by neurons in the gut and activates nearby vagal afferents, which in turn activate the DVC to reduce food intake. This route requires intense investigation, as there is not enough data to support or reject it. It is well known that enteric neurons secrete CCK, which also works as a neurotransmitter in the gut.[272–274] In addition, enteric neurons can be activated by CCK-8, -33 and CCK-58, oleate, and camostat,[217,218,262,263,275–278] and devazepide, a specific CCK_1 receptor antagonist, attenuates this activation.[264,279–283] As such, CCK may be secreted by enteric neurons to activate the vagal afferents that will signal the DVC to reduce food intake. This hypothesis requires significant further investigation.

8. INTERACTION BETWEEN CCK AND OTHER SATIETY PEPTIDES

Many reports have shown that CCK may work in conjunction with other peptides, including leptin, cocaine and amphetamine-regulated transcript (CART), and apolipoprotein-IV (ApoA-IV), to enhance the reduction of food intake, possibly to decrease body weight. There is less evidence for the involvement of the following molecules with CCK to reduce food intake: PYY, dopamine, and serotonin (5-hydroxytryptamine or 5-HT).

It has been shown that CCK-8 enhances long-term satiety effects (24 h following food intake) in rats, mice, and fish.[284–287] The reduction in food intake by leptin and CCK-8 was accompanied by an increase in Fos-LI within the feeding control areas of the forebrain and hindbrain.[286,288] In addition, the reduction of food intake by leptin and CCK-8 was attenuated by capsaicin treatments and pretreatment with the specific CCK_1 receptor antagonist devazepide.[286] Recently, it has been shown that leptin receptors are coexpressed with the low-affinity CCK_1 receptor, which mediates the reduction of food intake induced by CCK-8, in neurons of the nodose ganglia, which contain the cell bodies of the vagal sensory afferents that also mediate the reduction of food intake by CCK-8.[289] The mechanism by which leptin and CCK interact to enhance their satiety effects involves an increase in cytosolic calcium[290] and the phosphorylation of the signal transducer and activation of transcription 3 (STAT3) pathway, which causes a closure of K^+ channels that results in membrane depolarization and neuronal activation.[291]

Recently, it has been shown that CCK-8 reduces food intake synergistically with CART[292] and that CCK increases the expression of this neuromodulator.[293] The combination of both peptides also increased Fos expression in the NTS of the brainstem, the paraventricular nucleus, and the dorsomedial nucleus of the hypothalamus.[292] This effect was also attenuated by devazepide.[292]

The work of Lo and colleagues[294] has shown that a threshold dose of ApoA-IV, the protein that carries fat from the lymphatic vessels and the circulation, and a threshold dose of CCK-8 reduced food intake more than either peptide alone. This effect was also attenuated by lorglumide, a specific CCK_1 receptor antagonist.[294] In addition, they have shown that levels of dipeptidyl peptidase IV (DPP-IV) and aminopeptidase, enzymes responsible for destroying these peptides in the blood, were greater in plasma than in lymph.[295] Therefore, CCK-8 and ApoA-IV are more potent in reducing food intake when given intraperitonealy than intravenously. Furthermore, increased gastric motility in response to ApoA-IV in anesthetized rats was abolished by the CCK_1 receptor antagonist devazepide.[296]

9. THE POSSIBLE ROLE OF THE ENS OF THE GUT IN THE SHORT-TERM REGULATION OF FOOD INTAKE BY CCK

The GI tract[297] is a tube-like structure that extends from the mouth to the anus consisting of the mouth, esophagus, stomach, small intestine (duodenum, jejunum, and ileum), and the large intestine (cecum, colon, and rectum). Histologically (Fig. 8.8), starting from the lumen of the gut, the GI tract consists of four layers: (1) mucosa, including epithelium (enterocytes and endocrine cells), lamina propria, and muscularis mucosae; (2) submucosa; (3) two muscle layers, inner circular and outer longitudinal; and (4) serosa. Between the layers of the gut, there are two ganglionated nerve plexuses: the submucosal and the myenteric, which is sandwiched between the inner circular and the outer longitudinal muscle layers. These

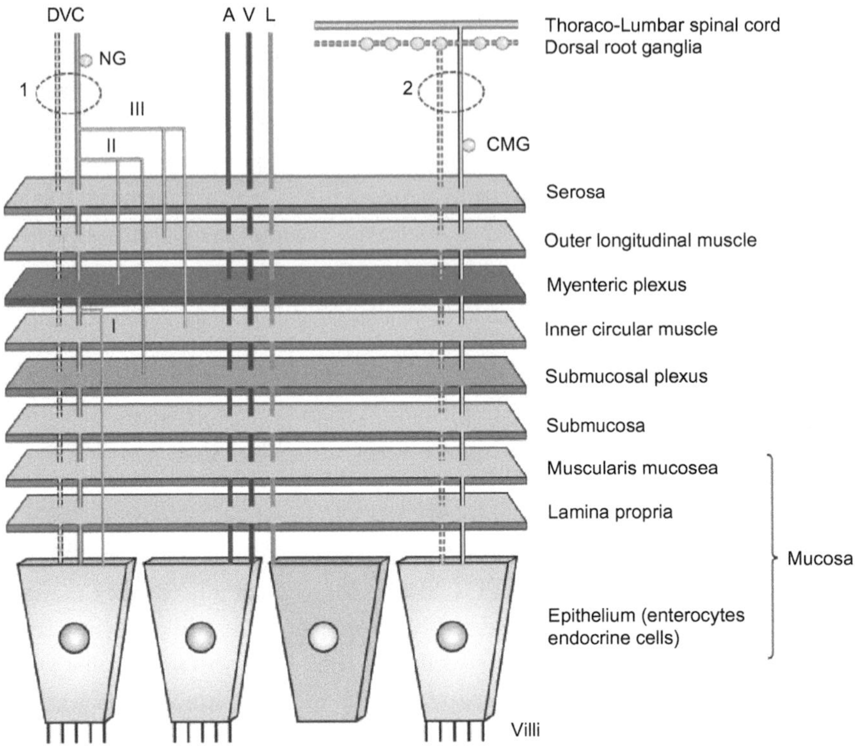

Figure 8.8 A diagram showing the various layers and innervation of the gastrointestinal tract; the layers include the mucosa (epithelium, lamina propria, and muscularis mucosea), submucosa, inner and outer muscle layers, and serosa. The myenteric and submucosal plexuses, which comprise the enteric nervous system, are located under the submucosa and between the two muscle layers, respectively. The extrinsic innervation of the gut consists of the splanchnic and vagus nerves. The vagus nerve (1) consists of two components. The motor component has cell bodies in the dorsal motor nucleus of the vagus located in the dorsal vagal complex (DVC) of the hindbrain. The cell bodies of the sensory portion of the vagus nerve reside in the nodose ganglia (NG). The sensory portion of the vagus consists of three known vagal afferents: intra villus arbors (I), intramuscular arrays (II), and interganglionic laminar endings (III). The splanchnic nerve (2) consists of two parts: sympathetic efferents and spinal afferents. The cell bodies of the sympathetic efferents reside in the thoracolumbar area of the spinal cord (presynaptic) and the celiacomesenteric ganglia (CMG, postsynaptic). The cell bodies of the spinal afferents reside in the dorsal root ganglia (DRG). A, artery; V, vein; L, lacteal (lymphatic vessel). (See Color Insert.)

plexuses collectively form the ENS, which is the intrinsic innervation of the gut and also controls most gut functions, for example, secretion, motility, and digestion. The gut is also supplied by extrinsic innervation through two nerves: the vagus nerve,[243–245,271] which consists of two portions, sensory and motor, and the splanchnic nerve, which also consists of two portions, sympathetic efferents and spinal afferents.[298–301] The cell bodies of the ENS are located in various ganglia throughout the gut. The cell bodies of the motor portion of the vagus nerve are located in the dorsal motor nucleus of the vagus in the hindbrain, and the cell bodies of the sensory portion of the vagus are located in the nodose ganglia.[241,302–306] The cell bodies of the sympathetic efferents are located in the thoracolumbar region of the spinal cord (presynaptic) and the celiacomesenteric ganglia (postsynaptic) between the cranial mesenteric and celiac arteries.[307,308] The cell bodies of the spinal afferents are located in the dorsal root ganglia along the spinal cord. All of these structures either contain CCK or CCK receptors or can be activated by CCK.[87,218,262–264,272,274,278,283,309–313]

The GI tract may play a primary role in the control of food intake by CCK.[314] This hypothesis is based on the following: (1) the major source of CCK in the periphery is the GI tract[297]; (2) the gut is the first system that is in direct contact with fat and protein, the main stimulus of CCK release[192,193]; (3) CCK-producing I-cells[21,22] and CCK_1 receptors[25,280] are heavily found in the gut; (4) the gut is supplied by the third component of the autonomic nervous system, the ENS,[297] which contains CCK[274] and CCK_1 receptors[25,280] and can be activated by all forms of exogenous and endogenous CCK[218,262–264,275,278,279,281–283,310,312,313] (Fig. 8.9); (5) the ENS

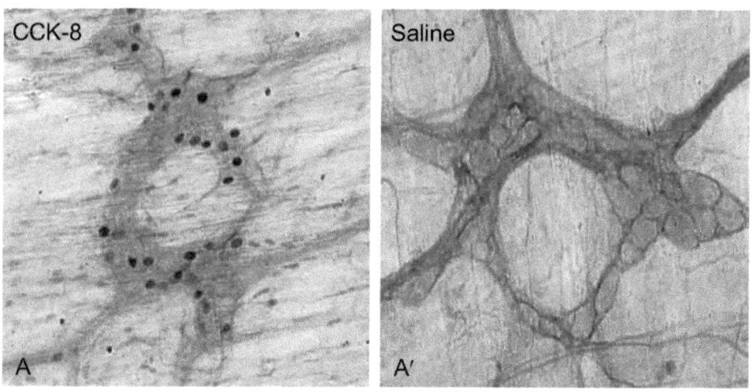

Figure 8.9 Activation of the myenteric neurons of the duodenum by CCK-8 (A) and saline vehicle (A′). The dark dots represent Fos-like immunoreactivity, a marker for neuronal activation, in this plexus.

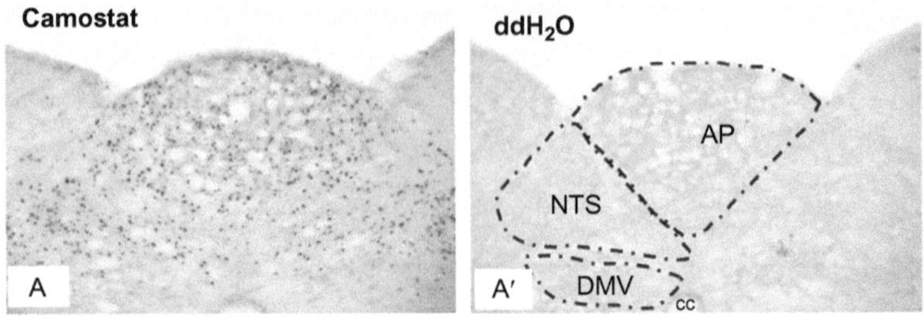

Figure 8.10 Activation of the dorsal vagal complex, area postrema (AP), nucleus tractus solitarius (NTS), and dorsal motor nucleus of the vagus (DMV) by camostat (A) and double-distilled water (ddH_2O) vehicle (A′). The pinpoint dots represent Fos-like immunoreactivity, a marker for neuronal activation, in these areas.

also communicates neurally with the feeding control areas in the DVC of the hindbrain via the vagus[243,244] and the splanchnic[88,315] nerves, which also contain CCK_1 receptors and can be activated by CCK[265,307,316,317] (Fig. 8.10); and (6) the gut contains vagal, capsaicin-sensitive, sensory afferents that also contain CCK_1 receptors, can be activated by CCK, and reside in close proximity to the I-cells that secrete CCK.[241] As such, the GI tract may have a role in the short-term regulation of food intake by CCK. This role could be direct or indirect, for example, through the modulation of GI functions that will ultimately affect food intake, such as the inhibition of gastric emptying, contraction of the gallbladder, and smooth muscle cells or an increase in pancreatic secretions.

Washington and colleagues have shown that the mRNA for CCK_1 receptors can be found in whole gut tissue of neonatal rats as young as 4 days old.[277] In adult rats, Lateef et al.[25] demonstrated that CCK_1 receptor mRNA exists in whole gut tissue in the duodenum >> jejunum > ileum. Lateef and colleagues[25] have also shown that the duodenum is the site of action for the reduction in MS by CCK-8, CCK-33 and camostat, and the prolongation of the IMI by CCK-33 and camostat[11,12,254,318]; CCK-8 increased the satiety ratio in diabetic rats to a greater extent than CCK-33.[319] This reduction and prolongation by CCK was mediated by the vagus and the splanchnic nerves, respectively.[254]

It has been shown that all forms of CCK,[217,218,262,263,278,280,310] oleate,[275] and camostat[217,218] activate specific enteric neurons in the rat gut.[275,276,320] Such activation was measured by quantifying the enteric neurons expressing Fos-LI.[321] By utilizing a double-labeling immunohistochemical technique of Fos and five different phenotypic markers,[320] it has been shown that CCK and oleate activate approximately 48% of the nitric oxide synthase-immunoreactive neurons in the myenteric plexus and 60% of the calbindin- and calretinin-immunoreactive neurons in the submucosal plexus.[275,276,320] These findings may suggest that CCK activates motor neurons in the myenteric plexus and sensory neurons in the submucosal plexus.

Furthermore, the activation of enteric neurons by CCK and oleate was mediated by the CCK_1 receptor.[217,264,280,283,313] However, unlike the activation of the DVC by CCK, the activation of enteric neurons by CCK was independent of the vagus nerve.[312] This finding indicates that the vagus nerve or vagal afferents may be necessary for transmitting the satiety signal to the DVC of the hindbrain, but they are not required for the activation of enteric neurons by CCK.

The complexity of the ENS, consisting of millions of neurons and tens of neurotransmitters, and the multiplicity of the layers of the gut along with extrinsic innervation through vagal and splanchnic spinal afferents has made our full understanding of a possible role for this system in the control of food intake by CCK difficult. The data described above do not confirm such a role, but only suggest that a role for the ENS in the control of food intake by CCK is possible. However, intense investigation is required to unveil such a role. A genetically engineered animal model that lacks the ENS, at least in one or more parts of the gut, could be a good tool for studying a possible direct role of the ENS in the control of food intake by CCK. In addition, the chemical ablation of the ENS, that is, using benzalkonium chloride or pharmacological blockers of enteric neurotransmission, might also be good tools for evaluating a possible direct role of the ENS on the satiation and satiety responses evoked by CCK.

ACKNOWLEDGMENTS

The author would like to thank Ms. Martha C. Washington, MS. and Dr. Shannon Raboin from the Gastroenterology Laboratory of Tuskegee University, College of Veterinary Medicine for their help in collecting some of the original data used in this work. Support: NIH 1SC1DK094972-01A1 and The Birmingham Racing Commission.

REFERENCES

1. Bermudez-Silva FJ, Viveros MP, McPartland JM, Rodriguez de Fonseca F. The endocannabinoid system, eating behavior and energy homeostasis: the end or a new beginning? *Pharmacol Biochem Behav* 2010;**95**:375–82.
2. Das UN. Obesity: genes, brain, gut, and environment. *Nutrition* 2010;**26**:459–73.
3. Bonomi AG, Westerterp KR. Advances in physical activity monitoring and lifestyle interventions in obesity: a review. *Int J Obes (Lond)* 2012;**36**:167–77.

4. Schwartz MW, Seeley RJ, Campfield LA, Burn P, Baskin DG. Identification of targets of leptin action in rat hypothalamus. *J Clin Invest* 1996;**98**:1101–6.
5. Elias CF, Aschkenasi C, Lee C, Kelly J, Ahima RS, Bjorbaek C, et al. Leptin differentially regulates NPY and POMC neurons projecting to the lateral hypothalamic area. *Neuron* 1999;**23**:775–86.
6. Baker RA, Herkenham M. Arcuate nucleus neurons that project to the hypothalamic paraventricular nucleus: neuropeptidergic identity and consequences of adrenalectomy on mRNA levels in the rat. *J Comp Neurol* 1995;**358**:518–30.
7. Elmquist JK. Hypothalamic pathways underlying the endocrine, autonomic, and behavioral effects of leptin. *Physiol Behav* 2001;**74**:703–8.
8. Schwartz MW, Seeley RJ, Woods SC, Weigle DS, Campfield LA, Burn P, et al. Leptin increases hypothalamic pro-opiomelanocortin mRNA expression in the rostral arcuate nucleus. *Diabetes* 1997;**46**:2119–23.
9. Cowley MA, Smart JL, Rubinstein M, Cerdan MG, Diano S, Horvath TL, et al. Leptin activates anorexigenic POMC neurons through a neural network in the arcuate nucleus. *Nature* 2001;**411**:480–4.
10. Korner J, Savontaus E, Chua Jr SC, Leibel RL, Wardlaw SL. Leptin regulation of Agrp and Npy mRNA in the rat hypothalamus. *J Neuroendocrinol* 2001;**13**:959–66.
11. Lateef DM, Washington MC, Sayegh AI. The short term satiety peptide cholecystokinin reduces meal size and prolongs intermeal interval. *Peptides* 2011;**32**:1289–95.
12. Washington MC, Coggeshall J, Sayegh AI. Cholecystokinin-33 inhibits meal size and prolongs the subsequent intermeal interval. *Peptides* 2011;**32**:971–7.
13. Washington MC, Wright SA, Sayegh AI. Gastrin releasing peptide-29 evokes feeding responses in the rat. *Peptides* 2011;**32**:241–5.
14. Bayliss WM, Starling EH. The mechanism of pancreatic secretion. *J Physiol* 1902;**28**:325–53.
15. Ivy AC, Oldberg EA. A hormone mechanism for gallbladder contraction and evacuation. *Am J Physiol* 1928;**86**:599–613.
16. Harper AA, Raper HS. Pancreozymin, a stimulant of the secretion of pancreatic enzymes in extracts of the small intestine. *J Physiol* 1943;**102**:115–25.
17. Greengard H, Grossman MI, Woolley JR, Ivy AC. A confirmation of the presence of pancreozymin in the duodenal mucosa. *Science* 1944;**99**:350–1.
18. Mutt V, Jorpes E. Hormonal polypeptides of the upper intestine. *Biochem J* 1971;**125**:57P–58P.
19. Mutt V, Jorpes JE. Structure of porcine cholecystokinin-pancreozymin. 1. Cleavage with thrombin and with trypsin. *Eur J Biochem* 1968;**6**:156–62.
20. Polak JM, Bloom SR, Rayford PL, Pearse AG, Buchan AM, Thompson JC. Identification of cholecystokinin-secreting cells. *Lancet* 1975;**2**:1016–8.
21. Buffa R, Solcia E, Go VL. Immunohistochemical identification of the cholecystokinin cell in the intestinal mucosa. *Gastroenterology* 1976;**70**:528–32.
22. Larsson LI, Rehfeld JF. Distribution of gastrin and CCK cells in the rat gastrointestinal tract. Evidence for the occurrence of three distinct cell types storing COOH-terminal gastrin immunoreactivity. *Histochemistry* 1978;**58**:23–31.
23. Larsson LI, Rehfeld JF. Localization and molecular heterogeneity of cholecystokinin in the central and peripheral nervous system. *Brain Res* 1979;**165**:201–18.
24. Larsson LI, Rehfeld JF. Peptidergic and adrenergic innervation of pancreatic ganglia. *Scand J Gastroenterol* 1979;**14**:433–7.
25. Lateef DM, Washington MC, Raboin SJ, Roberson AE, Mansour MM, Williams CS, et al. Duodenal myotomy blocks reduction of meal size and prolongation of intermeal interval by cholecystokinin. *Physiol Behav* 2012;**105**:829–34.
26. Lundberg JM, Hokfelt T, Nilsson G, Terenius L, Rehfeld J, Elde R, et al. Peptide neurons in the vagus, splanchnic and sciatic nerves. *Acta Physiol Scand* 1978;**104**:499–501.

27. Rehfeld JF, Lundberg JM. Cholecystokinin in feline vagal and sciatic nerves: concentration, molecular form and transport velocity. *Brain Res* 1983;**275**:341–7.
28. Dockray GJ, Gregory RA, Tracy HJ, Zhu WY. Transport of cholecystokinin-octapeptide-like immunoreactivity toward the gut in afferent vagal fibres in cat and dog. *J Physiol* 1981;**314**:501–11.
29. Hokfelt T, Rehfeld JF, Skirboll L, Ivemark B, Goldstein M, Markey K. Evidence for coexistence of dopamine and CCK in meso-limbic neurones. *Nature* 1980;**285**:476–8.
30. Hokfelt T, Skirboll L, Rehfeld JF, Goldstein M, Markey K, Dann O. A subpopulation of mesencephalic dopamine neurons projecting to limbic areas contains a cholecystokinin-like peptide: evidence from immunohistochemistry combined with retrograde tracing. *Neuroscience* 1980;**5**:2093–124.
31. Takahashi Y, Kato K, Hayashizaki Y, Wakabayashi T, Ohtsuka E, Matsuki S, et al. Molecular cloning of the human cholecystokinin gene by use of a synthetic probe containing deoxyinosine. *Proc Natl Acad Sci USA* 1985;**82**:1931–5.
32. Dockray GJ. Immunochemical evidence of cholecystokinin-like peptides in brain. *Nature* 1976;**264**:568–70.
33. Rehfeld JF. Immunochemical studies on cholecystokinin. II. Distribution and molecular heterogeneity in the central nervous system and small intestine of man and hog. *J Biol Chem* 1978;**253**:4022–30.
34. Rehfeld JF. Accumulation of nonamidated preprogastrin and preprocholecystokinin products in porcine pituitary corticotrophs. Evidence of post-translational control of cell differentiation. *J Biol Chem* 1986;**261**:5841–7.
35. Rehfeld JF, Hansen HF, Larsson LI, Stengaard-Pedersen K, Thorn NA. Gastrin and cholecystokinin in pituitary neurons. *Proc Natl Acad Sci USA* 1984;**81**:1902–5.
36. Reubi JC. Targeting CCK receptors in human cancers. *Curr Top Med Chem* 2007;**7**:1239–42.
37. Zaccaro L, Del Gatto A, Pedone C, Saviano M. Peptides for tumour therapy and diagnosis: current status and future directions. *Curr Med Chem* 2009;**16**:780–95.
38. Baldwin GS, Shulkes A. CCK receptors and cancer. *Curr Top Med Chem* 2007;**7**:1232–8.
39. Mutt V. Further investigations of intestinal hormonal polypeptides. *Clin Endocrinol (Oxf)* 1976;(Suppl. 5):175S–183S.
40. Eysselein VE, Eberlein GA, Schaeffer M, Grandt D, Goebell H, Niebel W, et al. Characterization of the major form of cholecystokinin in human intestine: CCK-58. *Am J Physiol* 1990;**258**:G253–G260.
41. Eysselein VE, Reeve Jr JR, Shively JE, Hawke D, Walsh JH. Partial structure of a large canine cholecystokinin (CCK58): amino acid sequence. *Peptides* 1982;**3**:687–91.
42. Eberlein GA, Eysselein VE, Davis MT, Lee TD, Shively JE, Grandt D, et al. Patterns of prohormone processing. Order revealed by a new procholecystokinin-derived peptide. *J Biol Chem* 1992;**267**:1517–21.
43. Reeve Jr JR, Eysselein V, Walsh JH, Ben-Avram CM, Shively JE. New molecular forms of cholecystokinin. Microsequence analysis of forms previously characterized by chromatographic methods. *J Biol Chem* 1986;**261**:16392–7.
44. Eng J, Du BH, Pan YC, Chang M, Hulmes JD, Yalow RS. Purification and sequencing of a rat intestinal 22 amino acid C-terminal CCK fragment. *Peptides* 1984;**5**:1203–6.
45. Zhou ZZ, Eng J, Pan YE, Chang M, Hulmes JD, Raufman JP, et al. Unique cholecystokinin peptides isolated from guinea pig intestine. *Peptides* 1985;**6**:337–41.
46. Miller LJ, Jardine I, Weissman E, Go VL, Speicher D. Characterization of cholecystokinin from the human brain. *J Neurochem* 1984;**43**:835–40.
47. Eysselein VE, Reeve Jr JR, Shively JE, Miller C, Walsh JH. Isolation of a large cholecystokinin precursor from canine brain. *Proc Natl Acad Sci USA* 1984;**81**:6565–8.
48. Tatemoto K, Jornvall H, Siimesmaa S, Hallden G, Mutt V. Isolation and characterization of cholecystokinin-58 (CCK-58) from porcine brain. *FEBS Lett* 1984;**174**:289–93.

49. Eng J, Li HR, Yalow RS. Purification of bovine cholecystokinin-58 and sequencing of its N-terminus. *Regul Pept* 1990;**30**:15–9.
50. Christophe J, De Neef P, Deschodt-Lanckman M, Robberecht P. The interaction of caerulein with the rat pancreas. 2. Specific binding of [3H]caerulein on dispersed acinar cells. *Eur J Biochem* 1978;**91**:31–8.
51. Deschodt-Lanckman M, Robberecht P, Camus J, Christophe J. The interaction of caerulein with the rat pancreas. 1. Specific binding of [3H]caerulein on plasma membranes and evidence for negative cooperativity. *Eur J Biochem* 1978;**91**:21–9.
52. Jensen RT, Lemp GF, Gardner JD. Interaction of cholecystokinin with specific membrane receptors on pancreatic acinar cells. *Proc Natl Acad Sci USA* 1980;**77**: 2079–83.
53. Saito A, Sankaran H, Goldfine ID, Williams JA. Cholecystokinin receptors in the brain: characterization and distribution. *Science* 1980;**208**:1155–6.
54. Sankaran H, Goldfine ID, Deveney CW, Wong KY, Williams JA. Binding of cholecystokinin to high affinity receptors on isolated rat pancreatic acini. *J Biol Chem* 1980;**255**:1849–53.
55. Innis RB, Snyder SH. Distinct cholecystokinin receptors in brain and pancreas. *Proc Natl Acad Sci USA* 1980;**77**:6917–21.
56. Schmitz F, Goke MN, Otte JM, Schrader H, Reimann B, Kruse ML, et al. Cellular expression of CCK-A and CCK-B/gastrin receptors in human gastric mucosa. *Regul Pept* 2001;**102**:101–10.
57. Reubi JC, Waser B, Laderach U, Stettler C, Friess H, Halter F, et al. Localization of cholecystokinin A and cholecystokinin B-gastrin receptors in the human stomach. *Gastroenterology* 1997;**112**:1197–205.
58. Mantyh CR, Pappas TN, Vigna SR. Localization of cholecystokinin A and cholecystokinin B/gastrin receptors in the canine upper gastrointestinal tract. *Gastroenterology* 1994;**107**:1019–30.
59. Clerc P, Dufresne M, Saillan C, Chastre E, Andre T, Escrieut C, et al. Differential expression of the CCK-A and CCK-B/gastrin receptor genes in human cancers of the esophagus, stomach and colon. *Int J Cancer* 1997;**72**:931–6.
60. Ji B, Bi Y, Simeone D, Mortensen RM, Logsdon CD. Human pancreatic acinar cells lack functional responses to cholecystokinin and gastrin. *Gastroenterology* 2001;**121**:1380–90.
61. Williams JA, Sankaran H, Korc M, Goldfine ID. Receptors for cholecystokinin and insulin in isolated pancreatic acini: hormonal control of secretion and metabolism. *Fed Proc* 1981;**40**:2497–502.
62. Vigna SR, Steigerwalt RW, Williams JA. Characterization of cholecystokinin receptors in bullfrog (Rana catesbeiana) brain and pancreas. *Regul Pept* 1984;**9**:199–212.
63. Jensen RT, Gardner JD. Identification and characterization of receptors for secretagogues on pancreatic acinar cells. *Fed Proc* 1981;**40**:2486–96.
64. Fourmy D, Zahidi A, Fabre R, Guidet M, Pradayrol L, Ribet A. Receptors for cholecystokinin and gastrin peptides display specific binding properties and are structurally different in guinea-pig and dog pancreas. *Eur J Biochem* 1987;**165**:683–92.
65. Le Drean G, Le Huerou-Luron I, Gestin M, Desbois C, Rome V, Bernard C, et al. Exogenous CCK and gastrin stimulate pancreatic exocrine secretion via CCK-A but also via CCK-B/gastrin receptors in the calf. *Pflugers Arch* 1999;**438**:86–93.
66. Philippe C, Lhoste EF, Dufresne M, Moroder L, Corring T, Fourmy D. Pharmacological and biochemical evidence for the simultaneous expression of CCKB/gastrin and CCKA receptors in the pig pancreas. *Br J Pharmacol* 1997;**120**:447–54.
67. Morisset J, Julien S, Laine J. Localization of cholecystokinin receptor subtypes in the endocine pancreas. *J Histochem Cytochem* 2003;**51**:1501–13.

68. Karlsson S, Sundler F, Ahren B. CCK receptor subtype in insulin-producing cells: a combined functional and in situ hybridization study in rat islets and a rat insulinoma cell line. *Regul Pept* 1998;**78**:95–103.
69. Schweiger M, Erhard MH, Amselgruber WM. Cell-specific localization of the cholecystokininA receptor in the porcine pancreas. *Anat Histol Embryol* 2000;**29**: 357–61.
70. Holicky EL, Hadac EM, Ding XQ, Miller LJ. Molecular characterization and organ distribution of type A and B cholecystokinin receptors in cynomolgus monkey. *Am J Physiol Gastrointest Liver Physiol* 2001;**281**:G507–G514.
71. Steigerwalt RW, Goldfine ID, Williams JA. Characterization of cholecystokinin receptors on bovine gallbladder membranes. *Am J Physiol* 1984;**247**:G709–G714.
72. Poston GJ, Singh P, Maclellan DG, Yao CZ, Uchida T, Townsend Jr CM, et al. Age-related changes in gallbladder contractility and gallbladder cholecystokinin receptor population in the guinea pig. *Mech Ageing Dev* 1988;**46**:225–36.
73. von Schrenck T, Moran TH, Heinz-Erian P, Gardner JD, Jensen RT. Cholecystokinin receptors on gallbladder muscle and pancreatic acinar cells: a comparative study. *Am J Physiol* 1988;**255**:G512–G521.
74. Tokunaga Y, Cox KL, Coleman R, Concepcion W, Nakazato P, Esquivel CO. Characterization of cholecystokinin receptors on the human gallbladder. *Surgery* 1993;**113**:155–62.
75. Aoki T, Ueno T, Toyonaga A, Sakata R, Kimura Y, Gondo K, et al. Radiographic evidence of cholecystokinin octapeptide receptors in the hamster gallbladder. *Scand J Gastroenterol* 1991;**26**:1165–72.
76. Oliver AS, Vigna SR. CCK-A- and CCK-B-like receptors in the gallbladder and stomach of the alligator (Alligator mississippiensis). *Gen Comp Endocrinol* 1997;**105**:91–101.
77. Patterson LM, Zheng H, Ward SM, Berthoud HR. Immunohistochemical identification of cholecystokinin A receptors on interstitial cells of Cajal, smooth muscle, and enteric neurons in rat pylorus. *Cell Tissue Res* 2001;**305**:11–23.
78. de Weerth A, von Schrenck T, Gronewold M, Freudenberg F, Mirau S, Schulz M, et al. Characterization of CCK receptors in stomach smooth muscle: evidence for two subtypes. *Biochim Biophys Acta* 1997;**1327**:213–21.
79. Smith GT, Moran TH, Coyle JT, Kuhar MJ, O'Donahue TL, McHugh PR. Anatomic localization of cholecystokinin receptors to the pyloric sphincter. *Am J Physiol* 1984;**246**:R127–R130.
80. Rettenbacher M, Reubi JC. Localization and characterization of neuropeptide receptors in human colon. *Naunyn Schmiedebergs Arch Pharmacol* 2001;**364**:291–304.
81. Reubi JC, Waser B, Schmassmann A, Laissue JA. Receptor autoradiographic evaluation of cholecystokinin, neurotensin, somatostatin and vasoactive intestinal peptide receptors in gastro-intestinal adenocarcinoma samples: where are they really located? *Int J Cancer* 1999;**81**:376–86.
82. Motomura Y, Chijiiwa Y, Iwakiri Y, Nawata H. Direct contractile effect of cholecystokinin octapeptide on caecal circular smooth muscle cells of guinea pig via both CCK-A and CCK-B/gastrin receptors. *Life Sci* 1997;**60**:499–504.
83. Mazzocchi G, Malendowicz LK, Aragona F, Spinazzi R, Nussdorfer GG. Cholecystokinin (CCK) stimulates aldosterone secretion from human adrenocortical cells via CCK2 receptors coupled to the adenylate cyclase/protein kinase A signaling cascade. *J Clin Endocrinol Metab* 2004;**89**:1277–84.
84. Malendowicz LK, Nowak M, Gottardo L, Tortorella C, Majchrzak M, Nussdorfer GG. Cholecystokinin stimulates aldosterone secretion from dispersed rat zona glomerulosa cells, acting through cholecystokinin receptors 1 and 2 coupled with the adenylate cyclase-dependent cascade. *Endocrinology* 2001;**142**:4251–5.

85. Monstein HJ, Nylander AG, Salehi A, Chen D, Lundquist I, Hakanson R. Cholecystokinin-A and cholecystokinin-B/gastrin receptor mRNA expression in the gastrointestinal tract and pancreas of the rat and man. A polymerase chain reaction study. *Scand J Gastroenterol* 1996;**31**:383–90.
86. Lay JM, Jenkins C, Friis-Hansen L, Samuelson LC. Structure and developmental expression of the mouse CCK-B receptor gene. *Biochem Biophys Res Commun* 2000;**272**:837–42.
87. Sternini C, Wong H, Pham T, De Giorgio R, Miller LJ, Kuntz SM, et al. Expression of cholecystokinin A receptors in neurons innervating the rat stomach and intestine. *Gastroenterology* 1999;**117**:1136–46.
88. Broberger C, Holmberg K, Shi TJ, Dockray G, Hokfelt T. Expression and regulation of cholecystokinin and cholecystokinin receptors in rat nodose and dorsal root ganglia. *Brain Res* 2001;**903**:128–40.
89. Moriarty P, Dimaline R, Thompson DG, Dockray GJ. Characterization of cholecystokininA and cholecystokininB receptors expressed by vagal afferent neurons. *Neuroscience* 1997;**79**:905–13.
90. Lin CW, Miller TR. Both CCK-A and CCK-B/gastrin receptors are present on rabbit vagus nerve. *Am J Physiol* 1992;**263**:R591–R595.
91. Patterson LM, Zheng H, Berthoud HR. Vagal afferents innervating the gastrointestinal tract and CCKA-receptor immunoreactivity. *Anat Rec* 2002;**266**:10–20.
92. Kulaksiz H, Arnold R, Goke B, Maronde E, Meyer M, Fahrenholz F, et al. Expression and cell-specific localization of the cholecystokinin B/gastrin receptor in the human stomach. *Cell Tissue Res* 2000;**299**:289–98.
93. Tarasova NI, Romanov VI, Da Silva PP, Michejda CJ. Numerous cell targets for gastrin in the guinea pig stomach revealed by gastrin/CCKB receptor localization. *Cell Tissue Res* 1996;**283**:1–6.
94. Pisegna JR, de Weerth A, Huppi K, Wank SA. Molecular cloning of the human brain and gastric cholecystokinin receptor: structure, functional expression and chromosomal localization. *Biochem Biophys Res Commun* 1992;**189**:296–303.
95. Yu DH, Noguchi M, Zhou ZC, Villanueva ML, Gardner JD, Jensen RT. Characterization of gastrin receptors on guinea pig pancreatic acini. *Am J Physiol* 1987;**253**:G793–G801.
96. Le Meuth V, Philouze-Rome V, Le Huerou-Luron I, Formal M, Vaysse N, Gespach C, et al. Differential expression of A- and B-subtypes of cholecystokinin/gastrin receptors in the developing calf pancreas. *Endocrinology* 1993;**133**:1182–91.
97. Tang C, Biemond I, Lamers CB. Cholecystokinin receptors in human pancreas and gallbladder muscle: a comparative study. *Gastroenterology* 1996;**111**:1621–6.
98. Reubi JC, Waser B, Gugger M, Friess H, Kleeff J, Kayed H, et al. Distribution of CCK1 and CCK2 receptors in normal and diseased human pancreatic tissue. *Gastroenterology* 2003;**125**:98–106.
99. Saillan-Barreau C, Dufresne M, Clerc P, Sanchez D, Corominola H, Moriscot C, et al. Evidence for a functional role of the cholecystokinin-B/gastrin receptor in the human fetal and adult pancreas. *Diabetes* 1999;**48**:2015–21.
100. Glaser SS, Rodgers RE, Phinizy JL, Robertson WE, Lasater J, Caligiuri A, et al. Gastrin inhibits secretin-induced ductal secretion by interaction with specific receptors on rat cholangiocytes. *Am J Physiol* 1997;**273**:G1061–G1070.
101. Attoub S, Levasseur S, Buyse M, Goiot H, Laigneau JP, Moizo L, et al. Physiological role of cholecystokinin B/gastrin receptor in leptin secretion. *Endocrinology* 1999;**140**:4406–10.

102. Dornand J, Roche S, Michel F, Bali JP, Cabane S, Favero J, et al. Gastrin-CCK-B type receptors on human T lymphoblastoid Jurkat cells. *Am J Physiol* 1995;**268**: G522–G529.
103. Cuq P, Gross A, Terraza A, Fourmy D, Clerc P, Dornand J, et al. mRNAs encoding CCKB but not CCKA receptors are expressed in human T lymphocytes and Jurkat lymphoblastoid cells. *Life Sci* 1997;**61**:543–55.
104. Lignon MF, Bernad N, Martinez J. Pharmacological characterization of type B cholecystokinin binding sites on the human JURKAT T lymphocyte cell line. *Mol Pharmacol* 1991;**39**:615–20.
105. de Weerth A, Jonas L, Schade R, Schoneberg T, Wolf G, Pace A, et al. Gastrin/cholecystokinin type B receptors in the kidney: molecular, pharmacological, functional characterization, and localization. *Eur J Clin Invest* 1998;**28**:592–601.
106. von Schrenck T, Ahrens M, de Weerth A, Bobrowski C, Wolf G, Jonas L, et al. CCKB/gastrin receptors mediate changes in sodium and potassium absorption in the isolated perfused rat kidney. *Kidney Int* 2000;**58**:995–1003.
107. von Schrenck T, de Weerth A, Bechtel S, Eschenhagen T, Weil J, Wolf G, et al. Evidence for CCK(B) receptors in the guinea-pig kidney: localization and characterization by [125I]gastrin binding studies and by RT-PCR. *Naunyn Schmiedebergs Arch Pharmacol* 1998;**358**:287–92.
108. Samuelson LC, Isakoff MS, Lacourse KA. Localization of the murine cholecystokinin A and B receptor genes. *Mamm Genome* 1995;**6**:242–6.
109. Huppi K, Siwarski D, Pisegna JR, Wank S. Chromosomal localization of the gastric and brain receptors for cholecystokinin (CCKAR and CCKBR) in human and mouse. *Genomics* 1995;**25**:727–9.
110. Wank SA, Harkins R, Jensen RT, Shapira H, de Weerth A, Slattery T. Purification, molecular cloning, and functional expression of the cholecystokinin receptor from rat pancreas. *Proc Natl Acad Sci USA* 1992;**89**:3125–9.
111. Wank SA, Pisegna JR, de Weerth A. Brain and gastrointestinal cholecystokinin receptor family: structure and functional expression. *Proc Natl Acad Sci USA* 1992;**89**:8691–5.
112. Kopin AS, Lee YM, McBride EW, Miller LJ, Lu M, Lin HY, et al. Expression cloning and characterization of the canine parietal cell gastrin receptor. *Proc Natl Acad Sci USA* 1992;**89**:3605–9.
113. Lee YM, Beinborn M, McBride EW, Lu M, Kolakowski Jr LF, Kopin AS. The human brain cholecystokinin-B/gastrin receptor. Cloning and characterization. *J Biol Chem* 1993;**268**:8164–9.
114. Ito M, Matsui T, Taniguchi T, Tsukamoto T, Murayama T, Arima N, et al. Functional characterization of a human brain cholecystokinin-B receptor. A trophic effect of cholecystokinin and gastrin. *J Biol Chem* 1993;**268**:18300–5.
115. De Weerth A, Pisegna JR, Wank SA. Guinea pig gallbladder and pancreas possess identical CCK-A receptor subtypes: receptor cloning and expression. *Am J Physiol* 1993;**265**:G1116–G1121.
116. Dufresne M, Escrieut C, Clerc P, Le Huerou-Luron I, Prats H, Bertrand V, et al. Molecular cloning, developmental expression and pharmacological characterization of the CCKB/gastrin receptor in the calf pancreas. *Eur J Pharmacol* 1996;**297**:165–79.
117. Ghanekar D, Hadac EM, Holicky EL, Miller LJ. Differences in partial agonist action at cholecystokinin receptors of mouse and rat are dependent on parameters extrinsic to receptor structure: molecular cloning, expression and functional characterization of the mouse type A cholecystokinin receptor. *J Pharmacol Exp Ther* 1997;**282**: 1206–12.

118. Nakata H, Matsui T, Ito M, Taniguchi T, Naribayashi Y, Arima N, et al. Cloning and characterization of gastrin receptor from ECL carcinoid tumor of Mastomys natalensis. *Biochem Biophys Res Commun* 1992;**187**:1151–7.
119. Reuben M, Rising L, Prinz C, Hersey S, Sachs G. Cloning and expression of the rabbit gastric CCK-A receptor. *Biochim Biophys Acta* 1994;**1219**:321–7.
120. Langer I, Tikhonova IG, Travers MA, Archer-Lahlou E, Escrieut C, Maigret B, et al. Evidence that interspecies polymorphism in the human and rat cholecystokinin receptor-2 affects structure of the binding site for the endogenous agonist cholecystokinin. *J Biol Chem* 2005;**280**:22198–204.
121. Huang SC, Fortune KP, Wank SA, Kopin AS, Gardner JD. Multiple affinity states of different cholecystokinin receptors. *J Biol Chem* 1994;**269**:26121–6.
122. Solomon TE, Yamada T, Elashoff J, Wood J, Beglinger C. Bioactivity of cholecystokinin analogues: CCK-8 is not more potent than CCK-33. *Am J Physiol* 1984;**247**: G105–G111.
123. Reeve Jr JR, McVey DC, Bunnett NW, Solomon TE, Keire DA, Ho FJ, et al. Differences in receptor binding and stability to enzymatic digestion between CCK-8 and CCK-58. *Pancreas* 2002;**25**:e50–e55.
124. Huang SC, Yu DH, Wank SA, Mantey S, Gardner JD, Jensen RT. Importance of sulfation of gastrin or cholecystokinin (CCK) on affinity for gastrin and CCK receptors. *Peptides* 1989;**10**:785–9.
125. Chang RS, Lotti VJ. Biochemical and pharmacological characterization of an extremely potent and selective nonpeptide cholecystokinin antagonist. *Proc Natl Acad Sci USA* 1986;**83**:4923–6.
126. Makovec F, Peris W, Revel L, Giovanetti R, Mennuni L, Rovati LC. Structure-antigastrin activity relationships of new (R)-4-benzamido-5-oxopentanoic acid derivatives. *J Med Chem* 1992;**35**:28–38.
127. Gully D, Frehel D, Marcy C, Spinazze A, Lespy L, Neliat G, et al. Peripheral biological activity of SR 27897: a new potent non-peptide antagonist of CCKA receptors. *Eur J Pharmacol* 1993;**232**:13–9.
128. Lotti VJ, Chang RS. A new potent and selective non-peptide gastrin antagonist and brain cholecystokinin receptor (CCK-B) ligand: L-365,260. *Eur J Pharmacol* 1989;**162**:273–80.
129. Bock MG, DiPardo RM, Newton RC, Bergman JM, Veber DF, Freedman SB, et al. Selective non-peptide ligands for an accommodating peptide receptor. Imidazobenzodiazepines as potent cholecystokinin type B receptor antagonists. *Bioorg Med Chem* 1994;**2**:987–98.
130. Satoh M, Kondoh Y, Okamoto Y, Nishida A, Miyata K, Ohta M, et al. New 1,4-benzodiazepin-2-one derivatives as gastrin/cholecystokinin-B antagonists. *Chem Pharm Bull(Tokyo)* 1995;**43**:2159–67.
131. Makovec F, Revel L, Letari O, Mennuni L, Impicciatore M. Characterization of antisecretory and antiulcer activity of CR 2945, a new potent and selective gastrin/CCK (B) receptor antagonist. *Eur J Pharmacol* 1999;**369**:81–90.
132. Semple G, Ryder H, Rooker DP, Batt AR, Kendrick DA, Szelke M, et al. (3R)-N-(1-(tert-butylcarbonylmethyl)-2,3-dihydro-2-oxo-5-(2-pyridyl)-1H-1, 4-benzodiazepin-3-yl)-N'-(3-(methylamino)phenyl)urea (YF476): a potent and orally active gastrin/CCK-B antagonist. *J Med Chem* 1997;**40**:331–41.
133. Ursini A, Capelli AM, Carr RA, Cassara P, Corsi M, Curcuruto O, et al. Synthesis and SAR of new 5-phenyl-3-ureido-1,5-benzodiazepines as cholecystokinin-B receptor antagonists. *J Med Chem* 2000;**43**:3596–613.
134. Hughes J, Boden P, Costall B, Domeney A, Kelly E, Horwell DC, et al. Development of a class of selective cholecystokinin type B receptor antagonists having potent anxiolytic activity. *Proc Natl Acad Sci USA* 1990;**87**:6728–32.

135. Galas MC, Lignon MF, Rodriguez M, Mendre C, Fulcrand P, Laur J, et al. Structure-activity relationship studies on cholecystokinin: analogues with partial agonist activity. *Am J Physiol* 1988;**254**:G176–G182.
136. Lignon MF, Galas MC, Rodriguez M, Laur J, Aumelas A, Martinez J. A synthetic peptide derivative that is a cholecystokinin receptor antagonist. *J Biol Chem* 1987;**262**: 7226–31.
137. Murthy KS, Makhlouf GM. Functional characterization of phosphoinositide-specific phospholipase C-beta 1 and -beta 3 in intestinal smooth muscle. *Am J Physiol* 1995;**269**:C969–C978.
138. Paulssen RH, Fraeyman N, Florholmen J. Activation of phospholipase C by cholecystokinin receptor subtypes with different G-protein-coupling specificities in hormone-secreting pancreatic cell lines. *Biochem Pharmacol* 2000;**60**:865–75.
139. Piiper A, Stryjek-Kaminska D, Klengel R, Zeuzem S. CCK, carbachol, and bombesin activate distinct PLC-beta isoenzymes via Gq/11 in rat pancreatic acinar membranes. *Am J Physiol* 1997;**272**:G135–G140.
140. Sans MD, Xie Q, Williams JA. Regulation of translation elongation and phosphorylation of eEF2 in rat pancreatic acini. *Biochem Biophys Res Commun* 2004;**319**:144–51.
141. Stepan VM, Dickinson CJ, del Valle J, Matsushima M, Todisco A. Cell type-specific requirement of the MAPK pathway for the growth factor action of gastrin. *Am J Physiol* 1999;**276**:G1363–G1372.
142. Hocker M. Molecular mechanisms of gastrin-dependent gene regulation. *Ann N Y Acad Sci* 2004;**1014**:97–109.
143. Bragado MJ, Dabrowski A, Groblewski GE, Williams JA. CCK activates p90rsk in rat pancreatic acini through protein kinase C. *Am J Physiol* 1997;**272**:G401–G407.
144. Bragado MJ, Groblewski GE, Williams JA. p70s6k is activated by CCK in rat pancreatic acini. *Am J Physiol* 1997;**273**:C101–C109.
145. Bragado MJ, Groblewski GE, Williams JA. Regulation of protein synthesis by cholecystokinin in rat pancreatic acini involves PHAS-I and the p70 S6 kinase pathway. *Gastroenterology* 1998;**115**:733–42.
146. Bierkamp C, Kowalski-Chauvel A, Dehez S, Fourmy D, Pradayrol L, Seva C. Gastrin mediated cholecystokinin-2 receptor activation induces loss of cell adhesion and scattering in epithelial MDCK cells. *Oncogene* 2002;**21**:7656–70.
147. Daulhac L, Kowalski-Chauvel A, Pradayrol L, Vaysse N, Seva C. Src-family tyrosine kinases in activation of ERK-1 and p85/p110-phosphatidylinositol 3-kinase by G/CCKB receptors. *J Biol Chem* 1999;**274**:20657–63.
148. Ferrand A, Kowalski-Chauvel A, Bertrand C, Pradayrol L, Fourmy D, Dufresne M, et al. Involvement of JAK2 upstream of the PI 3-kinase in cell-cell adhesion regulation by gastrin. *Exp Cell Res* 2004;**301**:128–38.
149. Kowalski-Chauvel A, Pradayrol L, Vaysse N, Seva C. Gastrin stimulates tyrosine phosphorylation of insulin receptor substrate 1 and its association with Grb2 and the phosphatidylinositol 3-kinase. *J Biol Chem* 1996;**271**:26356–61.
150. Garcia LJ, Rosado JA, Tsuda T, Jensen RT. CCK causes rapid tyrosine phosphorylation of p125FAK focal adhesion kinase and paxillin in rat pancreatic acini. *Biochim Biophys Acta* 1997;**1358**:189–99.
151. Kiehne K, Herzig KH, Folsch UR. CCK-stimulated changes in pancreatic acinar morphology are mediated by rho. *Digestion* 2002;**65**:47–55.
152. Pace A, Garcia-Marin LJ, Tapia JA, Bragado MJ, Jensen RT. Phosphospecific site tyrosine phosphorylation of p125FAK and proline-rich kinase 2 is differentially regulated by cholecystokinin receptor type A activation in pancreatic acini. *J Biol Chem* 2003;**278**:19008–16.
153. Sjodin L, Gardner JD. Effect of cholecystokinin variant (CCK39) on dispersed acinar cells from guinea pig pancreas. *Gastroenterology* 1977;**73**:1015–8.

154. Yule DI, Tseng MJ, Williams JA, Logdson CD. A cloned CCK-A receptor transduces multiple signals in response to full and partial agonists. *Am J Physiol* 1993;**265**: G999–G1004.
155. Cordelier P, Esteve JP, Bousquet C, Delesque N, O'Carroll AM, Schally AV, et al. Characterization of the antiproliferative signal mediated by the somatostatin receptor subtype sst5. *Proc Natl Acad Sci USA* 1997;**94**:9343–8.
156. Cordelier P, Esteve JP, Rivard N, Marletta M, Vaysse N, Susini C, et al. The activation of neuronal NO synthase is mediated by G-protein betagamma subunit and the tyrosine phosphatase SHP-2. *FASEB J* 1999;**13**:2037–50.
157. Ahn SH, Seo DW, Ko YK, Sung DS, Bae GU, Yoon JW, et al. NO/cGMP pathway is involved in exocrine secretion from rat pancreatic acinar cells. *Arch Pharm Res* 1998;**21**:657–63.
158. DiMagno MJ, Hao Y, Tsunoda Y, Williams JA, Owyang C. Secretagogue-stimulated pancreatic secretion is differentially regulated by constitutive NOS isoforms in mice. *Am J Physiol Gastrointest Liver Physiol* 2004;**286**:G428–G436.
159. Brzozowski T, Konturek PC, Konturek SJ, Pajdo R, Drozdowicz D, Kwiecien S, et al. Acceleration of ulcer healing by cholecystokinin (CCK): role of CCK-A receptors, somatostatin, nitric oxide and sensory nerves. *Regul Pept* 1999;**82**:19–33.
160. Brzozowski T, Konturek PC, Konturek SJ, Pajdo R, Duda A, Pierzchalski P, et al. Leptin in gastroprotection induced by cholecystokinin or by a meal. Role of vagal and sensory nerves and nitric oxide. *Eur J Pharmacol* 1999;**374**:263–76.
161. Ferrand A, Kowalski-Chauvel A, Bertrand C, Escrieut C, Mathieu A, Portolan G, et al. A novel mechanism for JAK2 activation by a G protein-coupled receptor, the CCK2R: implication of this signaling pathway in pancreatic tumor models. *J Biol Chem* 2005;**280**:10710–5.
162. Cameron AJ, Phillips SF, Summerskill WH. Effect of cholecystokinin, gastrin, and glucagon on human gallbladder muscle in vitro. *Proc Soc Exp Biol Med* 1969;**131**:149–54.
163. Dyck WP, Hall FF, Ratliff CR. Hormonal control of intestinal alkaline phosphatase secretion in the dog. *Gastroenterology* 1973;**65**:445–50.
164. Dyck WP, Martin GA, Ratliff CR. Influence of secretin and cholecystokinin on intestinal alkaline phosphatase secretion. *Gastroenterology* 1973;**64**:599–602.
165. Dyck WP, Bonnet D, Lasater J, Stinson C, Hall FF. Hormonal stimulation of intestinal disaccharidase release in the dog. *Gastroenterology* 1974;**66**:533–8.
166. Moss S, Lobley RW, Holmes R. Enterokinase in human duodenal juice following secretin and pancreozymin and its relationship to bile salts and trypsin. *Gut* 1972;**13**:851.
167. Leroy J, Morisset JA, Webster PD. Dose-related response of pancreatic synthesis and secretion to cholecystokinin-pancreazymin. *J Lab Clin Med* 1971;**78**:149–57.
168. Rothman SS, Wells H. Enhancement of pancreatic enzyme synthesis by pancreozymin. *Am J Physiol* 1967;**213**:215–8.
169. Rosenfeld MG, Abrass I, Chang B. Hormonal stimulation of alpha-amylase synthesis in porcine pancreatic minces. *Endocrinology* 1976;**99**:611–8.
170. Debas HT, Grossman MI. Pure cholecystokinin: pancreatic protein and bicarbonate response. *Digestion* 1973;**9**:469–81.
171. Henriksen FW, Worning H. The mutual influence of gastrin and secretin on the external pancreatic secretion in dogs. *Acta Physiol Scand* 1969;**76**:67–72.
172. Hermansen K. Effects of cholecystokinin (CCK)-4, nonsulfated CCK-8, and sulfated CCK-8 on pancreatic somatostatin, insulin, and glucagon secretion in the dog: studies in vitro. *Endocrinology* 1984;**114**:1770–5.
173. Otsuki M, Sakamoto C, Yuu H, Maeda M, Morita S, Ohki A, et al. Discrepancies between the doses of cholecystokinin or caerulein-stimulating exocrine and endocrine responses in perfused isolated rat pancreas. *J Clin Invest* 1979;**63**:478–84.

174. Schultzberg M, Hokfelt T, Nilsson G, Terenius L, Rehfeld JF, Brown M, et al. Distribution of peptide- and catecholamine-containing neurons in the gastro-intestinal tract of rat and guinea-pig: immunohistochemical studies with antisera to substance P, vasoactive intestinal polypeptide, enkephalins, somatostatin, gastrin/cholecystokinin, neurotensin and dopamine beta-hydroxylase. *Neuroscience* 1980;**5**:689–744.
175. Gutierrez JG, Chey WY, Dinoso VP. Actions of cholecystokinin and secretin on the motor activity of the small intestine in man. *Gastroenterology* 1974;**67**:35–41.
176. Hedner P, Persson H, Rorsman G. Effect of cholecystokinin on small intestine. *Acta Physiol Scand* 1967;**70**:250–4.
177. Dinoso Jr VP, Meshkinpour H, Lorber SH, Gutierrez JG, Chey WY. Motor responses of the sigmoid colon and rectum to exogenous cholecystokinin and secretin. *Gastroenterology* 1973;**65**:438–44.
178. Thulin L, Samnegard H. Circulatory effects of gastrointestinal hormones and related peptides. *Acta Chir Scand Suppl* 1978;**482**:73–4.
179. Chey WY, Hitanant S, Hendricks J, Lorber SH. Effect of secretin and cholecystokinin on gastric emptying and gastric secretion in man. *Gastroenterology* 1970;**58**:820–7.
180. Debas HT, Farooq O, Grossman MI. Inhibition of gastric emptying is a physiological action of cholecystokinin. *Gastroenterology* 1975;**68**:1211–7.
181. Isenberg JI, Csendes A. Effect of octapeptide of cholecystokinin on canine pyloric pressure. *Am J Physiol* 1972;**222**:428–31.
182. Yamagishi T, Debas HT. Cholecystokinin inhibits gastric emptying by acting on both proximal stomach and pylorus. *Am J Physiol* 1978;**234**:E375–E378.
183. Gerner T. Pressure responses to OP-CCK compared to CCK-PZ in the antrum and fundus of isolated guinea-pig stomachs. *Scand J Gastroenterol* 1979;**14**:73–7.
184. Gibbs J, Young RC, Smith GP. Cholecystokinin decreases food intake in rats. *J Comp Physiol Psychol* 1973;**84**:488–95.
185. Houpt TR, Anika SM, Wolff NC. Satiety effects of cholecystokinin and caerulein in rabbits. *Am J Physiol* 1978;**235**:R23–R28.
186. Kissileff HR, Pi-Sunyer FX, Thornton J, Smith GP. C-terminal octapeptide of cholecystokinin decreases food intake in man. *Am J Clin Nutr* 1981;**34**:154–60.
187. Gibbs J, Falasco JD, McHugh PR. Cholecystokinin-decreased food intake in rhesus monkeys. *Am J Physiol* 1976;**230**:15–8.
188. Parrott RF, Baldwin BA. Operant feeding and drinking in pigs following intracerebroventricular injection of synthetic cholecystokinin octapeptide. *Physiol Behav* 1981;**26**:419–22.
189. Della-Fera MA, Baile CA. Cholecystokinin octapeptide: continuous picomole injections into the cerebral ventricles of sheep suppress feeding. *Science* 1979;**206**:471–3.
190. Greenberg D, Smith GP, Gibbs J. Intraduodenal infusions of fats elicit satiety in sham-feeding rats. *Am J Physiol* 1990;**259**:R110–R118.
191. Olsen O, Schaffalitzky de Muckadell OB, Cantor P. The significance of plasma CCK and secretin in the oleate-stimulated pancreatico-biliary secretion in man. *Int J Pancreatol* 1986;**1**:363–72.
192. Liddle RA, Goldfine ID, Rosen MS, Taplitz RA, Williams JA. Cholecystokinin bioactivity in human plasma. Molecular forms, responses to feeding, and relationship to gallbladder contraction. *J Clin Invest* 1985;**75**:1144–52.
193. Liddle RA, Green GM, Conrad CK, Williams JA. Proteins but not amino acids, carbohydrates, or fats stimulate cholecystokinin secretion in the rat. *Am J Physiol* 1986;**251**:G243–G248.
194. Yonezawa H. Discrepancy between the potency of various trypsin inhibitors to inhibit trypsin activity and the potency to release biologically active cholecystokinin-pancreozymin. *Jpn J Physiol* 1984;**34**:849–56.

195. Iwai K, Fukuoka S, Fushiki T, Tsujikawa M, Hirose M, Tsunasawa S, et al. Purification and sequencing of a trypsin-sensitive cholecystokinin-releasing peptide from rat pancreatic juice. Its homology with pancreatic secretory trypsin inhibitor. *J Biol Chem* 1987;**262**:8956–9.
196. Spannagel AW, Green GM, Guan D, Liddle RA, Faull K, Reeve Jr JR. Purification and characterization of a luminal cholecystokinin-releasing factor from rat intestinal secretion. *Proc Natl Acad Sci USA* 1996;**93**:4415–20.
197. Herzig KH, Schon I, Tatemoto K, Ohe Y, Li Y, Folsch UR, et al. Diazepam binding inhibitor is a potent cholecystokinin-releasing peptide in the intestine. *Proc Natl Acad Sci USA* 1996;**93**:7927–32.
198. Antin J, Gibbs J, Holt J, Young RC, Smith GP. Cholecystokinin elicits the complete behavioral sequence of satiety in rats. *J Comp Physiol Psychol* 1975;**89**:784–90.
199. Falasco JD, Smith GP, Gibbs J. Cholecystokinin suppresses sham feeding in the rhesus monkey. *Physiol Behav* 1979;**23**:887–90.
200. Stacher G, Bauer H, Steinringer H. Cholecystokinin decreases appetite and activation evoked by stimuli arising from the preparation of a meal in man. *Physiol Behav* 1979;**23**:325–31.
201. Anika SM. Ontogeny of cholecystokinin satiety in rats. *Eur J Pharmacol* 1983;**89**:211–5.
202. Lorenz DN, Goldman SA. Vagal mediation of the cholecystokinin satiety effect in rats. *Physiol Behav* 1982;**29**:599–604.
203. Melville LD, Smith GP, Gibbs J. Comparison of the satiating potencies of cholecystokinin-33 and cholecystokinin-8. *Pharmacol Biochem Behav* 1993;**45**:85–7.
204. Glatzle J, Raybould HE, Kueper MA, Reeve Jr JR, Zittel TT. Cholecystokinin-58 is more potent in inhibiting food intake than cholecystokinin-8 in rats. *Nutr Neurosci* 2008;**11**:69–74.
205. Overduin J., Reeve JRJ., Strohmayer AJ., Gibbs J. (2000). *Soc Neurosc* 1531.
206. Raboin SJ, Reeve Jr JR, Green GM, Cox JE, Gibbs J, Perez LJ, et al. Differential effects of exogenous CCK-8 and CCK-58 on food intake. *Appetite* 2007;**49**:322.
207. Gores GJ, LaRusso NF, Miller LJ. Hepatic processing of cholecystokinin peptides. I. Structural specificity and mechanism of hepatic extraction. *Am J Physiol* 1986;**250**: G344–G349.
208. Gores GJ, Miller LJ, LaRusso NF. Hepatic processing of cholecystokinin peptides. II. Cellular metabolism, transport, and biliary excretion. *Am J Physiol* 1986;**250**: G350–G356.
209. Hoffmann P, Eberlein GA, Reeve Jr JR, Bunte RH, Grandt D, Goebell H, et al. Comparison of clearance and metabolism of infused cholecystokinins 8 and 58 in dogs. *Gastroenterology* 1993;**105**:1732–6.
210. Keire DA, Solomon TE, Reeve Jr JR. NMR evidence for different conformations of the bioactive region of rat CCK-8 and CCK-58. *Biochem Biophys Res Commun* 2002;**293**:1014–20.
211. McLaughlin CL, Peikin SR, Baile CA. Trypsin inhibitor effects on food intake and weight gain in Zucker rats. *Physiol Behav* 1983;**31**:487–91.
212. Yox DP, Ritter RC. Capsaicin attenuates suppression of sham feeding induced by intestinal nutrients. *Am J Physiol* 1988;**255**:R569–R574.
213. Yox DP, Brenner L, Ritter RC. CCK-receptor antagonists attenuate suppression of sham feeding by intestinal nutrients. *Am J Physiol* 1992;**262**:R554–R561.
214. Brenner L, Yox DP, Ritter RC. Suppression of sham feeding by intraintestinal nutrients is not correlated with plasma cholecystokinin elevation. *Am J Physiol* 1993;**264**: R972–R976.
215. Brenner L, Ritter RC. Peptide cholesystokinin receptor antagonist increases food intake in rats. *Appetite* 1995;**24**:1–9.
216. Voigt JP, Wenz D, Voits M, Fink H. Does increased endogenous CCK interact with serotonin to reduce food intake in rats? *Peptides* 2000;**21**:1895–901.

217. Sullivan CN, Raboin SJ, Gulley S, Sinzobahamvya NT, Green GM, Reeve Jr JR, et al. Endogenous cholecystokinin reduces food intake and increases Fos-like immunoreactivity in the dorsal vagal complex but not in the myenteric plexus by CCK1 receptor in the adult rat. *Am J Physiol Regul Integr Comp Physiol* 2007;**292**:R1071–R1080.
218. Raboin SJ, Reeve Jr JR, Cooper MS, Green GM, Sayegh AI. Activation of submucosal but not myenteric plexus of the gastrointestinal tract accompanies reduction of food intake by camostat. *Regul Pept* 2008;**150**:73–80.
219. Reeve Jr JR, Green GM, Chew P, Eysselein VE, Keire DA. CCK-58 is the only detectable endocrine form of cholecystokinin in rat. *Am J Physiol Gastrointest Liver Physiol* 2003;**285**:G255–G265.
220. Yamamoto M, Reeve Jr JR, Keire DA, Green GM. Water and enzyme secretion are tightly coupled in pancreatic secretion stimulated by food or CCK-58 but not by CCK-8. *Am J Physiol Gastrointest Liver Physiol* 2005;**288**:G866–G879.
221. Corwin RL, Gibbs J, Smith GP. Increased food intake after type A but not type B cholecystokinin receptor blockade. *Physiol Behav* 1991;**50**:255–8.
222. Moran TH, Ameglio PJ, Schwartz GJ, McHugh PR. Blockade of type A, not type B, CCK receptors attenuates satiety actions of exogenous and endogenous CCK. *Am J Physiol* 1992;**262**:R46–R50.
223. Moran TH, Ameglio PJ, Peyton HJ, Schwartz GJ, McHugh PR. Blockade of type A, but not type B, CCK receptors postpones satiety in rhesus monkeys. *Am J Physiol* 1993;**265**:R620–R624.
224. Kawano K, Hirashima T, Mori S, Natori T. OLETF (Otsuka Long-Evans Tokushima Fatty) rat: a new NIDDM rat strain. *Diabetes Res Clin Pract* 1994;**24**(Suppl): S317–S320.
225. Moran TH. Unraveling the obesity of OLETF rats. *Physiol Behav* 2008;**94**:71–8.
226. Funakoshi A, Miyasaka K, Shinozaki H, Masuda M, Kawanami T, Takata Y, et al. An animal model of congenital defect of gene expression of cholecystokinin (CCK)-A receptor. *Biochem Biophys Res Commun* 1995;**210**:787–96.
227. Bi S, Ladenheim EE, Schwartz GJ, Moran TH. A role for NPY overexpression in the dorsomedial hypothalamus in hyperphagia and obesity of OLETF rats. *Am J Physiol Regul Integr Comp Physiol* 2001;**281**:R254–R260.
228. Moran TH, Katz LF, Plata-Salaman CR, Schwartz GJ. Disordered food intake and obesity in rats lacking cholecystokinin A receptors. *Am J Physiol* 1998;**274**: R618–R625.
229. Schwartz GJ, Whitney A, Skoglund C, Castonguay TW, Moran TH. Decreased responsiveness to dietary fat in Otsuka Long-Evans Tokushima fatty rats lacking CCK-A receptors. *Am J Physiol* 1999;**277**:R1144–R1151.
230. Covasa M, Ritter RC. Attenuated satiation response to intestinal nutrients in rats that do not express CCK-A receptors. *Peptides* 2001;**22**:1339–48.
231. McLaughlin CL, Baile CA. Cholecystokinin, amphetamine and diazepam and feeding in lean and obese Zucker rats. *Pharmacol Biochem Behav* 1979;**10**:87–93.
232. Covasa M, Ritter RC. Rats maintained on high-fat diets exhibit reduced satiety in response to CCK and bombesin. *Peptides* 1998;**19**:1407–15.
233. Balint JA, Fried MB, Imai C. Ileal uptake of oleic acid: evidence for adaptive response to high fat feeding. *Am J Clin Nutr* 1980;**33**:2276–80.
234. Goda T, Takase S. Effect of dietary fat content on microvillus in rat jejunum. *J Nutr Sci Vitaminol (Tokyo)* 1994;**40**:127–36.
235. Spannagel AW, Nakano I, Tawil T, Chey WY, Liddle RA, Green GM. Adaptation to fat markedly increases pancreatic secretory response to intraduodenal fat in rats. *Am J Physiol* 1996;**270**:G128–G135.
236. French SJ, Murray B, Rumsey RD, Fadzlin R, Read NW. Adaptation to high-fat diets: effects on eating behaviour and plasma cholecystokinin. *Br J Nutr* 1995;**73**:179–89.

237. Covasa M, Grahn J, Ritter RC. High fat maintenance diet attenuates hindbrain neuronal response to CCK. *Regul Pept* 2000;**86**:83–8.
238. Ohlsson B, Borg K, Mulder H, Rehfeld JF, Axelson J, Sundler F. Continuous infusion of cholecystokinin leads to down-regulation of the cholecystokinin-A receptor in the rat pancreas. *Scand J Gastroenterol* 2000;**35**:612–8.
239. Otsuki M, Williams JA. Amylase secretion by isolated pancreatic acini after chronic cholecystokinin treatment in vivo. *Am J Physiol* 1983;**244**:G683–G688.
240. Brenner LA, Ritter RC. Type A CCK receptors mediate satiety effects of intestinal nutrients. *Pharmacol Biochem Behav* 1996;**54**:625–31.
241. Berthoud HR, Patterson LM. Anatomical relationship between vagal afferent fibers and CCK-immunoreactive entero-endocrine cells in the rat small intestinal mucosa. *Acta Anat (Basel)* 1996;**156**:123–31.
242. Corp ES, McQuade J, Moran TH, Smith GP. Characterization of type A and type B CCK receptor binding sites in rat vagus nerve. *Brain Res* 1993;**623**:161–6.
243. Moran TH, Smith GP, Hostetler AM, McHugh PR. Transport of cholecystokinin (CCK) binding sites in subdiaphragmatic vagal branches. *Brain Res* 1987;**415**:149–52.
244. Zarbin MA, Wamsley JK, Innis RB, Kuhar MJ. Cholecystokinin receptors: presence and axonal flow in the rat vagus nerve. *Life Sci* 1981;**29**:697–705.
245. Moran TH, Norgren R, Crosby RJ, McHugh PR. Central and peripheral vagal transport of cholecystokinin binding sites occurs in afferent fibers. *Brain Res* 1990;**526**:95–102.
246. Joyner K, Smith GP, Gibbs J. Abdominal vagotomy decreases the satiating potency of CCK-8 in sham and real feeding. *Am J Physiol* 1993;**264**:R912–R916.
247. Smith GP, Jerome C, Cushin BJ, Eterno R, Simansky KJ. Abdominal vagotomy blocks the satiety effect of cholecystokinin in the rat. *Science* 1981;**213**:1036–7.
248. Schwartz GJ, Berkow G, McHugh PR, Moran TH. Gastric branch vagotomy blocks nutrient and cholecystokinin-induced suppression of gastric emptying. *Am J Physiol* 1993;**264**:R630–R637.
249. Eisen S, Phillips RJ, Geary N, Baronowsky EA, Powley TL, Smith GP. Inhibitory effects on intake of cholecystokinin-8 and cholecystokinin-33 in rats with hepatic proper or common hepatic branch vagal innervation. *Am J Physiol Regul Integr Comp Physiol* 2005;**289**:R456–R462.
250. Yox DP, Stokesberry H, Ritter RC. Vagotomy attenuates suppression of sham feeding induced by intestinal nutrients. *Am J Physiol* 1991;**260**:R503–R508.
251. MacLean DB. Abrogation of peripheral cholecystokinin-satiety in the capsaicin treated rat. *Regul Pept* 1985;**11**:321–33.
252. Ritter RC, Ladenheim EE. Capsaicin pretreatment attenuates suppression of food intake by cholecystokinin. *Am J Physiol* 1985;**248**:R501–R504.
253. South EH, Ritter RC. Capsaicin application to central or peripheral vagal fibers attenuates CCK satiety. *Peptides* 1988;**9**:601–12.
254. Brown TA, Washington MC, Metcalf SA, Sayegh AI. The feeding responses evoked by cholecystokinin are mediated by vagus and splanchnic nerves. *Peptides* 2011;**32**:1581–6.
255. Moran TH, Shnayder L, Hostetler AM, McHugh PR. Pylorectomy reduces the satiety action of cholecystokinin. *Am J Physiol* 1988;**255**:R1059–R1063.
256. Cox JE, McCown SM, Bridges JM, Tyler WJ. Inhibition of sucrose intake by continuous celiac, superior mesenteric, and intravenous CCK-8 infusions. *Am J Physiol* 1996;**270**:R319–R325.
257. Cox JE, Perdue GS, Tyler WJ. Suppression of sucrose intake by continuous near-celiac and intravenous cholecystokinin infusions in rats. *Am J Physiol* 1995;**268**:R150–R155.
258. Washington MC, Sayegh AI. Gastrin releasing peptides increase Fos-like immunoreactivity in the enteric nervous system and the dorsal vagal complex. *Peptides* 2011;**32**:1600–5.

259. Wright SA, Washington MC, Garcia C, Sayegh AI. Gastrin releasing peptide-29 requires vagal and splanchnic neurons to evoke satiation and satiety. *Peptides* 2012;**33**:125–31.
260. Edwards GL, Ladenheim EE, Ritter RC. Dorsomedial hindbrain participation in cholecystokinin-induced satiety. *Am J Physiol* 1986;**251**:R971–R977.
261. Day HE, McKnight AT, Poat JA, Hughes J. Evidence that cholecystokinin induces immediate early gene expression in the brainstem, hypothalamus and amygdala of the rat by a CCKA receptor mechanism. *Neuropharmacology* 1994;**33**:719–27.
262. Cooper MS, Reeve Jr JR, Abdalla MO, Moyer L, Raboin SJ, Green GM, et al. Cholecystokinin-33 is more effective than cholecystokinin-8 in inhibiting food intake and in stimulating the myenteric plexus and dorsal vagal complex. *Brain Res* 2008;**1205**:27–35.
263. Cooper MS, Reeve Jr JR, Raboin SJ, Abdalla MO, Green GM, Sayegh AI. Cholecystokinin-58 and cholecystokinin-8 produce similar but not identical activations of myenteric plexus and dorsal vagal complex. *Regul Pept* 2008;**148**:88–94.
264. Gulley S, Covasa M, Ritter RC, Sayegh AI. Cholecystokinin1 receptors mediate the increase in Fos-like immunoreactivity in the rat myenteric plexus following intestinal oleate infusion. *Physiol Behav* 2005;**86**:128–35.
265. Raybould HE, Gayton RJ, Dockray GJ. Mechanisms of action of peripherally administered cholecystokinin octapeptide on brain stem neurons in the rat. *J Neurosci* 1988;**8**:3018–24.
266. Greenberg D, Smith GP, Gibbs J. Infusion of CCK-8 into hepatic-portal vein fails to reduce food intake in rats. *Am J Physiol* 1987;**252**:R1015–R1018.
267. Greenberg D, Smith GP. Hepatic-portal infusion reduces the satiating potency of CCK-8. *Physiol Behav* 1988;**44**:535–8.
268. Pandya PK, Huang SC, Talkad VD, Wank SA, Gardner JD. Biochemical regulation of the three different states of the cholecystokinin (CCK) receptor in pancreatic acini. *Biochim Biophys Acta* 1994;**1224**:117–26.
269. Talkad VD, Patto RJ, Metz DC, Turner RJ, Fortune KP, Bhat ST, et al. Characterization of the three different states of the cholecystokinin (CCK) receptor in pancreatic acini. *Biochim Biophys Acta* 1994;**1224**:103–16.
270. Weatherford SC, Laughton WB, Salabarria J, Danho W, Tilley JW, Netterville LA, et al. CCK satiety is differentially mediated by high- and low-affinity CCK receptors in mice and rats. *Am J Physiol* 1993;**264**:R244–R249.
271. Simasko SM, Wiens J, Karpiel A, Covasa M, Ritter RC. Cholecystokinin increases cytosolic calcium in a subpopulation of cultured vagal afferent neurons. *Am J Physiol Regul Integr Comp Physiol* 2002;**283**:R1303–R1313.
272. Schutte IW, Hollestein KB, Akkermans LM, Kroese AB. Evidence for a role of cholecystokinin as neurotransmitter in the guinea-pig enteric nervous system. *Neurosci Lett* 1997;**236**:155–8.
273. Ekblad E, Arnbjornsson E, Ekman R, Hakanson R, Sundler F. Neuropeptides in the human appendix. Distribution and motor effects. *Dig Dis Sci* 1989;**34**:1217–30.
274. Nemeth PR, Zafirov DH, Wood JD. Effects of cholecystokinin, caerulein and pentagastrin on electrical behavior of myenteric neurons. *Eur J Pharmacol* 1985;**116**:263–9.
275. Sayegh AI, Covasa M, Ritter RC. Intestinal infusions of oleate and glucose activate distinct enteric neurons in the rat. *Auton Neurosci* 2004;**115**:54–63.
276. Sayegh AI, Ritter RC. Cholecystokinin activates specific enteric neurons in the rat small intestine. *Peptides* 2003;**24**:237–44.
277. Washington MC, Murry CR, Raboin SJ, Roberson AE, Mansour MM, Williams CS, et al. Cholecystokinin-8 activates myenteric neurons in 21- and 35-day old but not 4- and 14-day old rats. *Peptides* 2011;**32**:272–80.

278. Webb T, Gulley S, Pruitt F, Esdaile AR, Sharma SK, Cox JE, et al. Cholecystokinin-8 increases Fos-like immunoreactivity in myenteric neurons of the duodenum and jejunum more after intraperitoneal than after intravenous injection. *Neurosci Lett* 2005;**389**:157–62.
279. Gulley S, Raboin SJ, Henley SC, Chan WC, Sharma SK, Billups LH, et al. Chemical sympathectomy attenuates myenteric but not dorsal vagal complex Fos-like immunoreactivity induced by cholecystokinin-8 in the rat. *Brain Res* 2005;**1064**: 119–25.
280. Gulley S, Sharma SK, Mansour M, Sullivan CN, Moran TH, Sayegh AI. Strain differences in myenteric neuron number and CCK1 receptor mRNA expression may account for differences in CCK induced c-Fos activation. *Brain Res* 2005;**1058**:109–19.
281. Raboin SJ, Gulley S, Henley SC, Chan WC, Esdaile AR, Jackson CA, et al. Effect of sympathectomy and demedullation on increased myenteric and dorsal vagal complex Fos-like immunoreactivity by cholecystokinin-8. *Regul Pept* 2006;**134**:141–8.
282. Raboin SJ, Gulley S, Henley SC, Chang WC, Esdaile AR, Jackson CA, et al. Effect of adrenalectomy on cholecystokinin-8-induced Fos-like immunoreactivity in myenteric neurons and the dorsal vagal complex in rats. *Am J Vet Res* 2006;**67**:1552–6.
283. Webb T, Gulley S, Esdaile AR, Pruitt F, Sharma SK, Williams CS, et al. Effects of cholecystokinin-receptor antagonists on Fos-like immunoreactivity stimulated by sulfated cholecystokinin-8 in neurons of the myenteric plexus and hindbrain of rats. *Am J Vet Res* 2005;**66**:1308–13.
284. Wang L, Barachina MD, Martinez V, Wei JY, Tache Y. Synergistic interaction between CCK and leptin to regulate food intake. *Regul Pept* 2000;**92**:79–85.
285. Volkoff H, Canosa LF, Unniappan S, Cerda-Reverter JM, Bernier NJ, Kelly SP, et al. Neuropeptides and the control of food intake in fish. *Gen Comp Endocrinol* 2005;**142**:3–19.
286. Barrachina MD, Martinez V, Wang L, Wei JY, Tache Y. Synergistic interaction between leptin and cholecystokinin to reduce short-term food intake in lean mice. *Proc Natl Acad Sci USA* 1997;**94**:10455–60.
287. Matson CA, Reid DF, Cannon TA, Ritter RC. Cholecystokinin and leptin act synergistically to reduce body weight. *Am J Physiol Regul Integr Comp Physiol* 2000;**278**: R882–R890.
288. Wang L, Martinez V, Barrachina MD, Tache Y. Fos expression in the brain induced by peripheral injection of CCK or leptin plus CCK in fasted lean mice. *Brain Res* 1998;**791**:157–66.
289. Li Y, Wu X, Zhou S, Owyang C. Low-affinity CCK-A receptors are coexpressed with leptin receptors in rat nodose ganglia: implications for leptin as a regulator of short-term satiety. *Am J Physiol Gastrointest Liver Physiol* 2011;**300**:G217–G227.
290. Peters JH, Karpiel AB, Ritter RC, Simasko SM. Cooperative activation of cultured vagal afferent neurons by leptin and cholecystokinin. *Endocrinology* 2004;**145**:3652–7.
291. Heldsinger A, Grabauskas G, Song I, Owyang C. Synergistic interaction between leptin and cholecystokinin in the rat nodose ganglia is mediated by PI3K and STAT3 signaling pathways: implications for leptin as a regulator of short term satiety. *J Biol Chem* 2011;**286**:11707–15.
292. Maletinska L, Maixnerova J, Matyskova R, Haugvicova R, Pirnik Z, Kiss A, et al. Synergistic effect of CART (cocaine- and amphetamine-regulated transcript) peptide and cholecystokinin on food intake regulation in lean mice. *BMC Neurosci* 2008;**9**:101.
293. de Lartigue G, Dimaline R, Varro A, Dockray GJ. Cocaine- and amphetamine-regulated transcript: stimulation of expression in rat vagal afferent neurons by cholecystokinin and suppression by ghrelin. *J Neurosci* 2007;**27**:2876–82.

294. Lo CM, Zhang DM, Pearson K, Ma L, Sun W, Sakai RR, et al. Interaction of apolipoprotein AIV with cholecystokinin on the control of food intake. *Am J Physiol Regul Integr Comp Physiol* 2007;**293**:R1490–R1494.
295. Lo CM, Xu M, Yang Q, Zheng S, Carey KM, Tubb MR, et al. Effect of intraperitoneal and intravenous administration of cholecystokinin-8 and apolipoprotein AIV on intestinal lymphatic CCK-8 and apo AIV concentration. *Am J Physiol Regul Integr Comp Physiol* 2009;**296**:R43–R50.
296. Glatzle J, Darcel N, Rechs AJ, Kalogeris TJ, Tso P, Raybould HE. Apolipoprotein A-IV stimulates duodenal vagal afferent activity to inhibit gastric motility via a CCK1 pathway. *Am J Physiol Regul Integr Comp Physiol* 2004;**287**:R354–R359.
297. Sayegh AI, Reeve Jr JR, Lampley ST, Hart B, Gulley S, Esdaile AR, et al. Role for the enteric nervous system in the regulation of satiety via cholecystokinin-8. *J Am Vet Med Assoc* 2005;**226**:1809–16.
298. Gibbins IL, Furness JB, Costa M. Pathway-specific patterns of the co-existence of substance P, calcitonin gene-related peptide, cholecystokinin and dynorphin in neurons of the dorsal root ganglia of the guinea-pig. *Cell Tissue Res* 1987;**248**:417–37.
299. Grothe C, Unsicker K. Neuron-enriched cultures of adult rat dorsal root ganglia: establishment, characterization, survival, and neuropeptide expression in response to trophic factors. *J Neurosci Res* 1987;**18**:539–50.
300. Ju G, Hokfelt T, Brodin E, Fahrenkrug J, Fischer JA, Frey P, et al. Primary sensory neurons of the rat showing calcitonin gene-related peptide immunoreactivity and their relation to substance P-, somatostatin-, galanin-, vasoactive intestinal polypeptide- and cholecystokinin-immunoreactive ganglion cells. *Cell Tissue Res* 1987;**247**:417–31.
301. Yaksh TL, Michener SR, Bailey JE, Harty GJ, Lucas DL, Nelson DK, et al. Survey of distribution of substance P, vasoactive intestinal polypeptide, cholecystokinin, neurotensin, Met-enkephalin, bombesin and PHI in the spinal cord of cat, dog, sloth and monkey. *Peptides* 1988;**9**:357–72.
302. Berthoud HR, Kressel M, Raybould HE, Neuhuber WL. Vagal sensors in the rat duodenal mucosa: distribution and structure as revealed by in vivo DiI-tracing. *Anat Embryol (Berl)* 1995;**191**:203–12.
303. Czyzyk-Krzeska MF, Bayliss DA, Seroogy KB, Millhorn DE. Gene expression for peptides in neurons of the petrosal and nodose ganglia in rat. *Exp Brain Res* 1991;**83**:411–8.
304. Dun NJ, Wu SY, Lin CW. Excitatory effects of cholecystokinin octapeptide on rat nodose ganglion cells in vitro. *Brain Res* 1991;**556**:161–4.
305. Ghilardi JR, Allen CJ, Vigna SR, McVey DC, Mantyh PW. Cholecystokinin and neuropeptide Y receptors on single rabbit vagal afferent ganglion neurons: site of prejunctional modulation of visceral sensory neurons. *Brain Res* 1994;**633**:33–40.
306. Widdop RE, Krstew E, Mercer LD, Carlberg M, Beart PM, Jarrott B. Electrophysiological and autoradiographical evidence for cholecystokinin A receptors on rat isolated nodose ganglia. *J Auton Nerv Syst* 1994;**46**:65–73.
307. Hamaji M, Kawai Y, Kawashima Y, Tohyama M. An electron microscopic study on VIP-, BOM- and CCK-like immunoreactive terminals in the celiac-superior mesenteric ganglion complex of the guinea pig. *Brain Res* 1989;**488**:283–7.
308. Holle GE, Forssmann WG, Rehfeld JF, Holle F. Effects of superior mesenteric and coeliac ganglionectomy on peptide-producing cells of the small intestinal mucosa in the Hanford mini pig. II. Immunohistochemical study. *J Auton Nerv Syst* 1989;**26**:147–56.
309. Eastwood C, Maubach K, Kirkup AJ, Grundy D. The role of endogenous cholecystokinin in the sensory transduction of luminal nutrient signals in the rat jejunum. *Neurosci Lett* 1998;**254**:145–8.

310. Gulley S, Sharma SK, Moran TH, Sayegh AI. Cholecystokinin-8 increases Fos-like immunoreactivity in the brainstem and myenteric neurons of rats through CCK1 receptors. *Peptides* 2005;**26**:1617–22.
311. Richards W, Hillsley K, Eastwood C, Grundy D. Sensitivity of vagal mucosal afferents to cholecystokinin and its role in afferent signal transduction in the rat. *J Physiol* 1996;**497**(Pt 2):473–81.
312. Sayegh AI, Ritter RC. Vagus nerve participates in CCK-induced Fos expression in hindbrain but not myenteric plexus. *Brain Res* 2000;**878**:155–62.
313. Sayegh AI, Ritter RC. CCK-A receptor activation induces fos expression in myenteric neurons of rat small intestine. *Regul Pept* 2000;**88**:75–81.
314. Ritter RC, Brenner LA, Tamura CS. Endogenous CCK and the peripheral neural substrates of intestinal satiety. *Ann N Y Acad Sci* 1994;**713**:255–67.
315. Ghilardi JR, Allen CJ, Vigna SR, McVey DC, Mantyh PW. Trigeminal and dorsal root ganglion neurons express CCK receptor binding sites in the rat, rabbit, and monkey: possible site of opiate-CCK analgesic interactions. *J Neurosci* 1992;**12**:4854–66.
316. Funakoshi K, Nakano M, Atobe Y, Kadota T, Goris RC, Kishida R. Selective projections of cholecystokinin-8 immunoreactive fibers to galanin immunoreactive sympathetic preganglionic neurons in a teleost, Stephanolepis cirrhifer. *Neurosci Lett* 2001;**316**:111–3.
317. Knoper SR, Meehan AG, Purnyn S, Coggan JS, Anthony TL, Kreulen DL. CCKA receptors mediate slow depolarizations in cultured mammalian sympathetic neurons. *Eur J Pharmacol* 1993;**232**:65–9.
318. Metcalf SA, Washington MC, Brown TA, Williams CS, Strader AD, Sayegh AI. Ileal interposition attenuates the satiety responses evoked by cholecystokinin-8 and −33. *Peptides* 2011;**32**:1296–302.
319. Larsen CJ, Washington MC, Sayegh AI. Cholecystokinin-8 increases the satiety ratio in diabetic rats more than cholecystokinin-33. *Physiol Behav* 2010;**101**:649–52.
320. Sayegh AI, Ritter RC. Morphology and distribution of nitric oxide synthase-, neurokinin-1 receptor-, calretinin-, calbindin-, and neurofilament-M-immunoreactive neurons in the myenteric and submucosal plexuses of the rat small intestine. *Anat Rec A Discov Mol Cell Evol Biol* 2003;**271**:209–16.
321. Dragunow M, Faull R. The use of c-fos as a metabolic marker in neuronal pathway tracing. *J Neurosci Methods* 1989;**29**:261–5.

CHAPTER NINE

Adiponectin Receptors in Energy Homeostasis and Obesity Pathogenesis

Benson T. Akingbemi
Department of Anatomy, Physiology and Pharmacology, Auburn University, Auburn, Alabama, USA

Contents

Abstract

Adipokines, that is factors secreted by adipose tissue, act through a network of autocrine, paracrine, and endocrine pathways to regulate several aspects of physiology, including glucose and lipid metabolism, neuroendocrine function, reproduction, and cardiovascular function. In particular, adiponectin, a 30-kDa protein, is associated with the regulation of insulin sensitivity, and its levels in serum are affected by altered metabolic homeostasis. Adiponectin effects are mediated by adiponectin receptors, which occur as two isoforms (AdipoR1 and AdipoR2). Transcriptional regulation of adiponectin is by the peroxisome proliferator-activated receptor-gamma (PPAR-γ). However, acting through AdipoR1 and AdipoR2, adiponectin enhances 5′ adenosine monophosphate-activated protein kinase (AMPK) and the PPARα-mediated pathways in the liver and skeletal muscles. Adiponectin receptors mediate a wide spectrum of metabolic reactions, including gluconeogenesis and fatty-acid oxidation. Altogether, adiponectin deficiency and/or decreased adiponectin receptor-mediated activity possibly contribute to insulin resistance in metabolic syndromes, coronary heart disease, and liver disease.

Progress in Molecular Biology and Translational Science, Volume 114
ISSN 1877-1173
http://dx.doi.org/10.1016/B978-0-12-386933-3.00009-1

1. INTRODUCTION

Although considered initially as an inert fat storage organ, adipose tissue is now known to secrete several proteins and bioactive peptides (i.e., adipokines) that serve locally as autocrine or paracrine regulators or systemically as hormonally active substances. Adiponectin is probably the most abundantly secreted adipokine (Fig. 9.1), forming about 0.05% of serum proteins and measuring 3–30 μg/ml in humans and 3–6 μg/ml in rodents.[1] There appears to be sexual dimorphism in adiponectin secretion because several reports have shown that adiponectin levels are lower in males than in females.[2,3] However, despite high levels in plasma, adiponectin is cleared rapidly by the liver and has only a half-life of ~75 min, suggesting a rapid turnover and indicating that a high synthetic rate is required in adipocytes in order to maintain high blood levels.[4]

There is now general agreement that adiponectin exerts many of its cellular effects primarily by binding to two receptor isoforms, which exhibit distinct distribution patterns. The two receptor isoforms (AdipoR1 and AdipoR2) are seven-transmembrane receptors and have a 66.7% homology in structure. However, AdipoR1 and AdipoR2 are structurally and functionally distinct from G protein-coupled receptors because their N-terminus is intracellular, whereas the C-terminus is extracellular.[5] Adiponectin receptors are expressed in a wide spectrum of tissues, including metabolically

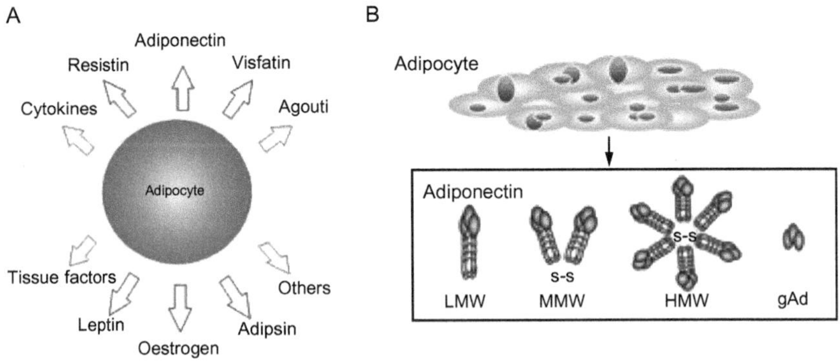

Figure 9.1 Adipose tissue is an endocrine organ secreting hormones, cytokines, and other tissue-active factors (A). Adiponectin is a major cytokine produced by adipocytes and is present in plasma as multimer forms. LMW, low molecular weight; MMW, medium molecular weight; HMW, high molecular weight[121]. (See Color Insert.)

active organs such as adipose tissue, liver, skeletal muscles, as well as in the cardiovascular system, reproductive tract, and brain. Of the metabolically active organs, AdipoR1 and AdipoR2 are both expressed in adipose tissue while AdipoR1 is the predominant form expressed in skeletal muscle and AdipoR2 in the liver.[6]

2. EXPRESSION

Adiponectin is an autocrine regulator of adipose tissue, which expresses both AdipoR1 and AdipoR2.[7] However, expression levels of AdipoR1 and AdipoR2 are decreased in conditions of obesity, which has the effect of diminishing adiponectin sensitivity and causing insulin resistance in adipocytes[8] (Fig. 9.2). On the other hand, AdipoR2 is expressed at higher levels than AdipoR1 in the liver,[6] but expression levels of both receptor subtypes in the liver are higher in insulin-resistant subjects than in control animals.[10] The higher level of adiponectin receptor expression associated with insulin resistance is probably a compensatory mechanism to mitigate the effects of decreased plasma adiponectin levels. Unlike the liver, which expresses mostly AdipoR2, skeletal muscle expresses predominantly AdipoR1.[6] Since both liver and skeletal muscles are metabolically active, differential expression of adiponectin receptors (AdipoR2 vs. AdipoR1) in these organs seems to suggest that both receptor isoforms have similar capacity to mediate metabolic activity. Also, deficiency in expression levels of one receptor subtype possibly affects expression of the other because AdipoR1 null mice have increased AdipoR2 mRNA levels in brown adipose tissue (BAT) while expression levels are unaffected in white adipose

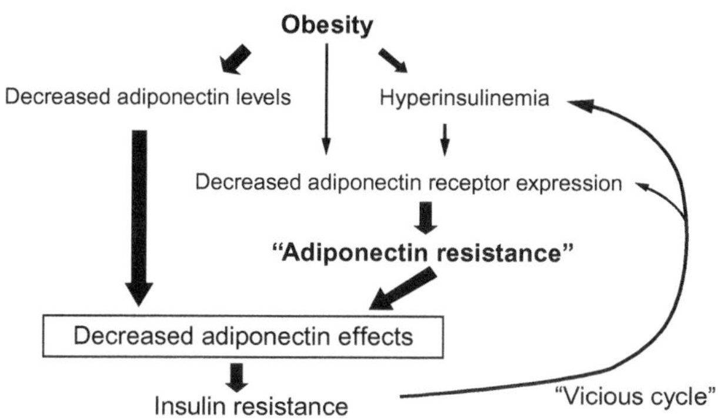

Figure 9.2 Adiponectin of adipose tissue origin signals to other peripheral tissues and represents the link between obesity and insulin resistance.[9]

tissue (WAT), skeletal muscles, and brain. On the other hand, AdipoR1 expression levels were increased in the livers of AdipoR2 null mice.[11]

The heart is an organ with high metabolic activity, and both cardiomyocytes and epicardial adipose tissue have the capacity to secrete adiponectin.[12,13] There is evidence that cardiac fibroblasts express AdipoR1,[14,15] and several pieces of evidence suggest that both AdipoR1 and AdipoR2 are expressed at high levels in the brain.[11] Besides robust expression levels in adipose tissue, liver, skeletal muscle, reproductive tract, cardiovascular system, and the brain, AdipoR1 and AdipoR2 are expressed in several other tissues, including the lungs,[16] placentae,[17] osteoblasts,[18] kidneys,[19] and pancreatic β-cells.[20]

The hypothalamus–pituitary–gonadal (HPG) axis is the primary regulator of reproductive activity in both sexes, and several reproductive organs are known to express adiponectin receptors. Although it remains to be clarified whether adiponectin has the capacity to cross the blood–brain barrier, both AdipoR1 and AdipoR2 are localized to several hypothalamic nuclei.[21] Interestingly, mRNA and protein for adiponectin, AdipoR1 and AdipoR2 are present in the rat testis.[17] AdipoR1 appears to be the predominant receptor in the spermatogenic epithelium, whereas AdipoR2 is the isoform expressed by interstitial Leydig cells.[11] Interestingly, a pubertal increase in AdipoR2 protein expression is evident in rat Leydig cells[22], but it is not clear whether higher expression levels translate to greater sensitivity to adiponectin stimulation. AdipoR1 and AdipoR2 mRNA have also been localized to the chicken testis.[23] In the female, there is evidence that AdipoR1 and AdipoR2 are expressed in pig[24] and human ovaries,[25,26] and the oviduct of cycling rats produces adiponectin localized to secretory epithelial cells.[27] Although there is no information on hormonal regulation of AdipoR1 and AdipoR2 expression, adiponectin expression was increased by gonadotropin stimulation in the human ovary.[28] While the physiological relevance of adiponectin and its receptors in reproductive tract tissues has received only little attention, expression of AdipoR1 and AdipoR2 in the reproductive tract suggests a role in reproductive activity.

3. PHYSIOLOGY

The preponderance of evidence in the literature indicates that adiponectin is biologically relevant across the life cycle. For example, adiponectin levels in cord blood of full-term infants are much greater than in adult subjects and correlate positively with birth weight and body mass index (BMI) but these levels are reversed later in childhood.[29,30] Incidentally, there is no information on age-dependent roles of adiponectin in the literature. However, there is a consensus that multiple environmental factors affect energy homeostasis by causing changes to adiponectin secretion. The American Obesity Association estimated in 2009 that obesity affects nearly one-third of the adult American population, that is, approximately 60 million individuals with annual costs in health care delivery of over $100 billion.[1] Insulin resistance, which is characterized by high insulin levels in blood (hyperinsulinaemia), is a feature of several metabolic syndromes. Circulating adiponectin levels are correlated with hyperinsulinaemia and insulin resistance,[31] and mice overexpressing adiponectin conversely show reduced insulin resistance.[32] Similarly, AdipoR1-deficient mice have increased adiposity associated with decreased glucose tolerance and energy expenditure, whereas adipoR2-null mice were lean and resistant to high fat diet-induced obesity as a result of improved glucose tolerance and energy expenditure.[11] Also, expression levels of AdipoR1 and AdipoR2 may influence fat deposition because AdipoR2 null mice have smaller reproductive and perirenal WAT and BAT compared to wild-type (WT) litter-mates.[11]

Liver: The liver is not a significant source of adiponectin but its role in mediating adiponectin effects on metabolism has been known for a long time. The hepatic actions of adiponectin are similar to those in skeletal muscles and are mediated mostly by AdipoR2 activation of AMPK- and PPARα-mediated pathways[33] (Fig. 9.3). Adiponectin and AdipoR1-induced AMPK activation was found to decrease expression of genes encoding hepatic gluconeogenic enzymes (e.g., glucose-6-phosphatase and phosphoenolpyruvate carboxykinase) and lipogenic genes.[34] In contrast, AdipoR2 has no effect on expression of enzyme proteins involved in gluconeogenesis although its overexpression nevertheless increases expression of the gene encoding PPARα and its targets, including acyl-CoA oxidase (*Acox1*) and uncoupling protein 1 (*Ucp1*).[35] Adiponectin may also regulate glucose homeostasis by increasing fatty-acid oxidation and reducing fatty-acid influx and oxidation to triglycerides in the liver[36] (Fig. 9.4). Plasma adiponectin levels are negatively correlated with insulin, glucose, and triglyceride levels while deficiencies in adiponectin action correlate with steatosis, hepatomegaly, and inflammation that are present in liver disease. Thus, simultaneous disruption of AdipoR1 and AdipoR2 in mice abolishes adiponectin binding and actions, resulting in increased tissue triglyceride content, inflammation, and oxidative stress as well as insulin resistance and glucose intolerance.[35]

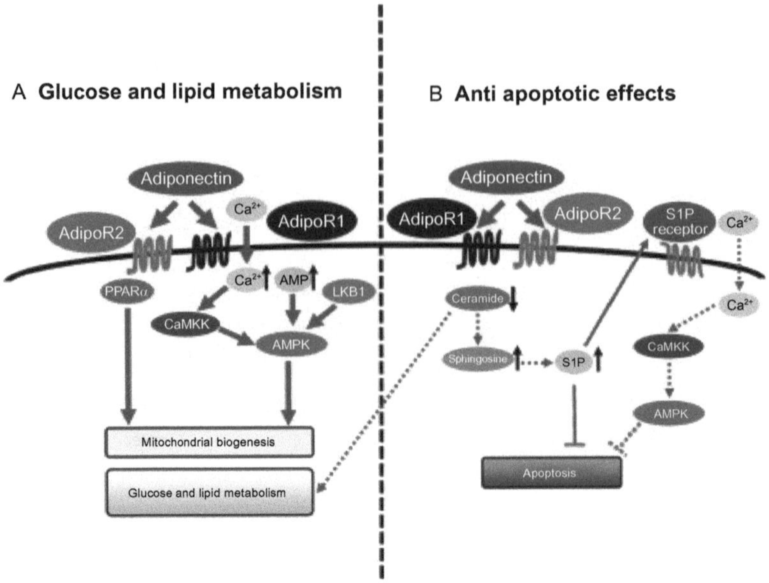

Figure 9.3 Adiponectin signals through both AdipoR1 and AdipoR2 (A and B), and by activation of AMPK and induction of extracellular Ca^{2+} influx and stimulation of PPARα, which are involved in glucose and lipid metabolism[122]. (See Color Insert.)

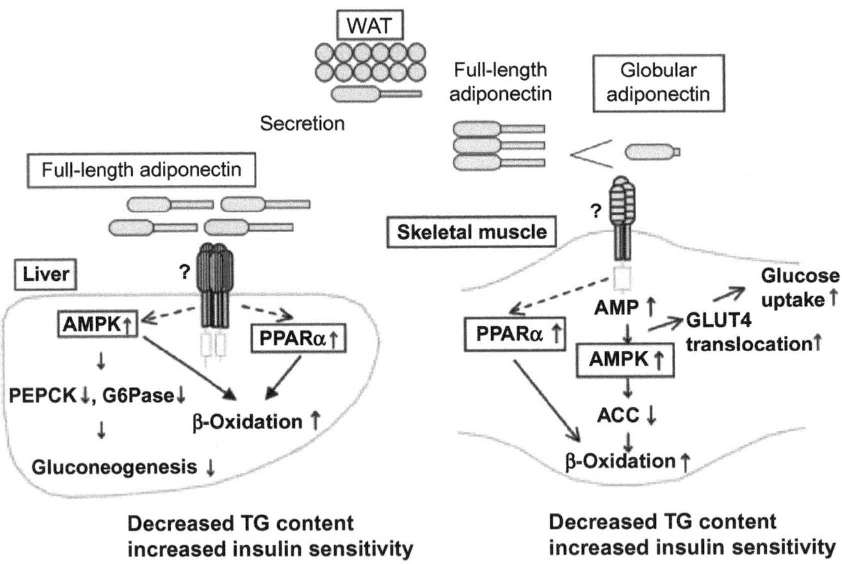

Figure 9.4 Adiponectin activates AMPK and cause fatty-acid oxidation and glucose uptake or PPARα to stimulate fatty-acid oxidation in skeletal muscle. In the liver, adiponectin activates AMPK to promote gluconeogenesis and fatty-acid oxidation. WAT, White adipose tissue; PEPCK, phosphoenolpyruvate carboxykinase; G6Pase, glucose-6-phosphatase G6Pase.[9] (See Color Insert.)

Skeletal muscle: Adiponectin plays a major role in the regulation of skeletal muscle metabolism. In this regard, the results of *in vitro* studies have shown that adiponectin-enhanced basal and insulin-stimulated glucose uptake by skeletal muscle fibers occurs by induction of GLUT4 translocation to the cell membrane,[37] increased fatty-acid uptake, and oxidation.[38] Other reports have suggested that adiponectin increases mitochondrial mass and oxidative capacity in skeletal muscle, at least in part, by stimulating extracellular Ca^{2+} influx and subsequently activating the Ca^{2+}/calmodulin-dependent protein kinase kinase-β PPARγ coactivator-1alpha (PGC1α) pathway.[39] Furthermore, adiponectin is expressed and secreted by skeletal muscle fibers.[40] Therefore, adiponectin may function as a locally acting myokine to enhance insulin-stimulated glucose uptake.[41] In agreement with these observations, adiponectin increased insulin-stimulated tyrosine phosphorylation of insulin receptors (IRs), insulin receptor substrate (IRS)-1, and protein kinase B or Akt phosphorylation in skeletal muscles of lipoatrophic mice.[32]

Lungs: There is a possibility that adiponectin may have a role in the regulation of respiratory function associated with its anti-inflammatory properties because respiratory dysfunction is a consistent feature in obese individuals, whereas adiponectin knockout mice exhibited progressive alveolar enlargement and endothelial cell apoptosis, which was attenuated by adenoviral adiponectin administration.[16] Also, chronic allergic airway inflammation and pulmonary vascular remodeling were found to be exacerbated in adiponectin-deficient mice.[42] The anti-inflammatory actions of adiponectin in the lungs are linked to inhibition of the cytokines IL-13 and IL-5.[43,44] Moreover, interactions between adiponectin and other inflammatory cytokines, for example, TNF-α, and IL-6, are thought to have a modulatory effect on insulin sensitivity and vascular atherogenicity.[45]

Cardiovascular system: The effect of adiponectin and AdipoR1- and AdipoR2-mediated activity in the regulation of cardiovascular function has received considerable attention. In general, plasma adiponectin levels negatively correlate with risk factors for ischemic heart disease, including dyslipidemia and high blood pressure.[46] Also, adiponectin-related impairment of heart function has been linked to cardiac fibrosis.[47] Adiponectin-deficient mice exhibited diet-induced insulin resistance and profound myocardial remodeling in response to pressure overload.[48] The view that adiponectin protects against heart disease is supported by clinical observations showing that plasma adiponectin levels have an inverse relationship with the incidence of cardiovascular diseases (CVDs) such as artherosclerosis, myocardial infarction, heart failure, endothelial dysfunction, and hypertension.[49]

Cardiac function is intricately tied to endothelial function. Early indications that adiponectin may protect against endothelial damage was provided by observations demonstrating the presence of adiponectin in catheter-injured vessel wall but not in the intact artery.[50] Adiponectin has the capacity to prevent macrophage adhesion to endothelium and was found to alleviate macrophage-induced injury[51] (Fig. 9.5). Adiponectin-induced inhibition of leukocyte adhesion to endothelium was attributed to decreased expression of adhesion molecules, including ICAM-1, vascular cellular adhesion molecule 1 and E-selectin, activation of Nuclear factor kappa beta (NFκB) and increased tumor necrosis factor (TNF) α-mediated activity.[8,52] Indeed, the size of artherosclerotic lesions was decreased by adiponectin overexpression in a mouse model of artherosclerosis.[53] These observations are supported by results of transgenic animal studies showing that cerebral infarct sizes are increased following ischemia-reperfusion, whereas supplementation with exogenous adiponectin decreases infarct size in wild-type (WT) and mutant mice.[54] Also, administration of adiponectin to Apolipoprotein (apo) E-deficient mice decreased aortic plaque formation and the size of lipid

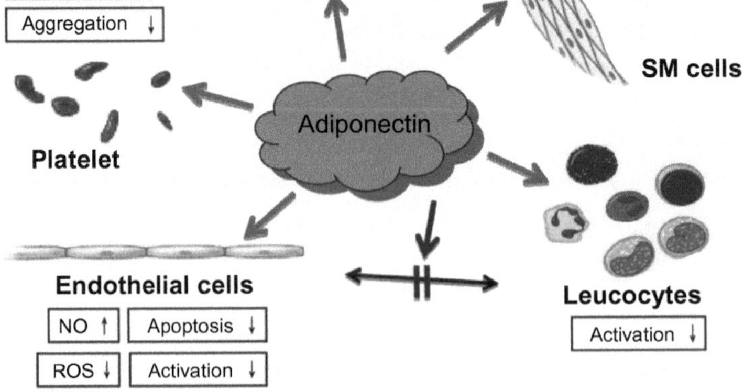

Figure 9.5 Adiponectin exerts its vasculoprotective activities through multiple mechanisms, including prevention of leukocyte adhesion to endothelium[121]. (See Color Insert.)

droplets compared to untreated mice.[53] Although serum parameters, for example, triglyceride, glucose, and insulin levels, were not changed, lesions in the aortic arch and descending aorta were smaller in globular adiponectin transgenic apoE-deficient mice than in control apoE-deficient littermates.[55] Taken together, these findings imply that overexpression of adiponectin results in marked reduction of atherosclerotic lesion formation.[53]

Oxidative stress and chronic inflammation are important pathogenic factors in Type 2 diabetes (T2D) and CVDs, and there is evidence that excessive oxidative stress may contribute to diabetes-induced artherosclerosis and coronary artery disease.[56] In this regard, elevated serum adiponectin levels are associated with increased activity of the antioxidant enzyme SOD-1 in WT and *db/db* mice.[57] Also, adenovirus-mediated administration of adiponectin stimulated eNOS phosphorylation during cerebral ischemia in WT and adiponectin knockout mice.[58] Altogether, the beneficial effects of adiponectin on cardiovascular function are attributed to increased lipoprotein lipase activity, reduced lipid levels, and increased NO production.[59] However, there are indications that adiponectin receptors may exert differential effects on antioxidant capacity because adenovirus-mediated expression of AdipoR2 in the liver of *db/db* mice, but not AdipoR1, increased catalase and SOD-1 enzyme activities.[35,60]

Reproductive tract: There appears to be a reciprocal relationship between adipose tissue and reproductive tract function because the P450 aromatase (enzyme that converts androgens into estrogens), steroid hormone receptors [estrogen receptors (ESRs) and androgen receptors (ARs)] and steroidogenic enzymes are expressed in adipose tissue.[61] Therefore, reproductive tract tissues are targeted by adipose tissue-secreted estrogen while estrogen and androgen of gonadal origin activate steroid hormone receptors to regulate adipose tissue development and function.[1] Relatedly, testosterone administration was found to reduce plasma adiponectin levels, which probably explains, at least in part, the sexual dimorphism in adiponectin levels seen in male and female subjects.[62] Taken together, these observations are in agreement with reports indicating that adiponectin levels in reproductive tract tissues are acutely regulated during different phases of the reproductive cycle, including sexual differentiation, puberty, gestation, and lactation.[63]

Several reports have shown that adiponectin exerts its effects at all levels of the HPG axis and thereby regulates hypothalamic GnRH and pituitary gonadotropin secretion as well as gonadal function. For example, adiponectin decreased basal and GnRH-stimulated LH, but not FSH,

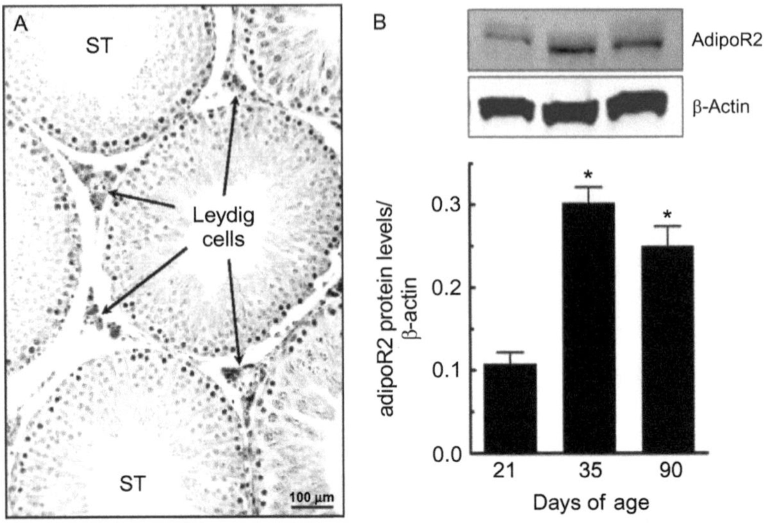

Figure 9.6 Adiponectin receptors are expressed in testicular cells (A), and there is a pubertal increase in Leydig cell expression of AdipoR2 in the postnatal period (B).[17,22] $^{*}p < 0.05$ compared to 21 days of age. (See Color Insert.)

secretion,[64] and long-term caloric restriction was manifested as increased circulating adiponectin levels.[65] Neonatal castration of male rats resulted in adiponectin levels similar to those observed in females, whereas circulating adiponectin levels in adult females were elevated following ovariectomy.[63] Both adiponectin and its receptors (AdipoR1/R2) are expressed in testicular cells[17,22] (Figs. 9.6 and 9.7). Phenotypic analysis of AdipoR1 and AdipoR2 knockout mice indicated that both are viable and fertile with normal body weights.[11,35] However, AdipoR2 mice exhibited reduced testis weights compared to WT littermates, which was associated with atrophy of the seminiferous tubules and aspermia although plasma testosterone levels were unaffected.[11] While there is evidence that adiponectin regulates the steroidogenic pathway in Leydig cells,[22] the physiological role of adiponectin in Leydig cell proliferation and differentiation remains to be clarified.

In the ovary, AdipoR1 knockdown affects mostly cellular proliferation, survival, and apoptosis, whereas AdipoR2 knockdown disrupts the steroidogenic pathway associated with diminished responsiveness to adiponectin and FSH stimulation.[66] In particular, adiponectin increased progesterone and estradiol secretion by granulosa cells, which was linked to increased aromatase activity and protein kinase activation [extracellular regulated kinase or ERK, p38 mitogen activate protein kinase (MAPK)].[67] The physiological significance of these findings has not been described but fertility is impaired

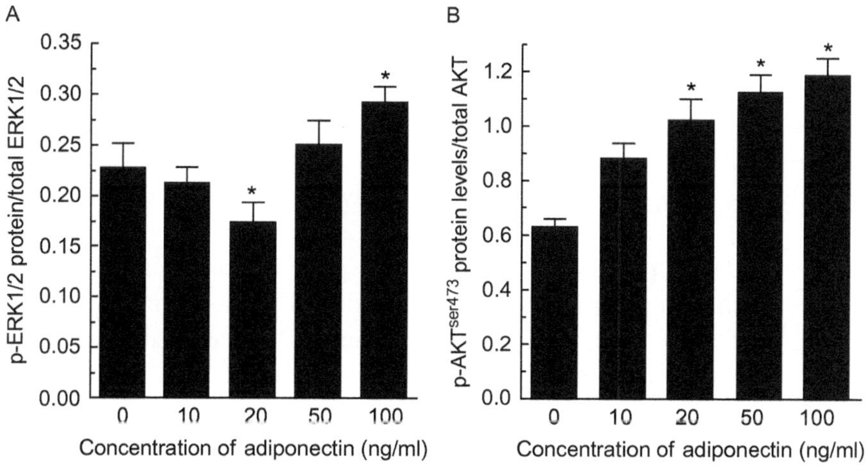

Figure 9.7 Adiponectin has the capacity to activate kinase cascades in testicular Leydig cells, for example, extracellular regulated kinase ERK (A) and protein kinase B or AKT (B).[22] *$p < 0.05$ compared to control.

in mice overexpressing adiponectin, though loss of adiponectin signaling appears to have no effect.[25] Thus, additional studies are required in order to fully elucidate the role of adiponectin, AdipoR1 and AdipoR2 in the regulation of ovarian development and function.

Brain: Multiple reports in the literature showed that intracerebroventricular administration of adiponectin decreased body weight and improves insulin sensitivity, which was related to increased energy expenditure and thermogenesis.[68] Adiponectin administration into the hypothalamus activated several kinase cascades, including phosphorylation of IRS-1, IRS-2, Akt, ERK, forkhead transcription factor-1 (FOXO1), JAK2, and signal transducer and activator of transcription 3, which mediate multiple signaling pathways.[69] Together, these observations imply that adiponectin causes centrally mediated effects, possibly acting in concert with leptin- and insulin-mediated pathways to regulate energy homeostasis. However, it is likely that inhibition of food intake by adiponectin is mediated specifically by hypothalamic AdipoR1[70] although a role for AdipoR2 has not been excluded.

Others: Information in the literature is scanty but adiponectin suppresses osteoclastogenesis and stimulates osteoblastogenesis, which has the effect of increasing bone mass.[71] Thus, it is likely that adiponectin exerts a protective effect against bone breakdown.[72] Adiponectin-induced osteoblast proliferation is probably mediated by AdipoR1, which activated the c-Jun

NH_2-terminal kinase pathway, whereas osteoblast differentiation is mediated by the p38 MAPK pathway.[18] A role for AdipoR2 in bone has not been advocated at this time but the effects of adiponectin and AdipoR1 provide some evidence that bone is a target for adiponectin action.

The high levels of serum 17β-estradiol in pregnancy were associated with decreased adiponectin levels.[63] It is likely that adiponectin exerts some control over placental metabolism because adiponectin activates the p38 MAPK-mediated pathway.[73] Furthermore, placental tissue isolated from women with severe preeclampsia have reduced adiponectin expression levels.[74] Although there is evidence that AdipoR1 may regulate angiogenesis in some tissues, it is not clear that AdipoR1 and AdipoR2 have similar roles in the placenta.[75,76]

Adiponectin levels are considered to be predictive for chronic kidney disease in male subjects because circulating adiponectin levels are increased in predialysis patients with end-stage renal disease.[77] The clinical data are supported by results of animal studies showing that adiponectin-deficient mice exhibit several histological alterations, including segmentally fused podocyte processes, albuminuria, and increased oxidative stress in the kidneys compared to WT litter-mates. AdipoR1 and AdipoR2 are both present in podocytes, and decreased AMPK-mediated activity has been associated with oxidative stress and podocyte dysfunction.[19]

There is evidence that AdipoR1/AdipoR2 are expressed in rat and human pancreatic β-cells, and pancreatic β-cell dysfunction is featured in conditions of hypoadiponectinemia. For example, deficiencies in AdipoR2-mediated signaling result in the failure of pancreatic β-cells to compensate for insulin resistance.[78] Adiponectin action in the pancreas involves protein kinase activation, that is, ERK and Akt phosphorylation, which increases glucose-stimulated insulin secretion in β-cells.[20] However, additional studies are necessary to better characterize the mechanisms of adiponectin action regulating pancreatic insulin secretion.

4. SIGNALING PATHWAYS

Adiponectin is secreted by adipose tissue either as a trimer (LMW), combination of two trimers (MMW) or as six trimers (HMW). Hepatic clearance is thought to occur at different rates such that HMW adiponectin is removed slowly while the trimer (LMW) is cleared at the fastest rate.[4] Interestingly, the two adiponectin receptor isoforms have different binding affinities for different adiponectin forms. AdipoR1 is thought to bind globular adiponectin with higher affinity than full-length adiponectin, whereas AdipoR2 appears to be an intermediate-affinity receptor. Nevertheless, AdipoR1 may serve as a receptor for globular adiponectin and AdipoR2 for the full-length adiponectin.[60] Although HMW adiponectin is thought to correlate better with insulin sensitivity than others, the relative contribution of different molecular forms of adiponectin to overall physiological action is compounded by the capacity for AdipoR1 and AdipoR2 to form homo- and heterodimers.[79]

AdipoR1 and AdipoR2 are considered to be the classical proteins mediating adiponectin action but there is a growing body of evidence that other adiponectin-binding proteins probably exist. For example, T-cadherin, which is expressed in endothelium and smooth muscles, is thought to have preference for HMW adiponectin.[80] Because simultaneous disruption of AdipoR1 and AdipoR2 was sufficient to almost completely abolish adiponectin binding to the liver (main organ mediating the metabolic effects of adiponectin), some have questioned whether T-cadherin has any role in mediating adiponectin effects.[35] On the other hand, the yeast homologue PHO_{36}, the receptor for osmotin, which has structural similarity albeit with no sequence homology with adiponectin, has the capacity to bind and activate AdipoR1 and AdipoR2.[81] Additional studies are warranted to fully elucidate the roles of T-cadherin and PHO_{36}, if any, in adiponectin biology.

In metabolically active organs such as the liver and skeletal muscles, it appears that AdipoR1 primarily activates the AMPK-mediated pathway, whereas AdipoR2 acts mostly through PPARα signaling. For example, adiponectin-induced inhibition of hypothalamic GnRH secretion is AMPK-mediated.[21] Although it was suggested that the anti-inflammatory effects of adiponectin results from inhibition of the inflammatory cytokine TNF-α,[82] its antifibrotic and anti-apoptotic effects in the liver are mediated at least in part by AMPK.[83,84] However, AMPK does not exclusively mediate adiponectin signaling and several other intermediates are involved in mediating adiponectin action in several tissues. For example, adiponectin protection against apoptosis in pancreatic β-cells was attributed to ERK and Akt, but not AMPK, phosphorylation.[20] Similarly, adiponectin increased progesterone and estradiol secretion in association with protein kinase activation (ERK, p38 MAPK).[67] Adiponectin also enhanced insulin-stimulated Akt phosphorylation in skeletal muscles.[32]

The local and tissue-specific effects of adiponectin are determined in part by changes in expression of its receptor isoforms and signaling intermediates. Although AdipoR1 and AdipoR2 may have differential binding affinities, several intermediate signaling molecules mediating adiponectin action have been identified. Perhaps the most extensively characterized factor is the adaptor protein containing pleckstrin homology domain, phosphotyrosine binding domain, and leucine zipper motif (APPL1). The APPL1 is able to bind AdipoR1 and AdipoR2 and is known to mediate adiponectin effects in the liver, skeletal muscles and endothelial cells. In contrast, other binding proteins, for example, ER protein 46, activated protein kinase C1 and protein kinase CK2β are thought to bind only to AdipoR1.[85,86] Thus, it appears that adiponectin-stimulated APPL1-AdipoR1 interactions represent a common mechanism for adiponectin-mediated action in several tissues.[87] It is likely that APPL1 activation of Akt promotes crosstalk between adiponectin- and insulin-mediated signaling pathways.

Altogether, available reports infer that targeted disruption of AdipoR1 results in abrogation of adiponectin-induced AMPK activation, whereas AdipoR2 deficiency causes a decrease in activity of PPARα-mediated pathways. Moreover, observations that AdipoR1 mRNA expression in adipose tissue, but not AdipoR2, is downregulated in adiponectin transgenic mice and is upregulated in adiponectin-deficient mice probably indicate that a feedback regulatory loop exists by which adiponectin regulates AdipoR1 expression.

5. REGULATION BY LEPTIN AND INSULIN

The plasma levels of adiponectin are about three orders of magnitude higher than leptin,[80] although adiponectin levels in cerebrospinal fluid of rats and humans measure at lower levels than levels found in the peripheral circulation.[68,88] Nevertheless, the possibility that adiponectin and insulin act in concert to cause centrally mediated effects on food intake and energy expenditure has been proposed. First, there is evidence that both adiponectin and leptin are insulin sensitizers and possibly signal through AMPK in skeletal muscles and liver to increase insulin sensitivity and stimulate β-oxidation and glucose uptake.[36,89] Second, reduced expression of gluconeogenic enzymes such as phosphoenolpyruvate carboxykinase and glucose-6-phosphatase is associated with elevated AMPK phosphorylation in the liver.[36] Therefore, AMPK activation is a probable mechanism by which insulin-sensitizing adipokines such as adiponectin and leptin increase insulin sensitivity.[59]

Several lines of evidence have demonstrated that leptin regulates adiponectin and adiponectin receptor function. For example, AdipoR1 and

AdipoR2 expression levels were decreased by ~65% and 55% in the liver of *db/db* mice compared to WT.[35] The latter observation implies that insulin resistance and diabetes in *db/db* mice are due in part to decreased expression of AdipoR1 and AdipoR2. The same effects are seen in hyperglycemic and hyperinsulinemic *ob/ob* mice, which have decreased AdipoR1 and AdipoR2 expression in skeletal muscle and adipose tissue.[90] In addition to regulating adiponectin receptors, changes in leptin signaling may affect adiponectin secretion directly because serum adiponectin levels are reduced in plasma, whereas liver clearance is prolonged in *ob/ob* mice.[4]

Insulin exerts similar effects as leptin on adiponectin signaling. For example, the levels of AdipoR1 mRNA is increased 2.5-fold in the skeletal muscle of streptozotocin (STZ)-diabetic mice, which are normalized following insulin administration,[91] suggesting that insulin has an inhibitory effect on AdipoR1 expression. Furthermore, adiponectin injection into the hypothalamus increases phosphorylation of several metabolic signaling molecules, including IRS-1, IRS-2, Akt, ERK, and signal transducer and activator of transcription 3.[69] Altogether, these observations imply that there is crosstalk between adiponectin, leptin-, and insulin-mediated pathways in the hypothalamus, liver, and skeletal muscle, cooperatively regulating multiple aspects of energy homeostasis.

6. REGULATION BY STEROID HORMONES

Receptors for steroid hormones (ESRs and ARs) and steroidogenic enzymes are present in adipocytes just as receptors for several adipokines, for example, leptin, adiponectin, have been localized to testicular cells.[22,92] There is a growing body of evidence showing that adipose tissue development is regulated by gonadal steroids.[93] Moreover, hypothalamic nuclei are sensitive to endocrine and metabolic signals and respond through effector pathways to regulate behavioral, endocrine, and metabolic changes affecting energy homeostasis. In this regard, ESRs are known to mediate hypothalamic regulation of locomotor activity and energy expenditure independently of the leptin receptor.[94] Indeed, deficiency in the *ESR1* or *aromatase* gene decreased physical activity and energy expenditure.[95] There is also evidence indicating that mice deficient in estrogen or androgen synthesis and/or signaling have altered adipose tissue accretion, affecting the number and size of adipocytes and resulting in altered metabolic homeostasis.[96,97] Thus, estrogen or compounds with estrogenic properties have the capacity to bind and activate ESRs and thereby regulate adiposity either directly, by modulation of lipogenesis, lipolysis, or adipogenesis, or indirectly, by modulating appetite and/or energy expenditure.[97]

The capacity of estrogen to act as a potent modulator of glucose homeostasis was indicated by findings in postmenopausal women who develop visceral obesity and insulin resistance and have an increased risk for diabetes; these metabolic aberrations were normalized by estrogen replacement therapy.[98] Individuals with mutations in the *aromatase* gene display truncal obesity, insulin resistance, and hyperlipidemia.[99] Similarly, mice lacking the *ESR1* or the *aromatase* gene have increased adiposity and fatty livers.[93,100] Estrogenic activity regulating adipose tissue has been ascribed to the ESR1 subtype and the role of ESR2 is much less clear because no clear metabolic phenotype is identified in mice lacking this subtype.[101] However, a more recent report indicated that feeding of a high-fat diet to βERKO mice increases their weight and adiposity to a greater extent than in WT littermates.[102]

Metabolic changes due to estrogenic compounds has led to suggestions that the recent increase in metabolic syndromes in the population is due in part to environmental factors because the rise in metabolic diseases such as obesity and diabetes was steepest in the last 40 years at the time of an astronomical rise in the global use of industrial chemicals.[103] Although still the subject of continuing debate, several exogenous estrogens, for example, dietary estrogens and the industrial chemical bisphenol A (BPA), are now thought to affect adipose tissue homeostasis and possibly contribute to and/or alleviate symptoms associated with metabolic syndromes.[104] A recent study demonstrated that BPA decreased adiponectin release from human adipose tissue.[105] The observations have led to the *obesogen hypothesis*, which states that altered developmental programming of adipose tissue results in metabolic syndromes in adulthood[103] (Fig. 9.8).

The role of androgens in adipose tissue development and function has received much less attention than for estrogens. However, low testosterone concentrations are consistent features in male subjects with T2D,[106,107] and epidemiological studies have suggested that low testosterone levels predict the occurrence of T2D.[108] Also, hypogonadism is associated with insulin resistance and metabolic syndromes. However, a cause and effect relationship has not been clearly described because low testosterone levels predispose to metabolic syndromes just as altered metabolic homeostasis increases the risk of hypogonadism.[109,110] Nonetheless, available evidence indicates that excessive estrogen secretion by adipose tissue in overweight and obese individuals has the effect of suppressing hypothalamic GnRH secretion, which in turn decreased pituitary gonadotropin and gonadal testosterone

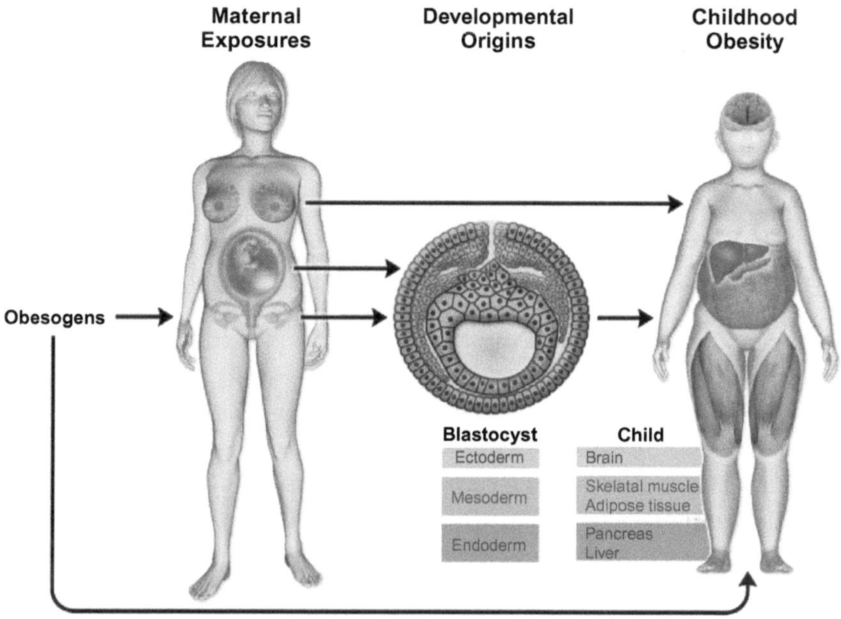

Figure 9.8 Embryonic tissues are targets for hormonally active environmental chemicals in the offspring, which cause changes to gene expression affecting differentiation in the postnatal period that lead to higher risk for metabolic syndromes in adulthood, and thereby acting as obesogens[123]. (See Color Insert.)

secretion.[111] Moreover, receptors for insulin and adiponectin are present in Leydig cells, which are the predominant source of androgens, and both factors are known to regulate androgen secretion.[22] Whether the absence of adiponectin signaling impairs development of testicular steroidogenesis is not known but excessive levels of adipokines affects the steroidogenic pathway in testicular Leydig cells.[22]

7. PATHOLOGY

The endocrine effects of adiponectin play a role in the regulation of many physiological processes and therefore contribute to the etiology of metabolic diseases. Obese individuals exhibit lower adiponectin levels (e.g., 3.7 vs. 8.9 μg/ml), which are negatively correlated to BMI, body fat mass, and fasting insulin and leptin levels.[112] Moreover, nonalcoholic fatty liver disease is associated with decreased expression of adiponectin receptors while several other liver diseases result in elevated serum adiponectin levels.[113] A role for adiponectin in multiple pathologies is further underscored by clinical signs associated with gene mutations that result in hypoadipoctinemia, including hypertension, hyperlipidemia, glucose intolerance, and coronary artery disease.[114] Because PPAR-γ is highly expressed in adipose tissue and stimulates adipocyte differentiation, thiazolidinediones, a class of compounds acting as PPAR-γ agonists, are found to improve insulin resistance and increased adiponectin mRNA levels in adipose tissue from both obese and lean individuals.[115]

Adiponectin may be involved in the pathogenesis of cancers because adiponectin levels are low in cancer patients.[1] In particular, several pathologies of the reproductive tract have been associated with altered serum adiponectin levels, including endometriosis, uterine leimyoma, breast cancer, and polycystic ovarian syndrome (PCOS).[116,117] Female subjects suffering from PCOS exhibit signs of clinical or biochemical hyperandrogenism in association with insulin resistance, which is related to decreased adiponectin levels as compared to healthy control subjects with similar BMI.[118] Altogether, the large number of reports associating low plasma adiponectin levels with several pathologies indicates that altered adiponectin levels represent independent risk factors for future development of metabolic syndromes, including diabetes and obesity.

Moreover, a genetic component to adiponectin-related metabolic disorders has been proposed. This is important because the 3q27 chromosomal region, which contains the adiponectin gene, has been linked to diabetes in humans.[119] In addition to relatively common single nucleotide polymorphisms (SNPs), a good number of mutations in the human adiponectin gene have been reported, many of which are related to diabetes.[9] Indeed, one report has confirmed that defects in the AdipoR2 gene is a major mechanism underlying the etiology of certain T2D.[78] The role of genetics in adiponectin regulation of energy homeostasis is perhaps best demonstrated in the severely obese Pima Indian population in the United States who have the highest known prevalence of obesity and T2D in the world. In the Pima Indian population, individuals with high adiponectin concentrations were 40% less likely to develop T2D than those with low concentrations even after adjustment for BMI.[120]

8. CONCLUSION AND FUTURE STUDIES

The finding that AdipoR1 and AdipoR2 double knockout mice are more insulin-resistant and glucose-intolerant than adiponectin-deficient mice raises the possibility that other ligands for the receptors exist and/or that AdipoR1 and AdipoR2 exert basal constitutive activities. If that were the case, it is likely that several physiological processes mediated by adiponectin have yet to be identified. Also, humans have mostly WAT while rodents have BAT and it is not clear to what extent differences in adipocyte biology impair extrapolation of findings collected in rodent studies to humans.

Thiazolidinediones such as rosiglitazone and pioglitazone are potent inducers of adiponectin expression. However, the metabolic and cardioprotective effects of PPAR-γ agonists are absent or decreased in adiponectin-deficient mice. Therefore, new therapeutic approaches to alleviate symptoms associated with T2D, obesity, and CVD will necessarily include strategies for development of adiponectin receptor agonists with capacity for tissue- or cell-specific delivery in order to upregulate expression of adiponectin receptors. Because variations in the *AdipoR2* gene are known to occur in the Amish and Caucasian population of T2D patients, development of pharmacogenetically specific strategies to treat such patients is needed.

The protective role of adiponectin in the brain is the least understood. Several observations have indicated that adiponectin, acting possibly in concert with estrogen, is an endogenous modulator of brain injury especially in response to acute ischemia. On the other hand, hypoadiponectinemia is known to exacerbate ischemic stroke. It is likely that adiponectin plays a larger role in the regulation of brain physiology than is presently conceived because expression levels for the orexigenic peptide agouti-related peptide and neuropeptide Y are increased by more than 150% and 60%, respectively, in the hypothalamus of AdipoR2-null mice.[11] Thus, further studies are warranted to identify AdipoR1 versus AdipoR2-mediated signaling pathways affecting brain function.

REFERENCES

1. Michalakis KG, Segars JH. The role of adiponectin in reproduction: from polycystic ovary syndrome to assisted reproduction. *Fertil Steril* 2010;**94**:1949–57.
2. Chandran M, Phillips SA, Ciaraldi T, Henry RR. Adiponectin: more than just another fat cell hormone? *Diabetes Care* 2003;**26**:2442–50.
3. Duncan BB, Schmidt MI, Pankow JS, Bang H, Couper D, Ballantyne CM, et al. Adiponectin and the development of type 2 diabetes: the atherosclerosis risk in communities study. *Diabetes* 2004;**53**:2473–8.
4. Halberg N, Schraw TD, Wang ZV, Kim JY, Yi J, Hamilton MP, et al. Systemic fate of the adipocyte-derived factor adiponectin. *Diabetes* 2009;**58**:1961–70.
5. Zhao T, Hou M, Xia M, Wang Q, Zhu H, Xiao Y, et al. Globular adiponectin decreases leptin-induced tumor necrosis factor-alpha expression by murine macrophages: involvement of cAMP-PKA and MAPK pathways. *Cell Immunol* 2005;**238**:19–30.

6. Dadson K, Liu Y, Sweeney G. Adiponectin action: a combination of endocrine and autocrine/paracrine effects. *Front Endocrinol (Lausanne)* 2011;**2**:62.
7. Beylot M, Pinteur C, Peroni O. Expression of the adiponectin receptors AdipoR1 and AdipoR2 in lean rats and in obese Zucker rats. *Metabolism* 2006;**55**:396–401.
8. Ouchi N, Kihara S, Arita Y, Okamoto Y, Maeda K, Kuriyama H, et al. Adiponectin, an adipocyte-derived plasma protein, inhibits endothelial NF-kappaB signaling through a cAMP-dependent pathway. *Circulation* 2000;**102**:1296–301.
9. Kadowaki T, Yamauchi T. Adiponectin and adiponectin receptors. *Endocr Rev* 2005;**26**:439–51.
10. Felder TK, Hahne P, Soyal SM, Miller K, Hoffinger H, Oberkofler H, et al. Hepatic adiponectin receptors (ADIPOR) 1 and 2 mRNA and their relation to insulin resistance in obese humans. *Int J Obes (Lond)* 2010;**34**:846–51.
11. Bjursell M, Ahnmark A, Bohlooly YM, William-Olsson L, Rhedin M, Peng XR, et al. Opposing effects of adiponectin receptors 1 and 2 on energy metabolism. *Diabetes* 2007;**56**:583–93.
12. Ding G, Qin Q, He N, Francis-David SC, Hou J, Liu J, et al. Adiponectin and its receptors are expressed in adult ventricular cardiomyocytes and upregulated by activation of peroxisome proliferator-activated receptor gamma. *J Mol Cell Cardiol* 2007;**43**: 73–84.
13. Bambace C, Telesca M, Zoico E, Sepe A, Olioso D, Rossi A, et al. Adiponectin gene expression and adipocyte diameter: a comparison between epicardial and subcutaneous adipose tissue in men. *Cardiovasc Pathol* 2011;**20**:e153–e156.
14. Huang D, Yang C, Wang Y, Liao Y, Huang K. PARP-1 suppresses adiponectin expression through poly(ADP-ribosyl)ation of PPAR gamma in cardiac fibroblasts. *Cardiovasc Res* 2009;**81**:98–107.
15. Fan D, Li L, Wang C, Cui XB, Zhou Y, Wu LL. Adiponectin induces interleukin-6 production and its underlying mechanism in adult rat cardiac fibroblasts. *J Cell Physiol* 2011;**226**:1793–802.
16. Nakanishi K, Takeda Y, Tetsumoto S, Iwasaki T, Tsujino K, Kuhara H, et al. Involvement of endothelial apoptosis underlying chronic obstructive pulmonary disease-like phenotype in adiponectin-null mice: implications for therapy. *Am J Respir Crit Care Med* 2011;**183**:1164–75.
17. Caminos JE, Nogueiras R, Gaytán F, Pineda R, González CR, Barreiro ML, et al. Novel expression and direct effects of adiponectin in the rat testis. *Endocrinology* 2008;**149**:3390–402.
18. Luo XH, Guo LJ, Yuan LQ, Xie H, Zhou HD, Wu XP, et al. Adiponectin stimulates human osteoblasts proliferation and differentiation via the MAPK signaling pathway. *Exp Cell Res* 2005;**309**:99–109.
19. Sharma K, Ramachandrarao S, Qiu G, Usui HK, Zhu Y, Dunn SR, et al. Adiponectin regulates albuminuria and podocyte function in mice. *J Clin Invest* 2008;**118**:1645–56.
20. Wijesekara N, Krishnamurthy M, Bhattacharjee A, Suhail A, Sweeney G, Wheeler MB. Adiponectin-induced ERK and Akt phosphorylation protects against pancreatic beta cell apoptosis and increases insulin gene expression and secretion. *J Biol Chem* 2010;**285**:33623–31.
21. Wen JP, Lv WS, Yang J, Nie AF, Cheng XB, Yang Y, et al. Globular adiponectin inhibits GnRH secretion from GT1-7 hypothalamic GnRH neurons by induction of hyperpolarization of membrane potential. *Biochem Biophys Res Commun* 2008;**371**: 756–61.
22. Pfaehler A, Nanjappa MK, Coleman ES, Mansour M, Wanders D, Plaisance EP, et al. Regulation of adiponectin secretion by soy isoflavones has implication for endocrine function of the testis. *Toxicol Lett* 2012;**209**:78–85.

23. Ocón-Grove OM, Krzysik-Walker SM, Maddineni SR, Hendricks III GL, Ramachandran R. Adiponectin and its receptors are expressed in the chicken testis: influence of sexual maturation on testicular ADIPOR1 and ADIPOR2 mRNA abundance. *Reproduction* 2008;**136**:627–38.
24. Lord E, Ledoux S, Murphy BD, Beaudry D, Palin MF. Expression of adiponectin and its receptors in swine. *J Anim Sci* 2005;**83**:565–78.
25. Ledoux S, Campos DB, Lopes FL, Dobias-Goff M, Palin MF, Murphy BD. Adiponectin induces periovulatory changes in ovarian follicular cells. *Endocrinology* 2006;**147**:5178–86.
26. Chappaz E, Albornoz MS, Campos D, Che L, Palin MF, Murphy BD, et al. Adiponectin enhances in vitro development of swine embryos. *Domest Anim Endocrinol* 2008;**35**:198–207.
27. Archanco M, Gomez-Ambrosi J, Tena-Sempere M, Fruhbeck G, Burrell MA. Expression of leptin and adiponectin in the rat oviduct. *J Histochem Cytochem* 2007;**55**: 1027–37.
28. Gutman G, Barak V, Maslovitz S, Amit A, Lessing JB, Geva E. Recombinant luteinizing hormone induces increased production of ovarian follicular adiponectin in vivo: implications for enhanced insulin sensitivity. *Fertil Steril* 2009;**91**:1837–41.
29. Kotani Y, Yokota I, Kitamura S, Matsuda J, Naito E, Kuroda Y. Plasma adiponectin levels in newborns are higher than those in adults and positively correlated with birth weight. *Clin Endocrinol (Oxf)* 2004;**61**:418–23.
30. Stefan N, Bunt JC, Salbe AD, Funahashi T, Matsuzawa Y, Tataranni PA. Plasma adiponectin concentrations in children: relationships with obesity and insulinemia. *J Clin Endocrinol Metab* 2002;**87**:4652–6.
31. Stefan N, Vozarova B, Funahashi T, Matsuzawa Y, Weyer C, Lindsay RS, et al. Plasma adiponectin concentration is associated with skeletal muscle insulin receptor tyrosine phosphorylation, and low plasma concentration precedes a decrease in whole-body insulin sensitivity in humans. *Diabetes* 2002;**51**:1884–8.
32. Yamauchi T, Kamon J, Waki H, Terauchi Y, Kubota N, Hara K, et al. The fat-derived hormone adiponectin reverses insulin resistance associated with both lipoatrophy and obesity. *Nat Med* 2001;**7**:941–6.
33. Polyzos SA, Kountouras J, Zavos C, Deretzi G. The potential adverse role of leptin resistance in nonalcoholic fatty liver disease: a hypothesis based on critical review of the literature. *J Clin Gastroenterol* 2011;**45**:50–4.
34. Woods A, Azzout-Marniche D, Foretz M, Stein SC, Lemarchand P, Ferre P, et al. Characterization of the role of AMP-activated protein kinase in the regulation of glucose-activated gene expression using constitutively active and dominant negative forms of the kinase. *Mol Cell Biol* 2000;**20**:6704–11.
35. Yamauchi T, Nio Y, Maki T, Kobayashi M, Takazawa T, Iwabu M, et al. Targeted disruption of AdipoR1 and AdipoR2 causes abrogation of adiponectin binding and metabolic actions. *Nat Med* 2007;**13**:332–9.
36. Yamauchi T, Kamon J, Minokoshi Y, Ito Y, Waki H, Uchida S, et al. Adiponectin stimulates glucose utilization and fatty-acid oxidation by activating AMP-activated protein kinase. *Nat Med* 2002;**8**:1288–95.
37. Ceddia RB, Somwar R, Maida A, Fang X, Bikopoulos G, Sweeney G. Globular adiponectin increases GLUT4 translocation and glucose uptake but reduces glycogen synthesis in rat skeletal muscle cells. *Diabetologia* 2005;**48**:132–9.
38. Yoon MJ, Lee GY, Chung JJ, Ahn YH, Hong SH, Kim JB. Adiponectin increases fatty acid oxidation in skeletal muscle cells by sequential activation of AMP-activated protein kinase, p38 mitogen-activated protein kinase, and peroxisome proliferator-activated receptor alpha. *Diabetes* 2006;**55**:2562–70.

39. Iwabu M, Yamauchi T, Okada-Iwabu M, Sato K, Nakagawa T, Funata M, et al. Adiponectin and AdipoR1 regulate PGC-1alpha and mitochondria by Ca(2+) and AMPK/SIRT1. *Nature* 2010;**464**:1313–9.
40. Christiansen T, Paulsen SK, Bruun JM, Ploug T, Pedersen SB, Richelsen B. Diet-induced weight loss and exercise alone and in combination enhance the expression of adiponectin receptors in adipose tissue and skeletal muscle, but only diet-induced weight loss enhanced circulating adiponectin. *J Clin Endocrinol Metab* 2010;**95**:911–9.
41. Liu Y, Chewchuk S, Lavigne C, Brule S, Pilon G, Houde V, et al. Functional significance of skeletal muscle adiponectin production, changes in animal models of obesity and diabetes, and regulation by rosiglitazone treatment. *Am J Physiol Endocrinol Metab* 2009;**297**:E657–E664.
42. Summer R, Fiack CA, Ikeda Y, Sato K, Dwyer D, Ouchi N, et al. Adiponectin deficiency: a model of pulmonary hypertension associated with pulmonary vascular disease. *Am J Physiol Lung Cell Mol Physiol* 2009;**297**:L432–L438.
43. Fasshauer M, Kralisch S, Klier M, Lossner U, Bluher M, Klein J, et al. Adiponectin gene expression and secretion is inhibited by interleukin-6 in 3T3-L1 adipocytes. *Biochem Biophys Res Commun* 2003;**301**:1045–50.
44. Shore SA, Terry RD, Flynt L, Xu A, Hug C. Adiponectin attenuates allergen-induced airway inflammation and hyperresponsiveness in mice. *J Allergy Clin Immunol* 2006;**118**: 389–95.
45. Bruun JM, Lihn AS, Verdich C, Pedersen SB, Toubro S, Astrup A, et al. Regulation of adiponectin by adipose tissue-derived cytokines: in vivo and in vitro investigations in humans. *Am J Physiol Endocrinol Metab* 2003;**285**:E527–E533.
46. Ouchi N, Kihara S, Funahashi T, Matsuzawa Y, Walsh K. Obesity, adiponectin and vascular inflammatory disease. *Curr Opin Lipidol* 2003;**14**:561–6.
47. Shimano M, Ouchi N, Shibata R, Ohashi K, Pimentel DR, Murohara T, et al. Adiponectin deficiency exacerbates cardiac dysfunction following pressure overload through disruption of an AMPK-dependent angiogenic response. *J Mol Cell Cardiol* 2010;**49**:210–20.
48. Shibata R, Sato K, Pimentel DR, Takemura Y, Kihara S, Ohashi K, et al. Adiponectin protects against myocardial ischemia-reperfusion injury through AMPK- and COX-2-dependent mechanisms. *Nat Med* 2005;**11**:1096–103.
49. Lee S, Park Y, Dellsperger KC, Zhang C. Exercise training improves endothelial function via adiponectin-dependent and independent pathways in type 2 diabetic mice. *Am J Physiol Heart Circ Physiol* 2011;**301**:H306–H314.
50. Okamoto Y, Arita Y, Nishida M, Muraguchi M, Ouchi N, Takahashi M, et al. An adipocyte-derived plasma protein, adiponectin, adheres to injured vascular walls. *Horm Metab Res* 2000;**32**:47–50.
51. Ouchi N, Kihara S, Arita Y, Nishida M, Matsuyama A, Okamoto Y, et al. Adipocyte-derived plasma protein, adiponectin, suppresses lipid accumulation and class A scavenger receptor expression in human monocyte-derived macrophages. *Circulation* 2001;**103**:1057–63.
52. Ouchi N, Kihara S, Arita Y, Maeda K, Kuriyama H, Okamoto Y, et al. Novel modulator for endothelial adhesion molecules: adipocyte-derived plasma protein adiponectin. *Circulation* 1999;**100**:2473–6.
53. Okamoto Y, Kihara S, Ouchi N, Nishida M, Arita Y, Kumada M, et al. Adiponectin reduces atherosclerosis in apolipoprotein E-deficient mice. *Circulation* 2002;**106**: 2767–70.
54. Nishimura M, Izumiya Y, Higuchi A, Shibata R, Qiu J, Kudo C, et al. Adiponectin prevents cerebral ischemic injury through endothelial nitric oxide synthase dependent mechanisms. *Circulation* 2008;**117**:216–23.

55. Yamauchi T, Kamon J, Waki H, Imai Y, Shimozawa N, Hioki K, et al. Globular adiponectin protected ob/ob mice from diabetes and ApoE-deficient mice from atherosclerosis. *J Biol Chem* 2003;**278**:2461–8.
56. Cao Y, Tao L, Yuan Y, Jiao X, Lau WB, Wang Y, et al. Endothelial dysfunction in adiponectin deficiency and its mechanisms involved. *J Mol Cell Cardiol* 2009;**46**:413–9.
57. Lee OH, Seo DH, Park CS, Kim YC. Puerarin enhances adipocyte differentiation, adiponectin expression, and antioxidant response in 3T3-L1 cells. *Biofactors* 2010;**36**:459–67.
58. Nishimura M, Takahashi H. Potential of plasma adiponectin concentrations as a predictive marker for in-stent restenosis after percutaneous coronary intervention in patients on hemodialysis. *Rinsho Byori* 2011;**59**:795–800.
59. Combs TP, Pajvani UB, Berg AH, Lin Y, Jelicks LA, Laplante M, et al. A transgenic mouse with a deletion in the collagenous domain of adiponectin displays elevated circulating adiponectin and improved insulin sensitivity. *Endocrinology* 2004; **145**:367–83.
60. Yamauchi T, Kamon J, Ito Y, Tsuchida A, Yokomizo T, Kita S, et al. Cloning of adiponectin receptors that mediate antidiabetic metabolic effects. *Nature* 2003;**423**:762–9.
61. Naaz A, Zakroczymski M, Heine P, Taylor J, Saunders P, Lubahn D, et al. Effect of ovariectomy on adipose tissue of mice in the absence of estrogen receptor alpha (ERalpha): a potential role for estrogen receptor beta (ERbeta). *Horm Metab Res* 2002;**34**:758–63.
62. Mohlig M, Wegewitz U, Osterhoff M, Isken F, Ristow M, Pfeiffer AF, et al. Insulin decreases human adiponectin plasma levels. *Horm Metab Res* 2002;**34**:655–8.
63. Combs TP, Berg AH, Rajala MW, Klebanov S, Iyengar P, Jimenez-Chillaron JC, et al. Sexual differentiation, pregnancy, calorie restriction, and aging affect the adipocyte-specific secretory protein adiponectin. Pregnancy, calorie restriction, and aging affect the adipocyte-specific secretory protein adiponectin. *Diabetes* 2003;**52**:268–76.
64. Rodriguez-Pacheco F, Martinez-Fuentes AJ, Tovar S, Pinilla L, Tena-Sempere M, Dieguez C, et al. Regulation of pituitary cell function by adiponectin. *Endocrinology* 2007;**148**:401–10.
65. Berg AH, Combs TP, Scherer PE. ACRP30/adiponectin: an adipokine regulating glucose and lipid metabolism. *Trends Endocrinol Metab* 2002;**13**:84–9.
66. Pierre P, Froment P, Negre D, Rame C, Barateau V, Chabrolle C, et al. Role of adiponectin receptors, AdipoR1 and AdipoR2, in the steroidogenesis of the human granulosa tumor cell line, KGN. *Hum Reprod* 2009;**24**:2890–901.
67. Chabrolle C, Tosca L, Dupont J. Regulation of adiponectin and its receptors in rat ovary by human chorionic gonadotrophin treatment and potential involvement of adiponectin in granulosa cell steroidogenesis. *Reproduction* 2007;**133**:719–31.
68. Qi Y, Takahashi N, Hileman SM, Patel HR, Berg AH, Pajvani UB, et al. Adiponectin acts in the brain to decrease body weight. *Nat Med* 2004;**10**:524–9.
69. Coope A, Milanski M, Araujo EP, Tambascia M, Saad MJ, Geloneze B, et al. AdipoR1 mediates the anorexigenic and insulin/leptin-like actions of adiponectin in the hypothalamus. *FEBS Lett* 2008;**582**:1471–6.
70. Kubota N, Yano W, Kubota T, Yamauchi T, Itoh S, Kumagai H, et al. Adiponectin stimulates AMP-activated protein kinase in the hypothalamus and increases food intake. *Cell Metab* 2007;**6**:55–68.
71. Tu Q, Zhang J, Dong LQ, Saunders E, Luo E, Tang J, et al. Adiponectin inhibits osteoclastogenesis and bone resorption via APPL1-mediated suppression of Akt1. *J Biol Chem* 2011;**286**(14):12542–53.
72. Tamura T, Yoneda M, Yamane K, Nakanishi S, Nakashima R, Okubo M, et al. Serum leptin and adiponectin are positively associated with bone mineral density at the distal radius in patients with type 2 diabetes mellitus. *Metabolism* 2007;**56**:623–8.

73. Chen J, Tan B, Karteris E, Zervou S, Digby J, Hillhouse EW, et al. Secretion of adiponectin by human placenta: differential modulation of adiponectin and its receptors by cytokines. *Diabetologia* 2006;**49**:1292–302.
74. Wang Y, Tao L, Yuan Y, Lau WB, Li R, Lopez BL, et al. Cardioprotective effect of adiponectin is partially mediated by its AMPK-independent antinitrative action. *Am J Physiol Endocrinol Metab* 2009;**297**:E384–E391.
75. Thyfault JP, Hedberg EM, Anchan RM, Thorne OP, Isler CM, Newton ER, et al. Gestational diabetes is associated with depressed adiponectin levels. *J Soc Gynecol Investig* 2005;**12**:41–5.
76. Jarvenpaa J, Vuoristo JT, Santaniemi M, Ukkola O, Savolainen ER, Jaaskelainen M, et al. Adiponectin induced placental cell apoptosis could be mediated via the ADIPOR1-receptor in pre-eclampsia with IUGR. *J Perinat Med* 2009;**37**:257–62.
77. Kollerits B, Fliser D, Heid IM, Ritz E, Kronenberg F. Gender-specific association of adiponectin as a predictor of progression of chronic kidney disease: the Mild to Moderate Kidney Disease Study. *Kidney Int* 2007;**71**:1279–86.
78. Liu H, Chen S, Zhang S, Xiao C, Ren Y, Tian H, et al. Adiponectin gene variation - 4522C/T is associated with type 2 diabetic obesity and insulin resistance in Chinese. *J Genet Genomics* 2007;**34**:877–84.
79. Pajvani UB, Hawkins M, Combs TP, Rajala MW, Doebber T, Berger JP, et al. Complex distribution, not absolute amount of adiponectin, correlates with thiazolidinedione-mediated improvement in insulin sensitivity. *J Biol Chem* 2004; **279**:12152–62.
80. Whitehead JP, Richards AA, Hickman IJ, Macdonald GA, Prins JB. Adiponectin—a key adipokine in the metabolic syndrome. *Diabetes Obes Metab* 2006;**8**:264–80.
81. Narasimhan ML, Coca MA, Jin J, Yamauchi T, Ito Y, Kadowaki T, et al. Osmotin is a homolog of mammalian adiponectin and controls apoptosis in yeast through a homolog of mammalian adiponectin receptor. *Mol Cell* 2005;**17**:171–80.
82. Sennello JA, Fayad R, Morris AM, Eckel RH, Asilmaz E, Montez J, et al. Regulation of T cell-mediated hepatic inflammation by adiponectin and leptin. *Endocrinology* 2005;**146**:2157–64.
83. Adachi M, Brenner DA. High molecular weight adiponectin inhibits proliferation of hepatic stellate cells via activation of adenosine monophosphate-activated protein kinase. *Hepatology* 2008;**47**:677–85.
84. Jung TW, Lee YJ, Lee MW, Kim SM. Full-length adiponectin protects hepatocytes from palmitate-induced apoptosis via inhibition of c-Jun NH2 terminal kinase. *FEBS J* 2009;**276**:2278–84.
85. Mao X, Hong JY, Dong LQ. The adiponectin signaling pathway as a novel pharmacological target. *Mini Rev Med Chem* 2006;**6**:1331–40.
86. Mao X, Kikani CK, Riojas RA, Langlais P, Wang L, Ramos FJ, et al. APPL1 binds to adiponectin receptors and mediates adiponectin signalling and function. *Nat Cell Biol* 2006;**8**:516–23.
87. Deepa SS, Dong LQ. APPL1: role in adiponectin signaling and beyond. *Am J Physiol Endocrinol Metab* 2009;**296**:E22–E36.
88. Neumeier M, Weigert J, Buettner R, Wanninger J, Schaffler A, Muller AM, et al. Detection of adiponectin in cerebrospinal fluid in humans. *Am J Physiol Endocrinol Metab* 2007;**293**:E965–E969.
89. Arita Y, Kihara S, Ouchi N, Takahashi M, Maeda K, Miyagawa J, et al. Paradoxical decrease of an adipose-specific protein, adiponectin, in obesity. *Biochem Biophys Res Commun* 1999;**257**:79–83.
90. Tsuchida A, Yamauchi T, Ito Y, Hada Y, Maki T, Takekawa S, et al. Insulin/Foxo1 pathway regulates expression levels of adiponectin receptors and adiponectin sensitivity. *J Biol Chem* 2004;**279**:30817–22.

91. Inukai K, Nakashima Y, Watanabe M, Takata N, Sawa T, Kurihara S, et al. Regulation of adiponectin receptor gene expression in diabetic mice. *Am J Physiol Endocrinol Metab* 2005;**288**:E876–E882.
92. Tena-Sempere M, Barreiro ML. Leptin in male reproduction: the testis paradigm. *Mol Cell Endocrinol* 2002;**188**:9–13.
93. Heine PA, Taylor JA, Iwamoto GA, Lubahn DB, Cooke PS. Increased adipose tissue in male and female estrogen receptor-alpha knockout mice. *Proc Natl Acad Sci U S A* 2000;**97**:12729–34.
94. Gao Q, Mezei G, Nie Y, Rao Y, Choi CS, Bechmann I, et al. Anorectic estrogen mimics leptin's effect on the rewiring of melanocortin cells and Stat3 signaling in obese animals. *Nat Med* 2007;**13**:89–94.
95. Ogawa S, Chan J, Gustafsson JA, Korach KS, Pfaff DW. Estrogen increases locomotor activity in mice through estrogen receptor alpha: specificity for the type of activity. *Endocrinology* 2003;**144**:230–9.
96. Robertson KM, O'Donnell L, Jones ME, Meachem SJ, Boon WC, Fisher CR, et al. Impairment of spermatogenesis in mice lacking a functional aromatase (cyp 19) gene. *Proc Natl Acad Sci U S A* 1999;**96**:7986–91.
97. Cooke PS, Naaz A. Role of estrogens in adipocyte development and function. *Exp Biol Med (Maywood)* 2004;**229**:1127–35.
98. Gambacciani M, Ciaponi M, Cappagli B, Piaggesi L, Genazzani AR. Effects of combined low dose of the isoflavone derivative ipriflavone and estrogen replacement on bone mineral density and metabolism in postmenopausal women. *Maturitas* 1997;**28**:75–81.
99. Carani C, Qin K, Simoni M, Faustini-Fustini M, Serpente S, Boyd J, et al. Effect of testosterone and estradiol in a man with aromatase deficiency. *N Engl J Med* 1997;**337**:91–5.
100. Jones ME, Thorburn AW, Britt KL, Hewitt KN, Misso ML, Wreford NG, et al. Aromatase-deficient (ArKO) mice have a phenotype of increased adiposity. *Proc Natl Acad Sci U S A* 2000;**97**(23):12735–40.
101. Ohlsson C, Hellberg N, Parini P, Vidal O, Bohlooly YM, Rudling M, et al. Obesity and disturbed lipoprotein profile in estrogen receptor-alpha-deficient male mice. *Biochem Biophys Res Commun* 2000;**278**:640–5.
102. Foryst-Ludwig A, Clemenz M, Hohmann S, Hartge M, Sprang C, Frost N, et al. Metabolic actions of estrogen receptor beta (ERbeta) are mediated by a negative cross-talk with PPARgamma. *PLoS Genet* 2008;**4**:e1000108.
103. Baillie-Hamilton P. Chemical toxins: a hypothesis to explain the global obesity epidemic. *J Altern Complement Med* 2002;**8**:185–92.
104. Cederroth CR, Nef S. Soy, phytoestrogens and metabolism: A review. *Mol Cell Endocrinol* 2009;**304**:30–42.
105. Ben-Jonathan N, Hugo ER, Brandebourg TD. Effects of bisphenol A on adipokine release from human adipose tissue: implications for the metabolic syndrome. *Mol Cell Endocrinol* 2009;**304**:49–54.
106. Barrett-Connor E, Khaw KT, Yen SS. Endogenous sex hormone levels in older adult men with diabetes mellitus. *Am J Epidemiol* 1990;**132**:895–901.
107. Anderson S. 7 care tips for managing patients with diabetes. *Am J Nurs* 1994;**94**:36–8.
108. Dandona P, Dhindsa S. Update: Hypogonadotropic hypogonadism in type 2 diabetes and obesity. *J Clin Endocrinol Metab* 2011;**96**:2643–51.
109. Laaksonen DE, Niskanen L, Punnonen K, Nyyssonen K, Tuomainen TP, Valkonen VP, et al. Testosterone and sex hormone-binding globulin predict the metabolic syndrome and diabetes in middle-aged men. *Diabetes Care* 2004;**27**:1036–41.
110. Laaksonen DE, Niskanen L, Punnonen K, Nyyssonen K, Tuomainen TP, Valkonen VP, et al. The metabolic syndrome and smoking in relation to hypogonadism

in middle-aged men: a prospective cohort study. *J Clin Endocrinol Metab* 2005;**90**: 712–9.
111. Pitteloud N, Villegas J, Dwyer AA, Crowley Jr WF, McPhaul MJ, Hayes FJ. Acute stress masking the biochemical phenotype of partial androgen insensitivity syndrome in a patient with a novel mutation in the androgen receptor. *J Clin Endocrinol Metab* 2004;**89**:1053–8.
112. Yamamoto Y, Hirose H, Saito I, Tomita M, Taniyama M, Matsubara K, et al. Correlation of the adipocyte-derived protein adiponectin with insulin resistance index and serum high-density lipoprotein-cholesterol, independent of body mass index, in the Japanese population. *Clin Sci (Lond)* 2002;**103**:137–42.
113. Buechler C, Wanninger J, Neumeier M. Adiponectin, a key adipokine in obesity related liver diseases. *World J Gastroenterol* 2011;**17**(23):2801–11.
114. Kondo H, Shimomura I, Matsukawa Y, Kumada M, Takahashi M, Matsuda M, et al. Association of adiponectin mutation with type 2 diabetes: a candidate gene for the insulin resistance syndrome. *Diabetes* 2002;**51**:2325–8.
115. Mori H, Ikegami H, Kawaguchi Y, Seino S, Yokoi N, Takeda J, et al. The Pro12 →Ala substitution in PPAR-gamma is associated with resistance to development of diabetes in the general population: possible involvement in impairment of insulin secretion in individuals with type 2 diabetes. *Diabetes* 2001;**50**:891–4.
116. Treeck O, Juhasz-Boess I, Lattrich C, Horn F, Goerse R, Ortmann O. Effects of exon-deleted estrogen receptor beta transcript variants on growth, apoptosis and gene expression of human breast cancer cell lines. *Breast Cancer Res Treat* 2008;**110**:507–20.
117. Gulcelik NE, Aral Y, Serter R, Koc G. Association of hypoadiponectinemia with metabolic syndrome in patients with polycystic ovary syndrome. *J Natl Med Assoc* 2008;**100**:64–8.
118. Toulis KA, Goulis DG, Kolibianakis EM, Venetis CA, Tarlatzis BC, Papadimas I. Risk of gestational diabetes mellitus in women with polycystic ovary syndrome: a systematic review and a meta-analysis. *Fertil Steril* 2009;**92**:667–77.
119. Maeda N, Shimomura I, Kishida K, Nishizawa H, Matsuda M, Nagaretani H, et al. Diet-induced insulin resistance in mice lacking adiponectin/ACRP30. *Nat Med* 2002;**8**:731–7.
120. Lindsay RS, Funahashi T, Hanson RL, Matsuzawa Y, Tanaka S, Tataranni PA, et al. Adiponectin and development of type 2 diabetes in the Pima Indian population. *Lancet* 2002;**360**:57–8.

CHAPTER TEN

The Role of Bombesin and Bombesin-Related Peptides in the Short-term Control of Food Intake

Ayman I. Sayegh
Gastroenterology Laboratory, Department of Biomedical Sciences, College of Veterinary Medicine, Tuskegee University, Tuskegee, Alabama, USA

Contents

Abstract

Bombesin (Bn) is a 14-amino acid peptide isolated from the skin of the frog *Bombina bombina*. The mammalian homologs of this peptide include three forms of gastrin-releasing peptide (GRP): GRP-10, GRP-27, and GRP-29, and a 10-amino acid peptide referred to as neuromedin-B (NMB). These peptides evoke a number of responses, including hyperthermia, bradycardia, inhibition of gastric emptying and inhibition of food intake, by activating one of three G protein-coupled receptors: an NMB-R or BB_1, a GRP-R or BB_2 and an orphan Bn receptor subtype-3 (BRS-3) or BB_3. Bombesin, GRP, and NMB have a role in the short-term control of food intake. These peptides reduce meal size (MS) and they prolong the intermeal interval (IMI), the time between the first and second meals. Studies have shown that the vagus and the splanchnic nerves in the upper gastrointestinal tract, which communicate with the feeding areas of the hindbrain, are necessary for reduction of MS and prolongation of the IMI by Bn, GRP, and NMB. In

Progress in Molecular Biology and Translational Science, Volume 114
ISSN 1877-1173
http://dx.doi.org/10.1016/B978-0-12-386933-3.00010-8

addition, one-tenth of the intraperitoneal dose of Bn, GRP, and NMB given in either the left gastric artery, which supplies the stomach, or the cranial mesenteric artery, which supplies the intestine, or the femoral vein, also reduces MS and prolongs the IMI. Thus, a potential neurocrine or an endocrine mode of action for these peptides requires further investigation.

1. INTRODUCTION

The control of food intake is a physiological process necessary in the fight against obesity, which is caused by excessive energy intake compared to energy expenditure. This physiological process consists of two mechanisms: a long-term mechanism, which governs energy homeostasis of the individual over long periods of time such as days or years, and a short-term mechanism, which governs two components: meal size (MS) and the intermeal interval (IMI, or time between two consecutive meals). In general, the long-term mechanism is controlled by two hormones, insulin and leptin. However, the short-term mechanism is controlled mostly by gut peptides/hormones. Such hormones/peptides may include the stomach peptide gastrin-releasing peptide (GRP) and the intestinal hormones/peptides cholecystokinin (CCK), peptide tyrosine tyrosine (PYY), and glucagon-like peptide-1 (GLP-1). The current review will focus on GRP as well as another similar but not identical peptide named neuromedin-B (NMB), and their amphibian homologue peptide bombesin (Bn).

The first gut peptide studied that regulates short-term control of food intake is CCK. However, the second class of peptide studied that regulates this control is Bn and its mammalian homologs GRP and NMB. Based on their structure, there are three groups of the amphibian peptides (Table 10.1): (1) the Bn group, (2) the ranatensin group and (3) the phyllolitorin group. The Bn group includes Bn, alytesin, and [pyroglutamic acid ^{1}]Bn_{6-14}. All of these peptides have a carboxyl terminus Gly-His-Leu-Met-NH_2. The ranatensin group consists of ranatensin, ranatensin R and C, litorin, rodheilitorin, and [Glu $(Ote)^2$ or $(Ome)^2$]litorin). These peptides have a carboxyl terminus Gly-His-Phe-Met-NH_2. The phyllolitorin group consists of phyllolitorin, [Leu^8]phyllolitorin and [Thr^5,Leu^8]phyllolitorin. These peptides have a carboxyl terminus Gly-Ser-Phe/Leu-Met-NH_2.[1,2]

In mammals, there are two Bn-related peptides (Table 10.1), GRP[3–5] and NMB.[6] Based on the number of amino acids in the peptide chain, the species of interest, and the most studied peptide, there are three forms

Table 10.1 The different forms of bombesin and bombesin-related peptides

Name of peptide	**Number of amino acids**
The mammalian forms of bombesin	
Gastrin-releasing peptide group	
1. Rat GRP	29
2. Avian GRP	27
3. Dog GRP	27
4. Pig GRP	27
5. Human GRP	27
6. GRP_{10} or GRP_{18-27} (also known as neuromedin C)	10
Neuromedin B group	
7. Neuromedin B	10
8. Neuromedin B32	32
The Amphibian Forms of Bombesin	
Bombesin group	
1. Bombesin	14
2. Alytesin	14
The ranatensin group	
3. Litorin	9
4. Ranatensin	11
5. Ranatensin C	11
6. Ranatensin R	17
7. rodhei-litorin	9
8. [Glu $(Ote)^2$ or $(Ome)^2$]litorin	9
The phyllolitorin group	
9. Phyllolitorin	9
10. [Leu^8]phyllolitorin	9
11. [Thr^5,Leu^8]phyllolitorin	9

of GRP: GRP-10 (or GRP_{18-27}, which was originally found in the porcine spinal cord and named NMC), GRP-27 (originally found in porcine non-antral tissues[4,5]), and GRP-29 (found only in the rat[7]). This review focuses on the role of Bn, GRP-10, GRP-27, GRP-29, and NMB in the short-term control of food intake.

2. HISTORICAL PERSPECTIVE

2.1. Bombesin

In 1970, Nakajima and colleagues isolated an undecapeptide (11 amino acid peptide) from the skin of the Northern Leopard Frog *Rana pipiens* and called it ranatensin. In the same year, and shortly thereafter, Erspamer *et al.* identified an amidated tetradecapeptide (14 amino acid peptide) with a structural similarity with ranatensin from the skin of two European frogs, *Bombina bombina* and *Bombina variegata*. They called this peptide Bn.[8–10] In 1972, a third peptide was isolated from the skin of the European frog *Alytes obstetricans* by the same group. This tetradecapeptide was named alytesin.[11] By 1978, similar peptides had been isolated from other amphibians: the nonapeptide (9 amino acid peptide) litorin, the undecapeptide ranatensin C, and the heptadecapeptide (17 amino acid peptide) ranatensin R.[12] These peptides have similar, but not identical, amino acid sequences[12] (see Table 10.1).

2.2. Gastrin-releasing peptides

In 1978, McDonald *et al.* identified a heptacosapeptide (27 amino acid peptide)[5] from porcine non-antral gastric tissue. The sequence of this peptide was identified by the same group in the next year,[4] and because of the most prominent function of this peptide, which is stimulation of gastric acid secretion due to the release of gastrin, the peptide was given the name gastrin-releasing polypeptide. The next year, the same group identified GRP in the chicken proventriculus as another 27-amino acid peptide.[3] In 1983, Reeve *et al.* characterized three forms of GRP (27, 23, and 10) from canine intestinal extracts.[13] Also in 1983, Minamino and colleagues[6] identified NMB (the mammalian equivalent of ranatensin and litorin[6]) in the porcine spinal cord; the next year, the same group identified GRP-10 in the porcine spinal cord, calling it NMC.[14] In 1984, GRP-27 was isolated and characterized from human pulmonary carcinoid cells.[15,16] Finally, in 1992, Nagalla and colleagues found that in the rat, the larger molecular form of GRP is GRP-29[7] (see Table 10.1).

Later on, similarity was found between the two classes of peptides, the amphibian peptides and GRP, especially at the biologically active COOH-terminal. However, because the NH_2-terminus was different between the amphibian and mammalian peptides, the name GRP was chosen for the mammalian peptide (see summary in Table 10.1).

3. DISTRIBUTION OF BOMBESIN AND BOMBESIN-RELATED PEPTIDES

Many studies have shown that Bn and GRP-like immunoreactivity and mRNA are widely distributed in the central and peripheral nervous systems as well as many other peripheral tissues.[17–19] In the rat stomach, bombesin-like immunoreactivity (Bn-LI) can be found at birth[20] on endocrine mucosal cells of the antrum. In adults, Bn-LI is found in the gut, lungs, and central and peripheral neurons.[21–24] In the human gut, Bn-LI is localized on endocrine cells.[25] However, in the rat gut, it has been shown that this immunoreactivity is localized to nerve fibers.[26] Bn-LI is found in the celiac ganglia[27] and the pancreas.[28] In addition, Bn-LI is found in the prevertebral, superior, and inferior mesenteric sympathetic ganglia of the rat and guinea pig.[29–31] It has also been found in several areas of the spinal cord,[32–34] especially laminae I-III of the dorsal horn,[35] which are responsible for afferent sensory pathways. Bn-LI has also been found in the hypothalamus (including arcuate nucleus)[23] and various other regions of the rat brain,[36] including the parabrachial nucleus, nucleus tractus solitarius (NTS), the vagus nerve[34,37] and the pituitary gland.[38] In addition, Bn-LI is found in the lungs.[24,39] Finally, Bn-LI is found in bovine milk.[40]

In the gut of the rat, GRP is produced by the myenteric neurons of the stomach, small intestine, and colon.[26] GRP-LI and mRNA are found in the hypothalamus, forebrain, medullary nuclei, and pituitary gland.[34,41–47] In contrast, NMB-LI and mRNA are found in the gut,[48] and in the central nervous system—the brain and spinal cord[49]—including in dorsal root ganglia.[17,50–53]

4. PHYSIOLOGICAL RESPONSES EVOKED BY BOMBESIN AND BOMBESIN-RELATED PEPTIDES

Bombesin increases plasma gastrin and gastric acid secretion,[54] a response that decreases following gastric antrectomy. These effects have been documented in dogs, humans, and pigs for GRP.[20,54–57] In addition to stimulation of gastrin release, Bn stimulates pancreatic polypeptide secretion in dogs,[56] stimulates secretion of CCK from the I cells of the small intestine, and causes gallbladder contraction.[58] Bombesin and GRP stimulate the secretion of various other peptides such as pancreatic polypeptide, enteroglucagon, pancreatic glucagon, motilin, vasoactive intestinal peptide, and insulin.[59,60] In humans and pigs, through rapid increase in intracellular calcium[61] and Ca^{+2}-dependent membrane hyperpolarization through opening K^{+} channels,[62] Bn stimulates pancreatic exocrine secretion.[63] These secretions are rich in amylase and bicarbonate but low in water. This same response was also documented with GRP-10 in rats.[64,65] In addition, Bn delays gastric emptying[66,67] and inhibits the myoelectric complex in the small intestine.[68] Bombesin injections cause antral gastrin cell proliferation[69] and acinar hyperplasia[70,71] in rats. Other effects of Bn include hyperthermia,[72] arterial hypertension, and bradycardia.[73]

GRP has effects on regulation of the circadian rhythm and regulation of anxiety and fear responses.[19,74–82] In many species, including rat, human, cat, dog, and pig, Bn and GRP cause smooth muscle contraction of several organs, such as the stomach, intestine, gallbladder, urinary bladder, and uterus, as well as stimulation of myenteric and submucosal neurons.[66,83–91] In addition, Bn and GRP cause reduction of MS (satiation) and IMI (satiety) in several species.[92–97]

GRP and NMB have potent effects on immune cells (e.g., leukocytes, macrophages, lymphocytes, and dendritic cells[98–103]), and they affect the growth and proliferation of various gastrointestinal (e.g., colon), respiratory (e.g., lung) and urogenital (e.g., prostate) cancers.[104–114] NMB, GRP and Bn evoke the responses described above by activating three G protein-coupled receptors, NMB-R (BB_1), GRP-R (BB_2), and BRS-3 (BB_3).

5. BOMBESIN AND BOMBESIN-RELATED PEPTIDES RECEPTORS

There are two different receptors and one orphan receptor for the mammalian Bn peptides NMB and GRP[115–121] (Table 10.2). These three seven-transmembrane-domain, G protein-coupled receptors are BB_1 (NMB-R), BB_2 (GRP-R), and BB_3 (BRS-3), respectively.

In 1990 and 1991, the GRP-R was cloned by various methods from murine Swiss 3T3 cells.[122–125] Also in 1990, the NMB receptor was cloned from a cDNA library of esophageal smooth muscle cells.[126,127] Both receptors were also characterized in small cell lung cancer,[128] CNS[120], and gastric

Table 10.2 The receptors of the mammalian bombesin-related peptides

	BB_1 (NMB-R)	BB_2 (GRP-R)	BB_3 (BRS-3)
Species	H, R, Mou, M	H, R, Mou, M, D, S	H, R, Mou, M, S
Gene Location	Chr 6p21	Chr Xp22	Chr Xq25
Structure	390 amino acids	384 amino acids	399 amino acids

H, human; R, rat; Mou, mouse; M, monkey; P, pig; S, sheep.

smooth muscle cells.[129] The amphibian peptide Bn has a high-binding affinity for both NMB-R and GRP-R, which makes it difficult to distinguish from the other peptide (see Table 10.3).

In 1992, a third receptor was cloned from guinea pig uterus,[131] with 52% amino acid sequence homology with the GRP receptor and 47% with the NMB receptor. This receptor was called the orphan receptor BRS-3 or BB_3. In 1995, Nagalla *et al.* found a fourth receptor, which has 61%, 56%, and 70% sequence homology with the GRP-R, NMB-R, and the BRS-3, respectively, and binds [Phe^{13}]-bombesin (frog skin) with high affinity but GRP and NMB with low affinity.[132] This receptor was called BB_4. A fifth receptor was cloned from chicken and has a 70% sequence homology with BB_4, 69% with BRS-3, 58% with GRP-R and 52% with NMB-R.[133] This receptor was referred to as chBRS-3.5 (see Table 10.3). Each of the three main receptors, BB_1, BB_2, and BB_3 will be discussed in detail.

5.1. The BB_1 (NMB-R) receptor

The human BB_1 receptor is a 72-kDa protein (43 kDa following deglycosylation)[134] that consists of 390 amino acids and has 89% sequence homology with the same receptor in the rat but only 55% with the human BB_2 receptor and 47% with the human BB_3 receptor.[128] The BB_1 receptor was also cloned in rats (63 kDa),[52] the frog *Bombina orientalis,*[132] mouse[135] and monkey.[136] The gene of the human BB_1 receptor is localized on chromosome 6p21-qter and in the mouse on chromosome 10[52,128] (see Table 10.2). The BB_1 receptor gene in human, mouse, and rat contains three exons with two introns.[52,128,135]

The BB_1 receptor mRNA is expressed in the CNS (e.g., the amygdala, caudate nucleus, hippocampus, hypothalamus, thalamus, brainstem, and spinal cord), as well as in the testicles, stomach, olfactory bulb, and esophagus.[52,120,135–138] In addition, the BB_1 receptor mRNA has been found on many tumors, including glioblastoma,[139] small cell and non-small

Table 10.3 Agonists and antagonists of the bombesin and bombesin-related peptide receptors[130]

	Affinity		
Agonist / antagonist	**BB_1 (nM)**	**BB_2 (nM)**	**BB_3 (nM)**
Natural Agonists			
GRP	440	18	>10,000
NMB	4	248	>10,000
Bn	34	4	>10,000
Litorin	7	6	>10,000
Ranatensin	13	2	>10,000
Alytensin	460	62	>10,000
Phyllolitorin	47	240	>10,000
NMC (GRP_{18-27})	140	20	>10,000
[Phe^{13}]bombesin	350	0.77	>10,000
Synthetic Agonists			
[D-Phe^6,β-Ala^{11},Phe^{13},Nle^{14}]Bn_{6-14}	0.36	0.99	4.2
[D-Try^6,(*R*)-Apa^{11},Phe^{13},Nle^{14}]Bn_{6-14}	7200	>1900	8.2
[D-Phe^6,Apa-$4Cl^{11}$,Phe^{13},Nle^{14}]Bn_{6-14}	2400	151	2.8
Ac-Phe,Trp,Ala,His (tBzl),Nip,Gly,Arg-NH_2	3800	5000	259
[D-Phe^6]Bn_{6-14}	14	2	>10,000
[D-Phe^6,D-Ala^{11},Leu^{14}]Bn_{6-14}	7600	13	>10,000
Antagonists			
[D-Phe^6]Bn_{6-13} methyl ester	7500	1.1	>10,000
N-Propionyl-[D-Ala^{11}]GRP_{20-26} methyl ester	13,660	3.4	>10,000
PD 168368	39	1300	1010
D-Nal,Cys,Tyr,D-Trp,Lys,Val,Cys, Nal-NH_2	59	2780	>10,000
[tyr^4,D-Phe^{12}]Bn_{6-14}	1900	>10,000	>10,000
Leu^{13},Ψ13-14, Leu^{14}]Bn_{6-14}	>10,000	430	>10,000
[D-Phe^6,Leu^{13},Cpa^{14},Ψ13-14]Bn_{6-14}	2700	42	6800
BW2258U89	>10,000	0.74	>10,000

Table 10.3 Agonists and antagonists of the bombesin and bombesin-related peptide receptors[130]—cont'd

	Affinity		
Agonist / antagonist	**BB_1 (nM)**	**BB_2 (nM)**	**BB_3 (nM)**
[D-Arg^1,D-Trp^{7,9},Leu^{11}]substance P	4100	11300	>10,000
JMV594	>10,000	2.2	>10,000
JMV641	1500	0.46	>10,000

cell lung cancer,[109,140] intestinal, thymic and bronchial,[141] ovarian,[142] and pancreatic cancers.[143–145]

Functionally, the BB_1 receptor mediates central responses such as fear, stress, spontaneous activities,[146,147] and inhibition of food intake.[148,149] In addition, this receptor mediates peripheral responses such as contraction of smooth muscle cells.[126,127,129,150–153] Studies from BB_1 receptor knockout mice demonstrated that this receptor also mediates thermoregulation, thyroxin release and function, and fear, stress, and anxiety.[74,154,155] It is known that the satiety effects mediated by the BB_1 receptor, unlike those mediated by the BB_2 receptor, are peripheral in location because they are attenuated by the neurotoxin capsaicin, which destroys afferent fibers.[148] Finally, stimulation of the BB_1 receptor or injections of NMB-stimulated growth in various cancers, including non-small cell lung cancer.[156]

5.2. The BB_2 (GRP-R) receptor

The human BB_2 receptor consists of 384 amino acids with 90% sequence homology with the mouse receptor, 55% with the human BB_2 receptor[128] and 51% with the human BB_3 receptor.[157] The minimum sequence required for activation of the BB_2 receptor is located in the first seven (heptapeptide) amino acids of the COOH-terminus, but the full activity of the receptor requires the first nine (nonapeptide) amino acids of the same terminus.[158–160] This receptor has been cloned from 21 species.[161] Although the molecular weight of the human BB_2 receptor is 60 kDa and the mouse is 82 kDa, the deglycosylated molecular weight for each of them is 43 kDa.[134,162–165] The human BB_2 receptor gene is localized on chromosome Xp22,[166–168] and the gene has three exons spanning more than 27 kb with intron 1 and intron 2 being 23 and 1.6 kb,[128,167] respectively.

Expression of the BB_2 receptor was studied in human, mouse, and monkey.[18,124,128,135,136] The highest concentrations of this receptor were found in the pancreas, stomach, prostate, skeletal muscle, and regions of the CNS including the hippocampus, hypothalamus, amygdala, and pons.[136,167] In the mouse brain, the BB_2 receptor is located in the hippocampus, hypothalamus, amygdala, and NTS.[169] In the gastrointestinal tract of the mouse, this receptor is present in the colon but not the stomach and small intestine.[124] In the rat,[18] the BB_2 receptor is found in the hypothalamus as well as other nuclei such as the suprachiasmatic and supraoptic nuclei.[19,120,137,138] However, in the gut (Fig. 10.1) of the rat,[170] BB_2 receptor is localized to the ileum > jejunum > stomach, on the antral G-cells[43] and D and parietal cells,[171] > cecum > rectum > duodenum > colon (Tenisha Johnson-Rouse, Martha C. Washington and Ayman I. Sayegh, 2012 unpublished data), and pancreatic acinar cells.[115,172] Finally, the BB_2 receptor can be found in many cancers, such as prostate, lung, gastrointestinal, uterine, and neuronal.[141,173–175]

Physiologically, the BB_2 receptor mediates a number of gastrointestinal functions, including regulation of gastric acid secretion through gastrin or somatostatin release[176,177] and regulation of gastrointestinal motility/gastric emptying, and gallbladder contraction.[178,179] In addition, the BB_2 receptor

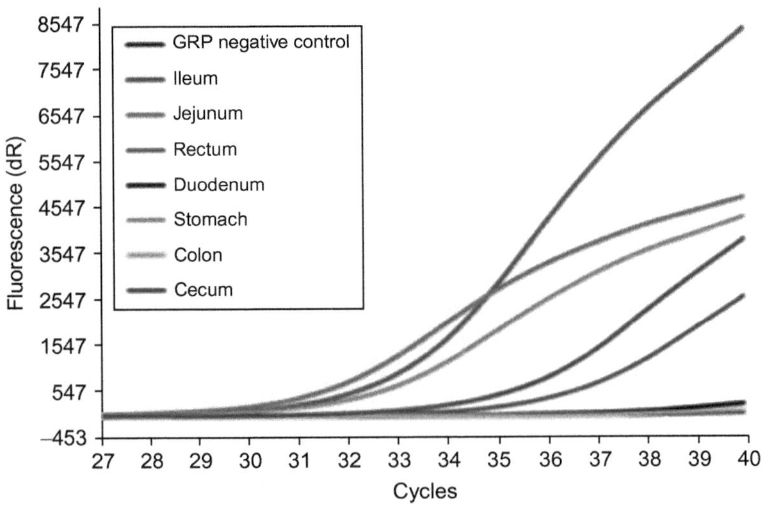

Figure 10.1 Gene expression of the gastrin-releasing peptide (GRP) receptor in the gastrointestinal tract of the rat. The expression of the GRP receptor in the gut of the rat as revealed by a real-time polymerase chain reaction test—the expression was greater in the ileum > jejunum > stomach > cecum > rectum > duodenum > colon (Tenisha Johnson-Rouse, Martha C. Washington and Ayman I. Sayegh, 2012 unpublished data). (See Color Insert.)

mediates stimulation of pancreatic exocrine secretion[180,181] and the release of insulin[182] and other hormones, including CCK, PYY, and glucagon.[60,183–185] This receptor also participates in immunological responses such as attracting peritoneal macrophages, monocytes, and lymphocytes.[98,99,102] The BB_2 receptor has also been shown to participate in lung development..[186] The central responses evoked by BB_2 receptors include satiety,[92,148,187–191] regulation of circadian rhythm,[192,193] memory,[194] stress, fear, and anxiety.[78,195–197] These and other responses, including insulin and GLP-1 secretion,[182] have also been demonstrated in BB_2 receptor knockout mice.[198] Finally, the BB_2 receptor has important effects on both normal cells, including bronchial epithelial cells,[199] chondrocytes,[200] and enterocytes[201]; and on neoplastic tissue,[202,203] including neuroblastoma,[204] pancreatic cancer,[205] colon cancer,[206] prostate cancer,[207,208] glioblastoma,[209] and non-small cell lung cancer.[210]

5.3. The BB_3 (BRS-3) receptor

The human BB_3 receptor is a 44.4 kDa, 399-amino acid protein[157] with 95% sequence homology with the rhesus monkey receptor and 80% with the rat receptor.[136,211] The gene for this receptor is located on chromosome Xq25,[157] and it contains two introns and three exons. In monkey, the greatest receptor expression was found in the CNS (e.g., hypothalamus including paraventricular nucleus and arcuate nucleus, thalamus, pituitary gland, amygdala, hippocampus, and caudate nucleus), testis, the tunica muscularis of the gastric antrum, duodenum, ileum and colon, and the smooth muscles, myenteric plexus, submucosal plexus, the interstitial cells of Cajal,[212] thyroid, and pancreas.[131,135,136,157] In addition, the BB_3 receptor, similar to BB_1 and BB_2 receptors, has been shown to be expressed in many tumors, including non-small cell lung tumors,[141,157] pulmonary carcinoids,[141,157] renal cell cancer,[141] pancreatic cancer,[213] pituitary cancer,[213] ovarian cancer,[142] and prostate cancer.[142,213]

Because there is no known natural ligand for the BB_3 receptor, its physiological and pathological effects are largely unknown. However, the BB_3 receptor knockout mouse[135] shows mild obesity, hypertension, impairment of glucose metabolism, a fivefold increase in leptin levels, a two- to threefold increase in insulin levels,[214] hyperphagia, and altered taste preferences.[215] The feeding alterations resulting from activation of the BB_3 receptor are possibly due to increases in melanin-concentrating hormone.[216] However, the BB_3 receptor is not a major cause of obesity because there is no mutation or polymorphism of this receptor in obese men.[217]

6. THE ROLE OF BOMBESIN AND BOMBESIN-RELATED PEPTIDES IN THE SHORT-TERM CONTROL OF FOOD INTAKE

After CCK, Bn is the most studied gut peptide that has a role in the regulation of the short-term control of food intake. Bn reduces food intake in rats,[92] sham rats,[218] obese mice,[219] cats,[220] fowl,[221] baboons,[222] and humans,[223] without producing any toxic or aversive effects/behaviors.[223–226] At first, because Bn releases endogenous CCK,[58] it was thought that reduction of food intake by Bn is due, at least in part, to CCK. However, this was soon found to be untrue. Bombesin reduces food intake in a different manner than CCK-8.[92] While Bn reduces both solid and liquid food intakes equally, CCK-8 reduces the intake of liquid food intake two- to fivefold more than solid food intake.[92]

Like Bn, intraperitoneal injections of GRP-10, GRP-27, GRP-29, and NMB have also been shown to reduce food intake,[95,97,227] as both reduction of MS (Fig. 10.2) and prolongation of the IMI (Fig. 10.3)[95] were observed. These effects are reversed with specific antagonists.[228,229] In addition, central injections of Bn and GRP reduce food intake.[188,230–232] GRP is more potent in decreasing food intake than NMB[229,232]; however, it is less potent than equimolar doses of Bn.[93,233] This effect occurs because Bn activates both GRP and NMB receptors.[93,94,233] Furthermore, the difference in potency between GRP and NMB is diet-dependent.[229] GRP decreases consumption of high-fat diet more than the low-fat diet.[93,97]

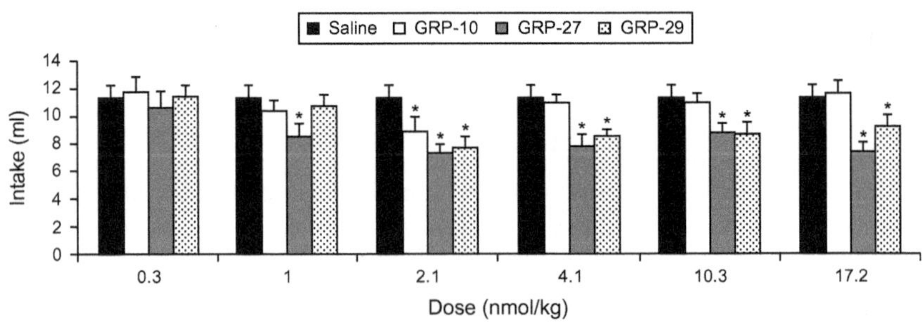

Figure 10.2 Effect of gastrin-releasing peptides (GRPs) on the size of the first meal. Food-deprived male rats ($n=17$) received an injection of GRP-10, -27, or -29 (0.3, 1, 2.1, 4.1, 10.3, and 17.2 nmol/kg) or saline i.p. (1 h into the light cycle) followed by presentation with a 10% sucrose test diet, after which the size of the first meal was determined. GRP-29 (2.1, 4.1, 10.3, 17.2 nmol/kg), GRP-27 (1, 2.1, 4.1, 10.3, 17.2 nmol/kg), and GRP-10 (2.1 nmol/kg) reduced the size of the first meal relative to saline (*). $p<0.05$.[95]

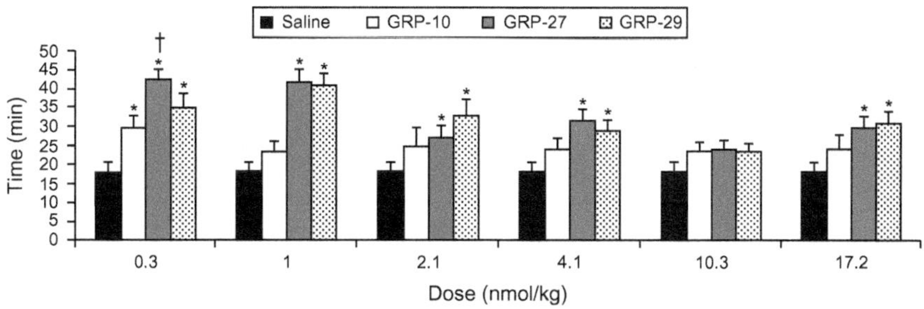

Figure 10.3 Effect of gastrin-releasing peptides (GRPs) on the length of the intermeal interval. Food-deprived rats ($n=17$) received an injection of GRP-10, -27, or -29 (0.3, 1, 2.1, 4.1, 10.3, and 17.2 nmol/kg) or saline i.p. 1 h into the light cycle. Immediately following the injection, rats were presented with a 10% sucrose test diet, and their minute-to-minute feeding behaviors were rated to determine the time between first and second meals or intermeal interval (IMI). GRP-10 (0.3 nmol/kg) and all doses of GRP-29 and GRP-27 except (10.3 nmol/kg) prolonged IMI relative to saline (⋆) and GRP-27 (0.3 nmol) prolonged it more than the same dose of GRP-10 (†). $p<0.05$.[95]

Due to their size, Bn, GRP, and NMB work in the periphery or in the circumventricular organs, which lack the blood–brain barrier. In addition, there are data supporting an upper gastrointestinal site of action for Bn to reduce food intake; this site is supplied by the celiac artery, a branch of the aorta.[234,235] We have shown that injections of Bn, GRP-10, GRP-27, GRP-29, and NMB intraperitoneally or through the left gastric artery, which supplies the lesser curvature of the stomach; the cranial mesenteric artery, which supplies the entire small intestine; or the femoral vein all reduce MS and prolonged the IMI (Martha C. Washington and Ayman I. Sayegh, 2012, unpublished data). These data suggest that the doses used in these studies require attenuation to determine the site of action and possibly the mode of action for these peptides, that is, whether it is neurocrine or endocrine. However, because GRP and NMB are peptides secreted by gastric and enteric neurons, it is plausible that the site of action for these peptides is gastrointestinal, a hypothesis that requires further testing.

The satiation (Fig. 10.4) and the satiety (Fig. 10.5) signals of Bn, GRP and NMB are transmitted to the feeding control areas of the hindbrain (i.e., area postrema, NTS, and dorsal motor nucleus of the vagus)[224] by the vagal and spinal afferents.[96,236] It has been shown that lesions of these areas attenuated the satiating effect of Bn, while lesions of the forebrain did not.[237–239]

We have shown that Bn, GRP-10, GRP-27, GRP-29, and NMB activate these areas (Fig. 10.6) as well as the enteric neurons (Fig. 10.7) of the

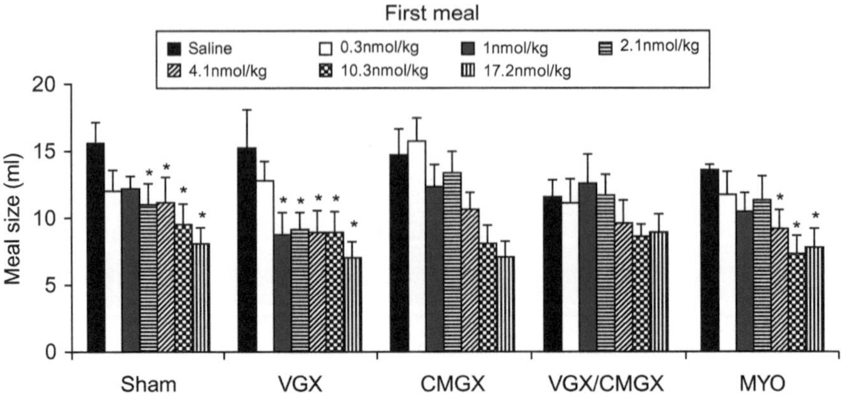

Figure 10.4 Effects of various surgical procedures on reduction of meal size by gastrin-releasing peptide (GRP)-29. Thirty-five rats were divided into five groups ($n=7$ rats per group): vagotomy (VGX), celiaco-mesenteric ganglionectomy (CMGX), combined VGX/CMGX, duodenal myotomy (MYO), and sham. Following recovery, the rats were given GRP (-29: 0.3, 1.0, 2.1, 4.1, 10.3, and 17.2 nmol/kg) or saline vehicle i.p. The behavior of the rats was rated every min, and the sizes of the first four meals (A, B, C, and D, respectively) were measured during a 120-min period. In sham-operated rats, GRP-29 (2.1, 4.1, 10.3, and 17.2 nmol/kg) reduced the size of the first meal relative to saline control (*). In the VGX group, GRP-29 (1.0, 2.1, 4.1, 10.3, and 17.2 nmol/kg) also reduced the size of the first meal (*). In the CMGX and combined VGX/CMGX group, GRP-29 failed to reduce food intake relative to saline control. In the MYO group, GRP-29 (4.1, 10.3, and 17.2 nmol/kg) reduced the size of the first meal relative to saline (*). GRP-29 failed to reduce the size of the second meal, third meal, and fourth meals, $p<0.05$.[96]

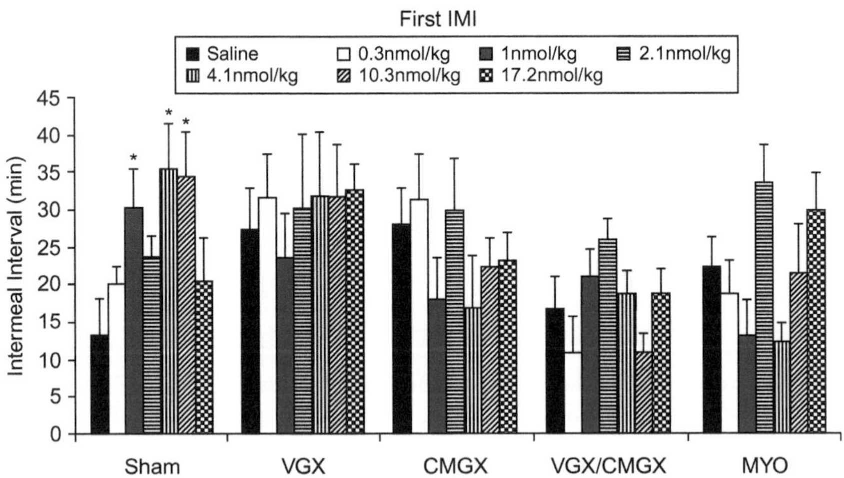

Figure 10.5 Effects of various surgical procedures on prolongation of the intermeal interval by gastrin-releasing peptide (GRP)-29. Thirty-five rats were divided into five groups ($n=7$ rats per group): vagotomy (VGX), celiaco-mesenteric ganglionectomy (CMGX), combined VGX/CMGX, duodenal myotomy (MYO), and sham. Following recovery, the rats were given gastrin-releasing peptide (GRP-29: 0.3, 1.0, 2.1, 4.1, 10.3, and 17.2 nmol/kg) or saline vehicle i.p. The behavior of the rats was rated every min for a total of 120 min and the time between the first four meals, the first three intermeal intervals (IMI) was determined. In sham-operated rats, GRP-29 (1, 4.1, and 10.3 nmol/kg) prolonged the first IMI relative to saline control (*). In the remaining groups, GRP-29 failed to prolong the IMI.[96]

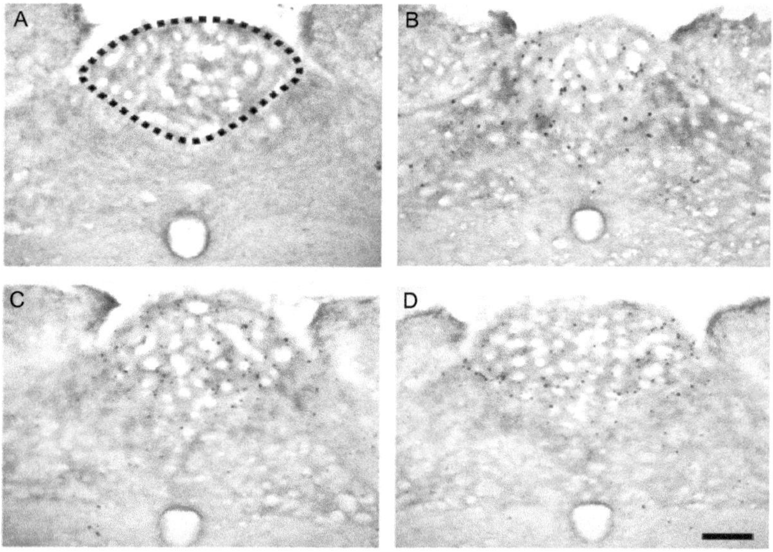

Figure 10.6 Photomicrographs of dorsal vagal complexes obtained from rats injected with saline (A), gastrin-releasing peptides (GRPs) 10 (B), 27 (C) and 29 (D). GRP increased Fos-like immunoreactivity, seen as dark pin-point dots, relative to saline in the area postrema (dotted shape) (bar = 100 μm).[91]

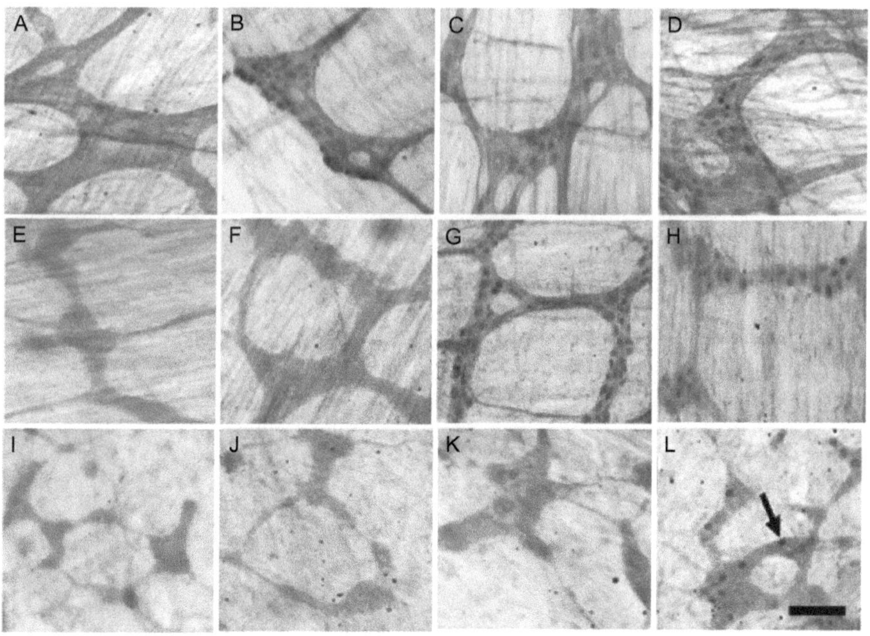

Figure 10.7 Photomicrographs of gastric (A–D) and duodenal (E–H) myenteric neurons, and duodenal submucosal neurons (I–L) captured from rats injected with saline (A, E, and I), gastrin-releasing peptide (GRP)-10 (B, F, and J), GRP-27 (C, G, and K), and GRP-29 (D, H, and L). Gastrin-releasing peptides 10, 27, and 29 increased Fos-like immunoreactivity in the myenteric neurons of the stomach (B–D), GRP-27 and 29 increased it in the myenteric plexus of the duodenum (G and H) and only GRP-29 increased it in the submucosal plexus of the duodenum (L). Arrow points at a Fos-positive nucleus (bar = 200 μm).[91]

stomach and intestine as measured by increased expression of Fos-like immunoreactivity, a marker for neuronal activation.[91] The vagus and the splanchnic nerves are necessary to carry the satiety and satiation signals of GRP-29 to the brain.[96] In addition, vagotomy by itself did not change the satiation effect of Bn.[240,241] However, combined vagotomy and thoracic spinal segment rhizotomy reversed these effects for Bn.[236] These pathways require further investigation for the remaining forms of GRP and NMB.

ACKNOWLEDGMENTS

The author would like to thank Dr. Robert Jensen from the NIH for providing valuable references for this work, and Ms. Martha C. Washington, MS, and Ms. Tanisha Johnson-Rouse, BS, from the Gastroenterology Laboratory of Tuskegee University, School of Veterinary Medicine, for their valuable help in preparing some of the data used in this chapter. SUPPORT: NIH 1SC1DK094972-01A1 and The Birmingham Racing Commission.

REFERENCES

1. Erspamer V. Discovery, isolation, and characterization of bombesin-like peptides. *Ann N Y Acad Sci* 1988;**547**:3–9.
2. Falconieri Erspamer G, Severini C, Erspamer V, Melchiorri P, Delle Fave G, Nakajima T. Parallel bioassay of 27 bombesin-like peptides on 9 smooth muscle preparations. Structure-activity relationships and bombesin receptor subtypes. *Regul Pept* 1988;**21**:1–11.
3. McDonald TJ, Jornvall H, Ghatei M, Bloom SR, Mutt V. Characterization of an avian gastric (proventricular) peptide having sequence homology with the porcine gastrin-releasing peptide and the amphibian peptides bombesin and alytesin. *FEBS Lett* 1980;**122**:45–8.
4. McDonald TJ, Jornvall H, Nilsson G, Vagne M, Ghatei M, Bloom SR, et al. Characterization of a gastrin releasing peptide from porcine non-antral gastric tissue. *Biochem Biophys Res Commun* 1979;**90**:227–33.
5. McDonald TJ, Nilsson G, Vagne M, Ghatei M, Bloom SR, Mutt V. A gastrin releasing peptide from the porcine nonantral gastric tissue. *Gut* 1978;**19**:767–74.
6. Minamino N, Kangawa K, Matsuo H. Neuromedin B: a novel bombesin-like peptide identified in porcine spinal cord. *Biochem Biophys Res Commun* 1983;**114**:541–8.
7. Nagalla SR, Gibson BW, Tang D, Reeve Jr JR, Spindel ER. Gastrin-releasing peptide (GRP) is not mammalian bombesin. Identification and molecular cloning of a true amphibian GRP distinct from amphibian bombesin in Bombina orientalis. *J Biol Chem* 1992;**267**:6916–22.
8. Anastasi A, Erspamer V, Bucci M. Isolation and structure of bombesin and alytesin, 2 analogous active peptides from the skin of the European amphibians Bombina and Alytes. *Experientia* 1971;**27**:166–7.
9. Erspamer V. Biogenic amines and active polypeptides of the amphibian skin. *Annu Rev Pharmacol* 1971;**11**:327–50.
10. Erspamer V, Erpamer GF, Inselvini M. Some pharmacological actions of alytesin and bombesin. *J Pharm Pharmacol* 1970;**22**:875–6.

11. Anastasi A, Erspamer V, Bucci M. Isolation and amino acid sequences of alytesin and bombesin, two analogous active tetradecapeptides from the skin of European discoglossid frogs. *Arch Biochem Biophys* 1972;**148**:443–6.
12. Erspamer V, Erspamer GF, Melchiorri P, Negri L. Occurrence and polymorphism of bombesin-like peptides in the gastrointestinal tract of birds and mammals. *Gut* 1979;**20**:1047–56.
13. Reeve Jr JR, Walsh JH, Chew P, Clark B, Hawke D, Shively JE. Amino acid sequences of three bombesin-like peptides from canine intestine extracts. *J Biol Chem* 1983;**258**:5582–8.
14. Minamino N, Kangawa K, Matsuo H. Neuromedin C: a bombesin-like peptide identified in porcine spinal cord. *Biochem Biophys Res Commun* 1984;**119**:14–20.
15. Orloff MS, Reeve Jr JR, Ben-Avram CM, Shively JE, Walsh JH. Isolation and sequence analysis of human bombesin-like peptides. *Peptides* 1984;**5**:865–70.
16. Spindel ER, Chin WW, Price J, Rees LH, Besser GM, Habener JF. Cloning and characterization of cDNAs encoding human gastrin-releasing peptide. *Proc Natl Acad Sci USA* 1984;**81**:5699–703.
17. Wada E, Way J, Lebacq-Verheyden AM, Battey JF. Neuromedin B and gastrin-releasing peptide mRNAs are differentially distributed in the rat nervous system. *J Neurosci* 1990;**10**:2917–30.
18. Battey J, Wada E. Two distinct receptor subtypes for mammalian bombesin-like peptides. *Trends Neurosci* 1991;**14**:524–8.
19. Moody TW, Merali Z. Bombesin-like peptides and associated receptors within the brain: distribution and behavioral implications. *Peptides* 2004;**25**:511–20.
20. Walsh JH, Wong HC, Dockray GJ. Bombesin-like peptides in mammals. *Fed Proc* 1979;**38**:2315–9.
21. Costello JF, Brown MR, Gray TS. Bombesin immunoreactive neurons in the hypothalamic paraventricular nucleus innervate the dorsal vagal complex in the rat. *Brain Res* 1991;**542**:77–82.
22. Timson CM, Polak JM, Wharton J, Ghatei MA, Bloom SR, Usellini L, et al. Bombesin-like immunoreactivity in the avian gut and its localisation to a distinct cell type. *Histochemistry* 1979;**61**:213–21.
23. Villarreal JA, Brown MR. Bombesin-like peptide in hypothalamus: chemical and immunological characterization. *Life Sci* 1978;**23**:2729–33.
24. Wharton J, Polak JM, Bloom SR, Ghatei MA, Solcia E, Brown MR, et al. Bombesin-like immunoreactivity in the lung. *Nature* 1978;**273**:769–70.
25. Polak JM, Bloom SR, Hobbs S, Solcia E, Pearse AG. Distribution of a bombesin-like peptide in human gastrointestinal tract. *Lancet* 1976;**1**:1109–10.
26. Dockray GJ, Vaillant C, Walsh JH. The neuronal origin of bombesin-like immunoreactivity in the rat gastrointestinal tract. *Neuroscience* 1979;**4**:1561–8.
27. Costa M, Furness JB, Yanaihara N, Yanaihara C, Moody TW. Distribution and projections of neurons with immunoreactivity for both gastrin-releasing peptide and bombesin in the guinea-pig small intestine. *Cell Tissue Res* 1984;**235**:285–93.
28. Knuhtsen S, Holst JJ, Jensen SL, Knigge U, Nielsen OV. Gastrin-releasing peptide: effect on exocrine secretion and release from isolated perfused porcine pancreas. *Am J Physiol* 1985;**248**:G281–G286.
29. Schultzberg M, Dalsgaard CJ. Enteric origin of bombesin immunoreactive fibres in the rat coeliac-superior mesenteric ganglion. *Brain Res* 1983;**269**:190–5.
30. Dalsgaard CJ, Hokfelt T, Schultzberg M, Lundberg JM, Terenius L, Dockray GJ, et al. Origin of peptide-containing fibers in the inferior mesenteric ganglion of the guinea-pig: immunohistochemical studies with antisera to substance P, enkephalin, vasoactive intestinal polypeptide, cholecystokinin and bombesin. *Neuroscience* 1983;**9**:191–211.

31. Schultzberg M. Bombesin-like immunoreactivity in sympathetic ganglia. *Neuroscience* 1983;**8**:363–74.
32. Massari VJ, Shults CW, Park CH, Tizabi Y, Moody TW, Chronwall BM, et al. Deafferentation causes a loss of presynaptic bombesin receptors and supersensitivity of substance P receptors in the dorsal horn of the cat spinal cord. *Brain Res* 1985;**343**:268–74.
33. Panula P, Hadjiconstantinou M, Yang HY, Costa E. Immunohistochemical localization of bombesin/gastrin-releasing peptide and substance P in primary sensory neurons. *J Neurosci* 1983;**3**:2021–9.
34. Panula P, Yang HY, Costa E. Comparative distribution of bombesin/GRP- and substance-P-like immunoreactivities in rat hypothalamus. *J Comp Neurol* 1984;**224**:606–17.
35. O'Donohue TL, Massari VJ, Pazoles CJ, Chronwall BM, Shults CW, Quirion R, et al. A role for bombesin in sensory processing in the spinal cord. *J Neurosci* 1984;**4**:2956–62.
36. Moody TW, Pert CB. Bombesin-like peptides in rat brain: quantitation and biochemical characterization. *Biochem Biophys Res Commun* 1979;**90**:7–14.
37. Roth KA, Weber E, Barchas JD. Distribution of gastrin releasing peptide—bombesin-like immunostaining in rat brain. *Brain Res* 1982;**251**:277–82.
38. Major J, Ghatei MA, Bloom SR. Bombesin-like immunoreactivity in the pituitary gland. *Experientia* 1983;**39**:1158–9.
39. Yoshizaki K, de Bock V, Solomon S. Origin of bombesin-like peptides in human fetal lung. *Life Sci* 1984;**34**:835–43.
40. Jahnke GD, Lazarus LH. A bombesin immunoreactive peptide in milk. *Proc Natl Acad Sci USA* 1984;**81**:578–82.
41. Houben H, Denef C. Evidence for the presence of gastrin-releasing peptide immunoreactivity in rat anterior pituitary corticotrophs and lactotrophs, AtT20 cells, and GH3 cells: failure to demonstrate participation in local control of hormone release. *Endocrinology* 1991;**128**:3208–18.
42. Larsen PJ, O'Hare MM, Vangsted A, Mikkelsen JD. Gastrin releasing peptide (GRP) is present in a GRP(1–27) form in anterior pituitary cells of the guinea pig. *Peptides* 1989;**10**:815–8.
43. Giraud AS, Smith AI, Rundle SE. Distribution and molecular forms of immunoreactive bombesin in the ovine median eminence. *Mol Cell Endocrinol* 1987;**53**:245–50.
44. Moody TW, Korman LY, O'Donohue TL. Neuromedin B-like peptides in rat brain: biochemical characterization, mechanism of release and localization in synaptosomes. *Peptides* 1986;**7**:815–20.
45. Chronwall BM, Pisano JJ, Bishop JF, Moody TW, O'Donohue TL. Biochemical and histochemical characterization of ranatensin immunoreactive peptides in rat brain: lack of coexistence with bombesin/GRP. *Brain Res* 1985;**338**:97–113.
46. Houben H, Vandenbroucke AT, Verheyden AM, Denef C. Expression of the genes encoding bombesin-related peptides and their receptors in anterior pituitary tissue. *Mol Cell Endocrinol* 1993;**97**:159–64.
47. Gundlach AL, Knobe KE. Preprogastrin-releasing peptide messenger ribonucleic acid: neuroanatomical localization in rat brain by in situ hybridization with synthetic oligodeoxynucleotide probes. *Neurosci Lett* 1992;**137**:123–8.
48. Spindel ER, Giladi E, Segerson TP, Nagalla S. Bombesin-like peptides: of ligands and receptors. *Recent Prog Horm Res* 1993;**48**:365–91.
49. Minamino N, Kangawa K, Matsuo H. Neuromedin B is a major bombesin-like peptide in rat brain: regional distribution of neuromedin B and neuromedin C in rat brain, pituitary and spinal cord. *Biochem Biophys Res Commun* 1984;**124**:925–32.
50. Lebacq-Verheyden AM, Krystal G, Sartor O, Way J, Battey JF. The rat prepro gastrin releasing peptide gene is transcribed from two initiation sites in the brain. *Mol Endocrinol* 1988;**2**:556–63.

51. Spindel ER, Zilberberg MD, Chin WW. Analysis of the gene and multiple messenger ribonucleic acids (mRNAs) encoding human gastrin-releasing peptide: alternate RNA splicing occurs in neural and endocrine tissue. *Mol Endocrinol* 1987;**1**:224–32.
52. Wada E, Way J, Shapira H, Kusano K, Lebacq-Verheyden AM, Coy D, et al. cDNA cloning, characterization, and brain region-specific expression of a neuromedin-B-preferring bombesin receptor. *Neuron* 1991;**6**:421–30.
53. Zoeller RT, Lebacq-Verheyden AM, Battey JF. Distribution of two distinct messenger ribonucleic acids encoding gastrin-releasing peptide in rat brain. *Peptides* 1989;**10**:415–22.
54. Bertaccini G, Erspamer V, Melchiorri P, Sopranzi N. Gastrin release by bombesin in the dog. *Br J Pharmacol* 1974;**52**:219–25.
55. Bunnett NW, Clark B, Debas HT, Del Milton RC, Kovacs TO, Orloff MS, et al. Canine bombesin-like gastrin releasing peptides stimulate gastrin release and acid secretion in the dog. *J Physiol* 1985;**365**:121–30.
56. Taylor IL, Walsh JH, Carter D, Wood J, Grossman MI. Effects of atropine and bethanechol on bombesin-stimulated release of pancreatic polypeptide and gastrin in dog. *Gastroenterology* 1979;**77**:714–8.
57. Delle Fave G, Kohn A, de Magistris L, Mancuso M, Sparvoli C. Effect of bombesin-stimulated gastrin on gastric acid secretion in man. *Life Sci* 1980;**27**:993–9.
58. Erspamer V, Improta G, Melchiorri P, Sopranzi N. Evidence of cholecystokinin release by bombesin in the dog. *Br J Pharmacol* 1974;**52**:227–32.
59. McDonald TJ, Ghatei MA, Bloom SR, Adrian TE, Mochizuki T, Yanaihara C, et al. Dose–response comparisons of canine plasma gastroenteropancreatic hormone responses to bombesin and the porcine gastrin-releasing peptide (GRP). *Regul Pept* 1983;**5**:125–37.
60. Ghatei MA, Jung RT, Stevenson JC, Hillyard CJ, Adrian TE, Lee YC, et al. Bombesin: action on gut hormones and calcium in man. *J Clin Endocrinol Metab* 1982;**54**:980–5.
61. Pandol SJ, Schoeffield MS, Sachs G, Muallem S. Role of free cytosolic calcium in secretagogue-stimulated amylase release from dispersed acini from guinea pig pancreas. *J Biol Chem* 1985;**260**:10081–6.
62. Pearson GT, Flanagan PM, Petersen OH. Neural and hormonal control of membrane conductance in the pig pancreatic acinar cell. *Am J Physiol* 1984;**247**:G520–G526.
63. Basso N, Giri S, Improta G, Lezoche E, Melchiorri P, Percoco M, et al. External pancreatic secretion after bombesin infusion in man. *Gut* 1975;**16**:994–8.
64. Deschodt-Lanckman M, Robberecht P, De Neef P, Lammens M, Christophe J. In vitro action of bombesin and bombesin-like peptides on amylase secretion, calcium efflux, and adenylate cyclase activity in the rat pancreas: a comparison with other secretagogues. *J Clin Invest* 1976;**58**:891–8.
65. Namba M, Ghatei MA, Ghiglione M, Bloom SR. Effects of decapeptide of mammalian bombesin and neuromedin B on pancreatic exocrine secretion in the rat. *Digestion* 1986;**34**:105–14.
66. Linari G, Linari MB. Effect of bombesin on pancreatic secretion and gall bladder motility of the chicken. *Eur J Pharmacol* 1975;**34**:305–10.
67. Scarpignato C, Bertaccini G. Bombesin delays gastric emptying in the rat. *Digestion* 1981;**21**:104–6.
68. Al-Saffar A, Hellstrom PM, Nylander G, Rosell S. Influence of fasting and bombesin-induced myoelectric activity on the transit of small-intestinal contents in the rat. *Scand J Gastroenterol* 1984;**19**:541–6.
69. Lehy T, Accary JP, Labeille D, Dubrasquet M. Chronic administration of bombesin stimulates antral gastrin cell proliferation in the rat. *Gastroenterology* 1983;**84**:914–9.
70. Lhoste E, Aprahamian M, Pousse A, Hoeltzel A, Stock-Damge C. Trophic effect of bombesin on the rat pancreas: is it mediated by the release of gastrin or cholecystokinin? *Peptides* 1985;**6**(Suppl. 3):89–97.

71. Lhoste E, Aprahamian M, Pousse A, Hoeltzel A, Stock-Damge C. Combined effect of chronic bombesin and secretin or cholecystokinin on the rat pancreas. *Peptides* 1985;**6** (Suppl. 3):83–7.
72. Brown M, Rivier J, Vale W. Bombesin:potent effects on thermoregulation in the rat. *Science* 1977;**196**:998–1000.
73. Fisher LA, Brown MR. Bombesin-induced stimulation of cardiac parasympathetic innervation. *Regul Pept* 1984;**8**:335–43.
74. Bedard T, Mountney C, Kent P, Anisman H, Merali Z. Role of gastrin-releasing peptide and neuromedin B in anxiety and fear-related behavior. *Behav Brain Res* 2007;**179**:133–40.
75. Mountney C, Sillberg V, Kent P, Anisman H, Merali Z. The role of gastrin-releasing peptide on conditioned fear: differential cortical and amygdaloid responses in the rat. *Psychopharmacology (Berl)* 2006;**189**:287–96.
76. Roesler R, Henriques JA, Schwartsmann G. Gastrin-releasing peptide receptor as a molecular target for psychiatric and neurological disorders. *CNS Neurol Disord Drug Targets* 2006;**5**:197–204.
77. Crawley JN, Moody TW. Anxiolytics block excessive grooming behavior induced by ACTH1-24 and bombesin. *Brain Res Bull* 1983;**10**:399–401.
78. Merali Z, McIntosh J, Kent P, Michaud D, Anisman H. Aversive and appetitive events evoke the release of corticotropin-releasing hormone and bombesin-like peptides at the central nucleus of the amygdala. *J Neurosci* 1998;**18**:4758–66.
79. Moody TW, Merali Z, Crawley JN. The effects of anxiolytics and other agents on rat grooming behavior. *Ann N Y Acad Sci* 1988;**525**:281–90.
80. Yamada K, Santo-Yamada Y, Wada E, Wada K. Role of bombesin (BN)-like peptides/receptors in emotional behavior by comparison of three strains of BN-like peptide receptor knockout mice. *Mol Psychiatry* 2002;**7**(113–7):6.
81. Yamada K, Santo-Yamada Y, Wada K. Stress-induced impairment of inhibitory avoidance learning in female neuromedin B receptor-deficient mice. *Physiol Behav* 2003;**78**:303–9.
82. Yamada K, Wada E, Wada K. Male mice lacking the gastrin-releasing peptide receptor (GRP-R) display elevated preference for conspecific odors and increased social investigatory behaviors. *Brain Res* 2000;**870**:20–6.
83. Endean R, Erspamer V, Falconieri Erspamer G, Improta G, Melchiorri P, Negri L, et al. Parallel bioassay of bombesin and litorin, a bombesin-like peptide from the skin of Litoria aurea. *Br J Pharmacol* 1975;**55**:213–9.
84. Mayer EA, Elashoff J, Walsh JH. Characterization of bombesin effects on canine gastric muscle. *Am J Physiol* 1982;**243**:G141–G147.
85. Mayer EA, Reeve Jr JR, Khawaja S, Chew P, Elashoff J, Clark B, et al. Potency of natural and synthetic canine gastrin-releasing decapeptide on canine antral muscle. *Am J Physiol* 1986;**250**:G581–G587.
86. Kantoh M, Takahashi T, Yamamura T, Ishikawa Y, Utsunomiya J. Bombesin evoked acetylcholine release from the guinea pig antrum. *Life Sci* 1985;**36**:2445–52.
87. Zafirov DH, Palmer JM, Nemeth PR, Wood JD. Bombesin, gastrin releasing peptide and vasoactive intestinal peptide excite myenteric neurons. *Eur J Pharmacol* 1985;**115**:103–7.
88. Zappia L, Molina E, Violini A, Bassani F. [Preliminary research on the action of the new polypeptide, bombesin, on the human appendix in vitro]. *Ateneo Parmense 1* 1972;**43**:169–74.
89. Corazziari E, Delle Fave G, Pozzessere C, Kohn A, de Magistris L, Anzini F, et al. Effect of bombesin on lower esophageal sphincter pressure in humans. *Gastroenterology* 1982;**83**:10–4.

90. Corazziari E, Habib FI, Delle Fave GF, Melchiorri P, Torsoli A, Carratu R. Gastrointestinal and gallbladder motor effects of bombesin in man. *Mater Med Pol* 1977;**9**:139–43.
91. Washington MC, Sayegh AI. Gastrin releasing peptides increase Fos-like immunoreactivity in the enteric nervous system and the dorsal vagal complex. *Peptides* 2012;**32**:1600–5.
92. Gibbs J, Fauser DJ, Rowe EA, Rolls BJ, Rolls ET, Maddison SP. Bombesin suppresses feeding in rats. *Nature* 1979;**282**:208–10.
93. Stratford TR, Gibbs J, Smith GP. Microstructural analysis of licking behavior following peripheral administration of bombesin or gastrin-releasing peptide. *Peptides* 1995;**16**:903–9.
94. Stratford TR, Gibbs J, Smith GP. Simultaneous administration of neuromedin B-10 and gastrin-releasing peptide(1–27) reproduces the satiating and microstructural effects of bombesin. *Peptides* 1996;**17**:107–10.
95. Washington MC, Wright SA, Sayegh AI. Gastrin releasing peptide-29 evokes feeding responses in the rat. *Peptides* 2012;**32**:241–5.
96. Wright SA, Washington MC, Garcia C, Sayegh AI. Gastrin releasing peptide-29 requires vagal and splanchnic neurons to evoke satiation and satiety. *Peptides* 2012;**33**:125–31.
97. Stein LJ, Woods SC. Gastrin releasing peptide reduces meal size in rats. *Peptides* 1982;**3**:833–5.
98. Ruff M, Schiffmann E, Terranova V, Pert CB. Neuropeptides are chemoattractants for human tumor cells and monocytes: a possible mechanism for metastasis. *Clin Immunol Immunopathol* 1985;**37**:387–96.
99. De la Fuente M, Del Rio M, Ferrandez MD, Hernanz A. Modulation of phagocytic function in murine peritoneal macrophages by bombesin, gastrin-releasing peptide and neuromedin C. *Immunology* 1991;**73**:205–11.
100. De la Fuente M, Del Rio M, Hernanz A. Stimulation of natural killer and antibody-dependent cellular cytotoxicity activities in mouse leukocytes by bombesin, gastrin-releasing peptide and neuromedin C: involvement of cyclic AMP, inositol 1,4,5-trisphosphate and protein kinase C. *J Neuroimmunol* 1993;**48**:143–50.
101. van Tol EA, Verspaget HW, Hansen BE, Lamers CB. Neuroenteric peptides affect natural killer activity by intestinal lamina propria mononuclear cells. *J Neuroimmunol* 1993;**42**:139–45.
102. Del Rio M, De la Fuente M. Chemoattractant capacity of bombesin, gastrin-releasing peptide and neuromedin C is mediated through PKC activation in murine peritoneal leukocytes. *Regul Pept* 1994;**49**:185–93.
103. Del Rio M, Hernanz A, de la Fuente M. Bombesin, gastrin-releasing peptide, and neuromedin C modulate murine lymphocyte proliferation through adherent accessory cells and activate protein kinase C. *Peptides* 1994;**15**:15–22.
104. Bologna M, Festuccia C, Muzi P, Biordi L, Ciomei M. Bombesin stimulates growth of human prostatic cancer cells in vitro. *Cancer* 1989;**63**:1714–20.
105. Flatt PR, Bailey CJ, Conlon JM. Somatostatin, gastrin-releasing peptide and gastrin in the stomach of rats with streptozotocin-induced diabetes and insulinoma. *J Nutr* 1991;**121**:1414–7.
106. Giacchetti S, Gauville C, de Cremoux P, Bertin L, Berthon P, Abita JP, et al. Characterization, in some human breast cancer cell lines, of gastrin-releasing peptide-like receptors which are absent in normal breast epithelial cells. *Int J Cancer* 1990;**46**:293–8.
107. Hajri A, Balboni G, Koenig M, Garaud JC, Damge C. Gastrin-releasing peptide: in vivo and in vitro growth effects on an acinar pancreatic carcinoma. *Cancer Res* 1992;**52**:3726–32.

108. Kawai K, Mukai H, Yuzawa K, Suzuki S, Kuzuya N, Fujii K, et al. Effects of neuromedin B and GRP-10 on gastrin and insulin release from cultured tumor cells of a malignant gastrinoma. *Endocrinol Jpn* 1990;**37**:857–65.
109. Moody TW, Carney DN, Cuttitta F, Quattrocchi K, Minna JD. High affinity receptors for bombesin/GRP-like peptides on human small cell lung cancer. *Life Sci* 1985;**37**:105–13.
110. Moody TW, Mahmoud S, Staley J, Naldini L, Cirillo D, South V, et al. Human glioblastoma cell lines have neuropeptide receptors for bombesin/gastrin-releasing peptide. *J Mol Neurosci* 1989;**1**:235–42.
111. Narayan S, Guo YS, Townsend Jr CM, Singh P. Specific binding and growth effects of bombesin-related peptides on mouse colon cancer cells in vitro. *Cancer Res* 1990;**50**:6772–8.
112. Narayan S, Spindel ER, Rubin NH, Singh P. A potent bombesin receptor antagonist inhibits bombesin-stimulated growth of mouse colon cancer cells in vitro: absence of autocrine effects. *Cell Growth Differ* 1992;**3**:111–8.
113. Vangsted AJ, Andersen EV, Nedergaard L, Zeuthen J. Gastrin releasing peptide GRP (14–27) in human breast cancer cells and in small cell lung cancer. *Breast Cancer Res Treat* 1991;**19**:119–28.
114. Viallet J, Minna JD. Gastrin-releasing peptide (GRP, mammalian bombesin) in the pathogenesis of lung cancer. *Prog Growth Factor Res* 1989;**1**:89–97.
115. Jensen RT, Moody T, Pert C, Rivier JE, Gardner JD. Interaction of bombesin and litorin with specific membrane receptors on pancreatic acinar cells. *Proc Natl Acad Sci USA* 1978;**75**:6139–43.
116. Moody TW, Pert CB, Rivier J, Brown MR. Bomebesin: specific binding to rat brain membranes. *Proc Natl Acad Sci USA* 1978;**75**:5372–6.
117. Jensen RT, Gardner JD. Identification and characterization of receptors for secretagogues on pancreatic acinar cells. *Fed Proc* 1981;**40**:2486–96.
118. Coy DH, Heinz-Erian P, Jiang NY, Sasaki Y, Taylor J, Moreau JP, et al. Probing peptide backbone function in bombesin. A reduced peptide bond analogue with potent and specific receptor antagonist activity. *J Biol Chem* 1988;**263**:5056–60.
119. Coy DH, Heinz-Erian P, Jiang NY, Taylor J, Moreau JP, Gardner JD, et al. Progress in the development of competitive bombesin antagonists. *Ann N Y Acad Sci* 1988;**547**: 150–7.
120. Ladenheim EE, Jensen RT, Mantey SA, McHugh PR, Moran TH. Receptor heterogeneity for bombesin-like peptides in the rat central nervous system. *Brain Res* 1990;**537**:233–40.
121. Jensen RT, Coy DH. Progress in the development of potent bombesin receptor antagonists. *Trends Pharmacol Sci* 1991;**12**:13–9.
122. Spindel ER, Gibson BW, Reeve Jr JR, Kelly M. Cloning of cDNAs encoding amphibian bombesin: evidence for the relationship between bombesin and gastrin-releasing peptide. *Proc Natl Acad Sci USA* 1990;**87**:9813–7.
123. Spindel ER, Giladi E, Brehm P, Goodman RH, Segerson TP. Cloning and functional characterization of a complementary DNA encoding the murine fibroblast bombesin/ gastrin-releasing peptide receptor. *Mol Endocrinol* 1990;**4**:1956–63.
124. Battey JF, Way JM, Corjay MH, Shapira H, Kusano K, Harkins R, et al. Molecular cloning of the bombesin/gastrin-releasing peptide receptor from Swiss 3T3 cells. *Proc Natl Acad Sci USA* 1991;**88**:395–9.
125. Feldman RI, Wu JM, Jenson JC, Mann E. Purification and characterization of the bombesin/gastrin-releasing peptide receptor from Swiss 3T3 cells. *J Biol Chem* 1990;**265**:17364–72.

126. von Schrenck T, Wang LH, Coy DH, Villanueva ML, Mantey S, Jensen RT. Potent bombesin receptor antagonists distinguish receptor subtypes. *Am J Physiol* 1990;**259**: G468–G473.
127. Von Schrenck T, Heinz-Erian P, Moran T, Mantey SA, Gardner JD, Jensen RT. Neuromedin B receptor in esophagus: evidence for subtypes of bombesin receptors. *Am J Physiol* 1989;**256**:G747–G758.
128. Corjay MH, Dobrzanski DJ, Way JM, Viallet J, Shapira H, Worland P, et al. Two distinct bombesin receptor subtypes are expressed and functional in human lung carcinoma cells. *J Biol Chem* 1991;**266**:18771–9.
129. Severi C, Jensen RT, Erspamer V, D'Arpino L, Coy DH, Torsoli A, et al. Different receptors mediate the action of bombesin-related peptides on gastric smooth muscle cells. *Am J Physiol* 1991;**260**:G683–G690.
130. Jensen RT, Battey JF, Spindel ER, Benya RV. International Union of Pharmacology. LXVIII. Mammalian bombesin receptors: nomenclature, distribution, pharmacology, signaling, and functions in normal and disease states. *Pharmacol Rev* 2008;**60**:1–42.
131. Gorbulev V, Akhundova A, Buchner H, Fahrenholz F. Molecular cloning of a new bombesin receptor subtype expressed in uterus during pregnancy. *Eur J Biochem* 1992;**208**:405–10.
132. Nagalla SR, Barry BJ, Creswick KC, Eden P, Taylor JT, Spindel ER. Cloning of a receptor for amphibian [Phe13]bombesin distinct from the receptor for gastrin-releasing peptide: identification of a fourth bombesin receptor subtype (BB4). *Proc Natl Acad Sci USA* 1995;**92**:6205–9.
133. Iwabuchi M, Ui-Tei K, Yamada K, Matsuda Y, Sakai Y, Tanaka K, et al. Molecular cloning and characterization of avian bombesin-like peptide receptors: new tools for investigating molecular basis for ligand selectivity. *Br J Pharmacol* 2003;**139**:555–66.
134. Benya RV, Kusui T, Pradhan TK, Battey JF, Jensen RT. Expression and characterization of cloned human bombesin receptors. *Mol Pharmacol* 1995;**47**:10–20.
135. Ohki-Hamazaki H, Wada E, Matsui K, Wada K. Cloning and expression of the neuromedin B receptor and the third subtype of bombesin receptor genes in the mouse. *Brain Res* 1997;**762**:165–72.
136. Sano H, Feighner SD, Hreniuk DL, Iwaasa H, Sailer AW, Pan J, et al. Characterization of the bombesin-like peptide receptor family in primates. *Genomics* 2004;**84**:139–46.
137. Ladenheim EE, Jensen RT, Mantey SA, Moran TH. Distinct distributions of two bombesin receptor subtypes in the rat central nervous system. *Brain Res* 1992;**593**:168–78.
138. Ladenheim EE, Jensen RT, Mantey SA, Taylor JE, Coy DH, Moran TH. Bombesin receptor antagonists differentiate receptor subtypes in rat brain. *Eur J Pharmacol* 1993;**235**:121–5.
139. Wang LH, Battey JF, Wada E, Lin JT, Mantey S, Coy DH, et al. Activation of neuromedin B-preferring bombesin receptors on rat glioblastoma C-6 cells increases cellular Ca2+ and phosphoinositides. *Biochem J* 1992;**286**(Pt 2):641–8.
140. Toi-Scott M, Jones CL, Kane MA. Clinical correlates of bombesin-like peptide receptor subtype expression in human lung cancer cells. *Lung Cancer* 1996;**15**:341–54.
141. Reubi JC, Wenger S, Schmuckli-Maurer J, Schaer JC, Gugger M. Bombesin receptor subtypes in human cancers: detection with the universal radioligand (125)I-[D-TYR (6), beta-ALA(11), PHE(13), NLE(14)] bombesin(6–14). *Clin Cancer Res* 2002;**8**:1139–46.
142. Sun B, Schally AV, Halmos G. The presence of receptors for bombesin/GRP and mRNA for three receptor subtypes in human ovarian epithelial cancers. *Regul Pept* 2000;**90**:77–84.

143. Bonora G, De Giorgio R, Toni R, Fanti MP, Cariani G, Vezzadini P. Bombesin-stimulated serum immunoreactive trypsin in the differential diagnosis between endocrine and exocrine tumors of the pancreas. *J Nucl Med Allied Sci* 1987;**31**:327–30.
144. Chung DH, Evers BM, Beauchamp RD, Upp Jr JR, Rajaraman S, Townsend Jr CM, et al. Bombesin stimulates growth of human gastrinoma. *Surgery* 1992;**112**:1059–65.
145. Lhoste EF, Longnecker DS. Effect of bombesin and caerulein on early stages of carcinogenesis induced by azaserine in the rat pancreas. *Cancer Res* 1987;**47**:3273–7.
146. Merali Z, Kent P, Anisman H. Role of bombesin-related peptides in the mediation or integration of the stress response. *Cell Mol Life Sci* 2002;**59**:272–87.
147. Merali Z, Moody T, Kateb P, Piggins H. Antagonism of satiety and grooming effects of bombesin by antiserum to bombesin and by [Tyr4, D-Phe12] bombesin: central versus peripheral effects. *Ann N Y Acad Sci* 1988;**547**:489–92.
148. Ladenheim EE, Knipp S. Capsaicin treatment differentially affects feeding suppression by bombesin-like peptides. *Physiol Behav* 2007;**91**:36–41.
149. Ladenheim EE, Ritter RC. Low-dose fourth ventricular bombesin selectively suppresses food intake. *Am J Physiol* 1988;**255**:R988–R993.
150. Regoli D, Dion S, Rhaleb NE, Drapeau G, Rouissi N, D'Orleans-Juste P. Receptors for neurokinins, tachykinins, and bombesin: a pharmacological study. *Ann N Y Acad Sci* 1988;**547**:158–73.
151. Kilgore WR, Mantyh PW, Mantyh CR, McVey DC, Vigna SR. Bombesin/GRP-preferring and neuromedin B-preferring receptors in the rat urogenital system. *Neuropeptides* 1993;**24**:43–52.
152. Parkman HP, Vozzelli MA, Pagano AP, Cowan A. Pharmacological analysis of receptors for bombesin-related peptides on guinea pig gallbladder smooth muscle. *Regul Pept* 1994;**52**:173–80.
153. Milusheva EA, Kortezova NI, Mizhorkova ZN, Papasova M, Coy DH, Balint A, et al. Role of different bombesin receptor subtypes mediating contractile activity in cat upper gastrointestinal tract. *Peptides* 1998;**19**:549–56.
154. Ohki-Hamazaki H, Iwabuchi M, Maekawa F. Development and function of bombesin-like peptides and their receptors. *Int J Dev Biol* 2005;**49**:293–300.
155. Ohki-Hamazaki H, Sakai Y, Kamata K, Ogura H, Okuyama S, Watase K, et al. Functional properties of two bombesin-like peptide receptors revealed by the analysis of mice lacking neuromedin B receptor. *J Neurosci* 1999;**19**:948–54.
156. Siegfried JM, DeMichele MA, Hunt JD, Davis AG, Vohra KP, Pilewski JM. Expression of mRNA for gastrin-releasing peptide receptor by human bronchial epithelial cells. Association with prolonged tobacco exposure and responsiveness to bombesin-like peptides. *Am J Respir Crit Care Med* 1997;**156**:358–66.
157. Fathi Z, Corjay MH, Shapira H, Wada E, Benya R, Jensen R, et al. BRS-3: a novel bombesin receptor subtype selectively expressed in testis and lung carcinoma cells. *J Biol Chem* 1993;**268**:5979–84.
158. Mazzanti G, Falconieri Erspamer G, Piccinelli D. Relative potencies of porcine bombesin-like heptacosapeptide (PB-27), amphibian bombesin (B-14) and litorin, and bombesin C-terminal nonapeptide (B-9) on in vitro and in vivo smooth muscle preparations. *J Pharm Pharmacol* 1982;**34**:120–1.
159. Heimbrook DC, Boyer ME, Garsky VM, Balishin NL, Kiefer DM, Oliff A, et al. Minimal ligand analysis of gastrin releasing peptide. Receptor binding and mitogenesis. *J Biol Chem* 1988;**263**:7016–9.
160. Lin JT, Coy DH, Mantey SA, Jensen RT. Comparison of the peptide structural requirements for high affinity interaction with bombesin receptors. *Eur J Pharmacol* 1995;**294**:55–69.
161. Baldwin GS, Patel O, Shulkes A. Phylogenetic analysis of the sequences of gastrin-releasing peptide and its receptors: biological implications. *Regul Pept* 2007;**143**:1–14.

162. Kris RM, Hazan R, Villines J, Moody TW, Schlessinger J. Identification of the bombesin receptor on murine and human cells by cross-linking experiments. *J Biol Chem* 1987;**262**:11215–20.
163. Rozengurt E. Bombesin-induction of cell proliferation in 3T3 cells. Specific receptors and early signaling events. *Ann N Y Acad Sci* 1988;**547**:277–92.
164. Huang SC, Yu DH, Wank SA, Gardner JD, Jensen RT. Characterization of the bombesin receptor on mouse pancreatic acini by chemical cross-linking. *Peptides* 1990;**11**:1143–50.
165. Benya RV, Fathi Z, Kusui T, Pradhan T, Battey JF, Jensen RT. Gastrin-releasing peptide receptor-induced internalization, down-regulation, desensitization, and growth: possible role for cyclic AMP. *Mol Pharmacol* 1994;**46**:235–45.
166. Maslen GL, Boyd Y. Comparative mapping of the Grpr locus on the X chromosomes of man and mouse. *Genomics* 1993;**17**:106–9.
167. Xiao D, Wang J, Hampton LL, Weber HC. The human gastrin-releasing peptide receptor gene structure, its tissue expression and promoter. *Gene* 2001;**264**:95–103.
168. Weber HC, Jensen RT, Battey JF. Molecular organization of the mouse gastrin-releasing peptide receptor gene and its promoter. *Gene* 2000;**244**:137–49.
169. Kamichi S, Wada E, Aoki S, Sekiguchi M, Kimura I, Wada K. Immunohistochemical localization of gastrin-releasing peptide receptor in the mouse brain. *Brain Res* 2005;**1032**:162–70.
170. Ferris HA, Carroll RE, Lorimer DL, Benya RV. Location and characterization of the human GRP receptor expressed by gastrointestinal epithelial cells. *Peptides* 1997;**18**:663–72.
171. Nakamura M, Oda M, Kaneko K, Akaiwa Y, Tsukada N, Komatsu H, et al. Autoradiographic demonstration of gastrin-releasing peptide-binding sites in the rat gastric mucosa. *Gastroenterology* 1988;**94**:968–76.
172. Jensen RT, Coy DH, Saeed ZA, Heinz-Erian P, Mantey S, Gardner JD. Interaction of bombesin and related peptides with receptors on pancreatic acinar cells. *Ann N Y Acad Sci* 1988;**547**:138–49.
173. Patel O, Shulkes A, Baldwin GS. Gastrin-releasing peptide and cancer. *Biochim Biophys Acta* 2006;**1766**:23–41.
174. Jensen RT, Niederle B, Mitry E, Ramage JK, Steinmuller T, Lewington V, et al. Gastrinoma (duodenal and pancreatic). *Neuroendocrinology* 2006;**84**:173–82.
175. Fleischmann A, Waser B, Gebbers JO, Reubi JC. Gastrin-releasing peptide receptors in normal and neoplastic human uterus: involvement of multiple tissue compartments. *J Clin Endocrinol Metab* 2005;**90**:4722–9.
176. Schubert ML, Hightower J, Coy DH, Makhlouf GM. Regulation of acid secretion by bombesin/GRP neurons of the gastric fundus. *Am J Physiol* 1991;**260**:G156–G160.
177. Hildebrand P, Lehmann FS, Ketterer S, Christ AD, Stingelin T, Beltinger J, et al. Regulation of gastric function by endogenous gastrin releasing peptide in humans: studies with a specific gastrin releasing peptide receptor antagonist. *Gut* 2001;**49**:23–8.
178. Degen LP, Peng F, Collet A, Rossi L, Ketterer S, Serrano Y, et al. Blockade of GRP receptors inhibits gastric emptying and gallbladder contraction but accelerates small intestinal transit. *Gastroenterology* 2001;**120**:361–8.
179. Yegen BC, Gurbuz V, Coskun T, Bozkurt A, Kurtel H, Alican I, et al. Inhibitory effects of gastrin releasing peptide on gastric emptying in rats. *Regul Pept* 1996;**61**:175–80.
180. Niebergall-Roth E, Singer MV. Central and peripheral neural control of pancreatic exocrine secretion. *J Physiol Pharmacol* 2001;**52**:523–38.
181. Nathan JD, Liddle RA. Neurohormonal control of pancreatic exocrine secretion. *Curr Opin Gastroenterol* 2002;**18**:536–44.
182. Persson K, Pacini G, Sundler F, Ahren B. Islet function phenotype in gastrin-releasing peptide receptor gene-deficient mice. *Endocrinology* 2002;**143**:3717–26.

183. Modlin IM, Lamers CB, Walsh JH. Stimulation of canine pancreatic polypeptide, gastrin, and gastric acid secretion by ranatensin, litorin, bombesin nonapeptide and substance P. *Regul Pept* 1981;**1**:279–88.
184. Modlin IM, Albert D, Crochelt R, Sank A, Jaffe BM. Evidence for an intestinal mechanism of pancreatic polypeptide release. *Dig Dis Sci* 1981;**26**:587–90.
185. Pettersson M, Ahren B. Gastrin releasing peptide (GRP): effects on basal and stimulated insulin and glucagon secretion in the mouse. *Peptides* 1987;**8**:55–60.
186. Subramaniam M, Sugiyama K, Coy DH, Kong Y, Miller YE, Weller PF, et al. Bombesin-like peptides and mast cell responses: relevance to bronchopulmonary dysplasia? *Am J Respir Crit Care Med* 2003;**168**:601–11.
187. Gibbs J, Smith GP. The actions of bombesin-like peptides on food intake. *Ann N Y Acad Sci* 1988;**547**:210–6.
188. Gibbs J, Kulkosky PJ, Smith GP. Effects of peripheral and central bombesin on feeding behavior of rats. *Peptides* 1981;**2**(Suppl. 2):179–83.
189. Flynn FW. Bombesin receptor antagonists block the effects of exogenous bombesin but not of nutrients on food intake. *Physiol Behav* 1997;**62**:791–8.
190. Ladenheim EE, Hampton LL, Whitney AC, White WO, Battey JF, Moran TH. Disruptions in feeding and body weight control in gastrin-releasing peptide receptor deficient mice. *J Endocrinol* 2002;**174**:273–81.
191. Hampton LL, Ladenheim EE, Akeson M, Way JM, Weber HC, Sutliff VE, et al. Loss of bombesin-induced feeding suppression in gastrin-releasing peptide receptor-deficient mice. *Proc Natl Acad Sci USA* 1998;**95**:3188–92.
192. Levine AS, Morley JE. Stress-induced eating in rats. *Am J Physiol* 1981;**241**:R72–R76.
193. Kraly FS, Miller LA, Gibbs J. Diurnal variation for inhibition of eating by bombesin in the rat. *Physiol Behav* 1983;**31**:395–9.
194. Roesler R, Luft T, Oliveira SH, Farias CB, Almeida VR, Quevedo J, et al. Molecular mechanisms mediating gastrin-releasing peptide receptor modulation of memory consolidation in the hippocampus. *Neuropharmacology* 2006;**51**:350–7.
195. Iguchi A, Kunoh Y, Gotoh M, Miura H, Uemura K, Tamagawa T, et al. Relative contribution of nervous system and hormones to CNS-mediated hyperglycemia is determined by the neurochemical specificity in the brain. *Physiol Behav* 1991;**50**:1019–25.
196. Brown MR, Gray TS. Peptide injections into the amygdala of conscious rats: effects on blood pressure, heart rate and plasma catecholamines. *Regul Pept* 1988;**21**:95–106.
197. Shumyatsky GP, Tsvetkov E, Malleret G, Vronskaya S, Hatton M, Hampton L, et al. Identification of a signaling network in lateral nucleus of amygdala important for inhibiting memory specifically related to learned fear. *Cell* 2002;**111**:905–18.
198. Wada E, Watase K, Yamada K, Ogura H, Yamano M, Inomata Y, et al. Generation and characterization of mice lacking gastrin-releasing peptide receptor. *Biochem Biophys Res Commun* 1997;**239**:28–33.
199. Willey JC, Lechner JF, Harris CC. Bombesin and the C-terminal tetradecapeptide of gastrin-releasing peptide are growth factors for normal human bronchial epithelial cells. *Exp Cell Res* 1984;**153**:245–8.
200. Hill DJ, McDonald TJ. Mitogenic action of gastrin-releasing polypeptide on isolated epiphyseal growth plate chondrocytes from the ovine fetus. *Endocrinology* 1992;**130**:2811–9.
201. Chu M, Nielsen FC, Franzen L, Rehfeld JF, Holst JJ, Borch K. Effect of endogenous hypergastrinemia on gastrin receptor expressing human colon carcinoma transplanted to athymic rats. *Gastroenterology* 1995;**109**:1415–20.
202. Moody TW, Staley J, Zia F, Coy DH, Jensen RT. Neuromedin B binds with high affinity, elevates cytosolic calcium and stimulates the growth of small-cell lung cancer cell lines. *J Pharmacol Exp Ther* 1992;**263**:311–7.

203. Endo T, Fukue H, Kanaya M, Mizunuma M, Fujii M, Yamamoto H, et al. Bombesin and bradykinin increase inositol phosphates and cytosolic free Ca2+, and stimulate DNA synthesis in human endometrial stromal cells. *J Endocrinol* 1991;**131**:313–8.
204. Kim S, Hu W, Kelly DR, Hellmich MR, Evers BM, Chung DH. Gastrin-releasing peptide is a growth factor for human neuroblastomas. *Ann Surg* 2002;**235**:621–9 Discussion 629–630.
205. Murphy LO, Abdel-Wahab YH, Wang QJ, Knezetic JA, Permnert J, Larsson J, et al. Receptors and ligands for autocrine growth pathways are up-regulated when pancreatic cancer cells are adapted to serum-free culture. *Pancreas* 2001;**22**:293–8.
206. Chave HS, Gough AC, Palmer K, Preston SR, Primrose JN. Bombesin family receptor and ligand gene expression in human colorectal cancer and normal mucosa. *Br J Cancer* 2000;**82**:124–30.
207. Plonowski A, Nagy A, Schally AV, Sun B, Groot K, Halmos G. In vivo inhibition of PC-3 human androgen-independent prostate cancer by a targeted cytotoxic bombesin analogue, AN-215. *Int J Cancer* 2000;**88**:652–7.
208. Plonowski A, Schally AV, Varga JL, Rekasi Z, Hebert F, Halmos G, et al. Potentiation of the inhibitory effect of growth hormone-releasing hormone antagonists on PC-3 human prostate cancer by bombesin antagonists indicative of interference with both IGF and EGF pathways. *Prostate* 2000;**44**:172–80.
209. Sharif TR, Luo W, Sharif M. Functional expression of bombesin receptor in most adult and pediatric human glioblastoma cell lines; role in mitogenesis and in stimulating the mitogen-activated protein kinase pathway. *Mol Cell Endocrinol* 1997;**130**:119–30.
210. Siegfried JM, Krishnamachary N, Gaither Davis A, Gubish C, Hunt JD, Shriver SP. Evidence for autocrine actions of neuromedin B and gastrin-releasing peptide in non-small cell lung cancer. *Pulm Pharmacol Ther* 1999;**12**:291–302.
211. Liu J, Lao ZJ, Zhang J, Schaeffer MT, Jiang MM, Guan XM, et al. Molecular basis of the pharmacological difference between rat and human bombesin receptor subtype-3 (BRS-3). *Biochemistry* 2002;**41**:8954–60.
212. Porcher C, Juhem A, Peinnequin A, Bonaz B. Bombesin receptor subtype-3 is expressed by the enteric nervous system and by interstitial cells of Cajal in the rat gastrointestinal tract. *Cell Tissue Res* 2005;**320**:21–31.
213. Schulz S, Rocken C. Immunohistochemical detection of bombesin receptor subtypes GRP-R and BRS-3 in human tumors using novel antipeptide antibodies. *Virchows Arch* 2006;**449**:421–7.
214. Matsumoto K, Iijima H. Sibutramine sensitivity assay revealed a unique phenotype of bombesin BB3 receptor-deficient mice. *Eur J Pharmacol* 2003;**473**:41–6.
215. Yamada K, Wada E, Imaki J, Ohki-Hamazaki H, Wada K. Hyperresponsiveness to palatable and aversive taste stimuli in genetically obese (bombesin receptor subtype-3-deficient) mice. *Physiol Behav* 1999;**66**:863–7.
216. Maekawa F, Quah HM, Tanaka K, Ohki-Hamazaki H. Leptin resistance and enhancement of feeding facilitation by melanin-concentrating hormone in mice lacking bombesin receptor subtype-3. *Diabetes* 2004;**53**:570–6.
217. Hotta K, Matsukawa Y, Nishida M, Kotani K, Takahashi M, Kuriyama H, et al. Mutation in bombesin receptor subtype-3 gene is not a major cause of obesity in the Japanese. *Horm Metab Res* 2000;**32**:33–4.
218. Martin CF, Gibbs J. Bombesin elicits satiety in sham feeding rats. *Peptides* 1980;**1**:131–4.
219. Taylor IL, Garcia R. Effects of pancreatic polypeptide, caerulein, and bombesin on satiety in obese mice. *Am J Physiol* 1985;**248**:G277–G280.
220. Bado A, Lewin MJ, Dubrasquet M. Effects of bombesin on food intake and gastric acid secretion in cats. *Am J Physiol* 1989;**256**:R181–R186.

221. Denbow DM. Centrally and peripherally administered bombesin decreases food intake in turkeys. *Peptides* 1989;**10**:275–9.
222. Woods SC, Stein LJ, Figlewicz DP, Porte Jr D. Bombesin stimulates insulin secretion and reduces food intake in the baboon. *Peptides* 1983;**4**:687–91.
223. Lieverse RJ, Jansen JB, Masclee AA, Lamers CB. Bombesin reduces food intake after a preload in man by a cholecystokinin-independent mechanism. *Clin Sci (Lond)* 1993;**85**:277–80.
224. Flynn FW. Caudal brain stem systems mediate effects of bombesin-like peptides on intake in rats. *Am J Physiol* 1992;**262**:R39–R45.
225. Gutzwiller JP, Drewe J, Hildebrand P, Rossi L, Lauper JZ, Beglinger C. Effect of intravenous human gastrin-releasing peptide on food intake in humans. *Gastroenterology* 1994;**106**:1168–73.
226. Muurahainen NE, Kissileff HR, Pi-Sunyer FX. Intravenous infusion of bombesin reduces food intake in humans. *Am J Physiol* 1993;**264**:R350–R354.
227. Figlewicz DP, Stein LJ, Woods SC, Porte Jr D. Acute and chronic gastrin-releasing peptide decreases food intake in baboons. *Am J Physiol* 1985;**248**:R578–R583.
228. Kirkham TC, Gibbs J, Smith GP, Geary N. Meal pattern analysis in rats reveals partial agonist activity of the bombesin receptor antagonist BW2258U89. *Pharmacol Biochem Behav* 1995;**52**:101–6.
229. Kirkham TC, Perez S, Gibbs J. Prefeeding potentiates anorectic actions of neuromedin B and gastrin releasing peptide. *Physiol Behav* 1995;**58**:1175–9.
230. Fekete E, Vigh J, Bagi EE, Lenard L. Gastrin-releasing peptide microinjected into the amygdala inhibits feeding. *Brain Res* 2002;**955**:55–63.
231. Ladenheim EE, Taylor JE, Coy DH, Carrigan TS, Wohn A, Moran TH. Caudal hindbrain neuromedin B-preferring receptors participate in the control of food intake. *Am J Physiol* 1997;**272**:R433–R437.
232. Ladenheim EE, Taylor JE, Coy DH, Moore KA, Moran TH. Hindbrain GRP receptor blockade antagonizes feeding suppression by peripherally administered GRP. *Am J Physiol* 1996;**271**:R180–R184.
233. Ladenheim EE, Wirth KE, Moran TH. Receptor subtype mediation of feeding suppression by bombesin-like peptides. *Pharmacol Biochem Behav* 1996;**54**:705–11.
234. Greenberg D, Smith GP. Hepatic-portal infusion reduces the satiating potency of CCK-8. *Physiol Behav* 1988;**44**:535–8.
235. Kirkham TC, Gibbs J, Smith GP. Satiating effect of bombesin is mediated by receptors perfused by celiac artery. *Am J Physiol* 1991;**261**:R614–R618.
236. Stuckey JA, Gibbs J, Smith GP. Neural disconnection of gut from brain blocks bombesin-induced satiety. *Peptides* 1985;**6**:1249–52.
237. West DB, Williams RH, Braget DJ, Woods SC. Bombesin reduces food intake of normal and hypothalamically obese rats and lowers body weight when given chronically. *Peptides* 1982;**3**:61–7.
238. Geary N, Smith GP, Gibbs J. Pancreatic glucagon and bombesin inhibit meal size in ventromedial hypothalamus-lesioned rats. *Regul Pept* 1986;**15**:261–8.
239. Bellinger LL, Bernardis LL. Suppression of feeding by cholecystokinin but not bombesin is attenuated in dorsomedial hypothalamic nuclei lesioned rats. *Peptides* 1984;**5**:547–52.
240. Smith GP, Jerome C, Gibbs J. Abdominal vagotomy does not block the satiety effect of bombesin in the rat. *Peptides* 1981;**2**:409–11.
241. Taylor IL, Garcia R, Elashoff J. Effects of vagotomy on satiety induced by gastrointestinal hormones in the rat. *Physiol Behav* 1985;**34**:957–61.

INDEX

Note: Page numbers followed by "*f*" indicate figures, and "*t*" indicate tables.

T

V

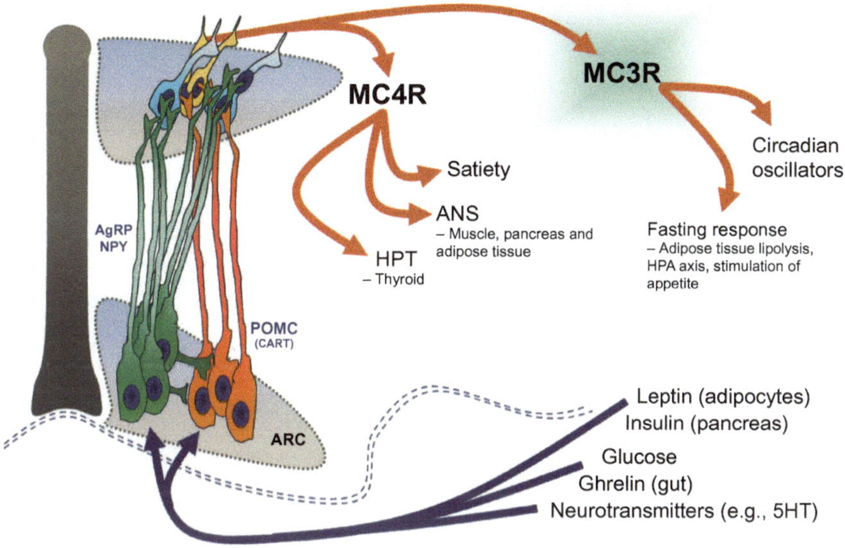

BEGRICHE *ET AL.*, FIGURE 4.1

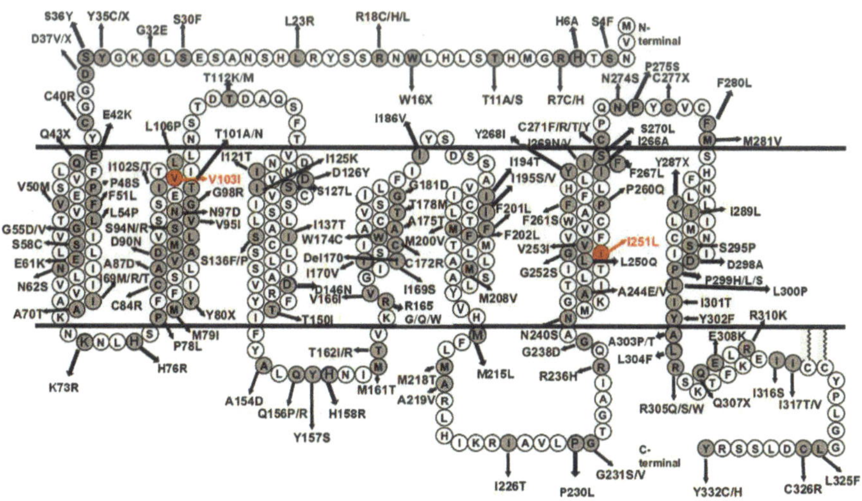

HINNEY *ET AL.*, FIGURE 5.1

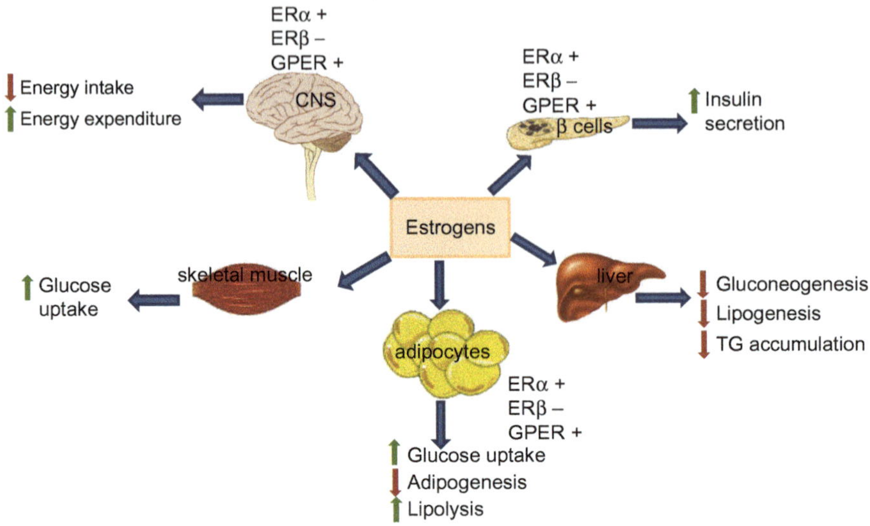

Shi *et al.*, Figure 6.1

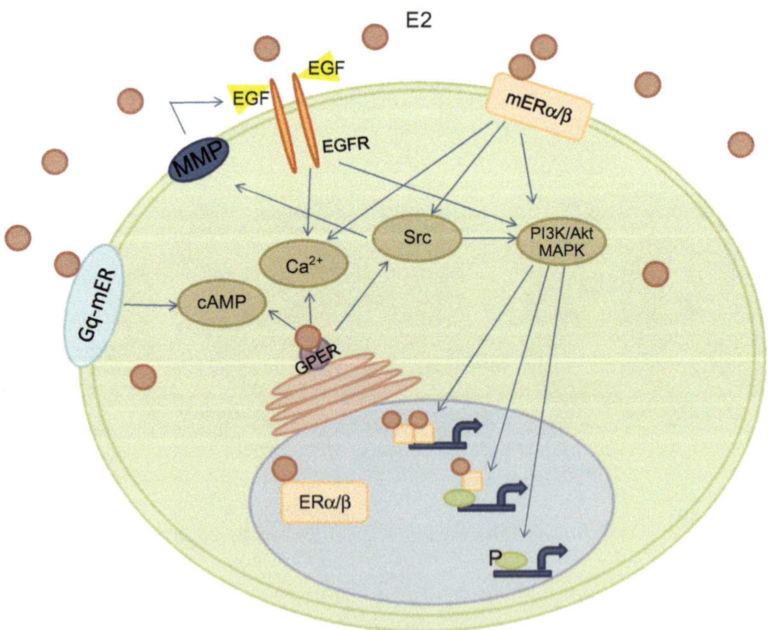

Shi *et al.*, Figure 6.2

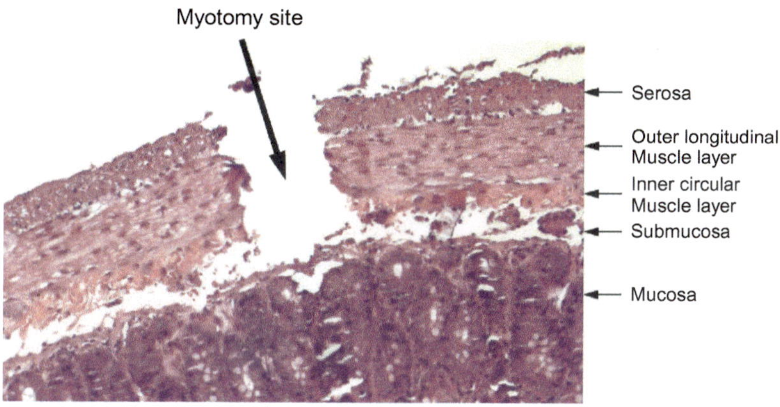

SAYEGH, FIGURE 8.6

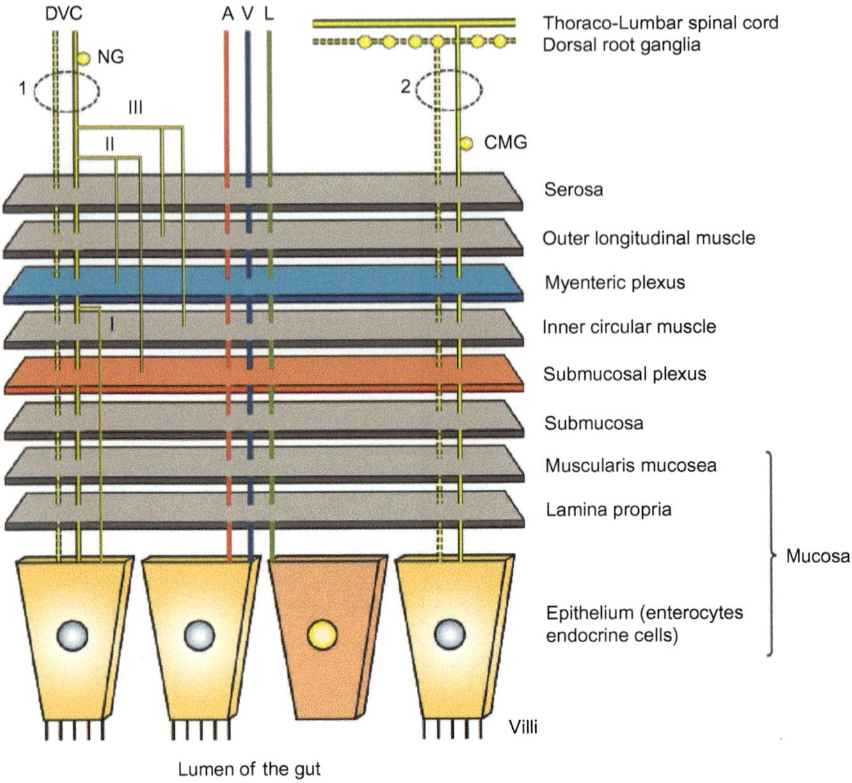

SAYEGH, FIGURE 8.8

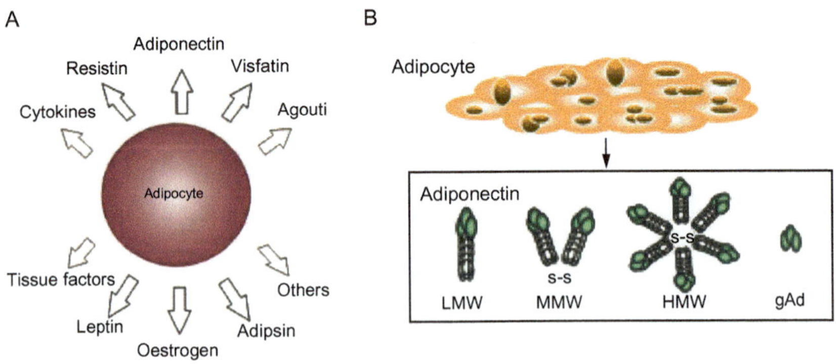

AKINGBEMI, FIGURE 9.1

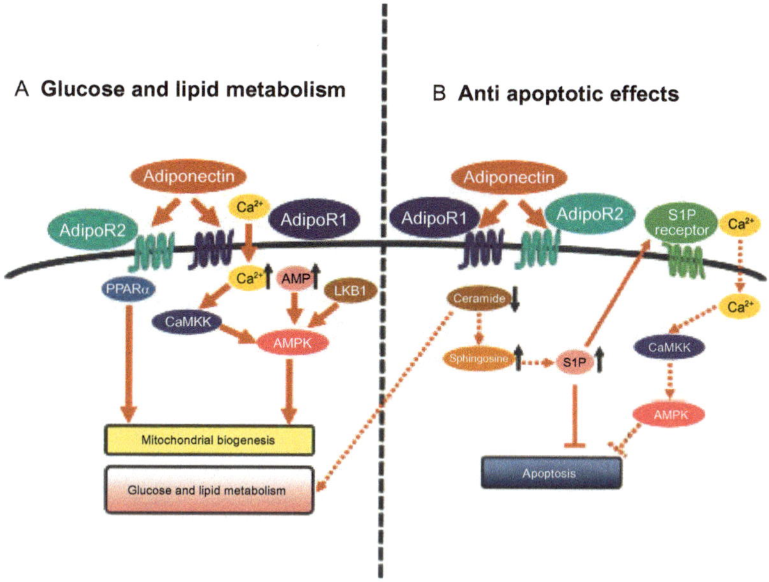

Akingbemi, Figure 9.3

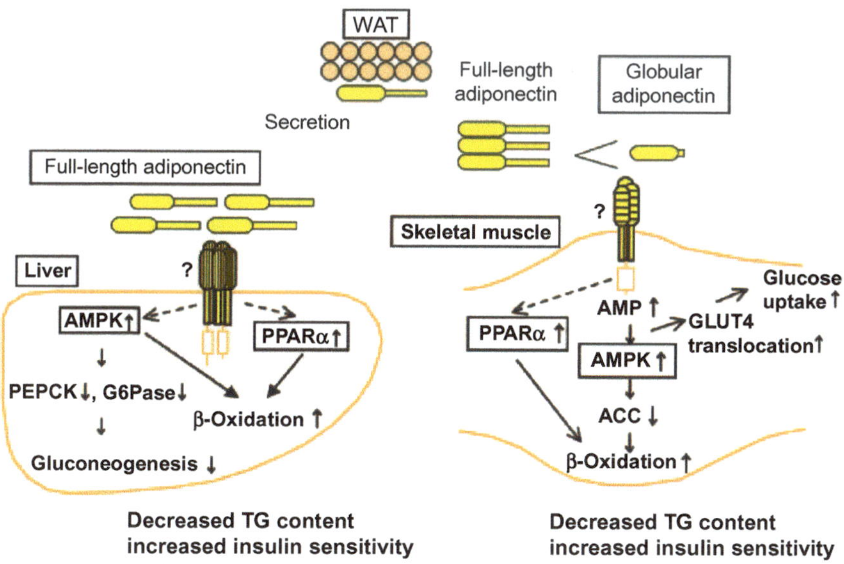

Akingbemi, Figure 9.4

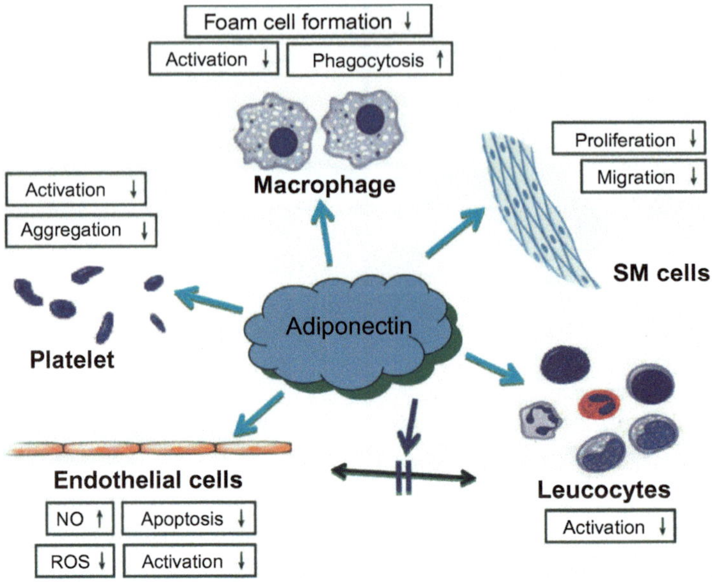

Akingbemi, Figure 9.5

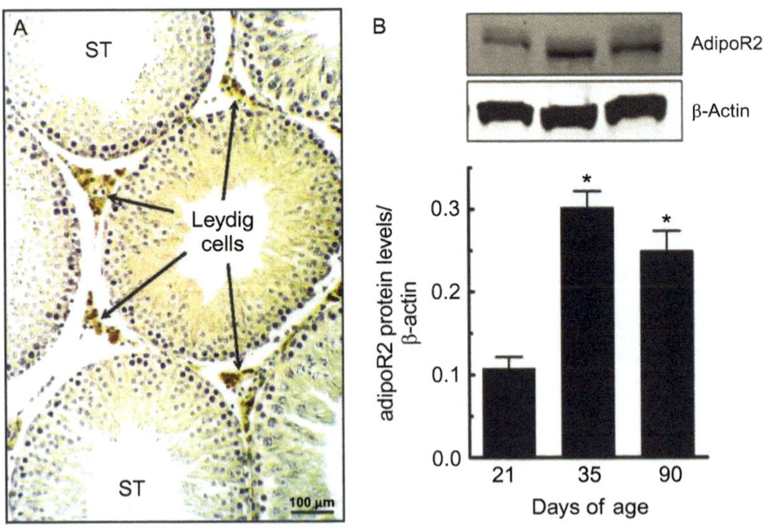

Akingbemi, Figure 9.6

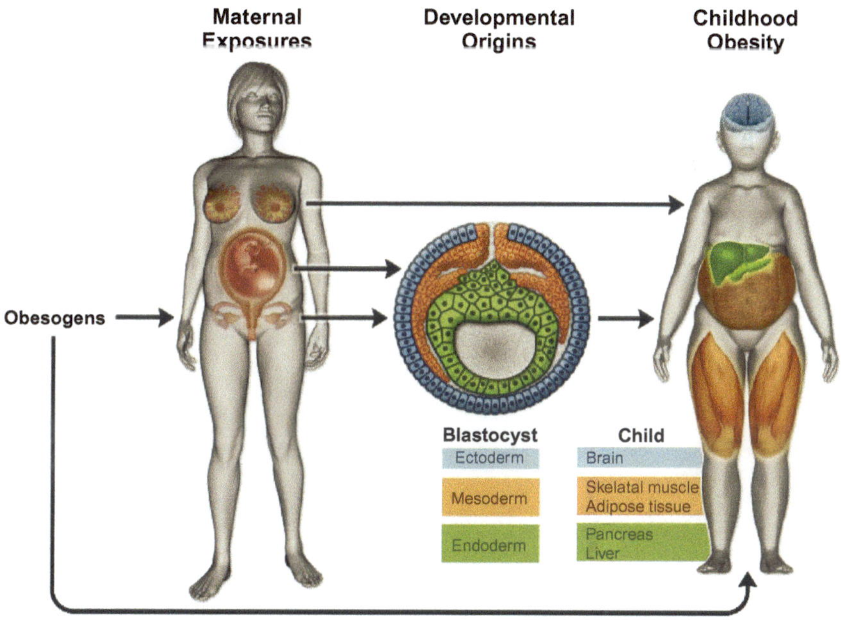

Akingbemi, Figure 9.8

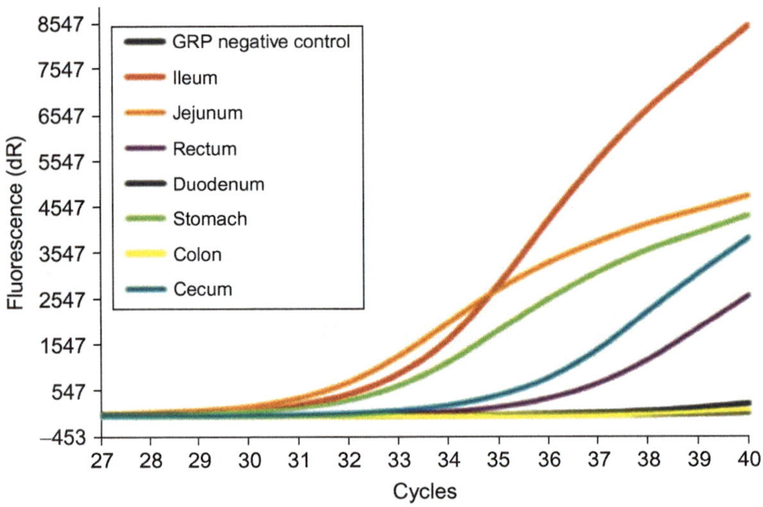

SAYEGH, FIGURE 10.1

FSC
www.fsc.org
MIX
Papier | Fördert gute Waldnutzung
FSC® C083411

Kolibri 360 GmbH
Ferdinand-Jühlke-Straße 7
99095 Erfurt, Deutschland
produktsicherheit@kolibri360.de